Sustainability

SUSTAINABILITY

A Philosophy of Adaptive Ecosystem Management

Bryan G. Norton

The University of Chicago Press
Chicago and London

The University of Chicago Press, Chicago 60637
The University of Chicago Press, Ltd., London

Printed in the United States of America

14 13 12 11 10 09 08 07 2 3 4 5

ISBN: 0-226-59519-6 (cloth)
ISBN: 0-226-59521-8 (paper)

Library of Congress Cataloging-in-Publication Data

Norton, Bryan G.
Sustainability : a philosophy of adaptive ecosystem management / Bryan G. Norton.
p. cm.
Includes bibliographical references and index.
ISBN 0-226-59519-6 (cloth : alk. paper) — ISBN 0-226-59521-8 (pbk. : alk. paper)
1. Environmental policy. 2. Environmental management—Decision making.
3. Interdisciplinary research. 4. Communication in science. 5. Sustainable development. I. Title.
GE170.N67 2005
333.72—dc22
2005007412

♾ The paper used in this publication meets the minimum requirements of the American National Standard for Information Sciences—Permanence of Paper for Printed Library Materials, ANSI Z39.48-1992.

CONTENTS

PART II: VALUE PLURALISM AND COOPERATION

PART III: INTEGRATED ENVIRONMENTAL ACTION

PREFACE: BEYOND IDEOLOGY

John Muir, in an early issue of the *Sierra Club Bulletin,* wrote: "The battle we have fought and are still fighting for the forests is a part of the eternal conflict between right and wrong, and we cannot expect to see the end of it." This inspired rhetoric could stand as the motto for the first hundred-plus years of the American conservation movement—what may in the future be called "The Age of Ideology." With all the practical and economic arguments arrayed against him, Muir appealed to higher principles. His rhetoric was matched on the other side by Gifford Pinchot, who loudly advocated a materialist approach to deciding conservation policy, using human (economic) utility to determine resource use. "There are two kinds of things in the world," Pinchot was fond of saying, "people and natural resources."

Today we are learning that this poignant rhetoric causes no end of trouble. Federal environmental policy whiplashes back and forth between rhetorical flourishes and weak action by Democrats, on the one side, and the Republican appointment of economic zealots who want to dismantle the system of environmental protections and turn resource management over to the private sector on the other. I refer to this rhetoric-driven process as *ideological.* It was polarizing in the days of Muir and Pinchot and remains so today. By calling the process ideological, I highlight the importance of preexperiential commitments of participants in the debate—to moral imperatives and to uncritical endorsement of economic efficiency as the only goal for policy, for example. A priori commitments profoundly influence the way people understand environmental problems. Ideologies shape not only a participant's understanding of problems faced, but also what people experience while addressing those problems. Outbursts of ideologically motivated rhetoric are unlikely to result in improved environmental policies.

Recent changes of federal administrations have resulted in about-faces equivalent to reversing a tanker at sea. Environmental policy has in this sense been held hostage to ideology, to right-versus-left and environment-versus-free-marketeering agenda shifts driven by the whim of electoral politics. Since ideological environmentalism, growing out of the historic battles between Muir and Pinchot, poses questions in right-or-wrong, all-or-nothing terms, ideological environmentalism seems to enforce an either-or choice on us today. It seems as if we must choose between moralism and economism—between doing right and doing well.

This formulation of the problem, however, presents us with false alternatives. Since environmental problems are characterized by ideologically committed advocates on both sides, the problems are experienced as zero-sum competitions. The two sides in this polarized rhetoric, confused by miscommunication, resort to enmity and name-calling, and very little gets done to protect the environment. Seldom do environmental agencies and departments pursue coherent goals; and almost nothing is done efficiently once goals are chosen. This book is premised on the possibility of rejecting ideological environmentalism altogether and of developing a new vocabulary that will allow and encourage a new formulation of environmental problems. My goal is to explore ways to avoid posing environmental problems in ideological terms.

There is an alternative to the either-or whiplash associated with ideological environmentalism. I propose to construct a new vocabulary, a new way of formulating and posing environmental choices. The centerpiece of the new vocabulary will be the much-used, and much-maligned, terms *sustainability* and *sustainable development.* I intend to explore how these terms, properly refurbished and semantically connected to other important scientific and evaluative terms, might serve to reformulate environmental problems as matters of more or less—of proper balances among competing social values—rather than as clashes of right and wrong.

As I begin my exploration of the role of language in environmental problem solving (or failures thereof), I will adopt a rather strong working hypothesis: *The way we Americans talk and write about the environment is a major cause for the failure of our governments at all levels to achieve rational actions to protect environmental public goods.* This hypothesis and the associated corollary, that *success and failure in environmental problem-solving is often determined by the way a problem is formulated and discussed in public discourse,* function together and guide the development of my argument. A central aspect of this argument focuses on communication and, especially in the diagnostic parts of the book, on failures of communication in public discourse about environmental goals and policies. For reasons that are explained throughout the book, I choose *sustainability* and *sustainable development* as examples of terms that, though used imprecisely today, could become effective tools of communication. The choice of these terms as our initial philosophical and linguistic beachhead implies that they have not yet been given clear enough meaning to be useful when precise communication is required.

Indeed, it is often said that the terms *sustainability* and *sustainable development* mean all things to all people, that they have become banners around which environmentalists rally only to find that the banners are of all colors and that to follow these banners is to go in many directions simultaneously. Similarly, it has been argued persuasively that sustainability and sustainable development currently function as "essentially contested concepts"; that is, these concepts, because of their vagueness and ambiguity, can be claimed by many movements and by diverse actors, each one interpreting the key terms differently. In this case the contest involves social, political, and especially disciplinary battles. A key aspect of the disciplinary battles, accordingly, has been the contest to define *sustainable* and *sustainable development,* in order to determine on whose turf, and in whose terms, the content of sustainable policy will be debated. One simply cannot overemphasize the key role of problem formulation in seeking cooperation and success in environmental management. And at the heart of problem formulation is the question of how we talk about—how we articulate and discuss—environmental problems.

There are of course lots of books, papers, speeches, and reports about sustainable development. Why should anyone read this book in particular? It's a good question and deserves an answer. The short answer is that this book, a philosophical exploration of the language of environmentalism, is unique in explicitly examining many assumptions normally taken for granted in environmental debates. These assumptions, usually invisible because they are implicitly embedded in the ways we talk, create the confusing context in which most environmental problems are encountered today. Using methods developed by philosophers of language—methods that are sometimes referred to as the pragmatics of the language of science—this book explores how we talk about the environment and how failures of communication across incommensurate conceptual frameworks result in social traps and confusing discourse. Since these assumptions shape the usual discussions about sustainable development, a philosophical examination of language can provide a uniquely neutral, cross-disciplinary viewpoint from which to examine environmental policy problems and current approaches to managing them. I hope that this fresh perspective will encourage interest in my radical project of reframing environmental problems by developing a new vocabulary for discussing them.

Most writers on environmental policy, even academic writers, mainly accept our current, dysfunctional discourse as something that is beyond our power to change. Most of us are linguistic fatalists in this respect; when trapped in a web of confusing terms and failures of communication, we struggle on, lamenting the failures but falling nonetheless into fruitless debates and confused rhetoric. Such pacifism is a recipe for more ideology and more whiplash, with no end in sight. This book is emphatically not fatalistic about the confusions and ambiguities that beset environmental policy discussions today; on the contrary, it sets out to do something about them by analyzing the causes of communicative failure and then proposing an alternative vocabulary designed to avoid such failure. I look at the problem from the unusual perspective of a philosopher of language who sees the ambiguity and confusion surrounding sustainability discussions as an opportunity to create better tools of communication and better formulations of problems.

My approach is to construct a set of terms and concepts that can be tested for their usefulness by employing them in discussions of real environmental problems. Relying on a study of the pragmatics of the language of environmental science, I examine that science as an *activist* discipline. In order to give concrete form to this activism, I situate my discussion within the tradition called adaptive management—hence the subtitle of the book, "A Philosophy of Adaptive Ecosystem Management." The advantage of this strategy, which places policy discussion in the context of decision making rather than in various academic disciplines, is that it allows discipline-independent tests of definitions and measures. More importantly, this strategy encourages definitions that are extradisciplinary in their scope. What is needed to improve environmental policy discourse is a system of bridge concepts that can serve to link the various scientific disciplines and to relate environmental science with social values in the search for rational policies.

Adaptive management is science-based management that assumes we usually do *not* know enough to choose what is absolutely best to do; adaptive managers adopt the attitude that actions to correct environmental problems must simultaneously be actions that reduce uncertainty in the future, allowing correction of our uncertain course in later decisions. Adaptive management is thus experimental management. Given the premise that even the experts do not know what is best, adaptive management is humble management—not management by experts, but management through political participation and social learning. Adaptive management places experts in the role of teachers and facilitators in a broader democratic process.

Adaptive management has a long and distinguished history. I will argue, for example, that Aldo Leopold, the much-revered philosophical forester of the first half of the twentieth century, was the first adaptive manager (though he never used the name). Leopold laid out the basic principles of adaptive management by embodying them in his famous simile "thinking like a mountain." More or less explicitly, adaptive managers have recognized Leopold's foresights; and adaptive management, named by the ecological theorist C. S. Holling in 1978, has become an important watchword of contemporary environmental policy formation. It is important here because the action orientation of adaptive management ensures that proposed linguistic innovations will be tested not against preexperiential ideologies, but against real-world problems and case studies, including controlled experiments whenever possible.

In providing a philosophy of adaptive management, I have also proposed an important amendment to adaptive management as it is currently practiced. Most of the practitioners and advocates of adaptive management, self-avowed or not, are scientists by training, and they adhere sharply to the positivist precept that factual and normative discourse are best separated. Writing as an environmental philosopher-ethicist with a strong pragmatic bent, I argue (1) that in all situations affecting policy, facts and values must be discussed together: environmental science must abandon the myth that it can be pursued as value-neutral science; and (2) that norms and environmental values should, like scientific hypotheses, be considered open for revision through testing in the face of expanding experience—experience that can be guided by intentional and controlled experiments in many cases. So I also follow the pragmatist idea of the unity of inquiry, which holds that there is only one way to improve both empirical understanding and normative judgment: experience. That method must, in turn, be supported by a commitment to norms of procedure and by acceptance of an obligation to communicate and give reasons when we disagree. This method—the "scientific method"—is applicable no less to moral problems than to physical ones. My analysis of environmental problems therefore responds to scientific claims and value claims in pretty much the same way: "Show me the evidence!" This amendment in the usual scope of adaptive management has the advantage of making values and goals endogenous to the adaptive management process; no less than scientific hypotheses, our norms and stated goals are held up to the standard of experience and corrected through ongoing adjustment in the face of new experience about their viability in an evolving situation.

This project originated in countless "experiments" in the use of concepts and ideas in various activist and academic contexts. For about fifteen years, I have been experimenting with different ways of talking about and evaluating the environment as a participant in policy discussions and conferences. After the less-than-satisfactory experiences I had trying to work as a consultant for the Environmental Protection Agency between 1985 and the late 1990s (as sketched in chapter 1), I formed allegiances with colleagues from other fields and professions in an attempt to devise a better approach to environmental policy problems and a better vocabulary for discussing them. I have participated in many interdisciplinary conferences on sustainable development, conferences that looked at sustainability from the more or less unique perspectives of the practitioners of multiple disciplines. As a philosopher of the language of science, I communicated with, and learned from, academic scientists, philosophers, activist scientists, and social activists. I found that if I would just listen and discuss possible policies with others of diverse backgrounds—often in an action-, policy-oriented context—I could try out many vocabularies, suggesting concepts, getting reactions, and reconsidering their communicative value.

This book should be read by everyone who is tired of rehearsing, again and again, the same old ideological disagreements about how to protect the environment. This book is different from others in that by concentrating on development of a vocabulary that can encourage communication, I present environmental problems in a new light, expanding opportunities for cooperation among former opponents and clearing the way for a new, post-ideological age of environmental progress.

Three notes on the reading of the text may be helpful to the reader: one is stylistic, one has to do with content, and one is about the referencing and footnoting of sources. First, I should clarify a point of unusual style. In self-reference I have sometimes used the first-person singular pronoun and sometimes the first-person plural pronoun. These switches are not random. Since the book both attempts a neutral analysis of linguistic-communicative problems in environmental discourse and advocates particular solutions to environmental problems, I have found it useful to employ the first-person plural, *we,* to express reasoning I expect to be noncontroversial, likely accepted by most readers. In other contexts, when I am advocating partisan positions that I expect will be controversial for many readers, or when I am simply exploring a line of reasoning that may or may not have widespread support, I use the first-person singular, *I,* to signal that I am speaking from an individual point of view.

Second, I should say that although integration is my theme, I have consciously steered clear of one aspect of policy formation and management: the role of political and economic power relationships, which often limit attempts to achieve a rational environmental policy. The reader should be assured that I am well aware of this confounding variable that affects official decisions so pervasively. To deal with that aspect, however, would have led to a different—and even longer—book. What I have tried to do is to discuss a *rational* process that is *possible*—but hardly guaranteed. If members of any community or government participate in bad faith, subverting the adaptive management process to their own selfish ends, this is a different problem from the problem of what we *should do as a community,* which is the challenge of this book. This book concerns both policy and management, and it sets out to provide as seamless as possible a connection among these separable aspects of what must become an integrative process. But the book examines a process that presupposes good faith and a cooperative spirit on the part of participants. My goal has not been to force selfish and uncooperative people to become less selfish; rather, I have set out to illustrate that, for those individuals and communities that are committed to transcending divisiveness and paralysis and engaging in cooperative action, there is a clear and optimistic road forward.

Third, a note on referencing and footnotes: since my multidisciplinary search for insights about sustainability has taken me to countless conferences, I have spoken with and listened to thousands of people talking about sustainability; I have also read a good deal in many related disciplines. I cannot hope to thank all of the people from whom I have learned in this ongoing pilgrimage across the borders of the disciplines of public knowledge. A few individuals are featured in the text because what I learned from them really changed the direction of my thought, or because their particular way of making a point seems apt, but these are a tiny minority of those from whom I have learned. Since I have learned from so many, I concluded that an attempt to fully reference all of my arguments would detract from my main task of integration. I am sensitive to the charge that this writing strategy may seem ungrateful to the many authors and speakers from whom I have gained insights, so I here refer the reader to the papers I have published in journals and in anthologies over the past decade. These more detail-oriented publications, which are heavily referenced, are better places to unearth the considerable intellectual debts I owe to others. In this book, the emphasis has to be on the big picture.

Accordingly, I must content myself with recognizing here those who have supported my research or contributed to the writing, revising, and production of this text. First, much of the research for this book was supported by three grants: one from the Program of Exploratory Research in the Office of Research and Development at the U.S. Environmental Protection Agency (which allowed me to work with Michael Toman of Resources for the Future on defining sustainability); one from the Methodology, Measurement, and Statistics Program of the National Science Foundation (SBR9729229); and one from the Environmental Protection Agency, Water and Watershed Program (R825758), on which I worked with my colleague Anne Steinemann and also with a research group at the University of Georgia, led by Bruce Beck. I also received support from the Georgia Institute of Technology in the form of a sabbatical for the academic year 1999–2000, which I spent as a Visiting Fellow at the Army Environmental Policy Institute. I would also like to thank the staff and researchers at two research institutes who hosted me during my work on the

Dutch approach to environmental politics and management. I am especially happy to mention the excellent hospitality shown by, and the wonderful discussions I had at, the Institute for Evolutionary and Ecological Studies at the University of Leiden, where I worked with—and became friends with—Cees Musters, Hans de Graf, and Wim ter Keurs; and the International Center for Integrative Studies of the University of Maastricht, where I had excellent discussions with Jan Rotmans, Marjolein van Asselt, and Dale Rothman as well as helpful interactions with other members of the staff. I also want to thank Strachan Donnelly for including me in many of his projects, first at the Hastings Center and later at the Center for Humans and Nature. These informal meetings served as opportunities to try out many of the ideas that became part of this book.

Finally, I would like to mention and thank my many friends and colleagues who assisted me in the completion of the manuscript. First, earlier versions were read and commented on by Stan Carpenter, Elizabeth Corley, Paul Hirsch, Richard Howarth, Bruce Hull, Ben Minteer, Michael Toman, Andy Ward, Clark Wolf and his students in an honors seminar at the University of Georgia, and Asim Zia, as well as several anonymous reviewers. From these readers I received much good advice, most of which I heeded; remaining failures are due to my own obstinacy. I also received technical and other kinds of assistance from Elizabeth Corley, Paul Hirsch, Sharon Mills, and especially Michael Waschak, who prepared the figures and the index.

A NOTE TO THE BUSY READER: SOME SHORTER PATHS

The purpose of this book is to integrate the many elements of a philosophy of environmental policy and management so that readers can compare a complete approach—complete in the sense that it brings together work from the various sciences summarized as "environmental science" with work in economics and the social sciences and environmental ethics, and with emerging work on public participation in collaborative management projects—with the other management theories and approaches they can find in the literature. I have, in the process, erred on the side of inclusion, hoping to anticipate many of the problems and issues that may arise in the search for a better understanding of sustainable living. I also know that the book is long and that many readers will be interested in only some aspects of the overall argument. Accordingly, I have outlined below several shortcuts and several more limited pathways through the book. In order to support this selectivity, I have tried to provide enough cross-references and signposts in the text and the notes so that "chapter-surfers" will know where they are in the larger argument (even if they are passing by some of the details of some parts of the argument).

First, a comment on the appendix. It describes and defends the philosophical methodology that is employed in the rest of the book, but the material in the appendix is not intended to advance the primary argument of the book. In the appendix I show that the methods developed by linguistic philosophers in the twentieth century can be understood as providing a rigorous method by which to address interdisciplinary failures of communication, such as those exemplified in cross-disciplinary disputes. For readers who are interested, reading the appendix between chapters 2 and 3 will provide the best connections. Readers interested in the path to better environmental policy may find this careful justification of my method tedious or unnecessary and may choose to skip the appendix.

Second, section 3.6 is a bit of a tangent—an interesting one, I hope, but it is not essential to the main arguments of the book; also, some might find section 10.1 more than they want to know about decision analysis; going from chapter 9 to section 10.2 will not interrupt the argument unduly.

More generally, the book has several main themes, and some readers might like to follow somewhat shorter paths that explore particular themes. Of these paths, the central one is an exploration of the meaning of sustainability—one that draws on many disciplines and shows how, though the terms *sustainable* and *sustainable development* are currently unclear and confusing, we might offer a comprehensive approach to understanding and evaluating intertemporal normative relationships. Readers mainly interested in this theme should concentrate on chapters 1, 2, 3, 4, 6, 8, 9, 11, and 12.

Also, this volume embeds the new approach to sustainability—which it is argued must be developed by each community that resolves to live sustainably—within a philosophy of adaptive management. This "book-within-a-book" can be followed by reading chapter 1, sections 2.1 and 2.2 of chapter 2, and chapters 3, 4, and 7–12.

I have also tried to develop and demonstrate a comprehensive approach to interdisciplinary studies, an approach that places a great deal of weight on problems of communication and that employs the tools of the philosophy of science and the philosophy of language to develop better cross-disciplinary and integrative communication. Those wishing to follow this argument should read chapters 1–4, 6, and 7; sections 8.6, 8.7, and 10.5; chapters 11–12; and the appendix.

Finally, some readers may find the division of the book into three parts useful in choosing what to read. Part 1 develops the main ideas of an adaptive approach to management, part 2 provides a small "book" on environmental ethics and values, and part 3 directs attention toward integrated action.

1

AN INNOCENT AT EPA

1.1 The Old EPA Building

At first it just seemed to be a large, ugly building composed of cubes and rectangles piled side by side and one on top of another, architecturally typical of the worst buildings of the 1960s. It was the "old" Environmental Protection Agency Headquarters building in southwest Washington, which housed EPA federal headquarters from 1971 until the midnineties. Eventually the old building came to symbolize the many confusions and frustrations I experienced in trying to save the world.

This isn't really a book about bad architecture, nor one about my experiences at EPA. It's about constructively improving policy discourse by clarifying the meaning of the term *sustainability* and other related terms. This first chapter, however, starts with my experiences with EPA projects—and with the old building—because those experiences in that place have shaped this book by forcing upon me the questions the book sets out to answer. In this chapter I introduce those questions as they came to me, in the context of struggling to evaluate EPA policies and regulations. I hope this little bit of biography will help you to understand the perspective from which I write and the complexity of the problem of improving communication and understanding of environmental policy formation and implementation.

Although EPA's struggle to form a more coherent environmental policy provides an excellent entrée into the problems I wish to address, the quandaries I encountered at EPA are really only symptoms of deeper and broader problems, problems that go to the heart of environmental policy formation and implementation today. They are most basically problems of communication, the failure to develop effective vocabularies for relating the sciences to each other and to the broader context of social values and policy goals. So this is not an exposé of EPA—I do not mean to disparage the sincere good efforts of the people who work in that beleaguered agency. Rather, I use EPA as an example of the broader problems with environmental policy discourse and as a starting point for a constructive approach to an integrated environmental policy.

The story of how I hoped to reform the goals of EPA, and do my bit to save the world, begins with a few of my early visits—in the mideighties—to the old EPA building in Washington, DC. At first I thought EPA employees were remarkably hospitable and gracious people, because whenever I had a meeting at the old EPA building, even when I was meeting someone fairly high up in the administrative hierarchy, the person would tell me to wait at the main entrance. When I arrived, a receptionist called upstairs, and the EPA employee came down to meet me and led me to his or her office. Gracious, if not very efficient, I thought. Only later, when I was teaching courses in a graduate program for EPA employees, did I learn the real reason for this practice. The students in my class—this was, incidentally, as sharp and highly motivated a group of graduate students as I have encountered—were all EPA employees, and they were working in an executive degree program to obtain a masters degree in public affairs. Some of them invited me to one of their offices in the old EPA building after class one day; following some open-ended discussion, it was suggested that I might want to talk to another employee with common interests in another part of the building. After calling ahead, my host began to give directions and then stopped. He began again, and stopped again, exhibiting signs of frustration. Finally he said, "Come on. I'll lead you there." On the way to the nearest elevators, he explained that it was almost impossible to direct someone within the building, so we set out on a hike; I followed him through a warren of hallways and elevators, going down and up, and eventually—it seemed like fifteen minutes later—I was delivered to my appointment.

Rumor has it that the building was originally conceived as residential, and that it got its bizarre geography from an intention to create clusters of residential units, which might have made sense of the separate towers and clusters as smaller "communities" in the vast complex. In fact, I learned from talking to officials in the EPA real estate office—and the developer Charlie Bresler—that despite the plausible rumors, the building was from the beginning planned mainly for office space. The eccentric traffic patterns and inaccessible hallways were actually a result of the rapid, but Balkanized, growth of the agency itself. As I will develop in more detail presently, EPA grew as a result of separate legislation to protect air and water and manage toxics and solid waste. No organic act was ever completed, so the agency grew as fiefdoms. The government, in 1971, rented more than 525,000 square feet of almost completed office space in Waterside Mall and eventually moved the headquarters and central offices there. But that original contract was followed in quick succession by several expansions of the EPA mandate and by attendant, often desperate, cries for new blocks of office space. As new departments were added to EPA, new chunks of office space were requested. Given overlapping projects and other complications, the mall grew like the proverbial Topsy; towers were built around it, creating more and more disjointed workspaces. In addition to the Main Mall, there was a Northeast Mall and a Southeast Mall, some areas of which were devoted to EPA. These malls have mismatched floors and flank the Main Mall. The data center occupied a huge area that seemed to exist in another dimension, tucked in behind a large supermarket. Eventually, a west tower (with twelve floors) and an east tower (with twelve floors) were added as the developer responded to new requests for space, which came frequently during the 1970s and 1980s. Each of a series of locations was built up and out on the large property, and the chunkiness of the projects contributed to the creation of separate fiefdoms. The result was a building with towers sprouted around a core, just as EPA mandates to protect air and water quality, manage pesticides, and so forth, sprouted around a central agency composed of existing offices that had been uprooted from other departments and agencies. Each tower, it turned out, had its own elevators, and the connective tissue—which was to have been provided by the central mall—was simply inadequate to provide a reasonable traffic pattern connecting elements of the agency.

Therefore, "You can't get there from here" was too often the answer to a question about navigating the structure. At last I learned why administrators meet their guests at the building door: unescorted guests could get so lost that they would become missing persons and get discovered, days later, wandering shell-shocked through corridors that go mainly in circles. At some point in my relationship with EPA, after a Kafkaesque frustration dream in which I was struggling to get somewhere, I realized I was wandering a maze of hallways, circling the disconnected towers of Babel in the old EPA building. It was then that I realized that the old building was a perfect metaphor for the way environmental policy is discussed and made in the United States.

I had lived in Washington off and on during the early and middle eighties, on leave from my professorship and working at think tanks in the capital area, mainly on policy to protect biological resources. Since falling by good fortune into a wonderful project on the values and rationale behind the Endangered Species Act in the early 1980s, I had published a couple of books, and on this basis had become one voice in an emerging coherent and scientifically informed public discourse about the importance of protecting vulnerable species and biodiversity. Since I'd had some success analyzing and evaluating endangered species policy, I thought I might apply those insights more broadly to other policy problems. As the eighties progressed, I expanded the focus of my environmental policy studies and began to think seriously about whether EPA, the agency best poised to develop a comprehensive and rational national policy to protect the environment, was up to the task. Soon I had an excellent chance to explore this question; by the end of the decade, I was invited to join various EPA policy panels, and I gradually began to learn the ropes at EPA. It seemed clear, even from casual observation, that the agency lacked a central, unifying vision—indeed this was just as obvious to EPA employees and to the agency's strongest supporters. As an environmental philosopher, I really thought it would be possible to both help EPA develop a more coherent approach to evaluation of environmental change *and* bring those values to bear upon policy by strengthening the contribution of science to environmental decision making.

I admit I was naive—foolish, even—to think a philosopher might bring rationality to national environmental policy; perhaps I still am. But back then I was so naive that I thought I could work from within the agency, helping various parts of it to develop some new methods and a more coherent and comprehensive approach to evaluation. I guess I thought that a better method, once proposed, would catch on. So the story of my experiences at EPA, out of which I came to see the questions of environmental policy in the way that I do today, begins as a story of disillusionment. It tells how I came to believe that the solution to the environmental problems of the twenty-first century will require new ways of thinking as well as new organizational structures. In this book I recount the terms of my disillusionment—the reasons I finally saw that working to reform EPA from within was unlikely to work—but I also look more generally at environmental problems, and especially at the public discourse that passes for rational analysis of environmental problems in advanced industrial societies like ours. First, however, it will be helpful to know something about the internal barriers, both structural and conceptual-communicative barriers, that make it so difficult, at EPA, to get "there" from "here." I came to understand those barriers when I was an innocent philosopher accepting invitations to serve on committees to improve the analysis of environmental problems within EPA.

My timing, at least, was right. William Reilly, George Bush's choice to lead EPA, decided to broaden the focus of the agency and urged the Science Advisory Board (SAB), which had been formed in 1978 to provide the agency with "scientific and technical" guidance, to review and compare risks. EPA had become highly concentrated on human health—a strategy pursued especially in the Carter administration, one that survived until Reilly argued that EPA should focus equally on threats to human health and threats to ecological processes and natural systems. Reilly followed up the SAB report with action and directed the agency to develop a methodology for "ecological risk assessment," which led to lots of policy discussions in many of my meetings. Reilly favored expansion of the risk assessment methodology to broader areas, including risk to ecological systems and processes, because one of his personal priorities as administrator was to systematize information-gathering across the agency, and he considered the risk assessment/risk management (RA/RM) approach the one most likely to provide a comprehensive framework for collecting information and comparing risks to identify effective strategies for the agency.

So I came to the Science Advisory Board at a time when a lot was happening and deep questions about how to evaluate risks were to be debated in a highly charged political atmosphere. My strategy was to involve myself in as many policy-relevant panels and policy discussions at EPA as possible. Through good fortune and a handful of kind contacts who apparently thought EPA might benefit from a bit of philosophical input, I got my chances, over the decade of the nineties, to serve on a variety of EPA studies and panels. These studies were designed to integrate science into policy, to evaluate the success of given programs, and to recommend new courses of action. I met some wonderful friends, got an education in policy formation and development—in the kitchen, so to speak—and I even got paid a modest consulting fee. In the process I learned about a lot of good things that go on at EPA; but I also learned of problems. If I emphasize the negative in this chapter, it is because I am preparing to be more constructive in subsequent chapters. I found in these meetings that people with the best of intentions often spoke past each other; indeed I sometimes felt as if I were at a multilingual conference without translators. What worried me most was that people from different backgrounds and disciplines continued to interact, carry on conversations, and do their jobs, hardly noticing that they spoke languages without available translations. They, like unattended visitors at the EPA building, wandered into blind corridors and, when they asked how one might get to a more rational environmental policy, were too often told, "You can't get there from here."

In the early and middle 1990s, I served simultaneously on several EPA panels, panels that appeared to have very similar goals—to oversee the day-to-day application of science to EPA tasks and regulatory duties. These panels were all, to varying degrees, influenced by Reilly's attempt to broaden the agency purview, but they continued into the Clinton administration. I will rely heavily on these and subsequent experiences I had while consulting with EPA as I try to carefully characterize the problems of environmental discourse today. In this chapter, we concentrate on three such panels and committees.

First, I served as a charter member of the Environmental Economics Advisory Committee (EEAC), a new addition to the SAB, which had been instituted in 1974 and was given a legislative mandate by Congress in 1978. I (the only noneconomist on the committee) and my colleagues struggled to understand how economic science was being used, and should be used, in the formation of agency policy. Not long after joining the SAB, I was also invited to work on the Risk Assessment Forum, an internal EPA initiative supported by another section of the EPA bureaucracy, which had the task of improving risk analysis and management, in this case by preparing scientific background papers for the first-ever protocols for "ecological risk assessment." The general assignment of the forum was to examine the scientific background of ecological risk; our subcommittee's specific task was to write a paper on how to determine the "ecological significance" of changes in ecological processes and systems. About this time, I was also serving in a less formal discussion group called the Ecosystem Valuation Forum, which brought together a panel of leading scholars and practitioners to discuss the problem of how to evaluate changes in ecological systems. More will be said about the Ecosystem Valuation Forum below. Let's start by comparing the discussions I participated in as a member of the EEAC with those I encountered in the Risk Assessment Forum. My experience on these two panels provided ideal laboratories for the analysis of communication at the nexus where science and social values intersect in the formation of policy, and it was on these panels that I learned the extent of the Balkanization, according to offices and according to disciplines, that occurs in the policy development process.

The first thing I noticed was that I was about the only person that was on both of these panels, with the exception of my friend Bill Cooper, an ecologist from the Ecological Processes and Impacts Committee of the Science Advisory Board, who joined us occasionally on the EEAC as a liaison member. Bill was a veteran of the EPA science wars and had chaired an important subcommittee leading up to EPA's important *Reducing Risk,* a report published in 1990. This report was used by Reilly to justify broadening the mission of EPA.[1] The Risk Assessment Forum was designed to improve the use of risk assessment in EPA's policy processes and further Reilly's goal of developing a systematic, agency-wide procedure for gathering, interpreting, and comparing information in the regulatory process. If ecological risk was going to be integrated into the RA/RM process, it would be necessary to extend the techniques and models of toxicologists—key players in traditional risk assessment—to apply to the broader problem of "ecological risk." It seemed a stretch to me, but I was encouraged because I was joined on the subcommittee on "ecological significance" by Cooper and some other excellent ecologists. Our specific subcommittee task was to summarize what is known about how to identify "ecologically significant" processes and endpoints and to show how this knowledge could guide risk assessors in determining degrees of ecological risk. I knew Cooper was as interested as I was in developing a basis on which ecological science and environmental valuation could be made more congruent and commensurate, so I undertook the subcommittee work with enthusiasm.[2]

To my surprise, of the thirty or so scientists invited by the Risk Assessment Forum to look at the problem of ecological risk, I was the only one who had more than a passing acquaintance with valuation—the task of evaluating changes and their impacts on social values. This meant that unless I forced the issue, the group of ecologists and aquatic toxicologists assembled to guide EPA policy on ecological risk would never raise the question of how ecological risk assessment relates to social values. If ecological risk assessment was going to conform to the practice in analysis of human health risk, we would apparently have to complete our discussion of "ecological significance" without appealing to broader social values. The latter questions must be, on the assumption of a sharp separation of science and values, left to "risk managers." Still, since I was a member of other panels, I thought I could make an important contribution by integrating what I was learning about environmental valuation with the EEAC and other groups into the process of risk assessment and management. It was at least disconcerting, however, to find myself on a panel of thirty highly trained specialists, researchers who represented the cream of the crop when it came to "assessing risks," and learn that none of the other twenty-nine, though they could write equations around me on the topic of risk, seemed the least bit interested in deciding what was a "good" or a "bad" outcome of an action under uncertainty. They could, with unquestioned technical facility, tell you the likelihood—to as many decimal places as you would like—that a given outcome would occur, given specified empirical conditions; but they had no professional interest in deciding which outcomes were "good" or "bad." As risk analysts, they expected someone else to tell them which outcomes were "bad" (and therefore requiring a probabilistic assessment of the risk that they would happen).

Almost everyone else on the ecological risk committee was either an environmental toxicologist or a "risk assessor," and many of them had done research for EPA or for the National Institutes of Health or in one of the national laboratories. There were also M.D.'s, especially epidemiologists, and a few other disciplines I couldn't pronounce were represented. Virtually all of these scientists represented an "artificial science" called "risk assessment," which resulted from demands by EPA and other agencies that they have a rigorous methodology for assessing risks of various chemical exposures. This group, whose members were mostly located in national laboratories, especially Oak Ridge National Laboratory, and a few universities, formed the core group of researchers who tested various chemicals for toxicity under contract to EPA. They often used the mindless "kill-'em-and-count-'em" technique for determining the level of concentration at which a chemical is fatally poisonous to 50 percent of a population of fathead minnows, an unfortunate fish species that shares its fate with laboratory mice. Although it would be possible to question the usefulness of such data for most decisions, risk assessors had succeeded in developing elaborate "models" to judge exposure risks, the risk of transport by ecological means, and a variety of other predictors of morbidity and mortality.

This science of risk assessment developed at close quarters with EPA; since most risk assessments were done for EPA, the agency had shaped the research in the field by its requests for proposals and the terms of its research contracts. The methodologies developed were quite narrow, best suited for modeling the potency of suspected carcinogens in humans, because human health—especially protecting humans from the risk of cancers—was the main focus of EPA for its first three decades. On the RA/RM model, risk assessment referred to the "hard" science of computing probabilities of harm from a chemical exposure, whereas risk management referred to the difficult-to-quantify value judgments that were to be made by "managers," not "scientists." We will see later that the RA/RM separation is itself an arbitrary artifact of an exaggerated belief in the separability of scientific measurement from value judgments. Nonetheless, it was in this stilted conceptual context that EPA managers had to deal with Reilly's new directive to pay equal attention to ecological risks and human health risks.

What I found was a subdiscipline, environmental toxicology, including especially ecotoxicology, in considerable disarray. The professionals on the forum, except for me, had made a career of testing chemicals under contract to EPA; they had a comfortable working relationship with the agency—most had at some time been colleagues in labs or in group research projects, and many had worked for EPA in the past. Given the personnel at the meetings, it appeared that the intention was to institute ecological risk assessment by continuing to hire these same experienced contractors but to broaden their use of risk assessment models. Hence the need for scientific input into the risk assessment process, and our charge to provide scientific background papers for the writing of regulations regarding ecological risks. The implicit goal of our panel, I realized, would be to show how the methods of the toxicologists and the risk assessors could be expanded to apply to broader ecological problems affecting natural systems as well as the health of human individuals. I also understood that a lot was at stake for this assembled group. Each of them had developed a career, and had contributed to a discipline, based on the original policy orientation of EPA toward human health risk; if they could not respond with effective scientific models that applied to the new goal of measuring "eco-risk" scientifically and factoring it into their technical risk models, they would be left out. If EPA really did shift half of its attention to eco-risk, they stood to lose half of their research support and their consulting business. Most of my colleagues on the panel, it seemed, weren't sure what ecological risk was, but they wanted to be sure that the scientific tasks associated with it would include a role for their expertise in risk assessment.

Two problems made it highly problematical to integrate ecological risk into EPA's day-to-day business with familiar contractors. First, the health-based risk models developed in the first two decades of EPA's operations were designed to measure the likelihood of an increased incidence among human individuals of quite specific diseases, usually cancers, as a result of varied exposure levels. These models had no obvious application to more diffuse effects on ecological systems. Risk assessment science apparently could not be directly applied to ecological risks because that science counts increments of risk to individuals, whereas ecological systems must be regarded as more than simple collections of individuals. Ecosystems are self-organizing systems that unfold on many scales and at many speeds; indeed, ecosystems exist on all scales from microhabitat to eco-region, so it is apparently irrelevant to ecological risks to identify at-risk individuals and count risks to them. This problem vexed us as we tried to talk about "ecological significance"—significance would have to reflect more than the likelihood that an exposed individual would become ill. The idea of risk was in the case of risk to humans anchored to the value our society places on the health of individuals. In ecological risk, there is no such clear anchor.

The expected expansion of risk assessment models to apply to eco-risk also raised a related, even more important, problem about environmental and social values. As long as human health was the focus of the agency, value questions could be muted and even hidden, since nobody doubted the value of reductions in cancer incidence, nor was there serious disagreement with the principle that risk to various individuals should count equally. A commitment to protect the life and well-being of the population sidesteps many questions about values and fairness. Risk assessors could, given the wide agreement on these principles, avoid value questions and maintain the fiction that they only did "scientific" assessments of cause and effect. It was obvious, but conveniently not noticed, that the very act of valuing the health of humans as individuals implied unquestioned value axioms—the sanctity of human life and the obligation to avoid disease and death among the human population under EPA's care. Reilly's agenda expansion, however, seemed to require appeal to alternative or expanded value axioms by focusing on ecological as well as human health risk. Since ecosystems are not individuals, ecological risk assessment requires new value premises that cannot rest on the usual assumption of EPA's RA/RM model, that risk to illness of individual humans is the risk to be avoided. One can only reasonably speak of the risk to an ecological system if one has some idea of the aspects or elements of the system that are socially valued. Ecological risk assessment thus undermined the fiction that risk assessment could be completed prior to the exercise of evaluation or judgment by risk managers; and yet we were instructed to define ecological significance with no guidance from risk managers, who required that scientific input on ecological significance be gathered *before* they would make value judgments.

Both EPA and its contractors were comfortable with expanding the research portfolio of their common work, but the problem was that the methods available to risk assessors for estimating risks of cancer in human populations—developed over decades to provide data to support regulation of carcinogens—were simply inapplicable to questions like risks of food chain simplification. Now that Reilly had advocated a broader mission for EPA, including the protection of ecosystems and watersheds as well as human health, EPA—and the artificial science of risk assessment—faced a real quandary. Reilly was also pushing for the broader use of risk analysis and comparative risk calculations as the best hope to develop a more unified framework for analyzing data and preparing the information base for risk managers. Risk managers, as noted, were to decide on regulatory policy. But instructing agency employees to apply this risk assessment framework in ecological risk assessment only exacerbated—and made painfully obvious—the failure of assessment lingo as a comprehensive vernacular for the discussion of environmental policy. How could EPA and its usual contractors, specialists in human individuals' exposure to chemicals and the risk of cancer, transfer their risk equations and models from the narrower topic of human health risks to ecological risks? Should EPA and its risk assessors recognize that eco-risk was a whole new ball game and expect a new field, with new concepts, measures, and models to emerge to serve EPA's demand for ecological risk assessments?

We can begin to understand the politics of the research situation better by examining a book that was published about that time by one of my colleagues on the Risk Assessment Forum, Glenn Suter. Suter was then a respected risk analyst at Oak Ridge, and he had taken the lead among risk assessors in addressing the challenge members of that field faced as they were asked to broaden the scope of risks that could be comprehended in their models. The question for the discipline was, How could modelers in chemical toxicology retool their methods so that they could continue to provide the requisite science base for decisions to be made by risk managers who were charged to reduce ecological risk? This was the question Suter apparently sets out to answer in his book, *Ecological Risk Assessment*. The reader learns from the preface that the book is to be read both as a text and as a "manifesto." Suter acknowledges at the outset that "as of this writing, the U.S. Environmental Protection Agency has no agency guidance for ecological risk assessment" and that the lack of such guidance is "in large part due to the absence of formal ecological assessment methods for which to provide guidance." To remedy the situation, Suter recommends that those who use the results of assessments "provide assessors with more time and support to develop the needed assessment tools." This, he thinks, is preferable to a shift in "terminology" for assessing risks.[3]

Suter states that his treatment is not balanced, in that "more ecological risk assessments have been performed on aquatic ecosystems" and because "the body of data and models is far richer for aquatic systems." He bemoans the lack of studies of terrestrial system risk but acknowledges that this imbalance in his book "reflects the current state of environmental toxicology, chemistry, and assessment." Also, "risk assessments that predict the effects of new chemicals and effluents are emphasized," again because the standard risk assessment paradigm is designed for predictive assessments and because "most of the currently available tests and models were designed to support predictive assessments." This brings us to the manifesto aspect of the preface. Suter has admitted that risk assessors in fact have no methods by which to assess risks to ecological systems (except in the particular cases of some risks arising from exposure of fish populations to chemicals) and that the risk assessors have hardly any experience in examining chemical or any other risks to terrestrial systems. These are hardly strong recommendations for the services of the profession, and yet Suter ends the preface with an unabashed general advertisement for the services of risk assessors: "A hoped for benefit [of the book and of the method it presents] is recognition that ecological risk assessment, like other types of risk assessment, is an important and complex applied science that is worthy of support by private industry, governmental agencies, and environmental advocacy groups, and is worthy of the best efforts of the best students and environmental scientists." Aside from the job advertisement aspect, this paragraph emphasizes the need to establish ecological risk assessment as a *science*. Suter recognizes that it is strongly in the interest of risk assessors to carve out a "scientific" task in the ecological risk assessment/risk management process; otherwise, they and their labs will lose access to research dollars and jobs. But Suter admits that in ecological risk assessment the objectives are not nearly so clear as in human health studies, where the policy objective—reduction in risk of death or illness of human individuals—is obvious.[4]

Given the rigid separation of risk assessment and risk management, the risk assessors themselves cannot judge which values are important or which indicators best reflect valued characteristics of an ecosystem. All value judgments must be made by risk managers, so the device of an "endpoint" is used in order to sequester value decisions, keeping these independent from the scientific assessment. An endpoint, defined as "a formal expression of the environmental values to be protected," is furnished by the risk manager to the risk assessor as the starting point of a risk assessment. Defining an assessment endpoint requires two steps, according to Suter: "(1) identifying the valued attributes of the environment that are considered to be at risk, and (2) defining these attributes in 'operational terms.'" Although statutes usually lay out the broad values to be pursued, they are not usually sources of operational definitions. So "the second step is left to the judgment of the responsible agencies." Risk managers, in other words, must interpret statutes and turn these into operational and measurable goals.[5]

On what basis are these crucial decisions made? "In the past," Suter acknowledges, "little systematic thought has been given to this critical step."[6] Such an admission might give pause to inquirers less committed to the discipline, but Suter is unaffected. The problem that risk assessors are working with haphazardly and thoughtlessly prepared "formalizations of the environmental values to be protected" does not bother him very much. As a risk assessor, he has no responsibility for the quality of the endpoint choices made by the mysterious risk managers who work for the agency. Consequently, after one-half page of discussion (in a book of 538 pages), the problem of social values and how we formulate social values that are open to risk is set aside as someone else's problem. As a scientist, Suter is concerned with testing hypotheses. If someone asks him to test a hypothesis for which he lacks a method, he will oblige by writing a funding proposal to create such methods. The manifesto for a science turns out to be a shill for a large and open-ended research program, at public expense.

But this is not surprising in the bigger picture. The science of risk assessment is an artifact, created by EPA itself, of the RA/RM model that insists that the data for risk management be developed by scientists who never concern themselves with value or policy issues. In the early years of EPA, when the agency concentrated on human health only, it was easier to separate value questions from predictions of cancer rates associated with exposures. By pursuing narrow goals of cancer prevention, EPA narrowed and sequestered the value judgments that support this concentration on human health risks at the expense of all other risks and at the same time created a myth that all risk is risk to human health. As soon as an expansion of agency mandate and responsibilities was even considered, this value-neutral stance became untenable. It is necessary to assign values and priorities in order to assess ecological risks as risks to social values.

The science, risk assessment, that grew up in the research hole defined by EPA was thus too narrow for the broader tasks that Reilly envisioned. A clash inevitably occurred between Reilly's two priority goals for the agency: to expand the purview of the agency to include ecological as well as human concerns and to increase the formalization of decision support by expanding risk analysis to apply more broadly. Risk assessors, as Suter readily admitted, had no ready methods to offer such analyses. This quandary was especially evident when the Subcommittee on Human Health, composed of toxicologists, physicians, and chemists, which Reilly formed for the "Reducing Risk" study, refused to compare risks such as cancer risks and noncancer health risks because the "implicit value judgments and ethical issues" necessary to construct such an index were beyond the group's authority and confidence.[7] This reticence is the result of deeper incoherencies in the dominant EPA decision strategies: the decision strategies are deficient in ways of articulating, discussing, and deliberating about values, goals, and objectives. Agency managers have tried since the beginning to have it both ways. On the one hand, they want to insist that their contract scientists are "objective" and insulated from the political process; they believe this insistence is politically necessary so that they can refer to the evidence provided by these contractors as "objective" reasons to support the policies they advocate. On the other hand, in order to create political cover for themselves and their superiors, they have repeatedly pushed the problem of index and indicator construction off upon scientists, a task that cannot be accomplished if the scientists remain in the role of value-neutral hypothesis-testers and if the management of EPA continues to fail to address value questions and difficult trade-offs explicitly and openly.

By insisting that risk analysis can be undertaken in relatively insulated inquiries consisting of risk assessment (a purely formal, quantitative and predictive science) and risk management (a process of balancing risks and threatened values), the current strategy virtually ensures that there will be no open public discussion and deliberation of value issues. Political appointees and their appointers have every incentive to treat difficult decisions and trade-offs as purely "technical" or "scientific." And they are most comfortable when they can say a decision was "dictated by science." The problem was that EPA managers constantly shaped the science they received by the questions they asked; these, in turn, were affected by their value assumptions. EPA's rules for acceptable risk assessment explicitly required sticking to technical questions; but managers were themselves unwilling to take and defend any controversial value positions either, acting as if values were too subjective to discuss openly and rationally. The RA/RM process thus left no room for input about values, let alone providing a forum for deliberation regarding values and goals.

In the end, we will see, this inability to place priorities and value judgments up for rational discussion and debate left EPA open to political deal-cutting. Indeed, one of the reasons for EPA's creation was to form an administrative buffer against unpopular decisions regarding regulation of polluters. So politically appointed decision makers encouraged the myth that the science of risk assessment was untainted by political motives and decisions and provided solid data to support decisions. Risk assessors, for their part, had every reason to defend their work as scientific and above bias by political interference, thereby protecting their access to research funds even when changes in agency emphasis were motivated by political pressures. The risk assessors' research could thus be insulated from the political storms regarding what values, goals, and objectives to pursue. But a huge chasm—referred to as "risk management"—was created, across which communication, especially about values and priorities, was virtually impossible. Risk managers are the ones who have to act, but their actions, according to agency myth, must be based on "hard science." Since the hard science available grossly underdetermines what we ought to do in any particular case, this chasm is labeled "judgment" or "risk management." Scientists refer to the chasm as "uncertainty," but what actually occurs in the chasm is the balancing of political pressures from various interest groups. Value questions, at best, are simply resolved by splitting political differences; at worst, they are resolved by raw political power or by financial contributions to the campaigns of unscrupulous politicians. As a result, no intelligent discourse focused on environmental values, goals, and objectives has grown up around the agency. Into the vacuum created by lack of discourse about shared, community-wide values, of course, rush specialized individual and corporate interests.

Meanwhile, back at the Science Advisory Board, discussions of the economics of environmental policy in the Environmental Economics Advisory Committee were carried on in the jargon of economists, and as far as I could see, there was little interest in exploring social values that could not easily be understood as affecting aggregated individual welfare. The EEAC was well run, and the founding cochairmen, Allen Kneese and V. Kerry Smith, encouraged wide-ranging as well as technical discussions—I could not imagine a better education in the arts of environmental economics. But I was the only noneconomist among the regular members of the committee, and the disciplinary boundaries were drawn quite sharply; for example, though many members of the EEAC did research as risk analysts, the committee included, as far as I know, nobody who was a risk assessor. Economists worked to integrate risk assessment into economic science by using their traditional and newly developed techniques to estimate the public's aggregated willingness to pay for various risk reductions at various increments. They were interested, in other words, in measuring the economic value of risks, once characterized, but they viewed the actual assessment of risk from a chemical to be entirely outside their professional interests, better left to chemists, toxicologists, and epidemiologists.

The EEAC was in my view seriously impeded by its congressional mandate, which required that the EEAC advise the agency on issues having to do with *science and technology* but expressly forbade any SAB committee to address *policy* questions. This mandate may at first seem reasonable, but the most important questions to be answered in this area, I submit, are questions about which factual information is important information—important, that is, in the sense that the information affects social values. Questions such as this, however, cannot be reasonably discussed without reference to policy concerns. The artificial distinction between science and policy—a distinction that essentially parallels the artificial distinction between risk assessment and risk management in the discourse of risk analysis—was enforced on all SAB committees but seemed especially inappropriate for an economics committee. The EEAC could provide input on questions of whether a particular risk was "economically significant," but it could not ask whether that risk was "socially important" or make any policy recommendations regarding whether, or how, to avoid that risk. Application of this artificial distinction was crucial in setting the questions addressed by the EEAC, just as it had stilted the discussion of ecological risk in the Risk Assessment Forum.

The most important internal policy question facing the agency, Reilly's proposed expansion of the agency's purview to include ecological effects, was hardly discussed in the EEAC. In particular, we did not discuss the concept of ecological risk at all. Since they were restricted to discussing economic science, committee members could discuss only matters that could be measured, or at least characterized, using economic measures and models. So ecological risk was invisible to the economists on that committee, because they had not developed any vocabulary for evaluating systemic risks. I assert this on good authority: one of the subsequent cochairmen of the EEAC published, at about this time, a state-of-the-art book on economic methods for measuring environmental and resource goods, and in it he concluded that economic methods had not been developed to measure the value of ecological systems or to place a value on the risk of destroying ecologically based goods.[8] Ecological risk does not exist for economists because they do not have economic models that can bring such risk into their measurements and aggregations, which are aggregations of welfare effects on humans. By restricting SAB committees to "scientific" and "technical" questions in their discipline, Congress had in effect ensured that discipline-based science would be unable even to "see," much less measure, impacts for which no economic models or measures had been devised. The EEAC was formed as an SAB committee—rather late in the process of committee formation—only to oversee the "science" of economics.

The fiction was that the EEAC was convened to evaluate scientific data and theories and that the committee provided recommendations regarding economic science at EPA as science, but not in terms of its policy importance or implications for social values. Guidance suggested that SAB committees should all avoid appeal to values, except for the methodological norms of their discipline. This artificial requirement ensured that important questions such as whether to expand the idea of risk assessment to include ecological risk were left to risk managers, who were expressly forbidden to solicit advice and opinions from the SAB about what is "valuable." The fiction was maintained that even when commenting on economic implications of policies, the EEAC was evaluating scientific data and theories only with respect to their "scientific" validity; we were expected to be neutral in our policy positions.

And so it was that I, a philosopher, found myself wandering conceptual corridors that did not connect to each other, whether I was working for the Science Advisory Board or for the Risk Assessment Forum. Artificial distinctions, tracing back to the widely held assumption that science and values could be sharply separated, isolated the disciplines of scientists who were tasked to advise the agency. The different EPA offices thus solicited opinions from different groups of scientists of different disciplines; the problem of assessing risks in terms of ecological impacts of actions and policies was discussed by ecologists and toxicologists without input from economists, and economists couldn't even get the phenomenon on their radar screen because they were restricted to discussing ideas and concepts that had already been captured in their scientific models. There was nobody but the risk managers, whom I never met, and I to worry whether there were important questions of values involved in ecological risks and to ask whether there may be some social concerns that are not addressed within our current environmental discourse.

Again it was as if I were in a dream; in the dream, I felt great urgency to act, but each time I engaged someone in conversation, the person spoke another language. I would begin to pick up most of what was said by one respondent, beginning to understand the new language, only to encounter a whole new language as soon as I changed from one committee context to the next. When I talked to people on the various committees, trying to get consensus to act on our urgent common problems, I would repeat phrases and sentence fragments I had just learned on another committee and face the blank stares of incomprehension. To change contexts was to change languages; and as I encountered speakers of different languages, it became clear that they were not particularly interested in learning a new language, especially not a language from a foreign discipline. I the philosopher was left to ponder the multiplicity of languages I heard and to think about possible translations, which would at least initially have needed to be little more than hanging bridges among the otherwise impregnable towers of environmental science and policy.

1.2 Towers of Babel: The Structural Problems at EPA

We have already learned that the old EPA building grew in an ad hoc way, as new towers were created to house new programs in response to the growing legal mandate for the agency. Responding to the "chunkiness" of EPA's ad hoc growth, the developer created multiple towers, each with its own elevator shaft. The metaphor of a tower, which has relatively efficient vertical communications and transport but is intentionally impervious to horizontal interventions from occupants of other towers, is particularly relevant to EPA's problems in at least two ways: structural and conceptual.

EPA was created in 1970, and its current structure has been considered temporary by its administrators ever since. A brief history, which owes much to the more detailed account of Marc Landy, Marc Roberts, and Stephen Thomas,[9] will suffice to explain how it came to have, and maintain, its present structure, which—despite the physical movement of headquarters and most program offices out of the "old" building by 1995—continues to thwart all attempts to integrate environmental policy at the agency. According to Landy, Roberts, and Thomas, the creation of EPA and its original structure resulted from perceived political necessity, not rational planning. In the 1960s, both political parties were seeing their traditional power bases gradually eroded as the population shifted toward the suburbs and toward a style of life defined less by work relationships and more by leisure opportunities. A decline in union membership and the movement of population from city centers into the suburbs threatened the Democratic party, while a corresponding decline in numbers actively engaged in agriculture gradually emptied the small towns and farm communities, which were Republican strongholds. Meanwhile, news coverage expanded rapidly, and the national news was sprinkled with stories about poisoned water systems and pictures of oil-drenched birds; the media became a major force in increasing pressure for environmental action and cleanups. Both parties were thus anxious to appeal to the growing population in the suburbs, who exhibited greater awareness of environmental problems and their threats to the quality of American life.[10] Both parties, however, found strong resistance to many pro-environmental policy proposals from among their traditional supporters, and taking a strong and unqualified stand on environmental issues threatened both parties' traditional support base. For the Democrats, demands to clean up polluting industries in the decaying center cities alarmed labor by threatening to eliminate jobs; opposition to dams and large rural development projects was unpopular with small-scale farmers and rural people who had depended on such big projects for rural development, protection from drought and floods, and jobs. For the Republicans, too, direct appeals to suburban environmentalists were resisted by traditional supporters in the business community, who were concerned about increasing regulation and calls for spending on pollution controls.[11]

EPA was thus born as the parties tried to capture support of the growing suburban, leisure-oriented voters while simultaneously protecting their traditional, and threatened, constituencies. As Landy and coauthors put it: "Leaders in both parties sought to frame environmental programs to avoid these cleavages. A multibillion dollar grant program for the construction of local sewage plants allowed congressional Democrats to make pollution control palatable to the construction unions. Bold statutory preambles, which trumpeted the elimination of all pollution, were often joined to cumbersome enforcement mechanisms designed to appease both labor and management in target industries. The resulting incoherence in the design of environmental programs was therefore due in part to the ambivalent motivations that produced them."[12]

The Republican administration of Richard Nixon was goaded to action by growing success of congressional Democrats, led by Senator Edmund Muskie of Maine, in fashioning environmental legislation and appealing to environmentally aware voters in the new demographic situation. In early 1970 a task force, working under the direction of Richard Nixon's top domestic policy adviser, John Ehrlichman, produced a preliminary report that recommended establishment of a Department of Environment and Natural Resources (DENR). This new department would replace Interior and absorb the Forest Service and the pesticide program from Agriculture, the Army Corps of Engineers from the Department of Defense, and responsibility for Air Pollution Control from Health, Education, and Welfare. This proposal, which would of course have been very disruptive of traditional organizational structures in the federal bureaucracy, attracted support from two sources (despite the obvious entrenched interests favoring the status quo). First, some White House advisers —thinking politically—wanted an umbrella department, including agencies charged with both development and conservation responsibilities, allowing the department secretary to act as a balancer. In a supersized department, the department secretary would orchestrate compromises between advocates of development and of environmental protection and provide political cover for the White House by keeping the difficult trade-offs at arm's length. Second, Nixon had set up an independent advisory commission on government reorganization, which came to be known as the Ash Council, named after its chairman, Roy Ash. This commission at first toyed with the idea of a massive reorganization of the entire federal bureaucracy into four megadepartments;

the large scope proposed for DENR fit nicely into these plans, and the idea began to take on an independent life. At first it seemed as if the idea of a DENR had political legs. Nixon endorsed it and sent it forward to Ash's group for further planning.

In the end a smaller group, known informally as the "environmental protection group," was given responsibility by Ash for developing a detailed proposal for the reorganization, and this group did not believe the DENR plan was politically feasible or desirable. They objected, on political grounds, that the vast reorganization would never make it through Congress, given entrenched interests. Further, they did not think it was desirable to have environmental and development interests "traded off" against one another in a superdepartment—they feared the latter interests would too often outweigh the former and that no elected official could be held accountable. So they recommended the development of a new department, answering directly to the president. They thought the creation of an independent agency would make a statement that pollution control was important—cover needed by a president who feared having to run against Muskie in the next presidential election. Nixon therefore had an incentive to act decisively in favor of pollution control. But the DENR concept was savaged also by the rest of the Nixon cabinet, except for Walter Hickle, the secretary of interior, whose portfolio would have been expanded, and Secretary John Volpe, who would not have been affected. Nixon—lacking faith in Hickle and impressed with the intensity of the opposition, backed off from the DENR plan. It looked as if no reorganization would occur at all.

The political impetus to act remained strong, however, as Muskie shifted his position toward stronger and stronger environmental protection and created an expanding power base by skillfully using the congressional system to write more and more specific instructions to regulators. Seeing a compromise that stood halfway between the opposing forces—those who wanted a major reorganization and those who wanted no change—Nixon decided to add a new, noncabinet agency to address the problem of pollution and move a few pieces of other departments into it, disrupting the bureaucracy as little as possible but signaling aggressive action against pollution. The reorganization plan went into effect in September 1970, unopposed by Congress, and transferred more than a dozen agencies and offices from Interior, HEW, Agriculture, and the Atomic Energy Commission into the new Environmental Protection Agency. These offices and programs—including among others the Office of Research on Effects of Pesticides on Wildlife and Fish, the Bureau of Waste Management, the National Air Pollution Control Administration, and the Office of Pesticides Regulation—were moved into the new environment despite huge differences in mission and in agency cultures. For example, the Office of

Pesticides Regulation, previously under Agriculture, had an intimate relationship with the chemical pesticide industry it regulated, and its staff served mainly as marketing agents between pesticide manufacturers and farmers; strict regulation of agricultural chemicals went directly against the bureaucratic culture of the office. To merge this set of personnel into an "environmental" agency was to expect them to regulate the industry they currently served. And this is only one example of the difficulties faced in stitching together a quilt from such diverse fabrics.

The force of political necessity, amazingly, did join this group of already concretized, impervious units, wrested from their past institutional contexts, into a new agency. In the process, the worst of all possible outcomes came about: an agency charged to organize and integrate environmental policy but lacking the comprehensive scope of the proposed DENR and hence unable to address the whole range of environmental challenges. New departments were added to the agency, which was formed of already Balkanized and unintegrated units from elsewhere in the federal democracy, ensuring that there would be no common policy ground, no shared language of analysis, and no comprehensive discourse in which to discuss environmental problems. Instead, ownership of vocabulary was given over to a variety of disciplines and interest groups. The towers, it turns out, are the Towers of Babel, reminiscent of Jehovah's curse of multiple languages imposed upon humankind for the hubristic act of beginning construction of a tower to reach Heaven. He gave them multiple languages and dialects, dooming this and other projects that would require cooperation.

The combination of forces, and of multiple languages and conceptions, at the agency has led to incoherence of policy discourse and to real policy incoherence, opening the door to political pressure and political shenanigans. Worse, it has kept EPA from creating a common forum for deliberation, communication, and cooperative action in the face of uncertainty. The public came to see EPA itself, with its various offices and disciplines—what I have called policy towers—as a collection of interest groups.

The grab-bag nature of the agency was obvious to everyone, and it was expected that a major reorganization was necessary; the only debate on the matter within EPA was whether to reorganize first or to begin working, develop a track record, and reorganize while proceeding to regulate. The new administrator, William Ruckelshaus—who had a history of nailing polluters in complex enforcement actions—knew he had to move quickly to establish the agency as a force, to reduce its vulnerability to enemies on all sides; he also recognized that a major reorganization would consume huge amounts of energy and would pay off only in the long run. Nixon's executive order establishing the agency seemed to emphasize the need for reorganization, but

Nixon was preoccupied with other matters and there was no unanimity on the White House staff in giving Ruckelshaus his marching orders. Needing instant credibility, he did what he and his staff knew how to do—they went to court. In the first sixty days of its existence, EPA's enforcement offices brought five times as many enforcement actions against polluters as had ever been brought in a similar period by the precursors of the agency. Although this confrontational approach took the White House by surprise, Ruckelshaus was so successful that he immediately developed a public constituency that favored enforcement of pollution laws; Ruckelshaus emerged as the corporate dragon-slayer necessary to neutralize crosscutting interest groups and improve the environment.

Because he had acted largely independently of the Nixon administration, and because he developed an independent constituency among the public and environmental groups, Ruckelshaus and EPA became an independent force in the administration, at least somewhat buffered from business interests by their independence. In principle, Ruckelshaus acted at the behest of the president, but his success and independent constituency—not to mention internal disagreements in Nixon's inner circle—gave him considerable leeway and independence. In an additional, ironic twist, independence from Nixon—and the need for support in Congress—drove the agency into the arms of the effective cadre of environmental advocates assembled by Muskie around his Environment Subcommittee of the Senate's Public Works Committee. Ruckelshaus's aggressive approach to enforcement—in conjunction with Muskie's insistence on the independence of the agency from administration-based micromanagement, which he ensured by writing ever more detailed regulations into legislation—provided the new agency immediate credibility and stature; but it was a high-cost strategy. It angered business interests and development-oriented forces in the Nixon administration and, in an act of omission that would prove crucial for its future, delayed the possibility of a systematic reorganization and integration of agency tasks.

The original plan to reorganize the agency gave way to a three-stage plan advocated by Douglas Costle, who was the Ash Council staffer in charge of the transition and who would later be the administrator of EPA under President Jimmy Carter. In the first stage, each of the program areas that were moved in from other agencies would remain intact and begin to fulfill their traditional, or newly assigned, duties, as appropriate. Then, as a second stage, functional divisions would be added. Finally, the idea was to abolish the program offices and merge their efforts into the new, functionally defined units. Costle did not recommend a more comprehensive reorganization, fearing that the fledgling agency could not survive large bureaucratic changes. He argued that the agency should first prove its mettle; reorganization should await consolidation of a power base. Ruckelshaus completed the first two stages of the plan by dividing the EPA into five divisions, each headed by an assistant administrator but including three functional units (Planning and Management, Enforcement and General Counsel, and Research and Monitoring) and two program offices (Air and Water was one of them and Pesticides, Radiation, and Solid Waste was the other), which retained programmatic organizational structures.

When Jimmy Carter was elected president, Costle became the first head of the agency appointed by a Democrat, even though his background had been mainly serving under Republican administrations. Costle was a big supporter of Carter's goal of reforming the Civil Service, and he set out to make EPA more efficient—a model of good government. Although support for broad-based environmental policies remained strong, according to the polls, experienced political operatives insisted that the environmental movement—especially advocacy for wildlife and ecological protections—was "soft" and would fade. Costle, recognizing that there was still support for merging EPA into Interior and fearing a merger that would leave EPA as a minor actor in the huge Department of Interior, actively distanced his agency from wildlife issues, concentrating instead on pollution problems and their human health effects. He perceived that the high costs of pollution control programs would excite opposition and that the new agency would need a strong constituency to remain viable. Unsure how the fledgling agency would fare in competition with Interior and other cabinet departments if it competed with them to address wildlife issues, Costle saw that the area of environmental health—the impacts of pesticides and other chemicals on human health—was wide open for growth. He moved quickly to institute a human health agenda; EPA was the largest regulator in the area and the only agency with both regulatory and research agendas. In 1978 he announced, "EPA is a preventive health agency as well as an environmental agency," and EPA—still organized into crosscutting functional and program-oriented fiefdoms—began concentrating on chemicals that might cause cancer. The shift proved both effective and timely. With most environmental programs suffering cuts or at least freezes in the 1979 budget because of financial austerity, EPA was able to increase its budget by 25 percent. EPA retained its improbable and unworkable organizational structure but continued to grow, nonetheless, by responding to public concerns about the many chemicals being developed and released into the environment and about environmentally induced cancers.

As it turns out, then, EPA—like its old building—was in effect designed so as to thwart its central purpose of coordinating environmental policy, since additional space was added in a manner reflecting the chunk-by-chunk legislation despite the lack of a rational structure. Like its old building, the agency—consisting of many independent units inherited from other agencies that were perpetuated by disciplinary and administrative turf wars—was largely maintained as separate units. But these units took on more than historical identities, and it turned out that offices acted as representatives of, or were used as pawns of, powerful interest groups, which had their own perspectives. To concretize the idea, I have called such units "towers": they are policy fiefdoms that, though fairly effective in serving valued constituencies in some cases, are virtually impervious to information or cross-tower fertilization; these towers are practically unable to act in concert, even when their goals overlap. Here I have mainly repeated some elements of Landy and coauthors' much more detailed analysis of EPA's origins and development. Their story is, among other things, an account of lost opportunities to reorganize the agency and put it on a more rational and integrated footing. These opportunities were lost, usually, because of successive administrators' perception—often accurate—of the vulnerability of the agency to political enemies. So the old EPA building, with its division into horizontally impregnable, vertically integrated towers, stands as a fitting metaphor for the agency itself. For much of its thirty-plus years, the agency has been considered emblematic of inefficient government: characterized by failure of offices with similar tasks to cooperate, by policy reversals, and by the failure of offices to communicate in search of integrated environmental goals.

This, in a nutshell, is the structural problem with EPA; as just noted and as Landy and his coauthors clearly explain, the problem goes deeper than just unfortunate organizational structures. EPA's inefficient organizational structure and its inability to pursue integrated policies rest on a deeper cause. It will be impossible to unite environmental management under a single, coherent structure until a common conceptual framework emerges for articulating environmental goals perspicuously and for relating various environmental policies to each other.

1.3 The Costs of Not Being Able to Get There from Here (Conceptually)

We have seen that the organizational structure of the Environmental Protection Agency reflects the complex political forces that spurred its creation and development. Of course, these political forces reflect, on a deeper level, the attitudes and understandings of the society. Now we must examine this deeper social reality, exploring the emotional and intellectual forces that encourage Balkanization—what we have metaphorically called towering—in EPA. Towering occurs when bureaucrats and policymakers develop narrowly defined interest areas, respond only to other participants who share their own views and vocabularies for discussing those views, and insulate policy processes from open debate and challenges from critics. The old EPA building exhibited towering as a physical structure because—being designed willy-nilly around multiple elevator shafts—the building was so organized as to reduce horizontal interactions and cross-boundary relationships. This physical structure undoubtedly encouraged isolation and towering in day-to-day actions at EPA, exacerbating the political, disciplinary, and interest-group divisions that shaped the agency in its early years. I am not claiming, however, that the unfortunate physical structure of the old EPA building *created* isolation of EPA programs from each other or that the design *caused* failures of communication. Although it is hard to believe such concrete barriers did not complicate the communication problems within the agency, it would be more accurate to say that the two phenomena, one of physical space and the other of organizational structure, share a common origin in the fragmented legislative mandates that came from Congress. I claim only that the physical structure of the old EPA building provides an illuminating metaphor for the broader intellectual and social forces that deter open and effective communication regarding environmental problems and environmental policy goals in our society.

For the task of explaining the underlying reasons EPA has been so susceptible to conceptual towering, I am fortunate to be able to rely once again on the outstanding study by Landy, Roberts, and Thomas, who undertook a detailed analysis of EPA performance in the late 1980s. Their book was published in 1990 and again in an expanded edition in 1994. Its preparation involved careful observation, many interviews with EPA staff, and the benefit of close collaboration with Douglas Costle, who was an active participant in the research project while visiting for a semester at Harvard University's Public Health School. A lot has happened since that systematic study was completed, but my own observations and anecdotes—less systematic than their work, but more recent—strongly suggest that with respect to underlying causes and driving forces, little has changed at EPA. The basic problems described by Landy and colleagues remain the basic problems of today.

Because of the systematic and thorough nature of the Landy study, it is worth summarizing here the authors' conclusions and key parts of their analysis; they perfectly articulate the problems toward which this book is directed. Landy and colleagues argue that the performance of an agency can be rated on four standards: (1) fidelity to technical merits, (2) promoting civic education, (3) responsiveness to the public, and (4) building institutional capacity. After a detailed investigation, they conclude that EPA failed to live up to each of the four standards, and they explain why this is the case. In a section entitled "The Root of the Problem," the authors assert that their experience disproved the hypotheses that the failures of EPA resulted from lack of motivation, lack of commitment, or lack of hard work by employees. Their interviews and follow-up work convinced them that EPA employees were for the most part highly motivated, hardworking, and committed civil servants. Further, though they detail the structural problems that I summarized in section 1.2, they deny that these failings ultimately explain the most basic failures of EPA. In their words: "We find the source of the difficulty not in the motivation of public officials or in the obvious problems posed by organizational imperfections and limited data, but in something that lies behind all of these, the ways in which public managers think, reason, and decide."[13]

The underlying problem, Landy and his coauthors believe, has to do with how managers—and the whole society—think about and discuss environmental issues. I will let these authors speak for themselves:

> The crude models and theories people ordinarily employ are useful precisely because they embody simple distinctions, and thereby serve to facilitate routine decisions. But, a choice is often required between alternative simplifications, categories, or models, none of which convey all that there is to know about a particular situation. Choosing among such alternative formulations requires some pragmatic considerations. What aspects of a situation are most important to explain? . . . Does a given way of looking at a problem highlight critical choices by leaving out extraneous detail; or obscure the situation by oversimplifying and omitting relevant options and consequences?[14]

Landy and colleagues state that EPA works so badly because it has never evolved an effective way to encourage communication across the multiple disciplines and because diverse and incommensurate models for characterizing, understanding, and addressing environmental problems remain in use within and outside the agency. In particular, they fault EPA for not educating the public to understand the complexities, instead using oversimplifications to justify and expand their regulatory activities. They go on to assert that the failure to introduce a unifying vocabulary for communication among offices and disciplines undermines effective government at EPA: "These features of thinking in turn complicate the organization and operation of government. The simplification characteristic of all analyses means that multiple accounts of a situation are possible, indeed likely. Real problems almost never match the domain and perspective of any particular specialist. Different professional and functional groups are likely to see the same problem differently and to disagree in ways they find difficult to resolve."[15]

Applying these concerns to EPA, they note that EPA employees include engineers, scientists, and attorneys and that these professions have very different perspectives on environmental problems. "The EPA was and is organized in ways that reflect and reinforce these professional cleavages." But they also note that "the deeply held differences in society about environmental policy do not stop at the Agency's doorway,"[16] acknowledging that disagreements at the agency are expressive of society-wide conflicts. The problem, then, is that neither our society nor employees of EPA share a common model for understanding and addressing environmental problems. It is this lack of a shared conceptual model that prevents communication among segments of the public and encourages members of different disciplines to talk past participants from other disciplines. In short, the lack of a shared model sets the stage for ideological environmentalism, the situation in which preexperiential assumptions dominate because there is no shared framework for accounting for or testing reality.

Landy and colleagues further characterize the problems at EPA as an excess of "pluralism" and too much advocacy—by various offices and the agency as a whole—for environmental goals. I will not follow their terminology, since I wish to reserve the term *pluralism* for another use later in the book. But my analysis is directly responsive to the problems they pose so clearly. By pluralism, Landy and coauthors mean the tendency for an office or a program to become an interest group and adopt interest-based patterns of behavior, paying little attention to criticism from sources outside the group. This is the phenomenon I have called towering. The point is that our analyses correspond—though we use the term *pluralism* differently—and that I follow in the footsteps of their detailed analysis in identifying the most basic problem as one of communication and associated lack of coordination within the agency.

I wish I knew how to resolve the organizational problems at EPA. If I did know how to reorganize EPA and make it work, I promise you that I would write that book, too. But I think I'd better leave the EPA structural problems to my colleagues in organizational behavior and concentrate my efforts on the communication problem itself. I intend to address the lack of a unifying language or a shared set of linguistic conventions that, if present, would allow communication between the towers at EPA and across scientific and other policy disciplines in the broader policy debate. The problem is that neither EPA nor the society as a whole has established a unified discourse about environmental policy, a discourse in which all voices can be heard and in which communication, deliberation, and experimentation can take place in an open and inclusive public manner.

The fiction that the science of risk assessment could be separated from the process of evaluation encouraged the separation of scientific modeling from policy discussion. The presence of economists as virtually the *only* social scientists in the agency ensured that the ecological scientists would be marginal to the policy process. The tragic thing is that most ecologists—still anxious to earn their spurs as "hard" scientists by establishing generalizations and testing predictions based on hypotheses—welcomed the partitioning of science from values and policy issues. Within the agency, the staff of the Office of Policy, Planning, and Evaluation, dominated by economists, insisted that they were the ones to do analysis and to make recommendations; Congress, however, by writing requirements, often thwarted economists by choosing politically painless solutions over economically sound ones. Economics is a special case, of course, because much legislation requires that economic analysis play some specified role in rule-writing. Although there are a few—very few—noneconomist social scientists at EPA today, they have no unified voice in directing research or in affecting policy. My work with EPA has convinced me that there simply is no constituency to support noneconomic social science research, the kind of broader social science research that might help to develop broader methods of evaluation and to explore procedures for developing consensus in policy development and implementation. In essence, then, EPA is forbidden by legislation to seek advice about goals and policy options from scientists as employees on the Science Advisory Board, and its practice of strongly favoring economists, engineers, and attorneys over noneconomist social scientists creates a conceptual no-man's-land of value discourse, a land defined by the limits of economic analysis in expressing many environmental values and by the reticence of ecologists to become involved in discussions of social values and policy objectives.

Following in the footsteps of Landy and colleagues, we are now very close to the "root" problem at EPA. The absence of noneconomic social science input, coupled with the sharp separation of assessment of risk from "managing" risks, ensured that the discussion of environmental values would be politicized. Risk managers are, of course, political appointees, and their decisions are vulnerable to capture by interest groups. The tragedy is that the structure of EPA and its offices guarantees disciplinary imperialism and turf wars. The turf wars only exacerbate the communication problem and corroborate the conclusions of Landy's group, that the root of EPA's problem is the lack of a common framework for discussing and analyzing environmental problems. There is no common framework for articulating and comparing environmental values. Public discourse about environmental goals is thus doomed to towering and associated failures of communication across disciplines and interest groups. Towering was very evident in my experiences at the SAB and with the Risk Assessment Forum. In both experiences I saw well-meaning individuals formed into groups to compare careful reports that will never be read by the people who most need to read them. They will not read them because the reports are, in effect, written in the wrong language. In both cases the dialogue was captured by a particular discipline or interest perspective, essentially freezing out of the discourse other participants in the wider dialogue.

Now we can return briefly to the Ecosystem Valuation Forum, which was a modest effort to bring together a multidisciplinary group of scholars and practitioners with policy analysts from EPA; there was a professional facilitator leading the meetings to encourage creativity and capture insights. I was invited to join this group, once again fulfilling the quota of one philosopher, and found it the most pleasant and promising of all my activities at EPA. Some midlevel managers at EPA had brought together a great group of economists, ecologists, and ecological economists, many of whom I knew or at least knew of, professionally. Our assignment was to meet two or three times a year around a big square table and hammer out a framework for talking about how to evaluate changes in the conditions of ecosystems. We were also to explore ways to relate such changes to economic and other impacts on social values. At the first meeting, all the economists sat on one side of the big square table, and the ecologists sat on the other side; I and a few other renegades and antidisciplinarians sat at the end of the table, facing the facilitator and her large blank newsprint pads. The discussions were reasonably cordial, but it quickly became obvious that there were deep conceptual differences. The economists could not see how one could even talk about valuation unless one could somehow relate changes in ecosystem states to changes in individual welfare. They thought the only reasonable strategy was to discuss values within partial equilibrium models, using aggregated welfare as the way to measure social value. The ecologists insisted that models include dynamic processes, recognize discontinuities in natural systems, and accept the possibility of change on any and every scale of the system. Although the discussion was somewhat chaotic, the forum was intellectually stimulating and fun. By the second meeting, the disciplines were mixing up the seating arrangements and forming small interdisciplinary knots of informal working groups. I felt that if progress was to be made on the problem of developing a unified approach to environmental valuation, it would have to happen in a setting such as this, one where the disciplines were thrown together and would have no choice, if there was going to be a final report, but to develop some kind of a language to communicate the cross-disciplinary results.

After the first year, however, EPA withdrew the funding for the Ecosystem Valuation Forum, implying that it was not yielding results fast enough. Instead, a single field study of a single tributary of the Chesapeake Bay was supported. It was a great disappointment to me when the research design for the field study was published along with some other papers from the forum, contributed by the participants, in a special issue of *Ecological Economics*. The researchers who were chosen to do a field study had decided not to build a unified model or framework of analysis. Violating what I took to be the central goal of the Ecosystem Valuation Forum, the research was carried out by fairly separate teams of economists and ecologists, and they built two parallel models, one tracking land conversion and ecological effects in the watershed and the other tracking property values. These separate models were to be calibrated so that they could accept data from each other, but they remained separate economic and ecological models. So much for our efforts to address the problem of developing an integrative language for environmental policy head-on.[17]

So EPA killed the one forum that, in my experience, was making a creative effort to address the problems of interdisciplinary and communicative towering at the agency head-on. They short-circuited the attempt to bring the descriptive disciplines together with evaluative ones by failing to give the process time to work; meanwhile, the monolithic panels and groups—well supported by a single discipline or an office and united by ideology and interest—continue to thrive within their isolated towers, honing their narrow disciplinary skills and getting better and better at repelling all criticism from other disciplines and other interest-based groups.

This book is my best effort to understand and illuminate this problem of failure in public and bureaucratic communication. My analysis, which builds explicitly upon the empirical study of Landy and associates, is both broader and narrower than the target of the Landy study. Whereas Landy and colleagues looked at several case studies and analyses and a variety of problems of coordination at EPA, I concentrate on the linguistic and communication aspect of the problem. I do, however, examine this problem of communication in the broader context of environmental policy discourse, not in the context of actions and policy decisions within EPA alone. I intend to look hard at the failures of clarity and communication in the message environmentalists project, seeing EPA in its old building as a basic metaphor for the need to illuminate a most confusing area of public discourse.

1.4 Hijinks and Political Hijackings

Towers, reinforced by the use of discipline-based jargons into which foreign criticisms and ideas cannot be translated, protect the occupants of the towers from new ideas or criticisms from other disciplines, offices, and programs. Because towering is associated with the development of a specialized language or jargon, towering and its communication patterns encourage collusion between regulators within the agency and trained lobbyists and advocates from related interest groups. As revolving doors allow regulatory bureaucrats and lobbyists to move into and out of an agency tower, their movement facilitated by the development of a culture tied together by a common lingo, an artificial bias is created within the tower, just as EPA's demands for a "pure science" of risk created an artificial, narrow discipline defined by decisions within the tower. In this context, risk assessment evolved as a science so narrow that it cannot adapt to problems of ecological risk. By avoiding values discussion so assiduously, EPA bureaucrats and scientific risk assessors working under contract shared an interest in the creation of "pure" data on risks. The risk assessment community both within and outside the agency has not been able to respond to the more complex valuations required by a move from assessing human health risks to assessing ecological impacts on the full range of social values. Inhabitants of these towers, in conjunction with an external, interest-group community that speaks the same specialized language, create an exclusionary policy cabal, able to speak to and influence each other but deaf to criticism from other disciplines. The language used is in this sense impoverished, leaving members of the cabal trapped within their specialized discourse, unable to address complex value-laden questions about how to protect the full range of social values. These inside-outside liaisons and the resulting vacuum in evaluative discourse make agency policy vulnerable to interest-group manipulation of issues that should—but for the vacuum—be resolved by the interaction of scientists with advocates in an open search for explicit environmental goals.

The central, critical hypothesis of this book is that the vocabulary currently available for discussing policy goals and policy choices is so impoverished that it blocks communication and prohibits rational deliberation over the right questions. There is no accepted language for discussing, comparing, and prioritizing multiple values. Every interest group, clinging to its own values and trapped in its own specialized language, talks past others with differing interests and values. These linguistic/communication problems, symbolized by the old EPA building—but negatively affecting virtually all public discourse about the environment—will be shown to contribute heavily to policy failures.

Let us look at a case study: the failed attempts to protect America's wetlands, especially the administration of section 404 of the Clean Water Act, which has been interpreted to govern the permitting of dredge-and-fill operations in the surface waters—including wetlands—of the United States. The permitting process is jointly administered by EPA and the Army Corps of Engineers (ACE), with the cooperation of states. This case illustrates how towering leads to problems of communication among offices with overlapping, patchwork mandates. These communication problems appear as conceptual difficulties, as inhabitants of different towers have failed to connect across disciplines, preventing an approach to management that would integrate factual and evaluative information in the policymaking process. The case of the wetlands also reveals an even more specific area of communicative failure: EPA and ACE have failed to develop a plausible and rational method for evaluating wetland policies. Especially they have failed to develop an ecologically sound and transparent terminology for linking changes in wetland structure or functioning with important social values. Their failure has especially dire consequences because the regulatory framework developed under the Clean Water Act has established "banks" that create wetlands as substitutes for destroyed wetlands. The failure to develop methods to evaluate losses of function leaves permitters completely in the scientific dark in setting ratios and trade-offs. The door is thereby left open for political maneuvering, and as a result the politicians, far more than the ecologists, have controlled wetland policy. The case of wetland policy paradigmatically illustrates the phenomenon of policy towering by embodying all of its main aspects. Structural towering is evident in the way ecologists and policymakers at EPA have failed to work with engineers at ACE to develop an integrated policy of wetlands protection. This one case illustrates the linguistic and communicative failures that reinforce and are reinforced by towering.

That our wetlands policies have failed is no longer in doubt; the only question is, What will we do about it? A recently released report of an expert National Research Council panel assembled by the National Academy of Sciences concludes that the program has been badly administered and is failing to protect wetland functions and values.[18] The best that can be said is that by imposing bureaucratic costs on developers who choose to build on wetlands, the regulatory system encourages the choice of alternative, nonwetland parcels for development. But the program, as evaluated by the expert panel, has failed in its central goal, which is to ensure no net loss of wetland function. Few of the created wetlands that have been studied even come close to replacing the functions lost as a result of filling existing wetlands. This is true despite the usual requirement that replacement wetlands be built at a ratio greater than acre-for-acre. Even when constructed wetlands have fulfilled permit criteria, the artificial wetlands have failed to provide the functions—flood control, pollution abatement, and other functions—that were provided by the natural wetlands before they were destroyed. The panel found significant published support for this conclusion within the rather thin record. Worse yet, because of very lax follow-up, many contracted wetland projects have not even been started; others are left incomplete, suggesting that even the less stringent acre-for-acre requirement is not being fulfilled, largely because "performance expectations in section 404 permits have often been unclear, and compliance has often not been assured nor attained."[19] In summary, the panel concluded that we are failing to attain the widely accepted goal of wetland protection, which is "no net loss of wetlands." This case study therefore illustrates most of the problems that prevent the integration of scientific and evaluative information in policy formation.

Any analysis of the failures of wetlands mitigation policy must begin with the question, Why have professional ecologists—especially academic ecologists—played such a limited role in this huge ecological and social "experiment"? First some basic background. Federal wetlands policy has undergone a huge turnabout in the past fifty years. Not too many decades ago, the federal government was providing economic incentives for swamp-busting as a natural outgrowth of the widespread belief that wetlands—"swamps"—were either useless or positively harmful. As ecological information came in, however, it became obvious that wetlands perform many important, sometimes essential, functions in supporting social values: flood control, nutrient removal, wildlife habitat, and many others. The Clean Water Act and a hodgepodge of other legislation are interpreted to require permits to dredge and fill many U.S. waters. These acts evolved into the permitting process for dredge-and-fill operations now administered by EPA and ACE.

By the late 1980s, a consensus had formed around a slogan, endorsed by both major political parties and most agencies: "No net loss of wetlands." This ecologically vague but politically potent slogan became the centerpiece of wetlands management for at least two administrations. The problem was how to turn the vague slogan into effective policy, a problem that apparently is not yet solved. Ecological vagueness has encouraged prodevelopment politicians to battle environmentalists over how to count wetland losses. One question that was particularly vexing, as will be explained presently, has been, Should we count acres or measure functions when we insist on no net loss? This question is more complex than it might seem, because, as noted above, ACE's permitting process for dredging and filling has evolved to allow "mitigation banking"—the substitution of created or augmented wetlands for those filled under a permit. When a project threatens the public good, including environmental values—and explicitly including wetland values—the Corps can require either the elimination or the mitigation of impacts affecting the public interest. Mitigation has been interpreted to mean that if someone fills a wetland, the person (or other entity) can be required to build another wetland elsewhere as compensation. At first the expectation was that permittees would provide in-kind and on-site compensation—so that destruction of wetlands on one part of a property would be compensated by a similar wetland nearby. Gradually, however, large developers, rather than dealing with many such projects, would build one or a few large off-site wetlands and trade these acres as compensations for filling smaller sites. Eventually, by the mid-1990s, mitigation banking had become a business. Landowners began creating or restoring wetlands in order to sell credits to permittees who were required to provide mitigation for wetlands that they filled in development projects. Because constructed wetlands do not usually provide functions comparable to those provided by an undisturbed wetland, developers are usually required to support the creation of one and one-half to two acres for each acre of natural wetlands filled.

Although the number of mitigation actions is not huge (from 1993 to 2000, 24,000 acres were filled per year and 42,000 acres were required for compensation), one can see how this system highlights, rather than resolves, the question of whether to trade off functions for comparable functions or to count acres. Since science is inadequate at this time to relate structural requirements (percentage cover by various plants, for example) with desired functions, it is difficult to tell in advance which created wetlands will actually replace lost function. The problem is exacerbated by the long time required for a wetland to become established and stabilized—sometimes decades or more. Since permits are usually closed out after five or fewer years, performance requirements tied to function are difficult to enforce. The Corps's regulators can require only certain structural features, and they assume that function will follow. Unfortunately, the science isn't good enough to ensure that the desired functions will arise from the structural features, and given techniques now available, many wetlands that fulfill structural requirements do not develop adequate functioning.

The failure of wetland protection policies, with special emphasis on the mitigation banking aspect, is featured here because it so clearly illustrates three key aspects of towering and the way towering impoverishes public policy discourse about environmental problems and goals, leading to failed policies. Focusing sharply on these three aspects of towering in one case will help us recognize the pattern of maladies associated with the towering syndrome as it appears in the examples and case studies cited below.

The three aspects of towering that concern us most closely in this book are as follows. (1) Towering is often associated with an obsessive insistence on a sharp separation of science and values and with a clear segregation of processes of information-gathering from processes of policy choice. Too often, this means that the "right" studies are not done and that the best scientists create abstract models that cannot be adapted to particular decision situations. In the wetlands case, for example, the information that the engineers need—studies of what structural requirements are likely to result in a stable system that fulfills social values—has not been provided by ecologists, so enforcement of the no-net-loss rule has been severely compromised. (2) Towering increases miscommunication and creates blockage in the flow of important and relevant information, creating "blind spots." The wetlands case illustrates this towering failure also. Because neither ecologists nor policymakers had articulated the social values associated with wetlands, the engineers at the Army Corps wrote permit conditions not in terms of performance with respect to protection of values, but in terms of structural characteristics of proposed constructed wetlands. Since most advocates of the no-net-loss rule interpreted it to mean no net loss in socially valued functions, they assumed that the engineers were working for preservation of such functions. But alas, even when the engineers stated and enforced rigorous structural requirements on permittees, the criteria they created had no operational connection to the goals of policy. (3) Because towers tend to block the flow of information discordant with one's own beliefs, they prevent true learning from occurring over time. Good environmental management must be adaptive and experimental, but it is not a matter for specialists and experts alone. As John Dewey recognized, in a complex, technological society, democracy depends on the experts being teachers who guide and contribute to a many-sided public debate about what to do. But information must flow both ways: the scientists must also learn from the public discourse. The experts can learn which phenomena and which dynamics matter in important policy decisions, and this knowledge can guide their research toward questions that really matter in policy choices. In the next several pages, I will highlight these three aspects of towering by means of the wetlands case and sketch briefly the response I offer in this book to each aspect.

The first aspect of towering is that it is usually associated with a strong insistence upon the separation of science and policy. Further, we will learn, it is usually assumed in these contexts that the relevant scientific information can be gathered and packaged *before* direct engagement in value analysis or goal articulation. This is a view I will criticize as the "serial" view of science and policy. The serial view, though seldom stated explicitly or defended openly, is taken by many policy analysts as essential if science is to avoid being contaminated by social values. The serial view, and the underlying commitment to a separation of science and values, is aptly illustrated in wetlands policy.

Although ecologists did much in the past half century to change attitudes toward wetlands with studies showing their value to society, most informed observers would agree that ecological science has hardly contributed positively to the body of ecologically sensitive law and policy in the area of wetlands regulation. The study of the function of wetlands by professional and academic ecologists has, for the most part, resulted in complex models of "pristine" ecosystems. These models are generally too detailed and complicated to use in evaluating the functions or the contextual features of a given wetland that is threatened by development.[20] William J. Mitsch and James G. Gosselink, in their well-received study of wetlands, say, "Wetlands are now the focus of legal efforts to protect them but, as such, they are beginning to be defined by legal fiat rather than by the application of ecological principles."[21] Worse, there is no unified, specific national wetland law, and wetlands are not managed according to a coherent and comprehensive regulatory system, but rather according to a hodgepodge of water-quality laws and terrestrially based land-use law.[22] For example, even though President George H. W. Bush endorsed no net loss of wetlands, crucial questions, including how to define a wetland and how to evaluate and compare the social value of wetlands, ended up before Vice-President Dan Quayle's Competitiveness Council, which tried to institute a definition of wetlands that would favor economic development.

The reluctance of ecologists to become involved in value judgments or in social policy has greatly limited their impact on wetlands protection efforts. I concentrate on two aspects of this failure here: the inadequacy of current models to encompass contextual features of wetlands and the failure of academic ecologists to develop models useful in public decisions. By limiting their wetland classification methods to measurements of descriptive ecological values based mainly on site-specific judgments of capacity and function, ecologists have undercut efforts to include system-level ecological features as a part of the decision process.[23] Thus, although ecologists argue explicitly that the methods of economists ignore "ecosystem and global-level values relating to clean air and water and other 'life-support' functions,"[24] they have failed to expand the concept of "ecological value" of wetlands to include off-site features and the way these features affect public values. Dennis King insists upon a distinction between ecological *assessments* and ecosystem *valuations*, noting that the assessments normally undertaken by ecologists "provide only the front-end part of the analysis required to compare ecosystems on the basis of the services and values they provide."[25] Mitsch and Gosselink survey the techniques available for quantifying wetland values, after acknowledging that ecologists are usually not interested in "value to society" as much as in the ecological functions of "abstruse" wetlands processes.[26] They proceed to provide an overview of ecosystem-derived services such as recreation and other socially valued commodities, recognizing that many of these represent public goods that are often not captured by owners but are enjoyed by the public as a whole. These public values, since they are not owned and cannot be traded in markets, can be identified and measured only by use of contingent valuation (CV) techniques. CV studies, however, though they may illustrate particular public values, are too expensive to provide a comprehensive accounting of the vast number of public values affected by wetlands policy.[27]

King reports that ecologists have been slow to participate in attempts to establish a middle ground between economic, market valuation techniques on the one hand and structural, that is, morphological, analysis on the other.[28] Although long-term research continues on the analytical problems involved, it is unlikely to be useful in comparing wetlands in policy contexts because, as King argues, "in most legal and regulatory contexts where decision-makers are asked to compare ecosystems on the basis of their values, they are asked to compare their *expected future services and values* on the basis of information about their *current biophysical features and landscape contexts.*" But in order for comparisons to be useful in these situations, "there must be some way for the assessment of observable ecosystem features to be linked forward to expected ecosystem functions and services, and for the valuation of ecosystem services to be linked backwards to observable or measurable ecosystem features. Most of the ongoing research to improve the scientific basis of ecosystem assessment methods and the credibility of ecosystem valuation methods will not fill this gap." King hypothesizes that ecologists avoid evaluative classifications that truly rank wetlands according to social value in order to avoid becoming entangled in value judgments and to avoid having their work used to justify lesser protection for wetlands they describe as having lesser value.[29]

Since contextual ecological information is often crucial in making public decisions regarding what wetlands will be sacrificed to development and how to compare various sites, one might think that academic ecologists would be eager to work on these problems, as each engineering project is a potential experiment in the creation of wetlands. King, however, dividing relevant research into two categories, argues, "With a few exceptions the analytical tools that hold the most promise . . . are from the more practical ecosystem assessment literature, rather than from the more rigorous scientific and landscape modeling literature." He states that most studies in the latter category, designed to find out how ecosystems work, are too theoretical and that the cost of calibrating the landscape linkages necessary to apply the models is often prohibitive. He concludes, "At this time the methods of analysis described in the scientific literature dealing with ecosystems do not provide a practical basis for demonstrating that two ecosystems are equivalent or comparable in terms of services or values."[30]

The formula "No net loss of wetlands" and the institution of wetland mitigation banking thus provide an excellent example of the way inadequate language—and especially the poverty of our concepts for evaluating changes in the environment—can render EPA hostage to interest-group politics and unable to act decisively, even on such an important subject as protecting wetlands. The no-net-loss policy, in combination with pressure on EPA from development-oriented interests not to stand in the way of any projects, encouraged the creation of wetlands in return for rights to destroy wetlands and led to a virtual market in "acres" of wetlands, with created or restored wetlands being bartered against permitted losses. The entire idea of wetlands banking, however, depends upon being able to assign comparative values to restored or created wetlands and wetlands banking. If it is to be a real response to loss of public goods due to development, and not just an excuse for destruction of those goods, wetlands banking must be managed according to ecological principles.

Given the reluctance of ecologists to become involved in debates about values and policy, scientific progress in understanding the impacts of wetlands banking on socially valued processes has been slow in coming, although there are encouraging developments and some fragmentary data is coming in from some studies. So far, ecological principles have had little effect on these calculations. According to J. B. Zedler, "Rigorous research on wetland restoration construction is inadequate to provide simple formulas for constructing one wetland to compensate for functions lost in destroying another."[31] C. A. Simenstad and R. M. Thom similarly state: "The ability of wetland managers to assess compensatory-mitigation success over short-term (e.g., Regulatory) timeframes depends upon the selection of attributes that can predict long-term trends in the development of the restored/created system. However, we are hampered by a basic lack of long-term data sets describing the patterns, trends, and variability in natural wetland responses to disturbance, as well as natural variability in wetland attributes in presumably mature wetland communities." This lack of relevant science, though undeniable, is remarkable in itself, given its policy importance.[32] According to Mitsch and Gosselink, despite thousands of 404 dredge-and-fill permits issued to allow trade-offs between constructed and mature wetlands, ecologists have until recently published no relevant studies comparing undisturbed with constructed or restored wetlands. Considerable data has been published on wetland mitigation in the last few years, and at last the above-mentioned National Research Council study has brought together a respectable body of evidence. The news, however is mostly bad news, as noted above; required wetlands are often not built, and when they are, the requirements are neither enforced nor enforceable.

We see here, in wetlands policy and especially in wetlands mitigation banking, a serious failure of policy that can be traced directly to the phenomenon I have called towering. For decades two well-meaning agencies—trapped by the myth that facts and values must be discussed separately—failed to communicate about what is important to measure and what is important to save in protecting wetland ecosystems.

The second aspect of towering is failures of communication across disciplines and between disciplinary scientists and policymakers. It too has been evident in the development of wetlands policy. Lack of communication between responsible parties, especially regarding the nature of social values that should be targeted, monitored, and measured, has pervaded the entire history of wetlands policy. As was learned by the recent National Research Council/National

Academy study discussed above, the Army Corps was charged to oversee the permitting process, but almost no emphasis was placed on follow-up and enforcement. This was due partly to heavy workloads of individuals acting on permit applications, but it was also due to a huge failure of communication. Nobody made clear to the Army Corps what endpoint characteristics were the true goals of policy. Since the Corps did not have the ecological expertise to write performance standards in ecological terms, they fell back upon structural requirements, such as percentage of plant cover, as the basis for permitting and judging projects. Ecologists, for their part, provided no comprehensive, ecologically based tests of wetland function or of contribution to landscapes and watersheds, as was explained in detail above in the discussion of the first aspect of towering. What this meant was that, even where the structural requirements were monitored—overall, less than 50 percent of sites are inspected even once—no information was gathered to tell us which structural requirements are likely to be successful in protecting valued functions. This failing will be important in our discussion of the third aspect of towering—lack of learning.

Before turning to that discussion, I want to highlight one very specific form of linguistic poverty associated with towering that highlights a recognizable failure of our public discussions of environmental values: we have no generally recognized vocabulary that makes transparent the connections between ecological outcomes and social values. For convenience, we can say that what is lacking in policy discussions of wetlands, and what is lacking in most areas of public discourse about environmental policy, is that there are no clear "bridge" terms. Bridge terms are terms that have empirical, operational, and measurable descriptive content and therefore have a connection to the descriptive discourse and the literature of science; but bridge terms also connect to social values and our evaluative discourse by embodying or evoking important social values. For those who have already bought into the idea of a science sharply separated from values, the existence of such terms may seem impossible, or at least undesirable, because such terms cannot be purely scientific. This book, however, is premised on opposite assumptions—that factual material and evaluative material are routinely mixed in public discourse and that these factual-evaluative functions are as capable of being made precise as are descriptions of purely physical processes.

A simple analogy might be helpful. In medical science, they use a formula relating a person's height to body weight (easily measured quantities) to classify people as being of normal weight, overweight, or obese. Looked at from the scientific side, this index simply summarizes important descriptive information about an individual's height and weight. Given current understanding of obesity in causing disease (such as diabetes) and the apparently undeniable connection of obese and overweight conditions to morbidity and mortality, the category "obese" takes on a host of evaluative and normative associations. In this case, the index is a valued tool of communication in that (1) it provides a readily measurable, descriptive characteristic and also in that (2) it provides a direct linkage to a host of individual and social values. That a person seeking medical assistance is obese can be measured and treated as a descriptive fact; given scientific knowledge about causal relationships and risks that allows us to predict that a person so classified has higher risks of illness, we can also make important recommendations for the person. These recommendations already embody the value of health. Thus, in a context in which important causal hypotheses are accepted, the label "obese" both embodies empirical content and connects—through a network of scientific theories—a measurable condition with the value of human life and health. There is no reason that, by virtue of this mixture of both scientific and evaluative content, we should not treat the index as "objective" and precisely measurable. Precision and measurability can easily be achieved even when a bridge term such as this evokes evaluative content.

Medical science is a rich source of analogies that might guide us toward the choice of better bridge terms for communicating values and impacts of policies on values. Terms such as *healthy ecosystem* and *ecological integrity* have accordingly been suggested as bridge terms for ecological science and policy, but ecologists have generally scorned them as "too value-laden." I will not dispute this point on behalf of these particular terms; my point is more general. There are important connections between factual information about wetlands and other ecosystems and social values, connections that are missed—or at least become much more difficult to recognize—in the absence of bridge terms. Bridges can, under the right circumstances, connect towers. I hypothesize, however, that one of the problems in discourse about wetland and other environmental policies is that, lacking bridge terms, concentration on science (as separated from evaluation) leaves us with lots of precise numbers but unable to determine which of those numbers are important. So one of the main theses of this book is that if towering and its negative effects on environmental policy are to be avoided, we will need to develop some bridge concepts that are both scientifically precise and clearly expressive of values.

My analysis is that without accepted bridge concepts that encourage the association of scientific information with cherished human values, communication about policy will remain poor and the scientific information provided by ecologists and other scientists will not be brought to bear upon policy discourse in a useful way. My response to this problem—and the main burden of this book as a whole—is to develop a constellation of terms, each of which is appropriate *both* to introduce or summarize important descriptive information *and* to signal that a particular human value is associated with that description. At the center of this constellation is the term *sustainability,* which includes both a description (it says something about what will be left for people of the future) and an evaluative component (it expresses moral concern about whether our legacy is fair to future people). For the term *sustainability,* it turns out that I can offer only a "schematic" definition—a definition that includes a number of variables that must be turned into specifics by real communities that choose "important" indicators. If the people of a community choose indicators associated with values they hold dear, and use these indicators to state concrete sustainability goals with respect to their community, they will in effect be defining sustainability—for themselves. Every community might come up with a different definition. Even though sustainability is schematically defined, it nevertheless anchors the system of evaluative concepts being developed here. It does so by organizing our scientific information and our evaluative judgments to correspond to a world that unfolds on multiple temporal scales, or horizons. *Sustainability* thus both refers to systemic physical dynamics that will change the world humans encounter in the future and evokes a commitment to consider the important normative relationships that can develop in these dynamics, which today involve multigenerational impacts. If that sounds horribly abstract, we can bring the idea down to earth by referring to a simile of Aldo Leopold, one of the guiding lights of this book and of enlightened conservation. Leopold, who wrote an essay scolding himself for a shortsighted policy by which he had extirpated wolves from wilderness areas only to see the range destroyed by too many deer, explained his mistake by saying he had not yet "learned to think like a mountain."[33] The remainder of his essay and Leopold's graceful and profound book *A Sand County Almanac,* which includes the essay, are devoted to explaining how to come closer to thinking like a mountain.

What I have learned from Leopold is that constructing a working definition of sustainability requires thinking of time and the future in a way that is new to humans, a way of thinking in which understanding, love, and moral responsibility all expand beyond the bounds of a single life, unfolding against the backdrop of multiple temporal and moral horizons. And so in reaction to the positivists who still insist on a sharp separation of science and values, I will explore an alternative world, a world of practice in which members of communities are already engaged in communication and cooperative behavior, and in which the problem of how to secure the future is already a pressing problem. Even if we can rewrite ordinary discourse to artificially separate facts from values, science from society, creating formal, value-neutral languages, these languages cannot be the language of policy debate and deliberation about what to do. The language of public deliberation and of political discourse must be the "ordinary" public language of a community seeking to act cooperatively. I therefore try in the following chapters to encounter environmental problems in their natural habitats, which are real communities struggling with real issues and priority conflicts, and starting with available means of communication. Thinking adaptively, however, extends also to language, so starting with ordinary language is quite consistent with engaging in an explicit effort to create new descriptive and prescriptive languages to improve communication.

The third aspect of towering is a result of the first two: when science and policy are discussed within towers, little or no learning takes place. Again, wetlands policy illustrates the point. As noted, the engineers from ACE got no help from the ecologists, who refused to enter the debate about what functions and values of wetlands were crucial to save because they were afraid their science would be tainted by subjectivity and values. So they retreated to their pure-science ecological tower and created abstract models of the behavior of pristine ecosystems. EPA bureaucrats—having been told from day one of their employment that they were to make decisions based on "objective" (value-neutral) science, and who needed to cite objective science to justify their opinions, became coconspirators with the ecologists, delaying value discourse until all relevant science was contracted, gathered, and "characterized." The engineers at ACE, meanwhile, lacking information about the relationships between structural attributes of wetlands and wetland functioning on one hand and their relation to the surrounding habitat on the other, wrote permits based on structural and technological standards. For almost twenty years, nobody bothered to study which, if any, of the created wetlands in fact functioned as well as the ones they replaced, and nobody developed and tested hypotheses regarding which structural features were good predictors that a wetland would provide specified functions. Because the engineers and the ecologists did not communicate about what is important in wetlands, twenty precious years—hundreds of possible "experiments"—have gone unobserved and unrecorded. And as the NRC committee of experts reported, we still know little about these crucial relationships: "The committee noted instances where compensatory mitigation was having a positive result in watersheds and other cases that have problems. However, there is insufficient feedback to Corps' regulatory staff on whether the performance standards developed for a given project produced the expected results. As a result, the same performance standards are used repeatedly with uncertain results."[34]

And this is the greatest tragedy of towering. The artificial boundaries and exclusionary disciplinary jargons are so destructive of rational environmental policy because they limit our ability to learn. To learn we must be open to a wider realm of experience, which includes communicating with people who have different viewpoints and different experiences—and different values. Exclusionary disciplinary languages cut people off from criticism and suggestions that are motivated by different viewpoints and interests. The towering that separated the engineers at the Army Corps from professional ecologists and ecologically informed staff at EPA led to a massive loss of opportunities to learn what wetland policies would work, in what situations. This loss of information will make adaptation of our policies to the reality of real-world ecosystems impossible for the foreseeable future. Looking forward, we must learn from our mistakes and design and implement policies that will increase our knowledge and suggest new avenues to sustainable use of wetlands. We need, in response to this last aspect of towering—its ignorance-preserving capacity—to embrace adaptive management, management that operates to reduce uncertainty. For this reason a central theme of this book is the development of an adequate theory and practice of adaptive management. The experts from the NRC agree: "Designing restoration sites to help in learning which approaches work and why can greatly accelerate the learning curve. Unfortunately, there is no mechanism in place to build an experimental design or adaptive management process into mitigation projects in order to learn from these real-world tests of mitigation project design. Therefore, the committee recommends that the Corps establish a research program to study mitigation sites to determine what practices achieve long-term performance for creation, enhancement, and restoration of wetlands."[35]

In this chapter, and especially in this section, I have emphasized three aspects of the towering syndrome. Towering and the disciplinary isolationism that is associated with it have forced us into asking policy questions in an unrealistic context where the environmental sciences are artificially partitioned from the rest of public discourse by the forced application of the fact-value distinction. This unrealistic context for discussion and debate separates policy debate from the real, problem-oriented and value-permeated business of environmental management. Failures of communication have led in turn to failures of policies; and worse, because towers function to keep out criticism and alternative viewpoints, those who live and work in towers fail to learn and develop new and more satisfactory policies. They cannot adapt and learn.

The big loser in these failures is the public. The Corps's engineers accepted a responsibility to protect "the public interest." But the evidence is that they have not done so; many of the functions—water retention, habitat for wildlife, and water purification, for example—that were once performed by natural wetlands are now lost. They were lost, according to my analysis, as a result of towering, the failure of communication across disciplines and across offices and programs. The winners, once again, are the special interests, who work through their associates within the towers to place political and development opportunities ahead of healthy wetlands. As a result the permittees have destroyed wetlands and failed to provide adequate compensation to the public for those losses of public goods. They have absconded with the benefits of intact wetlands that once accrued to the public; and they have left behind either nothing or, in many cases, degraded systems requiring constant repair or further restoration. The moral is that, in the absence of a lively, interdisciplinary debate about what policies to pursue—a debate that is blocked by the many artificial boundaries known here as towering—the winners will be private interests as the vacuum in public values and environmental goals invites political manipulation. The losers, again, are members of the public.

In response to the problem of towering I will offer, first, some careful analysis of the failed communications that so often result, and I will suggest some new ways of talking about environmental values, new ways of talking that are associated with our central concept of sustainability. This concept, by the way, will remain under construction throughout most of the book. Most importantly, I suggest a whole new context in which to address environmental problems. Instead of thinking of environmental policy formation as a serial process, beginning with value-neutral science and progressing unidirectionally toward a policy judgment, we plunge our analysis into the management context, the context in which real people consider what management steps to take in the face of perceived problems. Being democratic in my leanings, I also try to open up this management process to include lively deliberation about environmental problems, environmental values, and what to do, as an important element of the management process. I call this an *adaptive management process,* a phrase I use quite broadly to include both the use of science in management and a collaborative process in which participation and social learning are an important part. It is important to realize at the outset, however, that when I talk about adaptive management, I refer to an activity that takes place in a context of action, complete with disagreements about what science is important and what policies should be pursued. Since action expresses our values, it is impossible to banish value discussion and deliberation of goals from the on-site development of policy. Adaptation in management must include a robust, evaluative vocabulary. Claimed environmental values are as much open to rebuttal as are facts and causal hypotheses. We need a discourse rich enough to express and disagree about values if we are to intelligently discuss and learn about our environmental goals and how to achieve them. The only languages known that are rich enough for this purpose are ordinary languages, languages that have evolved in real communities with real problems, real objectives, and real attempts to cooperate. The language of adaptive management, accordingly, must be ordinary language as it has evolved in specific communities facing problems and cherishing goals.

Acceptance of ordinary language is only a starting point, however; careful attention to the use of language in efforts at communication *can* lead to piecemeal improvements and clarifications of ordinary language. So the construction of sustainability must, simultaneously, be a struggle to learn how to act, to learn how to speak, and to learn our way to consensuses and compromises in the search for a better environmental tomorrow.

PART I

SETTING THE STAGE FOR ADAPTIVE MANAGEMENT

2

LANGUAGE AS OUR ENVIRONMENT

2.1 Introduction: The Importance of Language

Most conferences on the topic of sustainability or sustainable development—and believe me, I have attended more than my share of them—have at least one speaker who notes the many meanings of the term *sustainable.* At one conference not long ago, I heard an exchange in which several participants cited, one after the other as if in a bidding war, higher and higher numbers of distinct meanings of the term, as counted by various authors in recent publications. That the bidding began in double figures and continued into triple figures is hardly encouraging to those seeking a unified and tight definition of *sustainable.* It is typical, also, to bemoan the lack of precision attaching to the terms that have become, for better or worse, keystone concepts in the discourse surrounding environmental policy. Often, for example, one hears the fear expressed that *sustainable,* used by so many to evoke so much, has been rendered meaningless by the very inclusiveness that makes it a politically useful, large-umbrella characterization of environmentalists' goals and objectives.

It is always a good idea to know what we are talking about, so of course I share the attitude of skepticism that surrounds the use of ambiguous terms. My goal, however, is to go beyond skepticism and begin to sketch a general approach to defining sustainability and sustainable development that is reasonably inclusive—one that captures what most of us mean by the terms when we use them and yet does not leave these important terms devoid of determinate meaning. This task is daunting, given the number of scholars and expert practitioners who have tried to outline workable definitions before me; still, I hope I can do better because I have a different approach to definition than most of my predecessors. Most authors have chosen some definition and argued that it is *the* definition of sustainability, as if they had "found" the correct—but somehow hidden from most of us—definition. I, in contrast, see defining sustainability as one step—and probably not the first step—toward creating a broad theory of environmental values and science. Rather than taking on the task of capturing current usage, my approach will be to ask a bigger question: What language *should* we develop and use for discussing environmental goals, priorities, and policies? Once we have an acceptable start at answering this question, it will be possible to examine whether the term *sustainable* can play a useful role in such a linguistic array.

The susceptibility of the terms *sustainable* and *sustainable development* to so many definitions is itself a symptom of a more general problem, identified in chapter 1: we lack a unified, comprehensible vocabulary for discussing environmental problems as problems facing our democratic society. Especially, we lack a coherent set of terms for expressing environmental values and for explaining and justifying environmental goals. My objective in this chapter is to develop a new attitude toward these terms. I hope to convince you, the reader, that the problem is not so much a matter of choosing the "right" definition for sustainability but of proposing and refining—constructing, in other words—a new and improved language for discussion of environmental problems. Once such a language is in place, it may be possible to embed the troublesome terms in a constellation of effective and meaningful terms for communicating important scientific information and for expressing and defending social values. The premise supporting this enterprise, then, is the idea that what's wrong with terms such as *sustainable* is to be found not in their own inherent instability, but rather in the absence of clear connections to established bodies of theoretical knowledge, connections that would be more easily made if we had the linguistic thread available to weave richer connections between theories articulated in differing disciplines.

In this book I propose a more comprehensive and (I hope) more comprehensible language for describing and evaluating environmental change, and I do so by embodying that language in a general theory of environmental planning and management. This chapter focuses narrowly on the question of choosing a useful language for discussing environmental problems, goals, and policies. As will become clear later, however, my concern with language shapes the more substantive proposals developed later in the book.

One of the defining features of the pragmatist outlook on things is taking a problem-oriented approach to intellectual as well as practical dilemmas. Historically, of course, philosophers have addressed questions of great abstraction and generality, believing that if one gets first principles right, solutions to particular problems will fall out as corollaries of the general principles. Many philosophers believe that they confront special problems with special tools and that philosophy occupies a special intellectual space, with a form of access to truth not associated with the empirical methods of the other sciences. Indeed, in Western philosophy there is a venerable tradition, traceable back at least as far as Plato, that philosophy has a rational method that will allow penetration beyond the veil of language and experience, apprehending reality itself. Pragmatists, by contrast, doubt that philosophy has, or needs, a method that is independent of experience. Pragmatists aspire to an ideal of a unified conception of inquiry in which philosophers are one kind of workers in a larger enterprise. Pragmatists seek a unified method of inquiry—a method that is self-correcting, based in experience, but also involving interpretation and theory-building. Philosophy, on this view, differs only in degree from other sciences, all of which have truth as their ideal; philosophy may occupy the more abstract end of the continuum of knowledge, but it is a continuum, with all knowledge and wisdom—including definitions—ultimately answering to experience. Pragmatists therefore reject "higher intuitions" and prefer to deal with specific problems whenever possible.

What unifies inquiry, according to pragmatists, is a community's shared focus on a real-world problem. Aside from creating a healthy urgency, a problem orientation can go a long way toward setting a context, clarifying what values and interests are at stake in any question, and shaping disagreements as testable hypotheses. So what exactly is the problem to which this book offers the beginnings of a solution? The central problem, as we saw in chapter 1, is that there is a nearly complete breakdown in communication regarding environmental policy *right at the crucial nexus where the particular sciences are integrated with social values and translated into public policy.* This is where scientific data-gathering, model-building, and physical observations of scientists from many specialized disciplines—ecology, toxicology, economics, sociology, and so on—are brought together in the context of policy decisions. At this crucial point of unification, values are also brought into the process of policy study and formation, and this is precisely the locus—as we saw in our brief visit to the EPA building in chapter 1—of the deepest confusion and the most abysmal lack of adequate vocabulary for communicating about environmental problems, values, and goals.

Speaking generally, the problem featured in this book is the lack of a common language or shared discourse in which scientists and the public can discuss environmental problems, environmental goals, and possible environmental actions. Still speaking generally, our current language is inadequate because—as noted in the preface—it leads to polarization over environmental values and to ideological environmentalism. Speaking generally, however, is just that; a pragmatist is committed to looking at real cases. Speaking more specifically, lack of effective communication pervades and corrupts most contexts in which human communities are struggling to live within their environmental limits. And yet we cannot solve this problem on the most general scale until we have some idea of its manifestations at local levels and in real policy contexts. We must survey a number of specific environmental problems, looking for general features from our survey. To get the cards on the table, however, let me state at the outset a general hypothesis: at all levels of society, and in all kinds of places across our country, there are failures of communication in discussions of environmental problems. These failures are most basically due to the lack of an adequate language for integrating environmental science and environmental values, and as a result little true communication occurs in the process of formulating and discussing environmental policies. To be more specific, I will argue that in public policy debate regarding environmental choices, we lack a crucial type of term that can (1) encapsulate a great deal of information and (2) present this information in such a way that the its importance for widely held social values is transparent.

I believe this lack is illustrated by the comparative advantage economists have over ecologists in policy discussion. Public discourse contains terms such as *economic growth* and *savings,* which express social values widely held in the society but which have acquired more precise meanings by linkage to the theory and findings of economic science. For example, the term *Gross Domestic Product* has come, for better or worse—quite possibly for worse—to represent "economic growth" as defined within the assumptions and measures of economic science.

Although GDP is in many ways a flawed measure, it exerts a powerful force because, given economists' identification of strong growth and improved social welfare with upward trends in improvements in the GDP, the person on the street does not have to ask whether a severe drop in the GDP is good or bad news. It is a term that summarizes an immense amount of empirical data and, at the same time, wears its impact on widely shared social values on its sleeve, by connecting trends in scientific data with an unquestioned public good—a strong economy. Because the public understands the negative impact of a weak economy on their lives and lifestyles, the public can be very hard on incumbent candidates for office who preside over disappointing economic growth.

In contrast to economists' linguistic resources, discussions by ecologists of long-term impacts of current activities are often expressed in language that suggests no clear connection between scientific assessments and identifiable social values. For example, Kenneth Arrow and an interdisciplinary group of scholars, writing in a "Policy Forum" article, argue convincingly that achieving economic efficiency does not in many important cases guarantee adequate protection of resources. This same group, members of which are writing as distinguished representatives of economics, ecology, and related disciplines, then state not only that sustainable use of resources requires economic efficiency, but also that human activities protect the "resilience" of ecological systems.[1] Does the person on the street, or the average public servant, know the impact of having, or of losing, resilience in large-scale ecological systems? To their credit, Arrow and his colleagues set out to explain the importance of resilience, but it is questionable that their explanations will be clear or convincing to nonecologists, because they do little to link resilience to human values or to the day-to-day concerns of voters.[2]

The contrast with economic discourse is striking: whereas trends in the GDP are directly associated with matters that people care about, resilience has no such connection, unless that connection is laboriously made in each and every context and explained by appeal to anecdote and scientific generalization. Again, my point is not to praise the GDP measure as an ideal, or even to suggest that ecologists should seek a single, overall measure such as this. My point is that ecologists will have much more impact on policy if they use terms that transparently link ecological information and theory to widely favored social values and goals. Failure to employ language that helps nonecologists make connections between ecological trends and social values has a great cost: the public and the policymakers know whether trends in data are good or bad only if they are willing to learn a body of scientific information and its application to sectors of public interest. Ecologists, mostly unwittingly, have created a conceptual gulf between the information they gather and the social values people cherish, making it very difficult for participants in policy discussions to see the relationship between ecological science and public values. This is a serious problem for the discussion of environmental values and environmental policy.

This is my central contention—that policy discourse currently suffers because, whereas economic data is easily associated with the well-being of citizens in our democracy, ecological data has no such resonance. I contend that existing language is inadequate to establish a meaningful, multidirectional dialogue among scientific ecologists, policy analysts, policymakers, and the public because existing language does not clearly display the important connections between changes in the environment and changes in values enjoyed by the public.

2.2 Of Hedgehogs and Foxes

At a recent workshop on sustainability in environmental policy I was paired with fellow philosopher and friend Mark Sagoff to discuss the merits and demerits of the term *sustainability* as a guide to environmental policy formation.

Sagoff went beyond simply bemoaning the semantic sloppiness of the term, noted above, and provided a substantive—and provocative—argument that the term is doomed to meaninglessness. He introduced his argument by referring to the late Isaiah Berlin, who, in his famous book *The Hedgehog and the Fox,* quotes from the Greek poet Archilochus the statement "The fox knows many things but the hedgehog knows one big thing." Berlin interprets these words to define a chasm "between those, on one side, who relate everything to a single central vision, one system less or more coherent or articulate, . . . and, on the other side, those who pursue many ends, often unrelated and even contradictory." Berlin uses this distinction to explain different visions of history.[3]

Sagoff went on to deploy Archilochus's taxonomy to clarify the key difference between two opposing approaches we may take to understanding and solving environmental problems: "Those who share the hedgehog view attribute our environmental woes to one overwhelming cause: the expansion of the global economy. . . . These commentators believe that 'the Earth's ecosystems cannot sustain current levels of economic activity.' Relentless economic expansion world-wide continually gains momentum. 'An irresistible economy seems to be on a collision course with an immovable ecosphere.' Only by decreasing or at least containing the size of the global economy can we provide a sustainable environment for future generations."[4]

Environmental foxes, by contrast, believe that our environmental problems arise from many different sources: from poverty as well as from affluence, from economic contraction as well as from economic growth, and so forth. Rather than prescribing global economic retrenchment as a general solution, foxes recommend management strategies that "combine incentive and regulatory policies, recognize administrative constraints, and are tailored to specific problems."[5] They believe that the world must solve its environmental problems on a case-by-case basis. For example, by developing and deploying cleaner and more efficient technologies, rather than trying to reduce economic activity, foxes seek environmental improvement by such means as requiring industry to use resources ever more efficiently.

Sagoff associated the term *sustainable* with a particular "hedgehoggy" approach based in ecology and Malthusian theory, the reductionistic approach drawn from neo-Malthusian ideas of ecologists such as Paul Ehrlich. I call these ideas reductionistic because they reduce the manifold environmental problems to symptoms of a single cause: humans have exceeded their carrying capacity. This underlying cause is manifest in different ways as more and more limiting factors come into play and earth systems are progressively stressed. This theoretical approach is sometimes summarized in an equation, impact = population × affluence × technology (IPAT). Sagoff cited reasons to question the reductionistic approach, and he stated his agreement with those who emphasize the particular and local nature of many environmental problems. Rejecting the "just one thing" of the hedgehog, Sagoff decided to run with the foxes.

Although I do not agree with Sagoff's conclusion that there is no use for a term so general as *sustainability,* I find his application of Berlin's analogy useful because it focuses our attention on the second-order question of which concepts and terms are likely to be useful and which should be expunged from the vocabulary and discourse of environmental policy debate. In the process, Sagoff establishes a central enabling premise of this book: The question of what terms and vocabulary to use in discussing environmental problems—how to define them and in what terms—cannot be separated from substantive, theoretical beliefs about the nature, causes, and cures of those problems. The communication problem we hope to address occurs largely because in the interdisciplinary context of political discourse, there is no shared language or vocabulary to describe environmental problems and possible solutions. This lack reflects the absence of a theory shared by all parties in environmental debates that relates describable environmental change to impacts on social values. In chapter 1 I tried to show some of the damage done at EPA and in other institutions because of a lack of interdisciplinary communication, failures that are almost certainly accompanied by substantive disagreements about the nature of environmental problems. In this chapter I address Sagoff's question, emphasizing its linguistic nature: Is this particular term, *sustainable,* a useful one?

My question is not what the term currently means but what it *might* mean if it were embedded in a comprehensive but not-so-reductionistic theory of environmental management: not whether sustainability *is* a clear, unified, and unambiguous concept in current discourse, but whether it *could become* a useful concept to unify and guide environmental policy toward the future. So my first task is to explain my approach to the study of language and communication—my philosophy of language; this task begins in section 2.3 and continues in the appendix and in chapter 3.

Understandings of the nature of language range from "essentialist" views, which see language and its categories as reflective of natural realities and kinds, to more conventionalist views that language and its categories are bequeathed to us by our culture. I lean toward the conventional end of this continuum—a view I explain more fully in section 2.4. I also want to emphasize that language, being conventional, is highly dynamic and subject to change, especially under changing conditions. Once we reject the static view of language, the question arises whether a term such as *sustainability*—however suspect in current discourse—might acquire usefulness through creative redefinition. If my colleagues are right that on the basis of current usage the term *sustainable* is meaningless, another option—besides that of expunging it from our vocabulary—is to follow the advice of the philosopher W. V. O. Quine, who once allowed that of all terms, it is most seemly to assign meanings to the meaningless ones.

I part company with Sagoff when he implies that IPAT and the associated ecological reductionism is the only theory with which sustainability can be associated. By seeing our problem more constructively, as one of choosing and defining fruitful linguistic conventions for activist discourse, we may be able to turn what Sagoff saw as a liability—the association of the term with a questionable, faith-based reductionistic theory—into a positive direction for new research by raising the question whether the term can be given alternative meaning and function within a new and more acceptable theory of environmental management. It is this relationship to acceptable theory that can make the term *sustainable* nonvacuous and perhaps more useful than it is today. But this means we cannot substantially improve the use of the concept without creating a more satisfactory theory of environmental management, and especially of the moral relationships that obtain across generations. Therefore, the linguistic questions raised here at the beginning must pervade the entire project; the term *sustainable* can be given a stable and useful meaning only by building it into a comprehensive theory of environmental management.

Sagoff apparently is convinced that one must be either a hedgehog or a fox; he would have us decide between the single-minded carrying-capacity model, which reduces all environmental problems to one of growth in the scale of human activities, and his own position of having no theory at all, basking in haphazard many-mindedness. I share Sagoff's concern with reductionisms of all kinds and prefer, with him, some form of pluralism; but Sagoff refuses even to consider that there might be some plural but integrative theory that would make sense of varied environmental goals, relating them to each other and placing them in a coherent framework of understanding. He implies that we must choose between hedgehogs and hodgepodges. I, in contrast, fear that if we have no theory at all relating environmental change to human values, we will have no basis for deciding on general priorities or for deciding which questions to ask first. Environmental policy with no theory at all will remain ad hoc, adventitious, and riven by interest-group wrangling. These problems reflect the lack of a coherent vision of an environmentally better world. I suspect that there is no general understanding among the public—or among committed environmentalists themselves—about what policies environmentalists favor.[6]

I think it is unfortunate that Sagoff chooses as his only example of hedgehogs the ideologically loaded theory of carrying capacity applied at the global level. That makes it pretty attractive—but I think too easy—to embrace foxiness. If we consider some different types of theories of sustainable development, it may be possible to imagine a theoretically guided approach to the big picture in environmental problems without going all the way to the sort of physical reductionism embodied in the IPAT formula. Perhaps we should be looking for what might be called a "hedge-fox."

In short, it seems that Sagoff is pushing us into making an either-or distinction when in fact there are many alternatives and many ways to link theories—including moral theory, economic theory, scientific theory, and democratic theory—in an attempt to make sense of the myriad environmental problems we face and relate them to each other. If we could learn more about these linkages, we might find or create a powerful role for the term *sustainable*. Sagoff has also assumed that the only way to be a hedgehog is to ignore the concerns of foxes and force every moral concern into a single, reductionist theory, such as IPAT, that rests more on imagination and a priori reasoning than on careful observation and analysis of the facts of particular environmental problems and situations. Surely there is a lot of theoretical ground between IPAT and the chaos of basing policy on individual aesthetic tastes, which seems to be Sagoff's own proposal.

What we do know is that the way we think about the environment is all tied up with the language we use to describe it; I have suggested that in the case of environmental policy discussion, our language is disjointed at best and confused at worst. If our language for describing the environment is incoherent, our perceptions will likewise be incoherent, preventing even the clear statement of the problems faced. It is in this sense that the title of this chapter, "Language as Our Environment," is appropriate: language and our world are so intricately related that it is not inaccurate to say that language, in a very important sense to be explained here, *constitutes* our environment as we experience it. In the rest of this chapter and the rest of the book, I explore whether linguistic philosophy—which I take to be aggressively applying the tools of logic and the philosophy of language to public discourse about environmental policy—can bring some order to the chaotic and highly ineffective public discussions of environment policy.

The reader is forgiven some skepticism regarding this project; since the recent history of philosophy has not inspired confidence, especially regarding advice about public policy. As a partial palliative for the skeptical, I have undertaken in the appendix to retell the story of twentieth-century philosophy with a more positive and practical spin. If I spin it right, the story will end up with a new way forward for linguistic philosophy and with a new role for philosophers in public policy process. Relegating the technical and detailed philosophical analysis to the appendix, I will explain in the remainder of this chapter the general strategy of the book: to install linguistic philosophy as an important contributor to a new and more fruitful view of environmentalism and sustainability. To accomplish this, I will first (in section 2.3) explain how the basic approach of American pragmatists and fellow philosophical travelers in the late nineteenth and early twentieth centuries provided a rich ground for a more experience-based, adaptive approach to resource use and environmental protection. Then in section 2.4 I will show how this rich framework of ideas provides linguistic and methodological tools that can cut through the unfortunate polarization that has led to ideological environmentalism in America.

2.3 Progressivism, Pragmatism, and the Method of Experience

Ideological environmentalism, described in the preface as environmental advocacy based on preexperiential principles and moral commitments, is ineffective because it leaves no room for flexibility and for learning from experience. To make matters worse, opponents have gravitated toward polarized theories of environmental value, dogmas that block communication and make compromise actions more difficult. Two apparently opposed moral theories, one that restricts all value to the fulfillment of human preferences and one that extends moral value to the elements of nature itself, have polarized environmental thought. My ambitious strategy for dispelling ideological polarization from environmental thought and action has a historical aspect, and that is the topic of this section. I will show how the same intellectual strains and ferments that spawned the environmental movement, including the ideological splits embodied in the epic moral battles between Pinchot-style progressives and Muirian romanticists, also introduced a new "method," which provides an antidote for that very ideological polarization. My strategy also has a philosophical, or "logical," aspect (explored in section 2.4 and further in chapter 3). This method, which can be referred to as "the method of experience" or as "the pragmatic method," provides an alternative to rationalism and reliance on a priori values. Embracing a radical empiricism as one manifestation of a major shift away from authority and toward democracy, pragmatists trust only experience. This empiricism is not, however, based on the traditional, atomistic psychology of British empiricists such as Locke and Hume, who understood the senses as gateways for fragmentary images and impressions, but rather on a more action-oriented understanding of experience as active experimentation in the face of felt problems.

The main ingredients of the antidote mentioned above were prepared by American pragmatists and related thinkers during the social and intellectual ferment in post–Civil War America. As has been so clearly chronicled by

Louis Menand in his outstanding study *The Metaphysical Club*, a small group of creative and forward-thinking intellectuals responded to the earthquake caused by Darwin's theory of natural selection—and its scientific, social, and intellectual consequences—not as a disaster but rather as the inspiration for an epistemological breakthrough. They provide a method for debunking the widespread appeals of ideologues to a priori and theistic assumptions. Here I will sketch the historical context and the development of the pragmatic method and especially show how it bears particularly on the history of environmental thought and action.

Aldo Leopold, the great second-generation environmentalist who was thrust into this polarized situation as embodied by John Muir and Gifford Pinchot, is a central figure in this story. Most accounts of the development of the environmental movement in America—including some versions I myself have told—emphasize the bipolar ideas of environmental value motivating the first generation of environmentalists and the difficult dilemma this polarization posed for Leopold, as he followed in the footsteps of the giants of nineteenth-century conservation.[7] To summarize, Pinchot, the anthropocentric utilitarian, articulated a vision of human use of resources wherein that use was fettered only by limitations imposed by concern for future human consumption. Pinchot allied himself with progressive politics, and he emphasized fairness in the use of resources, including fairness to future generations; but he never expressed moral concerns beyond maximizing human welfare, limited by considerations of distributive fairness among persons born and unborn. This principle was utilitarianism: resources should be used for the greatest benefit to the greatest number of people over the longest period of time. Meanwhile, Muir and his followers—according to the widely accepted, bipolar caricature—set out to limit human consumption of resources in order to protect nature for its own sake, treating human consumption as the problematic, rather than the goal, of conservation. If one takes this bipolar caricature too seriously, it can seem as if ideological environmentalism and all its devastating effects are inevitable aspects of the intellectual and policy landscape on which environmentalism developed in America.

This bifurcated history can perhaps be a useful simplification of the moral terrain of early environmental thought, and many authors have found it useful to see Leopold as forced to choose between his intellectual forebears, the utilitarian manager Pinchot and the romantic preservationist Muir. For example, Max Oelschlager, in his epic treatment of the idea of wilderness in environmental thought, saw Leopold as having struggled between "Imperial ecology" (the use of ecological principles to efficiently control and exploit nature, à la Pinchot) and the "Arcadian ecology" of Muir, Thoreau, and others who "resonated with wilderness, with the birds and animals, the grass and the sky, feeling themselves to be a part of the larger and enveloping whole."[8] However useful as a simple representation of complex historical events, this account, which considers Leopold to have faced a stark choice between romanticism and economistic utilitarianism, misses a most important change in the intellectual context of American thought, a change that had huge impact on the American mind before Leopold began to develop his own approach to conservation. This "third force" altered the intellectual atmosphere in America and created an alternative to the utilitarianism, economism, and materialism that Pinchot had built into his approach to progressivism and science-based management. As a result of this third force, Leopold, I will argue—although he clearly saw the moral dilemma posed for him by Muir and Pinchot—was able to address the Muir-Pinchot dilemma in much less polarized and ideological terms than the titans of early conservation were able to. That third force could loosely be called "the pragmatic method." But I get ahead of the story.

To understand this third force it is necessary to examine, however briefly, the broader intellectual climate that emerged between the end of the Civil War and the early decades of the twentieth century; it was a time of great intellectual ferment and radical questioning of the foundations of knowledge and moral principle. Fortunately, a quick examination will suffice, and we can accomplish it by following Menand's wide-ranging and provocative account of intellectual life in New England in the post—Civil War era. Muir, who was born in Scotland in 1838, and Pinchot, born in the United States in 1865 and educated in Europe, perhaps never fully recognized the severity of the challenge that Darwinism posed for the moral and intellectual assumptions current in American thought, which remained through the Civil War under the domination of European problems and European ideas, even as those ideas were stretched beyond all recognition by the emerging democratic ideals of America. Since both Muir and Pinchot were mainly men of action and advocacy, often finding themselves in situations where they wished to persuade, they appealed to principles—Muir to God's law and Pinchot to the law of efficiency in the service of utilitarianism. Their appeals were consistent with the tradition of principle-based morality common at the time. Muir and Pinchot were both in favor of "principled" conservationism; they simply disagreed about which principle should apply, a spiritual one or an economic one.

Meanwhile, leaders of American thought, such as Oliver Wendell Holmes Jr., Charles Sanders Peirce, and William James, were engaged in a deep reconsideration of the most fundamental understanding of principles and rules. By the time Leopold headed east to study forestry at Yale University in 1905, the opposed rhetoric of conservationists and preservationists was at its height. It was highly polarized and ideologically virulent, with each pole firmly planted on its own principle. What had changed in the intellectual climate that Leopold found in New England was not the prominence of such appeals to higher and general principles by conservationists and other social reformers, but the understanding—in the broader intellectual community—of the nature of principles themselves. Darwin's theory had not just undermined arguments from design or facile ideas of human superiority; the idea of natural selection undermined a whole way of thinking and justifying actions.

The seeds of what I am calling the third force can actually be traced to a treatise on English Law published in 1863 by James Fitzjames Stephens, but since we are concentrating on its impacts in America, we can start with the statements of one member of a small discussion group that included Nicholas St. John Green, Oliver Wendell Holmes Jr., Charles Peirce, Chauncey Wright, and William James and met irregularly in the Cambridge area. Green, son of a four-time mayor of Cambridge and a professor of law at Harvard, was a friend of Holmes and a frequent contributor to the *American Law Review*, which Holmes edited. Green had a formative influence on Peirce and Holmes especially and was an outspoken opponent of "legal formalism," the view that legal principles capture and express a prior, unchanging, and determinate rule that can be codified. Green argued that legal concepts could vary according to the circumstances of their application and must be applied contextually. Just as Darwin had asserted that the concept *species*, rather than capturing some preexisting and fixed category of beings, served as a useful category in the study of biology, Green argued that categories like "embezzlement," "justifiable homicide," and "proximate cause" were likewise useful categories that got worked out in the practice of the law. Beliefs, for Green as for Stephens, were not made true by existing facts; rather, beliefs were tested as people acted upon them, and the test of a belief was its success in achieving actual goals of human actors in real contexts, just as the behavior of animals was tested by their ability to survive in their habitat.

Peirce, impressed with Green's idea—though disagreeing with one aspect of it—proposed calling the cluster of ideas shared by Green, Holmes, James, and himself "pragmatism," borrowing a term from Kant's *Critique of Pure Reason*. Kant had used the term to apply to contingent beliefs that are nevertheless used as a basis for action; Peirce used it to characterize all belief, effectively treating all beliefs as "bets" on the outcomes of proposed actions. Central to the new approach to knowledge was a clear recognition that experience must be the basis of all learning. Just as central, however, was a new understanding of experience as active rather than passive, as a guide to behavior, rather than as passive descriptions of an "outside" world. These ideas, which were explicitly recognized to have been shaped by Darwin's idea of natural selection, were the source of a huge and sweeping intellectual change, first in Boston and later in much of New England, Chicago, and throughout the country.[9]

These pragmatist ideas, while not entirely original, were made more than the sum of their parts by Peirce, who—objecting to the apparent "nominalist" tendencies of Green's formulation—insisted that the new concepts, created as tools of understanding and as guides to action, though conventional, must be anchored in something less ephemeral than individual speech and belief if they were to contribute to the larger search for truth and understanding. As a radical empiricist (accepting without question that all knowledge has its source in experience), Peirce did not conceive truth claims as comparing assertions to preexisting facts, nor was truth a matter of "correspondence" between a belief and some experience-independent "reality" out there. Rather, Peirce, who was a strong opponent of individualism, located the search for truth in a community process by locating the "conventions" and tools of language in a community of communicators, who use language in action. "Logic," he said, "is rooted in the social principle."[10] Peirce's emphasis on the role of community in anchoring our conventions was crucial in transforming the pragmatic method into a viable alternative to traditional epistemologies that treated assertions as representations of external reality and verification as involving a relation of correspondence between an isolated sentence and some fact existing in the world. Assertions, which are embedded by Peirce in settings demanding actions, are submitted to the constant and unrelenting test of being used by the community and its members to pursue felt goals. The test of truth is in action; and through linguistic communication each truth, once communicated, can be tested and retested throughout indefinite time in varied circumstances. The truth is that which emerges from this indefinite and open-ended process of communication and testing, testing not just by an individual, who is always susceptible to individual biases, but by countless individuals facing countless problematic situations in the future.

This epistemological breakthrough, which proves important in chapter 4, undermined the correspondence theory of truth. That theory apparently cannot be articulated without a hidden employment of rationalist assumptions that the world, as described by human assertions, *must* be prearranged so that it can correspond to the categories we use in our language. More importantly, Peirce articulated—however unclearly (Peirce was anything but a clear writer)—an idea that was eventually to mark the watershed between modern and postmodern epistemology. Modernism, which responded to problems formulated by René Descartes and the rationalists, was eventually undermined by the recognition that truth cannot be a relation between a linguistic item and a prelinguistic world out there, but rather that each truth—because it is inevitably expressed linguistically—exists in a complex relation among a problematic situation, an active asserter, and in the background, a community of communicators who lend stability to our concepts and eventually test such assertions in the open-ended extruding device of collective human experience. This "relational" approach to perception, verification, and knowledge, it turns out, has led to new fields of study, such as the pragmatics of language and communication and the study of semiotics, important both in the study of language in action and in the study of expressive arts. This activist, community-based approach to language, meaning, and assertion was carried to its logical conclusion by Peirce, who installed it as a pragmatist criterion of meaning: "Consider the effects, which might conceivably have practical bearings, we conceive the object of our conception to have. Then, our conception of these effects is the whole of our conception of the object."[11]

There is considerable confusion of terminology in this area, as some have begun using the term *postmodern* to refer to a particularly relativistic methodology that "deconstructs" all thought as perspectival and relative to viewpoint. More correctly, *postmodern* refers to what comes after modernism, and Peirce is historically important because he rejected not just the Cartesian solutions to skepticism and ontology but also the problems themselves, placing himself at the forefront of postmodernism. As of today, it is still not clear what will definitively characterize the postmodern period, but Peirce himself demonstrated that it is possible to dismiss the modernist worldview and at the same time embrace a strong form of realism. What is epistemologically key to pragmatists is that they believe there is a middle way between the rationalism, dogmatism, and authoritarianism of modernism, on the one hand, and unqualified relativism on the other. I believe pragmatists will play an important role in constructing whatever does become a dominant epistemological view in the postmodern period, because in rejecting both dualism and rationalism, they have shifted the epistemological ground to an emphasis on experience. They exhibit a reluctance to assign unquestioned truth to any claim; at the same time they see experience—as processed by a community of language users and actors—as the basis to build a common world.

Despite these shared ideas, it would be wrong to think of pragmatism as a unified set of philosophical principles or doctrines that are shared by a group of "disciples." Indeed, pragmatists were devoted to reducing the sway of pre-experiential "philosophical" or "moral" principles—and pragmatists maintained a variety of compatible and incompatible positions on many topics. For example, Peirce was a realist, declaring that the inquiries of the community will result in a single, unitary truth; James and Holmes were pluralists. What they all shared was the tendency to insist on experience over a priori principles, on a constructivist, conventionalist approach to concepts and to principles, on a belief that truth is tied to successful action, and on an experience-based debunking method based on these shared beliefs.

Mainly through the participation of Chauncey Wright in many discussions, these thinkers were affected by nineteenth-century positivism, the view that only assertions of scientific fact have meaning, as is evidenced by Peirce's pragmatist theory of meaning. Their more activist and relational conception of science and inquiry, however, led to one decidedly unpositivistic effect. Positivists, who clung to a belief that scientific assertions mirrored a preexperiential reality "out there," thought themselves well equipped to sort linguistic utterances into those that were "descriptive" and those that expressed not facts out there but normative judgments existent in the "inner" life of individuals. Because the pragmatists, on the contrary, saw *all* use of language, both in science and in advocacy, as immersed in community processes and involving claims and counterclaims that could be arbitrated only by experience, they had no use for a sharp epistemological dichotomy between factual description and moral urgings. All speech is embedded in the life of a community; all claims turn out to be useful, or not, in resolving problems. All claims can be challenged on the basis of experience. There is only one "method" of inquiry: the method of experience, with experience understood as active, rather than passive, and "truth" as reflecting the usefulness of beliefs in problematic situations.

Perhaps there is no better way to illustrate the spirit of pragmatic thinking than to recall Holmes's judicial realism, based on his "predictive theory of the law," which he described thus:

> The reason why . . . people will pay lawyers to argue for them or to advise them, is that in societies like ours the command of the public force is entrusted to the judges in certain cases, and the whole power of the state will be put forth, if necessary, to carry out their judgments and decrees. People want to know under what circumstances and how far they will run the risk of coming against what is so much stronger themselves, and hence it becomes a business to find out when this danger is to be feared. The object of our study, then, is prediction, the prediction of the incidence of the public force through the instrumentality of the courts.[12]

Holmes asserted that "No concrete proposition is self-evident, no matter how ready we may be to accept it,"[13] and went on to say that legal decisions must necessarily look backward to history but also forward to the ends of the society. On history, he says: "The rational study of law is still to a large extent the study of history. History must be a part of the study, because without it we cannot know the precise scope of the rules which it is our business to know." But history also has its limitations: "It is a part of the rational study, because it is the first step toward an enlightened skepticism, that is, towards a deliberate reconsideration of those rules."[14] And further: "We must beware of the pitfalls of antiquarianism, and must remember that for our purposes our only interest in the past is the light it throws on the future. I look forward to a time when the part played by history in the explanation of dogma shall be very small, and instead of ingenious research we shall spend our energy on a study of the ends sought to be attained and the reasons for desiring them."[15]

Holmes, speaking against formalism, argued that "the way to gain a liberal view of your subject is not to read something else, but to get to the bottom of the subject itself. The means of doing that are, in the first place, to follow the existing body of dogma into its highest generalizations by the help of jurisprudence; next to discover from history how it has come to be what it is; and, finally, so far as you can, to consider the ends which the several rules seek to accomplish, the reasons why those ends are desired, what is given up to gain them, and whether they are worth the price."[16] Thus, although Holmes urged young lawyers to study economics and statistics as training to make these choices, he clearly adopted a pluralist stance, recognizing that every legal decision involves a many-sided negotiation between the particular facts of the case and a complex set of social goals. Good decisions involve not the slavish application of a single, formal rule, but a creative reaction to local facts and a decision reached by balancing a number of competing rules. The goal of the judge—and what the legal advocate wishes to foresee—is a robust decision, a decision that responds as directly as possible to the multiple interests and considerations represented in the case. At the risk of oversimplification, one can say that the pragmatic attitude and method simply apply Holmes's reasoning about the common law to all questions in dispute, with current beliefs and knowledge analogous to the body of common law: the "truth" in any given situation can be approximated only by balancing a number of important (but none of them decisive) factors.

This pragmatic attitude is not relativistic—it expects to arrive at a justifiable decision in a particular situation—so it is better thought of as contextual. A contextual approach eschews one-size-fits-all solutions and rote applications of rules and formalisms to specific and unique situations, and hence it offers no dictates based on prior principles and rules but offers rather a method for subjecting claims to evaluation according to their effectiveness in problem solving.

Without getting too deeply into the details of pragmatist epistemology here (many of them can be delayed for a fuller treatment in chapter 3), let me explain how these philosophical developments affect conservation and environmental management. This "revolution" in American thought requires us to rethink the simple, bipolar view of environmentalism. Although Muir tended to lump all developers and progressives together as pursuing "Mammon," in fact the broader social and economic views of progressives were more diverse and continued to diversify in a burst of creative thought and experimental action in subsequent years. In other words, Pinchot's aggressive utilitarianism, along with his accompanying faith in economics, was not the only philosophical choice faced by young conservationists—such as Leopold—who were drawn, in the early years of the twentieth century, to progressive causes and to scientific management of resources. In fact, we know that Leopold was exposed to, and quoted, these pragmatist ideas and that his earliest account of a conservation ethic (written in 1923) embodied the pragmatist definition of truth. We also know that the pragmatists—and the related thinkers and movements that made up the revolution—were active and articulate critics of neoclassical economic calculations. Pragmatists and Darwinians were, in general, critical of individualism and emphasized community both in truth-seeking and in valuing cooperative behavior. Leopold, a convinced Darwinist, thus had access to this rich nest of ideas and did not really need to choose between Pinchot's materialist, atomistic, economic approach and Muir's religious fervor. Leopold chose, I believe, to base his land ethic on a human community's ability to evolve with and sustain the land. More of this later—the point is that the simple, sharp dichotomy between Pinchotists and Muirians comes nowhere near capturing the richness of the intellectual options available to second-generation conservationists. To assume that Leopold had to be either a Muirian romantic or a Pinchotist utilitarian is to neglect the most interesting options available to him and his generation of progressive conservationists.

To see the alternatives more clearly, we can note that advocates of the pragmatic approach, impressed by the Darwinist ideas of differential survival and natural selection, were strongly opposed to the views of Herbert Spencer and of the social Darwinists, who advocated Lamarckian evolution in the form of inheritance of acquired characteristics. The social Darwinists saw society as an economic competition in which the more fit survive and pass on their advanced characteristics to their offspring. This application of evolutionary thought to society, and especially to the competition between individuals, of course, became both a motivation and an apology for the industrial excesses and unlimited exploitation of nature characteristic of the Gilded Age in America. Because, following Peirce, the pragmatists recognized the importance of communities as arbiters of human action, they viewed humans as members of complex social networks, realizing that we do not face "nature" as individuals, but rather as members of a community guided by evolved tools, social norms, and institutions. They thus rejected individualism as a basis for action. Indeed, though Pinchot was careful (at least on some occasions) to distinguish his idea of conservation from that of unlimited exploitation—because of his concern for fair distribution and future human needs—his narrowly economistic version of utilitarianism was likewise rejected by the pragmatists as overemphasizing individual, consumptive desires at the expense of communal values and norms.

Leopold's most direct contact with pragmatists, as far as we know, was with Arthur Twining Hadley, who was president of Yale University when Leopold studied science, and then forestry, there from 1905 to 1909. Hadley, an immensely popular teacher at Yale, had studied political economy in Germany after graduating at the head of his class in 1876. He became the first lay president of Yale in 1899. He described himself as "a thoroughgoing pragmatist" and as a follower of William James.[17] Apparently, Hadley made an impression on Leopold (though the exact nature of his direct and indirect contact with Hadley has not been determined), because Leopold quoted—and attributed to Hadley—a version of the pragmatist conception of truth several times in his journals and in lectures, besides quoting this same definition in his first version of a conservation ethic. I infer from these attributions that Leopold certainly did not swallow Pinchotism whole, especially its narrow emphasis on economics and its appeal to materialism as the measure of individual fitness. The influence is clearly indicated in the way he mimics Hadley's discussion of social adaptation as a characteristic of cultures, not individuals, in an important passage in "Some Fundamentals of Conservation in the Southwest."

In that first attempt at articulating a conservation ethic, written in 1923 (but not published until thirty years after his death), Leopold considered nonanthropocentric moral principles as a basis for conservation, only to reject them as being "too intangible for most men of affairs to accept or reject as a guide to human conduct."[18] Further, he argued, despite the attractiveness of the nonanthropocentric position, that since it remained such a minority opinion, he would "not dispute the point." Instead, he asked, "Granting that the earth is for man—there is still a question: what man?" And he went on to describe how several previous cultures had lived on the fragile lands of the Southwest and how "our four predecessors [referring to particular cultures that had used the land, historically] . . . left the earth alive, undamaged." Then he turned the question back upon his own European-based grazing culture: "Is it possibly a proper question to ask what (subsequent generations) shall say about us? If we are logically (anthropocentric), yes." What is striking about this reasoning—aside from the fact that it apparently represents a clearly articulated sustainability principle, stated (minus the actual word *sustainable*) in 1923—is that it occurs immediately after an approving reference to Hadley's definition of truth as "'that which prevails in the long run'!" Leopold was here following directly on Hadley's criticism and correction of other evolutionists' misuse of Darwinian selection.[19]

J. Baird Callicott, who has publicly doubted that Leopold has pragmatist leanings, has also publicly ridiculed the idea that Leopold relied on the work of Hadley, arguing that the use of the pragmatic definition was "ironically" intended. This is a bit of an odd claim, in that people usually state things ironically when they think they are true! Nevertheless, given that Callicott is a respected scholar of Leopold's thought, one might ask whether this reference to Hadley was an isolated one. It was not. Curt Meine, Leopold's biographer, tells me that he has seen at least three or four references to Hadley's idea in Leopold's notes and papers, and my friend Ben Minteer has recently shown me an interesting quotation from a lecture by Leopold to the Women's Club of Albuquerque (when Leopold was serving as the secretary of the Albuquerque Chamber of Commerce in 1918):

> It is something of a platitude to say that in each succeeding century, human society has evolved one or more new ideas, which our ancestors then proceeded to write, speak, argue, fight, and die about, to the end that said idea might be proven or disproven,—adopted or discarded. When an idea has been tried by fire and adopted, it is known as Truth. So firmly has this evolutionary character of Truth been established that one of our modern philosophers, President Hadley of Yale—now defines the truth as "that which prevails in the long run."[20]

In 1913 Hadley published the McNair lectures, which were delivered at the University of North Carolina at Chapel Hill in 1912. He appended to these lectures one unpublished essay and a previously published paper that had appeared in the *Psychological Review* in 1909; the book was entitled *Some Influences in Modern Philosophic Thought.* (This was the year Leopold left New Haven for the Southwest—so the paper was likely under preparation when Leopold was on campus, increasing the likelihood that Leopold had seen it or heard some of its contents in a lecture.) This essay, "The Influence of Charles Darwin on Historical and Political Science," begins by noting that historians and political scientists found it less difficult than biologists to accept Darwin's theory of evolution through natural selection, because "all students of history accepted the idea of evolution in their own field of special study; most of them regarded historical evolution as the result of a process of natural selection."[21]

Hadley then asks what Darwinism in particular added to the well-established belief of historians in selective evolution of societies, and he answers by attributing two insights to Darwinians: "In the first place, they showed how natural selection was a means of developing, not only *individuals* of superior ability or intelligence, but *types* of superior adaptation to their surroundings; and they taught us further to regard this adaptation of the type to its surroundings as the thing which gave it its right to exist."[22] Hadley explained these contributions by reference to Malthus: the Darwinians taught us, he said, that Malthus's principle was valid, but at the level of "types." Pointing out that pre-Darwinian Malthusians assumed survival was determined by a direct struggle of individuals for food, and that the more productive workers "should" survive, and others be lost, Hadley said that Darwinians showed that individual struggle is not necessarily applicable to species that are highly socialized. "There is, in civilized communities at least, no habitual scarcity of food. This has been avoided by the development of certain institutions like the family and private property and certain motives which go with those institutions which prevent the scarcity that would otherwise exist."[23] For Hadley, the second contribution, then, was to offer a more defensible use of Malthusian models: "A generation ago the critics of Malthus thought that the nonexistence of the scarcity disproved the Malthusian theory. Today we see that it confirms it." Malthus was thus resurrected, but as applying to types, not individuals. Cultural practices, in other words, insulate the individual from selection in societies in which stable institutions allow orderly growth, based on past practices (cf. Holmes's comments on history, above, at note 14). Societies are thus seen as capable of "adapting" through a process of developing successful cultural modes of existence. It is possible, then, for cultures to develop practices and institutions that protect the individual from the forces of starvation and want; cultures can thus learn to "adapt" to their surroundings, avoiding the iron law of Malthus. This alteration, in turn, had several key implications for pragmatic thought: first, it suggested that it would be possible to compare cultures according to their success in adapting to their environment; and second, this new conception of group (cultural) selection introduced what was later called "social learning" by John Dewey. Once these two conceptualizations were in place, the context was set for applications of the pragmatist conception of truth at the societal, or cultural, level: the extent to which a society has discovered the "truth" about its immediate surroundings will be marked by the society's longevity. The ability of a social group to survive in a place, taking into account its available resources and the constraints embodied in the local environment, will manifest itself in the persistence of that community in the place.

This restoration of Malthus at the level of societies and civilizations, and especially the first implication just listed, was then used by Hadley to make a bold claim in favor of a Darwinian approach to the study of ethics: "It was Darwin who gave the historians and political thinkers the possibility of reaching objective results from their discussion which were previously unattainable. You like one kind of man and one kind of institution; I like another kind of man or another kind of institution. Very well; let us set to work to discover which, in the long run, is going to prevail over the other. That which will prevail in the long run must be right."[24]

Hadley's Darwinist approach to resolving ethical problems no doubt sounds somewhat brash to ethicists today, but Hadley added an important qualification: "Of course this is a doctrine that needs to be applied with great care"; and he points out that such a test of right is fraught with the "danger that we may take too short periods of history under our observation, and may think that an idea or an institution has won the race when it is riding most hurriedly toward its downfall." He went on to say that the main advantage of this Darwinist, pragmatist outlook was its ability to reduce dogmatism and focus debate on facts rather than opinion, helping "the man who is naturally objective become somewhat more so."[25]

The key point for our consideration of intellectual options available to Leopold, however, is Hadley's insistence that it was institutions and practices that matter at the all-important societal level. It was on this basis that he roundly criticized Darwin's contemporary Herbert Spencer for his inconsistency with the true spirit of Darwin: "It was far easier for popular writers to seize upon certain results of Darwin's thinking and try to apply them to history in the form of rhetorical analogies than it was to get at the Darwinian habit of mind in dealing with historical problems in general. Herbert Spencer's writings furnish a very marked instance of this error."[26] Hadley went on to praise other early pragmatist thinkers in political science and political economy, including the Englishman Walter Bagehot and the German economist Adolph Wagner, who, unlike Spencer, recognized the importance of communities and their social practices in mediating survival.

We would perhaps not share Hadley's optimism that capitalist economies have, as he says, "adapted [themselves] to [their] environment," thereby thwarting Malthus's law and avoiding scarcity; nor can we be comfortable with his implied justifications of European imperialism. However, if we think of the argument he sketched a bit more abstractly, it represents the "model" that Leopold used in his "anthropocentric" conservation ethic, as he sketched it in 1923. Societies, which embody—and are embodied by—cultural practices, behaviors, and various limits on behavior, represent experiments in group adaptation. The first two-thirds of the essay "Some Fundamentals" was devoted to a (very negative) assessment of the impacts of European culture's use of the fragile lands of the Southwest. Previous cultures, such as the cliff dwellers and the Pueblos, Leopold observed, apparently had practices that allowed them to live on the arid lands of the Southwest for many generations. (Here it is Leopold's turn to be overly optimistic—subsequent evidence suggests that these cultures were less than perfectly benign, environmentally.)

Those earlier cultures, by virtue of their longevity, were in this case better adapted and living closer to the "truth" than the current grazing culture. Whereas Hadley assumed that our later, "more advanced" culture would survive and thereby supersede the earlier, primitive cultures, Leopold—fresh from his pessimistic assessment of Forest Service land management in the Southwest—hypothesized that the primitive cultures had achieved greater longevity than we are likely to achieve on the land. By Hadley's criterion, then, applied at the level of cultures, the Pueblos and the cliff dwellers were apparently closer to the truth of sustainable living (survival) on the land than were the scientific land managers, including Leopold himself (who had been director of operations for the U.S. Forest Service since 1919). Despite the direct reversal of Hadley's prejudice in favor of modern, capitalist states, Leopold's argument is thus decidedly Hadleian.

Leopold based his argument on his own experience, and in this case he concluded that the institutions and managerial practices of European culture were not well adapted to the arid, local landscapes. Thus he reversed Hadley's judgment, which is just what Hadley apparently encouraged in his warning against finding "truth" in societies based on a too-short period of evaluation. Leopold, after looking at the evidence, concluded that the rapid economic expansion and high levels of exploitation of the fragile lands under his control represented an example of what Hadley had referred to when he cautioned that it is possible for a culture to consider itself about to "win the race" even when it is "riding most hurriedly toward its downfall." Leopold's field observations, recognizing Hadley's caution, convinced him that the imported culture was apparently taking "too short periods of history under . . . observation" and that in fact the European grazing culture was maladapted to the arid climate, characterized by periodic severe drought, of the Southwest. Leopold, employing the pragmatic method of experience and observation in the process of management actions, thus used Hadley's own model to upset the common prejudices—including those of Hadley himself—in favor of contemporary, growth-oriented economics. Leopold was engaging in real social learning, as a day-to-day manager, and it suggested to him that economics cannot possibly be the whole of the normative story in the treatment of land. He also recognized, based on his own observations, how important it is to examine impacts of our practices and institutions over longer, multigenerational periods of time. What he learned from Hadley was the pragmatic method and the importance of institutions and communities in regulating the relationship between individuals and society. But he went beyond even Hadley in rejecting unfettered capitalism itself and proceeding to express the ethical changes he proposed as matters of community evolution. As he said twenty-five years later in the final version of the land ethic, "this extension of ethics [to more and more humans], so far studied only by philosophers, is actually a process in ecological evolution. Its sequences may be described in ecological as well as in philosophical terms. An ethic, ecologically, is a limitation on freedom of action in the struggle for existence. An ethic, philosophically, is a differentiation of social from anti-social conduct. These are two definitions of one thing. The thing has its origin in the tendency of interdependent individuals or groups to evolve modes of co-operation."[27] Then, still emphasizing the effect of varied practices and institutions on the land, as Leopold makes his case for extending group-mandated ethics to the land, on the next page he says: "The extension of ethics [to the land] is, if I read the evidence correctly, an evolutionary possibility and an ecological necessity. . . . Individual thinkers since the days of Ezekiel and Isaiah have asserted that the despoliation of land is not only inexpedient but wrong. Society, however, has not yet affirmed their belief. I regard the present conservation movement as the embryo of such an affirmation."[28]

Based on Leopold's exposure to, and adoption of, Hadley's "model" for social adaptation and group learning, and based on the echoes of this early Darwinian approach to ethics in the final version of his land ethic, I conclude that Leopold did not see the choice between progressivism and romanticism as an either-or choice. By accepting Hadley's model of longevity as the test of whether a culture has learned the "truth" about its place, its habitat, Leopold could avoid challenging economists directly—by granting them anthropocentrism and acknowledging that some economic development is inevitable and good—while arguing that, in addition, each generation has an obligation to regulate the long-term impacts of its actions and policies. He saw his ethic-in-the-making as a matter of social learning. So our diversion into Hadley's particular applications of pragmatist principles to history and politics shows that Leopold—far from choosing between Muir and Pinchot—was actively searching for a way to integrate economics into a broader ethical framework. What Leopold borrowed from Hadley was the key idea that longevity over multiple generations provides a second, independent criterion of the success of a culture; can the culture survive for many generations with its current practices and institutions? Whereas the longevity criterion as applied to cultures was for most of human history a matter more of biology than culture, modern man has developed—within the constraints of biology, of course—extraordinary cultural adaptations. We are now living in the age of culture; humans today must learn very rapidly, because our impacts on nature are accelerating at the rapid pace of Lamarckian cultural evolution.

This criterion of longevity is applied to cultures and societies—to human communities—not to individuals; Leopold could, using it, sharply separate his own viewpoint from that of Pinchot, who measured all values in terms of the economics of individual consumption. And although Pinchot did say he acted to achieve "the greatest good for the greatest number over the longest period of time," this formula, especially when coupled with Pinchot's emphasis on individualistic economics, reduced conservation to a competition among individuals, with future individuals at a severe disadvantage in the competition (not being present at it!). Leopold could thus use Hadley's model to embrace economic development as normal and acceptable behavior without treating it as the ultimate arbiter of managerial success. Economic development is acceptable, provided it leaves the earth "alive and undamaged." By following Hadley in emphasizing the importance of cultural practices and institutions for mediating resource use and in showing the way out of the Malthusian tragedy, Leopold had found in the survival of cultures an alternative to Muir's theology and moralism and to Pinchot's narrow economism. The basic model was taken from evolutionary biology, coupled with emphasis on the highly social nature of human life, and it makes environmental problems a matter of identifying and altering—through socially conscious adaptation (i.e., social learning)—those practices and institutions that are having large impacts on ecological systems because of growth of human dominance over them. Thus, though I agree with Oelschlager that Leopold saw the limits of "imperial" ecology and that he recoiled from it, he never doubted that we are, henceforward, "captains of the venturing ship" of evolution and that the institutions we adopt, and the individual behaviors we encourage, will have a huge impact on both future generations of humans and on the systems of life itself—on "places." Leopold avoided the radical choice between Pinchot's economism and Muir's theological moralism by adopting a Darwinian approach to evaluating cultural institutions and practices in terms of their contribution to the multigenerational survival of a culture. For this reason, I think Leopold deserves to be considered "the first adaptive manager," not just the "Father of adaptive management."

But my friend Mark Sagoff will now perhaps say, "Gotcha!" He might think I have described Leopold as adopting a hard-edged, Ehrlichian ecological model. Have I made Leopold into a hedgehog? No, I do not think so, because Leopold, following Hadley, so heartily endorsed the relevant model as a model of *cultural* change in its context. It is not a simple physical calculation that tells us whether a given society is exceeding its carrying capacity. The key variable is not a physical one, but rather an institutional and behavioral one. In societies with low levels of technology, culture can change slowly because available technologies can have only limited impacts on the habitat—the evolutionary context. But with the advent of the industrial and information ages, the tools are more powerful, and the ability to change the ecological context of our culture is expanded. These cultural "adaptations" can outrun our ability to adapt—so the key characteristic of a culture of the future will be the ability to learn, even as we manage by the seat of our pants. Cultural flexibility and social learning become crucial; and some of what must be learned has to do with the physical vulnerabilities of the ecological systems in a place. In this sense, long-term ecological studies are the best way we have to gain experience about impacts of new technologies and new developments.

Although learning about such physical vulnerabilities of ecological systems is important, they are not at the center of Leopold's model, which is a management model; at the center of Leopold's model is the variable of culture and behavior; and in today's world of powerful technologies and huge populations in geographic motion, the key variable is the educability of cultures whose tools have so greatly outrun our ability to foresee the impacts of using them. The transformation of society into a system that changes its "context" on all levels of physical systems puts more pressure on human institutions and their ability to respond to environmental change and threats of catastrophic changes. If my interpretation of Leopold is correct, however, I think it follows—and this is of course consistent with my pragmatist interpretation—that Leopold was a pluralist about environmental values. The economics of any society or culture deal with individual values, whether consumptive or spiritual; but economic decisions are bounded by a larger cultural overlay, a multigenerational, living relationship between the land and the people who live on it. Leopold never considered reducing these cultural ideas and the "wisdom" of past cultures to a physical relationship *or* to an economic relationship. It is a cultural one; and the continuities a culture develops with its habitat have a value that transcends individual value; it is value that is stored in the wisdom and practices of multigenerational societies. It is noneconomic value.

Leopold, therefore, avoided falling into either of two narrow, reductionistic approaches. On the one hand, he avoided Pinchot's reductionistic economic model, by insisting on nonindividual—cultural—values attaching to cultural longevity. On the other hand, Leopold avoided "ecologizing" environmental problems—he did not treat them *simply* as matters of Malthusian calculations. He emphasized the variable of cultural institutions and practices, which do store the wisdom of the past but which today are constantly challenged by rapid cultural and technological change. The question, in other words, that Leopold's model sets out to address is whether a given culture has, or can develop, institutions and practices that will evolve responsively to changes in its ecological context. By making the main question of survival one of developing institutions that are stable enough to perpetuate our current social values, including love of nature, and at the same time flexible enough to respond to rapid change both in culture and especially in the ecological context of cultures, Leopold shifted the question out of economics and out of ecology, into the area of active management. As a manager, Leopold had to act, as Justice Holmes would say, in service of a variety of desired goods, including individual goods and also goods associated with the multigenerational community. But the effective variable in this balancing had to be an honest critique—such as Leopold made in his "reconnaissance missions" and accounting exercises in the Southwest—of long-term impacts of current behaviors, followed by the development of new institutions that would shape individual behavior so as to reduce the violence of human use of the land.

Leopold might be described as an "institutional economist"—an economist who believed that human economics is important but that long-term survival will be determined not by our ability to transform our environment quickly, but by our ability to quickly react to a more rapidly changing environment. The outcome will owe more to a society's stored wisdom, and increasingly to its ability to adapt quickly, than to any calculations of economists or of Malthusian ecologists. Based on his observations of the deterioration of the resource-producing systems of the Southwest, Leopold proposed—as an alternative to the economists' emphasis on individual accumulation as the basis of wealth and successful management of the environment—a Darwinian criterion related to the need (on penalty of extinction) to leave the land alive and undamaged for future generations as we engage in wealth-seeking.

Notice also that this approach represents a direct application to conservation of the arguments that Holmes and Green had applied to the common law: the best one can do is to look at history, examine the rules as they have been formulated, and choose an action that, while cognizant of history and principle, represents, in a given situation, the best means to achieve the multiple ends desired. For Leopold, survival over many generations was the only test available to evaluate such choices. In other words, Leopold continued to use economic success as one desideratum for action but recognized also that "leaving the earth alive, undamaged," also places a legitimate limitation on the action of members of each generation. We must, that is, in making difficult resource decisions (to quote Holmes, again), "consider the ends which the several rules seek to accomplish, the reasons why those ends are desired, what is given up to gain them, and whether they are worth the price." Leopold had been initiated into the mind-set of the pragmatists, accepting that conservation would always be a matter of experiment, a balancing of multiple desiderata, and carried out in uncertainty, because we can never know, no matter how careful our observation of impacts, whether we have examined the consequences of our actions over a sufficiently long period of time.

Therefore, although Leopold clearly felt the pull of both Muir's romanticism and Pinchot's materialistic utilitarianism, we can emphatically deny that he faced these alternatives in the stark form memorialized in the rhetoric of ideological environmentalism. I have shown that even as the rhetoric of environmentalism became more bipolar, as Muir decried the "evils" of development while Pinchot sang its praises in public battles over grazing in the forests and damming the Hetch Hetchy Valley for a reservoir, a powerful third force emerged, the pragmatic attitude, which espoused more pluralistic bases for judgment and rejected ideology and dogma in favor of unrelenting application of the method of experience. Leopold was not only exposed to this third force; evidence shows that he found it an ideal device for attacking prejudice and dogma on the basis of careful observation over expanding temporal frames.

To believe that Leopold was unaffected by this third force, one would have to believe that although the revolution in American thought ushered in by the pragmatic spirit and method had, by 1905, spread from Cambridge to Ann Arbor and Chicago, it had somehow missed New Haven. Even if such a lacuna were possible, we know it did not occur, because we know that Leopold somehow absorbed from Hadley—a sort of missionary for pragmatism—the essentials of the pragmatic habit of mind. It was this habit of mind—this suspicion of universal and preexperiential principles and a faith, ultimately, in experience alone as the arbiter of truth—that provided Leopold with an antidote for the polarization inherent in ideological environmentalism. Leopold did not, as suggested by Oelschlager and many others, start out a Pinchotist, on one extreme pole, and shift later in his life to the opposite, romanticist pole. Rather, Leopold was an integrator; and his guide toward the necessary integration was the pragmatic method.

I can now state the objectives I aim at and describe the strategy I will employ in pursuing them in this book. I have adopted three related objectives: (1) to show how environmentalism as a social movement can transcend its tendency toward ideological formulations of problems and options, (2) to show how to reduce (albeit gradually) the related phenomenon of towering—the inability of different agencies and interest groups, all with a tendency to use their own jargons and problem formulations, to cooperate in achieving widely shared social goals—and (3) to articulate, under the title "adaptive management," a positive, experience-based, mission-oriented approach to adaptive management science as social learning. My strategy, as may be obvious from the above account and my endorsement of pragmatism as Leopold's third way, is to argue that Leopold, insofar as he developed a consistent philosophy of management, laid out the general outlines of an adaptive theory of management and that he relied on experience and the use of Darwinian explanations of knowledge and ethics; moreover that these were in his day—as today—perfectly reasonable complements to a progressivist viewpoint. Since many Darwinists, including Hadley and Darwin himself, strongly rejected any form of economic reductionism, neither Leopold nor subsequent environmentalists need experience following Muir and Pinchot as an either-or choice between economic growth and nature preservation. Leopold—and I will suggest that today's environmentalists should enthusiastically follow him—substituted a Darwinian approach to truth and knowledge as well as ethics, thus embracing a standard of long-term survival as a noneconomic basis for evaluating a culture's practices and institutions.

Some of the implications of retelling the story of environmentalism as a matter of three paths from the past to the future are startling. My argument seems to suggest that in an important sense, environmental goods are "communal" or "community-level" goods, rather than individually experienced goods. This line of argument will lead me to reject one of the pillars of contemporary social sciences, the commitment of virtually all researchers to "methodological individualism" in accounting for social values, especially environmental values. A second apparent implication is that human and nonhuman cognition forms a continuum; as the cognitive scientist Henry Plotkin puts it, there is only a difference in kind, not in degree, between our knowledge that the desert is dry and the "knowledge" of the cactus, whose ability to store water efficiently is evidence that it has "learned."[29] A third, more practical implication seems to be that, although it is very interesting to pay attention to the languages of biology, physics, and economics, the relevant science—the science that can integrate our vocabularies—must be *management* science, not the special sciences, which are artificially and incompletely purged of evaluative judgments. Since environmental problems inevitably involve balancing competing social goals, only an activist science, such as adaptive management, can teach us to properly frame problems and to integrate science in our search for a balanced strategy of development and survival.

My strategy, then, is to provide a conception of sustainability that captures what is shared by Leopold's pragmatist-progressive ideas and by what we today call adaptive management. These and other implications of a pragmatist outlook on environmental problems are explored throughout this book; but the unifying objective is to understand, and to flesh out, the middle-ground, progressivist approach Leopold sketched in 1923 and to show how the pragmatist "habit of mind" that Leopold picked up from Hadley provides a way to avoid both ideology and towering in environmental policy discourse today. So much for history, objectives, and strategy, let us turn now to an examination of the "logic" of the pragmatic method and to the deep philosophical arguments that sway its proponents to accept experience, understood actively, as the ultimate arbiter of human knowledge.

2.4 Environmental Pragmatism and Action-Based Logic

Based on the argument of the previous section—that the histories of pragmatism and of environmental thought have been more deeply entwined than is usually recognized—three perhaps obvious questions arise.

1. What is pragmatism?
2. What is *environmental* pragmatism?
3. Why do its advocates find the pragmatic approach persuasive?

I have begun to answer the first question in section 2.3 and will return to it in chapters 3 and 4; here, I will push our understanding of pragmatism a bit deeper and explain what I mean by *environmental pragmatism,* mainly by answering question 3. The best way to understand pragmatism as a habit of mind—a habit of mind useful in guiding environmental thought—is to understand the pragmatists' central argument, which is an argument for a new way of thinking about language in relation to the world. To use an anatomical analogy, one might think of this argument, and the conception of language it implies, as the spine of pragmatism, with other connected beliefs, applications, and variations, some of them controversial even among pragmatists, as attached ribs. It is this argument that supports the pragmatist habit of mind; and it also serves as the nerve center for the multifaceted agreements and debates that make up pragmatism today. This argument about language and its relation to the world—which is not necessarily exclusively used by those who call themselves pragmatists—has turned out to have a huge impact on our understanding of philosophical method and its usefulness in practical endeavors.

Before introducing this argument, by referring to John Dewey's version of it, I first summarize what *I* mean by environmental pragmatism, which (I say with some pride) is now considered an important force in environmental philosophy. I hope my explanation will not seem preemptive of other versions, as I am aware that there are indeed a variety of formulations and approaches to pragmatism and neopragmatism, but it does seem fitting that I provide at the outset a sketch of how I view environmental pragmatism as a nascent movement.

In the mid-1980s, Anthony Weston and I, mainly independently, began exploring whether pragmatist ideas might have useful implications for environmental ethics and environmental thought. Weston emphasized, among other points, that environmental problems are often unproductively formulated and that the pragmatic method might help us to replace bad questions with better questions.[30] I observed that when one examines actual cases of important environmental legislation or other actions to protect species or natural systems, environmentalists with very different worldviews, who use different languages to explain and justify their policy objectives, can nonetheless agree on what those objectives should be.[31] Both approaches, I submit, are representative of the pragmatic habit of mind discussed by Hadley and other pragmatists early in the twentieth century.

In 1996 Andrew Light and Eric Katz published an anthology entitled *Environmental Pragmatism;* the seventeen essays making up the collection, by more than a dozen authors, demonstrated the growth of interest in a revival of Leopold's pragmatic analysis of environmental problems, as well as the diversity of approaches involved in such a revival. Some authors in that book, and others elsewhere, apparently use the term *environmental pragmatism* very broadly to include any problem-oriented perspective on environmental theory or practice. But I intend to use *environmental pragmatism* quite specifically to refer to the habit of mind referred to by Hadley and also to the particular understanding the pragmatists share about the nature of language and logic in relation to the world of experience. The habit of mind, the theory of language, and the adaptive, experimental approach to environmental action are to me three sides of the same coin. To articulate the argument for this intimate connection, I turn as promised to Dewey's particularly clear and compelling explanation of the not-yet-fully-understood consequences of Darwin's revolutionary idea.

John Dewey, who was both a philosopher and a social critic of immense influence in the first half of the twentieth century, sketched this view in a 1909 lecture that he called "The Influence of Darwinism on Philosophy."[32] His lecture comes closer than any other short piece to clearly exhibiting how Peirce and other earlier pragmatists drew insight from Darwin's theory of natural selection. In this lecture, which was later published as the title essay in a collection of Dewey's lectures and essays, Dewey recognized the revolutionary impacts of Darwin's dynamic selectionism and showed how those impacts would necessitate the reframing of the central philosophical questions facing a postmodern world. Dewey began by mentioning the impact of Darwinism, which many people thought incompatible with creationism, on religious thought and belief, but he spoke only briefly of the religious furor Darwin's books had ignited, noting that he doubted most people had yet seen the even deeper implications of Darwinism. Moving quickly past religious effects, and suggesting that we had not seen anything yet, Dewey went on to argue that Darwin's true impact on philosophy would be appreciated only when people grasped its implication that the categories—the taxonomy of kinds that we use in organizing and describing our world—starting with "the species, but including all categories and kinds," are not fixed but changing, changing in response to their use in describing and manipulating the world in which we survive as communicative, social beings. The language we speak does not get its meaning by reflecting an inert and passive world "out there," beyond experience, created and ordered by a benevolent, all-powerful being. Instead, language gains meaning from the dynamic relations emerging within a constantly changing and evolving culture composed of purposive individuals in linguistically cohesive communication. Language is thus integral to a complex set of behaviors that have evolved within a community's day-to-day practices. Meaningful speech is reflective of social relationships; social communication includes many exchanges of experience and gradually results in cultural adaptation. The same dynamism that undermined belief in fixed biological species and in the formalistic view of law critiqued by Green and Holmes, Dewey saw, must eventually undermine all commitments to a priori beliefs, to self-evident truths, and to preexperientially determinable categories of being. This was the crux of the matter.

In traditional philosophical jargon, the issue was "essentialism." Essentialism is the view that the categories of human thought answer to real, preexisting "kinds" or categories, a view that had seemed unquestionable in Western thought since Plato and Aristotle each articulated their own version of essentialism well over two thousand years ago. Essentialism, by assigning to nature a "real" structure that exists prior to experience or language, was able to support—for thousands of years—the view that at least on some deep, metaphysical level, *the* structure of the world is available for discovery. Darwin challenged this deep, metaphysical commitment to a single correct, prelinguistic, universal structure of nature, and Dewey explained how this challenge must undermine the belief that nature has one true rational structure, discoverable by humans.

Dewey realized that Darwin's dynamic, shifting categories of being, shaped by the hard, cold facts of natural selection, would eventually replace the old essentialist thinking everywhere. At the heart of this insight was the recognition that essentialism had become untenable after Darwin. Without essentialism, general words like *cat* and *tyrannosaurus* cannot be taken to *denote* or *refer to* a fixed and given kind unified by some "essence." Essences can fulfill their function in rationalizing the world of experience only if they exist independent of, and prior to, experience. When Darwin showed that the categories biologists used to describe nature were, in fact, ideas that were projected into a constantly changing world, the idea of a rational, unchanging reality was inevitably undermined.

Dewey saw that once the nose of the Darwinian camel—selectionism and dynamism of biological categories—was under the tent of modern thought, the tent's full upset was inevitable, only a matter of time. If the world we live within and adapt to is constantly changing and re-forming, if there are no "kinds" to correspond to the categories we use when we classify and describe nature, then the notion of essence and truth anchored in correspondence between sentences and sentence-sized chunks of reality becomes unanchored. If truth is to be the "correspondence" of a sentence to reality, and if a sentence uses its subject to identify some piece of the world and attribute a property to it, what corresponds to the "property"? Seen from the modernist viewpoint, this loss of essentialism is tragic; it forever blocks the way back to a fixed, unchanging, stable world of objects existing in preformed categories, as God created them. It denies the rationalist a rationally coherent world beyond, and independent of, our minds. The categories of being become, on these new assumptions, a function of language rather than the other way around.

From the vantage point of the modernist, authoritarian viewpoint, this consequence of Darwin's method "threatened" stability and rendered nature chaotic and unknowable. From the pragmatists' side of the modern-postmodern divide, however, the end of essentialism was the end of intellectual slavery—to fixed categories, to fruitless dichotomies, to unobservable "essences" and circular explanations. For the pragmatists, the way forward was to embrace experience as the measure of all things and forever abandon the failed Cartesian dream of describing a human-independent reality as it is, corresponding to our assertions about it. Every belief must eventually, and over and over, be tried before the jury of experience. On Peirce's side of the postmodern divide, Dewey's question must become dominant: if the categories and kinds we use to describe nature and its regularities do not come from nature—if they are not "discovered" by some combination of observation and reason—where do they come from? We cannot go forward without answering this question.

Pragmatists believe our descriptions of nature and the categories we use in these descriptions must be validated within communities that survive over time by learning, changing, and adapting in place. Our language and conceptions, now freed from the dead hand of religious orthodoxy and a priori doctrines about how nature *must* be, can now be seen as a part of the creative and positive processes of understanding and communication. Language, on this view, becomes a tool for communication, useful for achieving cooperative behavior in a larger social context.

In his monumental 1938 book *Logic: The Theory of Inquiry,* under a subsection entitled "Logic Is a Social Discipline," Dewey separates his views from those of symbolic logicians: "Inquiry is a mode of activity that is socially conditioned and that has cultural consequences. This fact has a narrower and a wider import. Its more limited import is expressed in the connection of logic with symbols. Those who are concerned with 'symbolic logic' do not always recognize the need for giving an account of the reference and function of symbols. While the relations of symbols to one another is important, symbols as such must be finally understood in terms of the function which symbolization serves."[33] Returning to this theme in a later chapter of the same book, on culture as the "existential matrix of inquiry," Dewey adds: "For man is social in another sense than the bee and ant, since his activities are encompassed in an environment that is culturally transmitted, so that what man does and how he acts, is determined not by organic structure and physical heredity alone but by the influence of cultural heredity, embedded in traditions, institutions, customs and the purposes and beliefs they both carry and inspire. . . . To speak, to read, to exercise any art, industrial, fine or political, are instances of modifications wrought *within* the biological organism by the cultural environment."[34]

Emphasizing the functional importance of language, Dewey refers to linguistic categories and concepts as "conventional," but he is quick to caution that "the physical sound or mark gets its meaning in and by conjoint community of functional use," not by any explicit convening in a "convention," and that "the particular existential sound or mark that stands for *dog* or *justice* in different cultures is arbitrary or conventional in the sense that although it has *causes* there are no *reasons* for it."[35] If Darwin showed that the categories of biology are not fixed but are developed as useful tools in the study of biology, then, accordingly, all language can be understood as tool-like: it must be evaluated on the basis of experience in using it. We can ask, Does this set of conventions that have evolved in cultural contexts actually conduce to improved communication and cooperative behavior? Language is an adaptation among other adaptations; and language is effective as long as it supports communication and forwards cooperative behaviors and effective social institutions. If the language we are using is not effective in communication and encouraging cooperation, it can be changed, and new forms of language can be tried. Choices of effective tools of communication become, therefore, testable hypotheses about what works in specific situations requiring action.

The categories of being to which we refer emerge from our communicative and cooperative behavior within a community. All language is thus symbolic, but Dewey also insists that this does not answer the key question for inquiry whether the symbols "are ready-made clothing for meanings that subsist independently, or whether they are necessary conditions for the existence of meanings—in terms often used, whether language is the dress of 'thought' or is something without which 'thought' cannot be."[36] Dewey, of course, clearly adopted the latter view, that the conventions of language create the categories and concepts essential to thought.

Our language in this sense creates the world as we know it. But creating the world as we know it is nothing like creating the world. Our beliefs and our linguistic tools for communicating those beliefs, and for testing them against our experience and others' experience, must of necessity be judged according to a criterion of functionality. A way of talking about the world, and an associated "ontology" of objects and kinds that "exist" because we say they do and for no better reason than our saying they do, helps us to organize and to learn from our experience.

Dewey emphasized, then, the situational nature of logic, the functional nature of language, and the conventional status of categories of being. In his *Logic,* Dewey described a "situation" as follows: "What is designated by the word 'situation' is *not* a single object or event or set of objects and events. For we never experience nor form judgments about objects and events in isolation, but only in connection with a contextual whole. This latter is what is called a 'situation.'"[37]

Dewey recognized that the very "situations" we face are shaped by the linguistic forms we use to communicate about our common experience. This shaping affects, of course, the facts we see and emphasize; but even more importantly, our language shapes whether we see a situation as one of complementary goals or of clashing interests. Intelligence evolved as a problem-solving ability, and Dewey wants societies of humans—who can learn much more quickly than can other species because of language and culture—to use that intelligence to improve the conduct of inquiry, to improve the efficiency of learning. For Dewey, the purpose of inquiry is to stabilize a situation and to make action possible. Dewey's functional logic drew only scorn from the "symbolic" logicians, and he was unfairly criticized by Bertrand Russell for his functional understanding of truth. Later in his life, to lay this controversy to rest, Dewey seldom referred to "truth" but rather characterized the epistemological problem as one of achieving "warranted assertibility," which is contextual and located in a situation demanding some form of resolution.[38]

We cannot, Dewey thought, rely on general, preexperiential principles to formulate problems, as occurs when ideological principles are used to force information and communication into particular symbolic forms. However, he does not leave the democratic process unguided: he offers, instead of fixed principles, situational intelligence, the ability to learn how to learn—what he called method, otherwise referred to here as the pragmatic habit of mind. Although learning is situational, we can inductively develop a method for looking at many particular situations. Logic—in a system of beliefs committed to action and relying on the method of experience—allows the gradual reformulation of issues and problems based on broadened experience and improved communication among participants with varied perspectives. Logic, for Dewey, was learning how symbolic behavior affects action in particular situations. He believed that our intelligence can also be turned upon linguistic forms and the function of symbolic behavior.

Dewey saw that linguistic characterizations of problems are often crucial in finding a cooperative solution; and he believed that since language is conventional and functional, there is no prohibition in creating new "models" and new linguistic forms for characterizing problems. If we are using outmoded, confusing concepts that encourage the formulation of disagreements ideologically—in terms of a priori principles and hemmed in by traditional categories of experience—prospects for cooperation are bleak. But perhaps, as one aspect of social learning, we can improve the ways we talk about shared experience and the ways we express our differences of opinion. Perhaps we can develop a language that encourages experimentation and careful observation of outcomes. In the process of developing more functional concepts, Dewey and the pragmatists would say, we are simultaneously reshaping our world and reinterpreting the problems we face in terms of new experiences.

I began this chapter with the controversial claim that language *is* our world; I hope that the specific meaning I give to this claim is now clearer. The world we experience is not the same world we would have experienced if we spoke a different language; but it is also true that languages are inseparable from the activities and practices of a people who use that language to communicate. And so the world we experience is not a matter of free choice—we cannot live in a world of make-believe. Reality will shape our world, because our practices are more or less direct responses to our world (our environment). Language is an adaptation, but language does not change without shifts in underlying practice; nor does language change independently of power relations and political pressures, especially when it is used to make important management decisions. Developing better language to communicate about and decide environmental policy proposals is, I submit, only one aspect of creating improved institutions and practices for managing the environment. It is, however, one manageable piece, and it is the piece that I take on in this book. As for the broader political and evaluative implications of our linguistic choices, once I admit that language is a politically sensitive adaptation, it is clear that my topic—the language of environmental decision making—cannot be sharply separated from either politics or policy.

Ultimately, the language in which these tools are chosen, the language in which important decisions are made, is a political matter. The language of politics—of power and negotiation—can in turn shape the formulation of environmental problems. But as careful analysis of environmental discourse in other countries has shown, the interaction of discourse and power in democratic societies is highly complex and depends on many "subpolitical processes," in which "discourse coalitions" form, gradually achieving agreement on the formulation of and possible solutions to environmental problems. In particular, I refer to the work of John Dryzek, a British political theorist, and to Maarten Hajer, who has done a careful empirical comparative case study of discourses concerning acid rain controversies in the United Kingdom and in the Netherlands. Dryzek and Hajer agree—drawing heavily on a few examples from Europe and Japan—that a new form of discourse, what they call "ecological modernization," has emerged as the dominant discourse coalition in environmental policy debate. They also suggest that the discourse of ecological modernization promises to become the dominant international discourse in environmentally aware countries. "Ecological modernization" is defined by Hajer as the assumption that "environmental problems can be solved in accordance with the workings of the main institutional arrangements of society. Environmental management is seen as a positive-sum game; . . . pollution prevention pays."[39]

Having granted ecological modernization a privileged role in public environmental discourse, both Dryzek and Hajer recommend—and practice—discourse analysis, which they originally associate with the discourse-theoretic perspectives of Michel Foucault. They (especially Hajer) modify the Foucauldian analysis somewhat along the way, rejecting the purely descriptive—and fatalistic—attitude of Foucault himself, and adopt a more positive approach according to which discourse, by becoming more "reflexive," can create "liberating practices based on reason and mutual respect."[40] Setting aside Foucault's fatalistic suggestion that the creation of new discourses simply replaces one "cage" with another, Hajer follows Ulrich Beck in envisioning a progressive clarification of problems through the development—by subpolitical processes—of discourse coalitions clustering around emergent problem formulations and shared conceptions of acceptable solutions.[41] To distinguish his approach from the more fatalistic approach of Foucault, Hajer introduces what he calls an "argumentative approach" to discourse analysis, an approach that "focuses on the level of the discursive interaction and argues that discursive interaction (i.e. language in use) can create new meanings and identities, i.e. it may alter cognitive patterns and create new cognitions and new positionings. Hence discourse fulfils a key role in processes of political change."[42] In particular, both Hajer and Dryzek introduce the mediating influence of "story lines," which are persuasive to the extent that they can provide, even temporarily, a coherent picture of environmental problems.

I plan to go much further than either Hajer or Dryzek in this respect, not only rejecting fatalism about linguistic forms but also encouraging self-conscious attention to both the captivating and the liberating aspects of our discourse. In particular, I will propose an idealized, but illustrative, process of public debate in which proposed heuristics—themselves, of course, open to reconsideration and adjustment—guide participants in policy processes toward better questions and away from disagreements based in ideology. I will substitute a form of "linguistic activism": explicitly experimenting with means of communication and judging them on the basis of their contribution (or lack thereof) to cooperative action. This linguistic activism, which is of course a characteristic of the pragmatic philosophy advocated in this section and the previous one, is in sharp contrast to Hajer's suggestion that through public policy discourse, scientific and disciplinary conceptualizations are "contaminated"—an apparently passive absorption of ideas across disciplinary boundaries. In this book I explicitly suggest that in addition to improving the first-order discourse in which we describe the world as problematic, we can also, more explicitly, shift into a "metadiscourse," in which we reflect not only about problems and solutions but also about the way we talk about problems and solutions.[43]

My goals in this book are similar to those of Dryzek and Hajer, and their work is methodologically complementary to my approach. I hope to go beyond their passive approach of recognizing the important role of language in environmental discourse, however, to begin the hard work of analyzing how conceptualizations change during the rough-and-tumble of public debate. Their analysis, of course, is based on a very different theoretical tradition. They relate the method of discourse analysis to Foucault's fatalistic method and make incremental alterations of that method by introducing a more liberal view from sociologists of science. Eventually, in order to escape the fatalism about discourse encouraged by Foucault's theory, they draw on the idea of Ulrich Beck that linguistic change can be liberating as well as captivating. Nevertheless, I believe neither Hajer nor Dryzek provides an adequate positive theory to support their more activist approach to discourse; most of their writings are simply descriptive of how language helps or, more often, hinders communication. What is most lacking in their analysis, in my view, is an explicit recognition that Peirce and the pragmatists presented a test that is applicable to action-oriented language—the test of the usefulness of modes of speech in real-world situations. Hajer, for example, explicitly distinguishes his work on discourse coalitions from Paul Sabatier's approach to analyzing policy development in terms of "advocacy coalitions," which Hajer criticizes for its "individualist ontology" (that is, for failing to see that modes of speech are used contextually and shift as individual actors shift from role to role) and for failing to account for the "constitutive" nature of language and the ways in which changes in discourse change opportunities for action.[44] I propose to combine the best of both discourse analytic approaches and analyses in terms of advocacy coalitions by firmly placing my linguistic analyses within the philosophical traditions of pragmatism, which, as noted in this and the previous section, sees language and action as intimately related in public discourse. A terminological note: the form of pragmatism I advocate is nothing like the "traditional pragmatism" attributed by Hajer to British conservatives, who adhered to the story line of conservatism and the no-action-on-acid-rain policy. Traditional pragmatism is equated by Hajer with an implacable refusal to act in the absence of unquestionable science. I want no part of what Hajer describes as inactive, political pragmatism. Real pragmatism stands for the direct opposite of what he calls traditional pragmatism. Real pragmatism stands for action, for experimentation—for relying on experience, not on preconceived principles.

A pragmatic attitude entails several modifications of the Foucauldian analyses of Dryzek and Hajer. First, and this may be largely a result of historical, cultural differences between European countries and the United States, the American pragmatist tradition thinks practice is more basic than patterns of discourse. Accordingly, I will show in chapter 3 that adaptive management, seldom mentioned in Europe, provides a better context for the analysis of environmental discourse and action, in America at least, and a more solid theoretical foundation than does Foucault's fatalistic analysis whereby discussions of action are constrained by inescapable linguistic "cages." By emphasizing the role of language as emergent *within a process of acting and managing,* it is possible to relate language and its uses directly to actions and to problems of communication in the search for consensus positions on what to do. Indeed, given the situation on the American side of the Atlantic, the discourse of "ecological modernization" has not emerged and is unlikely to do so, at least as long as environmental discourse in the United States is dominated by the bipolar rhetoric of ideological environmentalism. Once ideological environmentalism fades in America, there may be an opportunity for the emergence of a new "discourse coalition"; but I doubt it will be the discourse of ecological modernization; nor will the means of analysis be Foucauldian discourse analysis. I believe that in America, at least, the activist tradition of adaptive management and ecosystem management, including partnerships and public deliberation, will become dominant. As noted, these differences may be mainly cultural, but ecological modernism does not seem a fruitful discourse for American politics because adaptive management has already taken hold in many agencies and in watershed management partnerships, for example. It also seems highly unlikely that the idea of ecological modernization that Hajer describes in mainly political terms would have much resonance in America, where Leopold described ecology as the "fusion point of the sciences" and where ecology is taken seriously as providing a holistic, systematic model for integrating physical sciences and making them relevant to judging the quality of human habitats.[45]

Another major difference between European-style discourse analysis and pragmatism is that the former type of analysis still rests on a fairly explicit and unquestioned division of descriptive (scientific) and prescriptive (ethical) discourse. A more pragmatic analysis eschews this distinction as mainly artificial. Pragmatism and adaptive management, then, do not draw a sharp distinction between facts and values; they expect descriptive and prescriptive speech to be intermingled in policy discourse. Only time will tell whether my idea of adaptive management proves useful in Europe despite the fact that its ideas emerged from American thought and despite the fact that environmentalism in Europe has generally been unaffected by the ideas—powerful in the United States—of Aldo Leopold and the adaptive managers, most of whom have been North Americans. One might suggest that discourse analysis of environmental problems will reveal that the best approach to analysis will be highly country- or culture-specific. One might, alternatively, suggest that discourse analysis eventually may be reconstituted on a stronger, pragmatist philosophical base and that, internationally, there will (or at least should) be an eventual convergence of American-based pragmatist theory with the insights and methods, developed in Europe, of discourse analysis. In my view this would represent a wedding of the strongest theory with powerful techniques for teasing out the political relationships of power and subpolitical processes that determine the waxing and waning of discourse coalitions. Rather than speculate about such outcomes, I simply offer the pragmatic approach as an improved theoretical structure—by comparison to the overly intellectualized and fatalistic theory of Foucault.

Pragmatism promises that conscious attention to language in action can lead to improved ways of speaking and to reduced failures of communication. These linguistic innovations may range from quite informal suggestions of stipulative definitions for important terms already in use to highly formal ones, as when we introduce a mathematical model for representing some causal relationships used to track impacts of policies on outcomes. Today, however, environmental policy, at least in the United States, is paralyzed by the consequences of ideological environmentalism that affect the formulation and understanding of environmental problems. These effects in turn squelch any intelligent discussion of possible compromise actions within politically riven agencies like EPA and lead to towering in policy discussion and actions.

I have tried to show in this chapter how incorporation of pragmatist ideas into environmental management—by emphasizing the pragmatist roots of American conservationism, especially from the second generation forward—can liberate conservation thought from the paralysis of polarized valuations and practical stand-offs. The way beyond ideology to cooperative action is to develop a more neutral, and yet expressive, language that will allow formulation of problems in more-or-less terms. Once the moralism, theology, and economism are minimized, disagreements—and we certainly expect them to continue—should be stated as hypotheses that can be tested in a pilot project or evaluated by doing a case study of an existing program. My goal as a practical logician is to eliminate those value concepts that focus disagreement on preexperiential commitments—whether they be to market economies or to pantheism—and that focus attention on moral differences. Once those commitments are left behind, it is possible to think creatively, to wonder whether it is possible to create a new language—or fully refurbish, perhaps, some portion of presently used language and theory. In the remainder of the book, I set out to develop, and try out, a new framework language, including definitions and pragmatic heuristics, that may help communities to formulate their problems in less ideological terms and to begin a process of communication that can lead to cooperative solutions.

3

EPISTEMOLOGY AND ADAPTIVE MANAGEMENT

3.1 Aldo Leopold and Adaptive Management

Aldo Leopold, the great hero of American conservation, grew impatient with his colleagues who forever engaged in rhetorical debate about what ideas should guide conservation. He once wondered in exasperation why conservationists continually talk about ideas, "instead of going out on the land and giving them a trial."[1] This good advice illustrates both Leopold's affinity for American pragmatism and his pivotal role in American conservation. Leopold arrived on the conservation scene in the first decade of the twentieth century as a freshly trained forester, having been exposed to the experience-based philosophies of Hadley and the pragmatists while a graduate student at Yale. The conservation scene, as noted several times already, was divided by a bitter controversy between two wings of the conservation movement, one led by Gifford Pinchot, who advocated wise use of our national forests—understood as the universal application of economic utilitarianism—and the other inspired by John Muir, who argued on theistic grounds for setting aside many of the beautiful and pristine areas of mountains and forests. This broad conflict reeked of ideology and rhetoric, as noted above in the preface, but the disagreements, in particular contexts and in uses of particular pieces of land, were real and often bitter. Leopold, usually patient and thoughtful, lost patience when the disputants rested their case on rhetoric alone. He doubted that we would find the way by starting with universal principles and imposing these general solutions on local situations; on the contrary, he recognized that every conservation challenge is unique and should be viewed as an opportunity to try out new ideas and techniques in a new situation, learning in the process. He was, by inclination, in the terms of Isaiah Berlin's distinction discussed in section 2.2, more of a fox than a hedgehog, and his above-quoted, exasperated comment expresses, more or less consciously, the experience-based, experimental attitude of the philosophy of American pragmatism. In this chapter I show how adaptive management has developed into a promising way of addressing complex environmental problems of today and show along the way that an explicit infusion of pragmatism into adaptive management theory enriches both. For example, Leopold was known among his students for his ability to locate the origin of a handful of soil, brought into the classroom, from its touch, smell, look, and taste. He was thus held in awe for his powers of observation. Just as importantly, he also exercised the patience necessary to carefully record his observations for future reference, in order to allow observation of trends and patterns over time. What made his curiosity, experience, observations, and interpretations so revelatory was the *context* of his work: he sought the truth *while managing* U.S. Forest Service lands in the Southwest and *while serving* on a committee with responsibility for regulating the Wisconsin deer herd, to cite just two of many examples. In voluminous field notes and many notebooks of musings on the observations therein, Leopold recorded the details of the arrival of spring, marked by the arrival days of birds, and countless other daily occurrences. Leopold sought no shortcuts to environmental enlightenment. So it was ultimately his dogged curiosity, his willingness to learn his way to wisdom, that accounted for his particular genius. Ultimately, he placed his faith in nothing but his observations. But Leopold was also a complex and creative thinker, and it would be wrong to suggest that he confined himself to narrow questions of fact. Leopold loved philosophical speculation and engaged in it regularly in his popular essays. However, if I am right in saying Leopold was most basically a pragmatist philosophically, then this speculation should be understood as a guide to future thinking and research; the ultimate test of speculation must be experience in action.

Leopold was an amateur philosopher and a good one, but not a philosopher who read widely or joined a "school," even in the sense of settling on a "philosophy," understood as a set of first principles to be espoused and followed. Leopold was a philosopher in the pure sense—he was a lover of truth. But truth for Leopold was no abstraction; as he learned from Hadley, if a culture is not adapted to a place, if it does not understand its reality, it will not survive. Truth is forward-looking, and a society can claim to have found it only when the society's practices and institutions sustain its people indefinitely on the land it inhabits. And like the other pragmatists, Leopold accepted only one principle: rely on experience and only experience—formal experiments when possible, careful observation when control of variables is impossible. Experience is the only way to learning, and truth is a prediction about what will survive.

Like the professional pragmatists, Leopold also accepted the unity of inquiry. Experience was for him the measure of *all* truth, including scientific truth and moral rightness. This profound commitment has many implications for understanding Leopold and the depth of his contribution to conservation. Leopold's commitment to experience as the touchstone of all inquiry led him away from appeals to a priori principles and appeals to "foundational" ideas given by reason. First principles cannot trump experience. Again, the point is not that Leopold dismissed "the big questions" of metaphysics and morality; he simply thought these big questions, whether moral or "metaphysical," must be addressed in context, and using the same method—painstaking observations in many particular situations—that one uses to resolve day-to-day problems. Leopold believed that morality was tractably addressed by adopting the same sense of patient observation, coupled with speculation and theory-building as a complement; but he believed one must check speculatively proposed theories against real problems of existence and adaptation. This is the method that serves us well in science, more generally. Leopold did not, like the positivists, reduce all of life to the scientifically observable. Rather, he broadened his idea of observation so as to make life itself scientific. Leopold was, like Thoreau and the pragmatists before him, a naturalist in the sense that he believed morality to be subject to the same methods as he used in deciding how to manage a forest or a deer herd.

Leopold's activist role as a philosophical forester and wildlife manager also gave him a unique voice. Because he had served as director of operations for a Forest Service district, because he had designed and carried out management plans for game and other species, because he sat on a commission that managed Wisconsin's deer herd, Leopold awoke every morning with the expectation that he would be making decisions in real situations, decisions that would change the land and the environment. Action was forced because the lack of action in many cases can lead to real consequences for good or bad. Values permeated Leopold's every act and every observation. Values, in this sense, always shaped Leopold's thought; he was a manager first and a scientist second. It is in this broad sense that Leopold scientized life. He had no doubt that values are real—they have impacts. And so one can observe values being played out on farms and landscapes—the people's values have consequences for the land. Every management decision, for Leopold, tested a value as well as a fact. Values, Leopold believed, are no less, nor no more, subject to the regimens of observation than are scientific hypotheses.

Leopold thus followed the American pragmatists who emphasized action as the heart of their philosophies. For example, John Dewey argued that doubt takes on meaning only once it is experienced in a real situation with a real quandary. The practitioner, unlike the "pure" scientist and the academic, sees problems, including scientific quandaries, through a lens of values. The importance of a scientific hypothesis, the very interest in it, is conjured by its impact on the question, What ought we to do? When action is forced, facts come alive and are infused with meaning. Here Leopold's thinking is remarkably similar to that of Dewey, who argued for applying intelligence to real situations and emphasized social learning through the voluntary expansion of our experience and experimentalism in science, morals, and management.[2] Both Dewey and Leopold believed that actively expanding our experience—through experiments and observation under many conditions—was the path to social learning. Once first principles and intuitive, self-evident truths are rejected, experience comes to the fore. Epistemology is moved from the realm of the general, the abstract, and the a priori to the realm of local observation at the local level. Dewey's approach to dealing with uncertainty is especially effective in complex situations such as those that arise in the management of environmental systems. Both Dewey and Leopold eschewed appeals to general ideas as guides to action. They concentrated on amassing experience, experience gained in different situations and by different inquirers with differing motives and concerns. To justify management decisions, they would not appeal to preexperiential principles; they would, instead, attempt to resolve *this* particular problem in *this* particular place. Justification, in Dewey's sense, takes place in context, against a backdrop of shared assumptions and information but also of differing interests and motivations. Justification in this sense, the sense applicable to adaptive management, emphasizes local information and experiments to reduce specific uncertainties in specific situations. For the pragmatists—and here I include Leopold—the effective problem is not to deduce the correct managerial move from first principles but rather to see past such ideologies to reframe disagreements as predictions about future experience.

The Darwinian undermining of fixed essences and universal principles, as discussed in the last chapter, likewise affects the way we think of language. The shift from a world that is stable and organized from the top down to a random, changing, multifaceted world, Dewey recognized, represented a radical discontinuity in intellectual history. Truth was now to be found in the particular, and general principles must be built painstakingly on many particular observations. Since language exists in the concrete context of the human struggle to exist, linguistic forms, no less than specific beliefs about the way things are, must contribute to solving real human problems. Linguistic categories are not labels for transcendent essences but conventions reflecting human "choices" (often implicit); and they must be judged in action, functionally. They are tested against experience in the process of making choices about how to live. I hope to have shown that Leopold-the-manager operated with a method that had a distinctively pragmatist flavor and that for him the method of experience was the ultimate arbiter of both scientific truth and right action. Before going further, however, it is important to provide a more systematic explanation of what I mean by the term *adaptive management.*

3.2 What Is Adaptive Management?

I am arguing that Aldo Leopold, in the first half of the twentieth century, anticipated the core ideas of an increasingly popular approach to environmental management that today is called adaptive management. I have emphasized the importance of recognizing that a theory of environmental *management* must be a theory of *action.* The actions can be motivated only by social values, and all actions, including scientific study, are suffused with values. No system for managing the environment can be understood in purely physical terms. Understanding the physical systems involved is of course important: our theory of environmental action must embody an analysis of the physical mechanisms of environmental degradation. For convenience, we can refer to this component of our broader subject as the act of building a "model" of physical impacts of humans on natural systems—the habitats—that support human societies as their physical context. Since we seek a system of active *management,* our scientific models must be understood as embedded in a larger process of social discourse and political institutions. Management necessarily involves us in goal-directed activity; and our theory of management must therefore include a means of identifying, justifying, and/or legitimating goals by reference to some social value. Adaptive management, as understood here, is an approach to understanding, justifying, and implementing policies that affect the environment. This approach is worthy of being called adaptive by virtue of its intellectual pedigree and its increasingly important function in the life of human societies.

Although different advocates of adaptive management emphasize different aspects of the approach, I will here define adaptive management as management according to three key tenets.

1. Experimentalism. Adaptive managers emphasize experimentalism, taking actions capable of reducing uncertainty in the future.
2. Multiscalar analysis. Adaptive managers understand, model, and monitor natural systems on multiple scales of space and time.
3. Place sensitivity. Adaptive managers adopt local places, understood as humanly occupied geographic places, as the perspective from which multiscalar management orients.

The first and defining characteristic of adaptive management is experimentalism. The method used by adaptive managers, following Leopold, is a commitment to constantly use our experience to reduce uncertainty and also to adjust our goals and commitments. Experimentalism implies that we should take nothing for granted and that we should wherever possible replace assumptions with beliefs based on experimentation or careful observation. Taking nothing for granted means also that the goals and objectives set for policy, as well as physical models, are open to amendment. The very goal of sustainable living is a moving, changing target, to be defined as part of a process and refined as more experience pours in.[3]

Leopold was also the first to articulate the need to understand environmental problems as unfolding on multiple scales of time and space. Measurements, aggregations, and judgments, Leopold learned, must be considered tentative; but tentativeness is just what we would expect, given Hadley's historical caution that the Darwinian approach to identifying truth with permanence would lead us to look at some practices as permanent simply because we have examined their impacts within too short a frame of time. This warning was the operative concept, eventually, in Leopold's famous dictum "think like a mountain," a maxim that will become increasingly important as our analysis continues. According to this advice, we must pay attention to effects not only as they play out on individual and immediate scales but also on the scales of decades and generations. One important aspect of this second characteristic, then, is a commitment to open systems, to understanding nature and the environment as a complex and multiscalar interaction of parts. Since the parts change at different rates, multiscalar understanding introduces the possibility of emergent qualities, qualities of larger wholes that cannot be understood as the sum of actions of parts. The adoption of the multiscalar management model, treated as a method of systems analysis, provides the advantages of a more holistic viewpoint without the ontological commitments of organicism. It allows us to interpret impacts that emerge on different scales in terms of a single, integrated model and thus allows, at least in principle, a set of postulates for organizing space-time relations and perhaps even principles for integrating models over multiple levels. The second characteristic of adaptive management thus amounts to a commitment to build the formal apparatus necessary to follow the systematic consequences of our acts as they play out on different scales of the system.

The third characteristic of adaptive management is localism, a commitment to examine each problem in its particular context and to pay attention to differences that matter in a "place."[4] Darwinian adaptation is always local—one organism either survives or perishes in particular situations, and when

Darwin's principle is applied to societies, the relevant question is not whether the society has THE TRUTH (for all times and places) but rather whether the society has developed practices and institutions that are responsive to, and sustainable in, their local environment. Environmental management as community adaptation to a "place" is thus locally based. This is not to say, of course, that larger regional and global systems never impact local systems—but rather to say that the survival of the community takes place against the backdrop of changing systems on many scales, as these are viewed from the perspective of the local community. These complex, interlocking dynamics must be understood from a specific, local place, from a given perspective within a multiscalar system.

I have chosen to interpret the local aspect of adaptive management as signifying a local perspective not just in the physical sense but also in the social sense, involving a participatory component. Localism, as understood here, includes the idea of a community of people capable and willing to participate in decisions that affect their lives in their local context. This may be thought by some to go beyond the key ideas of adaptive management. For example, when my friend the philosopher and conservationist Peter Brown read an earlier version of this manuscript, he argued that I had unjustifiably conflated adaptive management with the idea of public participation in management practices. Using his own woodlot, which he manages for sustainable forestry, as a case study, he said: "I manage my land adaptively, discussing options with a forester from the local extension service—and we try various experiments, keeping track of results; but I don't consult the public in any way—adaptive management is scientific management, and has nothing conceptually to do with participatory governance." Especially if we apply the term to private woodlands such as Peter's, public process is hardly relevant. In this book, however, my goal is to examine environmental public policy and, at the community level, adaptive management. At that level, I believe public involvement is essential so I see his point. If one defines adaptive management narrowly as adaptive management that employs the scientific method to reduce uncertainty and guide management decisions, then the method of governance is simply irrelevant. Given that definition, I'd have to agree; but I have, despite Peter's excellent point, decided to define adaptive management more broadly to include also the goal-setting process that determines the direction of management inquiry.

I build this aspect into my definition by emphasizing the local nature of environmental values and by seeing localism as not just a geographic point but a "place," which is best thought of as a negotiation between the land and a human culture. In this sense the localism aspect of adaptive management—once it is supplemented in part 2 of this book with a pragmatist approach to values—entails an unavoidable interaction between adaptive managers, members of the public, politicians, and resource users.[5] Put simply, I understand the scientific aspect of adaptive management to be applicable to goal-setting and to social learning about community values as well as about physical processes, so defining a functional adaptive management system for a *public management process* (unlike Peter Brown's private use of adaptive management) requires also that the management be politically feasible and capable of reflecting community-based ("place-based") values. This decision, to use a broad, value-laden conception of adaptive management and hence to relate adaptive management to politics, is reinforced in chapter 4, where it is shown that many environmental problems reflect competition among multiple goods and that problem formulation requires iterative treatment of both science and values. Such iterative treatment requires some kind of political structure; and if the structure is to be supported by the public, the public must be involved and take ownership of the process.

On my understanding of the three principles of adaptive management, then, each principle has a goals-and-values aspect as well as a physical modeling aspect. Each of the principles has important implications for environmental values and valuation. Experimentalism as an attitude and a method—pragmatism—applies equally to factual and to evaluative claims. So a consistent application of the first principle requires an experimentalist approach to human values as well as to science. The second principle also has a normative aspect: the use of multiscalar physical models to describe impacts of humans on natural systems opens the door to a multiscalar analysis of environmental values. Might environmental values unfold on multiple scales? The place-based approach also has normative implications because it involves emphasizing local perspectives and locally articulated values, at least as a starting point. Adaptive management is thus committed to a place-based, contextual approach to evaluation as well as to modeling, and this stance disrespects "one-size-fits-all" solutions and implies favoring locally grounded values whenever possible. So adaptive management as presented in this book represents a philosophy of management; the same philosophy that governs the search for scientific understanding also governs the search for better management solutions and guides revisions of values and evaluations when observation and experience indicate the need for such revisions. Adaptive management is as much a search for the right thing to do as it is a search for the truth. Adaptive management, like medicine, is a normative science.

So far, this all may sound very pedestrian; adaptive management may seem like little more than common sense. There is, however, a payoff. Given this rather sparse set of assumptions and hypothetical premises, it is possible to provide a simple and elegant definition of *sustainability,* or rather what might better be called a definitional schema for sustainability definitions. Because of the place-based emphasis of adaptive management and the recognition of pervasive uncertainty, there is only so much that one can say about what is sustainable at the very general level of a universal definition. Speaking at this level of general theory, sustainability is best thought of as a cluster of variables; local communities can fill in the blanks, so to speak, to form a set of criteria and goals that reflect their needs and values. Although I recognize the importance of local details in particular determinations of sustainability, the three core characteristics of adaptive management go a long way toward specifying a schematic definition of sustainability. A schematic definition makes evident the *structure* and *internal relationships* that are essential to more specific, locally applicable definitions of sustainable policies.

First notice that the latter two principles of adaptive management can be represented in a very simple model of individual actors in a world encountered as a collection of resources. Each actor is treated as a chooser who acts upon observing her or his environment, which we can, in turn, represent as a mixture of opportunities and constraints; some of the chooser's choices result in survival; the chooser lives to choose again and has offspring who will also choose in the face of similar but changing environmental conditions. Choices of other opportunities lead to death with no offspring. This is the basic structure of an evolution-through-selection model that interprets the environment of a chooser as a mixture of opportunities and constraints; it contextualizes the "game" of adaptation and survival and can be represented as in figure 3.1.

This relationship is simply an expression of the relationship implicit in the second and third principles of adaptive management: the chooser is located at a space-time point within an environmental system, observing and acting from that perspective within the system. The actions of individuals, taken individually and collectively, moreover can be understood as experiments on two different scales. Survival of the individual depends, in the short run, on very local conditions of stability; but that local stability represents also a negotiation with slower-changing background conditions. The actions, once undertaken, will result in either survival or termination of the individual or the population over varying periods of time. Community-level success, in other words, requires success on two levels: at least some individuals from each generation must be sufficiently adapted to the environment to survive and reproduce, *and* for the population to survive over many generations, the collective actions of the population must be appropriate for (adaptive to?) its environment. Since humans are necessarily social animals (because of the long period of helpless infancy of individuals), individual survival depends also on reasonable levels of stability in the "ecological background," the stage on which individuals act. This environment normally changes much more slowly than

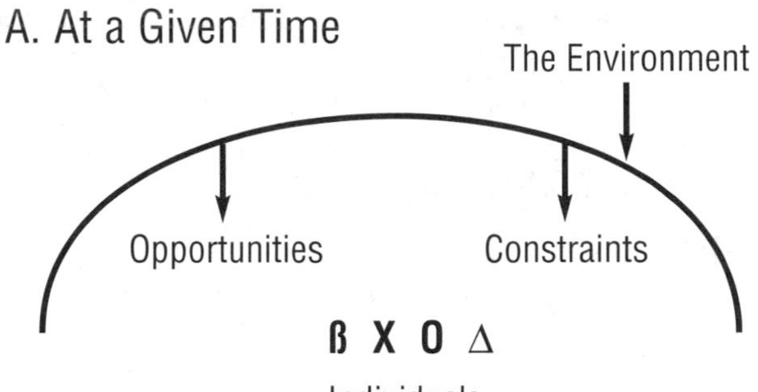

Individuals face their environment as a complex mix of opportunities and constraints as they adapt to their environment at any given time

B. The Cross-Scale Dynamic across Time

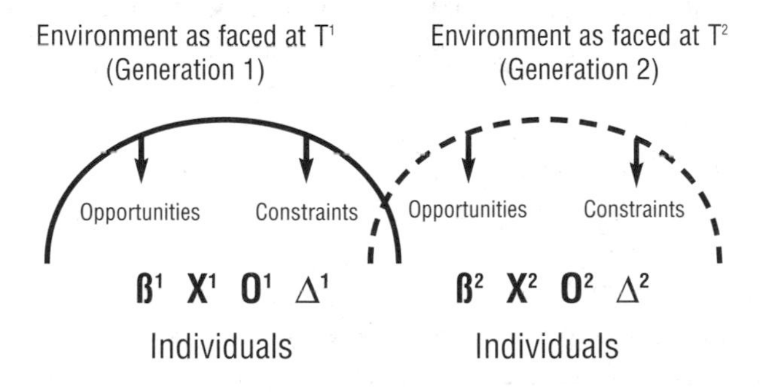

Choices made by members of an earlier generation can change the mix of opportunities and constraints faced by subsequent generations, limiting the latter's choices in their attempt to adapt

Figure 3.1 A simple hierarchical model

individual behaviors, permitting adaptation, over generations, to stable aspects of the environment. This simple model, if given a temporal expression, represents the relationship between individuals who live in an earlier generation and those who live later, as represented in figure 3.1B.

From this simple framework, a schematic definition of sustainability emerges: Individuals in earlier generations alter their environment, using up some resources, leaving others. If all individuals in the earlier generations overconsume, and if they do not create new opportunities, then they will have changed the environment that subsequent generations encounter, making survival more difficult. A set of behaviors is thus understood as sustainable if and only if its practice in generation *m* will not reduce the ratio of opportunities to constraints that will be encountered by individuals in subsequent generations *n, o, p*. Note that this simple model can be described, in its bare bones, as a natural-selection machine. No value judgments need be implied in the model—it can be viewed simply as representative of the relationship of individual and collective choices as they play out over generations. In each generation, individual actors make their choices *given* extant opportunities; looked at intergenerationally, aggregated choices of individuals may change the ratio of opportunities and constraints faced in subsequent generations.

Although the model has a "flat," schematic character, it could also be given a richer, normative-moral interpretation, as is surely hinted at by use of the terms *opportunities* and *constraints*. If we stipulate that the actors are human individuals, then the simple model provides a representation of intergenerational impacts of decisions regarding resources; our little model can thus be enriched to allow a normative interpretation or analogue. If we accept that having a range of choices is good for free human individuals, we can see the structure, in skeletal form, of the normative theory of sustainability. An action or a policy is not sustainable if it will reduce the ratio of opportunities to constraints in the future. Each generation stands in this asymmetric relationship to subsequent ones: choices made today could, in principle, reduce the range of free choices available to subsequent generations. Thus it makes sense to recognize impacts that play out on multiple, distinct scales. If we can agree that maintaining a constant or expanding set of choices for the future is good, and that imposing crushing constraints on future people is bad, our little model has the potential to represent, and relate to each other, the short- and long-term impacts of choices *and* to allow either a physical, descriptive interpretation or a normative one. This schematic definition, understood within the general model of adaptive management, captures two of our most important basic intuitions about sustainability: that sustainability refers to a relationship between generations existing at different times—a relationship having to do with the physical existence of important resources—and that this relationship has an important normative dimension. As my argument unfolds, it will become obvious that I believe any adequate conception of sustainability, one adequate to serve as a key term in public deliberation about what to do, will be explicitly normative. It is nonetheless interesting that this schematic definition can be viewed as flat, and descriptive of a Darwinian process, *or* as a richly normative expression of a community's values.

Thus we can tentatively put adaptive management—complete with a schematic definition of sustainability—forward as a useful model for environmental science and management. Its normative-moral aspect is the subject part 2 (chapters 5–9). Adaptive management is also based on a broad, multidisciplinary view of the physical aspects of the problems we face: it suggests that each community, located in a "place"—from its viewpoint within a complex, multiscalar system (its environment)—must make choices. These choices, especially when combined with similar choices made by others in the same generation, may have physically measurable impacts on the range of choices available to their successors. When conceived in physical terms, the model tracks the long-term adaptability of a population to a particular physical place, from the temporal viewpoint of a given generation. This schematic definition, in turn, directs attention to those impacts that are likely to affect the choice set available to future generations, linking the physical model to the realm of values associated with free choices. Given this dual nature of the schematic model, the multiscalar model can represent the possibility of intergenerational harm, so it begins to shape our notion of intergenerational responsibilities.

It is of course an empirical question whether at any given point in history, activities of one generation were, or are presently, harming future generations. Historians can reasonably disagree regarding when humans, through increasing populations and technological ability, gained enough power to significantly impact the range of actions and possibilities open to those who live in future generations. Wherever one draws that line, it certainly seems that we have passed it. Indeed, through slower processes ancient civilizations certainly degraded their land over generations; Plato discusses the deforestation and soil erosion that left the Attic peninsula largely barren. Today, expanded technological prowess, as well as growing human populations, enables individuals and societies to more rapidly, irreversibly, and pervasively change the context in which future generations will encounter their environment, and it seems certain that some of our choices today will change the set of choices available to future people in many ways. Among these are changes in the ecology of a "place," which is experienced by locals as a mixture of opportunities and constraints. Changes may be for the better, developing new options while holding others open. It is also at least possible that changes initiated today may impose serious hardships on future people by limiting the range of choices available to them. Leopold, immediately after acknowledging that change is natural and that not all change is bad, made this point eloquently in 1939: "Evolutionary changes, however, are usually slow and local. Man's invention of tools has enabled him to make changes of unprecedented violence, rapidity, and scope."[6] The adaptive management model becomes pertinent, then, in situations in which human populations have the technological and personal power to transform the opportunities of the present into constraints on future choices; it is thus the appropriate model for adaptive managers today and should become more apt as human impacts increase with population and growing consumption. Our physical model permits a moral interpretation by exhibiting the structure of intergenerational harms; threats to the future are represented as losses of significant options or opportunities. It therefore becomes important that we understand how to determine which options are significant in a place, a task that is central to the remainder of this book.

We now have a schematic model of what is at stake; but in order for people in earlier generations to have obligations to later ones, they must be able to foresee to some degree the possible impacts of their actions. Actions of an earlier generation can be judged morally only if the earlier generation had reason to anticipate negative future impacts and people failed to modify their present activities accordingly.

This is an example in which morality and moral responsibilities are highly dependent on the empirical facts of the situation. Although stone-age hunters with spears and stone axes affected their environment, their actions were less likely than ours to have large-scale and irreversible effects, given today's much larger populations wielding much more sophisticated technologies. Even when they were able to have huge and long-lasting impacts, as in the apparent destruction of the North American megafauna shortly after the arrival of stone-age hunters on the continent, it is unlikely that these actors had the conceptual tools and the necessary baseline data to judge that their activities would have these impacts. So any assertion of responsibility of our generation for impacts on future people rests on these two assumptions: that our choices have important impacts on future people and the choices they face, and that our scientific knowledge is sufficiently reliable that we can foresee and plan to avoid negative impacts and encourage benign ones. This latter assumption—that we have adequate knowledge to manage at all, even enough knowledge to start an experimental process—deserves very careful justification, because it is famously controversial; many politicians hide behind uncertainty in order to avoid tough decisions. In sections 3.4 and 3.5 I show that however limited, tentative, and uncertain our knowledge, it is adequate to undertake reforms of current policy and to begin a process of learning by doing as a community of communities. Before we take up that argument, however, it will be helpful to survey (in section 3.3) the problem of "uncertainty" and the range of available philosophical responses to it. Later, in something of a diversion from the practical arguments at hand, I explore (in section 3.6) why it is appropriate to describe adaptive management as adaptive in the traditions of evolutionary thought.

3.3 Uncertainty, Objectivity, and Sustainability

Perhaps no issue confounds environmental managers more than the "problem of uncertainty." On closer look the "problem" of uncertainty is really a grab bag of more or less related problems, all resulting from the fact that our finite knowledge will always fall short of any ideal of "full" knowledge upon which to base everyday decisions. Uncertainty, in this sense, is just a general label for all the failures of our scientific models. Speaking more precisely, experts have classified the types of uncertainty according to a number of taxonomies. For example, Granger M. Morgan and Max Henrion separate uncertainty into two classes: uncertainties about empirical quantities and uncertainties about the functional form of models.[7] Another classification, due to Sylvio O. Funtowicz and Jerome R. Ravetz, lists three categories of risk: technical uncertainties (concerning observations versus measurements), methodological uncertainties (concerning the right choice of analytical tools), and epistemological uncertainties (concerning the conception of a phenomenon).[8] In another useful taxonomy, Malte Faber, Reiner Manstetten, and John Proops speak of "open" and "closed" ignorance. The latter, what was called "ignorance of ignorance" by Plato, blocks inquiry; but people can shift to an open attitude if they recognize their ignorance. Open ignorance, then, can be understood as reducible if there is a personal or communal means to learn or irreducible if it involves chaos or true, unpredictable novelty.[9]

All of these types of uncertainty—and more—matter in real management situations in which managers face real dilemmas, and it is fair to ask, How can adaptive managers claim to "manage" in the face of all these kinds of uncertainty? Adaptive managers' experimental approach to management requires that they claim at least some faith in the reliability of the scientific models used to describe human impacts upon natural systems. So adaptive managers need a method by which they can explain and justify their proposals to decision makers and the public. (See the next two sections, which respond to the problem of uncertainty with an "epistemology" of adaptive management.)

There are several strategies for dealing with uncertainty in the uncertain world of management. One approach would be to go through the above list of types of uncertainty and try to address each of these types, providing some intellectually respectable solution or way around the problems presented by each type of uncertainty. I confess that I would not be able to accomplish this task (I doubt that anybody can), so another strategy will be necessary.

One way to begin to understand the experimental attitude developed by the pragmatists and used here as an epistemology of adaptive management is to explore the pragmatists' reasons for rejecting a priori first principles as their bulwark against uncertainty. Pragmatism and adaptive management similarly represent a rejection of the classical modernist philosophy of science, and with it the epistemological strategy of modern philosophy. At the very heart of modernism, eloquently posed—and inelegantly answered—by Descartes himself, was the question of certainty.

Descartes saw the problem of justifying actions as one of deriving justifications from fixed, unalterable, universal, and indubitable principles. This involved, for Descartes and subsequent rationalists, the articulation of self-evident principles derived from reason alone, followed by the application of these principles in an objective material world, a world that exists independent of human perception and ministrations. Descartes thus posed the apparently unavoidable problem of modernism: how can humans, in their finitude, trust their sensory knowledge of the real world? The question is terribly perplexing because a successful strategy, almost by definition, requires that we somehow perceive the (by hypothesis) unperceivable "reality" that lies beyond experience. How else would we be able to base our beliefs upon it? Since experience was denigrated by the rationalists as a necessarily flawed tool for ascertaining reality, Descartes rested his case on God's existence and veracity. He then based his proof of God's existence on the self-evident premise that every event must have a cause. Descartes's solution, one might say, has proved controversial.

Indeed, the history of modern philosophy since Descartes can be described, not inaccurately, as a series of failed attempts to respond to or avoid Descartes's epistemological question and its implications for action. The great British empiricists, for example, including John Locke, Bishop Berkeley, and David Hume, all struggled in some way to resolve Descartes's problem, as did Immanuel Kant. Centuries of attempts fell short; Descartes's question proved unanswerable, even by the greatest philosophical minds. Pragmatists, however, attacked Descartes's formulation of the question itself, arguing that his comprehensive doubts and appeals to universal truths confound our limited faculty of reason. Reason—pragmatists preferred to speak of "intelligence" and "logic," to avoid any association with rationalism and its problems—is a faculty that evolved within human communities, as a means to solve problems affecting survival of individuals and groups. Doubt occurs in real situations; it is individual and local, and it is encountered in a context in which a person or a group sets out to achieve some objective. Intelligence and logic function in concrete instances of real doubt, rather than at the universal level of systematic doubt, which can only be met with universal principles.

Furthermore, pragmatists reject the Cartesian implication that objectivity consists in a simple relationship between an assertion and some fact in the world. They recognize that ascertaining the truth by associating it with a chunk of reality beyond our experience, as Descartes attempted to do, is impossible. Pragmatists therefore seek an alternative means to create confidence in our beliefs and theories; that alternative is to submit them constantly to varied experimental tests. Finally, pragmatists seek objectivity not in relations of correspondence between assertion and unperceived reality, but rather in processes and methods that function within human experience, within, in particular, a community of truth-seekers. Pragmatism, then, represents an important epistemological break with the Western philosophical tradition in that it seeks the truth within everyday, constantly changing reality. Similarly, adaptive managers daily seek the truth without benefit of first principles. Can the pragmatist approach, in response to the pervasive uncertainty that is endemic to environmental problems, provide a sufficiently reliable epistemology for adaptive management?

Leopold, as noted above, absorbed some key pragmatist ideas from Arthur Twining Hadley, but we don't know in detail how much Leopold studied Hadley's ideas or how explicitly he adopted these as a basis for his scientific and managerial work. What we can show is that the ideas Leopold appealed to and used—ideas that have now been embodied in the methods of adaptive managers—emerge from a rich naturalist tradition in American intellectual history. Among other innovations, this tradition provides an alternative conception of truth and objectivity and a different approach to addressing problems of skepticism and uncertainty—by offering a fresh take on doing and learning.

We can begin to appreciate the radical nature of the pragmatists' epistemological departures from the modernist approach to epistemology—and also see how this new approach to epistemological uncertainty provides a useful strategy for adaptive managers—by starting with what is sometimes called the "pragmatic conception of truth." Leopold appealed to this principle, indicating Hadley as his source for it. Referring to Hadley, Leopold quoted the definition "Truth is that which prevails in the long run." As noted in section 2.3, this passage led Leopold directly into a concise but penetrating discussion of an ethic of sustainability based on broad anthropocentrism and the requirement that if we are "logically" anthropocentric, then we must care about the future of our culture and society. Although Leopold credited Hadley, it was C. S. Peirce who most clearly articulated what has come to be referred to as the "pragmatist conception of truth." Representative versions of Peirce's definition are (1) "Truth is that concordance of an abstract statement with the ideal limit towards which endless investigation would tend to bring scientific belief" and (2) truth is "the last result to which the following out of [the experimental] method would ultimately carry us." One reason to examine Peirce's definition, given its "forward-looking" temporal horizon, is that it provides an interesting analogy to problems in defining sustainability. Moreover, Peirce understood his notion of the search for truth as the defining pursuit of a community of inquirers who start with diverse viewpoints but are carried forward toward the truth "by a force outside of themselves to one and the same conclusion."[10] Surely any acceptable definition of sustainability must embody the idea of a forward-looking community that is normatively respectful of the pursuit, and also the perpetuation, of knowledge; so this approach to truth may be attractive to adaptive managers and advocates of sustainability.

Peirce's definition of truth expresses the philosophy of naturalistic epistemology that was emerging in the nineteenth century as a distinctively American alternative to the traditional philosophies of Europe. In *Walden*, Henry David Thoreau anticipates Peirce's definition of truth: "No face which we can give to a matter will stead us so well at last as the truth. This alone wears well."[11] The naturalism of Thoreau and Peirce addresses the problem of objectivity not in the usual terms of a time-bound relationship between thought and a chunk of the contemporaneous "external" world, but as an intertemporal relationship between present beliefs and future outcomes. Near the end of the explanatory chapter of *Walden*, "Where I Lived, and What I Lived For," Thoreau says: "Let us settle ourselves, and work and wedge our feet downward through the mud and slush of opinion, and prejudice, and tradition and delusion, and appearance, that alluvion which covers the globe, . . . through poetry and philosophy and religion, till we come to a hard bottom and rocks in place, which we can call *reality*." This passage emphasizes Thoreau's commitment to truth, not just opinion, and also links this idea to an experience-based process, a lifetime—even eternal—pursuit: "If you stand right fronting and face to face to a fact," Thoreau said, "you will see the sun glimmer on both its surfaces, as if it were a cimeter, and feel its sweet edge dividing you through the heart and marrow, and so you will happily conclude your mortal career. Be it life or death, we crave only reality."[12]

Thoreau anticipated two key aspects of the pragmatist's approach to truth and objectivity. Besides anticipating the temporal, forward-looking notion of truth, he also anticipated the idea that the struggle toward truth and objectivity takes place entirely within human experience, as we live, act, and observe within our world. In his journals Thoreau said, "I am not interested in mere phenomena, though it were the explosion of a planet, only as it may have lain in the experience of a human being."[13] Truth is not a matter of correspondence with an external reality.

As noted above, Western philosophy since Aristotle sought truth and objectivity in a correspondence between thought and a reality behind or beyond experience. Thoreau and later the pragmatists challenged the Cartesian dualistic world that separated human experience from the "real" world. For the new naturalists, truth is not a matter of correspondence to an element of reality located outside or beyond experience; it is rather a matter of struggling to separate reliable from unreliable bits of experience by seeking out more and more experience through time.

Thoreau, I believe, placed undue faith in what he called individual "genius," and it was thus left to Peirce to reconstrue the temporal relation more concretely as a *community* process, a process pursued by a very special community of scientific inquirers—the lovers of truth. This community has implicit norms and explicit methods for approximating the truth and can consciously study the "logic" of their enterprise. So I shall follow Peirce, not Thoreau, in construing the process of winnowing through many experiences as a community-based one, which points the way toward a pragmatist, community-based approach to knowledge.

3.4 A Pragmatist Epistemology for Adaptive Management

I have recommended that in order to respond to the pervasive uncertainty plaguing environmental action, adaptive managers explicitly adopt the pragmatist method of experience and that they focus attention on the way expanding experience—especially when experience is fortified with an explicit logic that governs experimentation and careful observation—can eventually reduce uncertainty and result in cooperative action by communities. One might, however, still reasonably ask about the problem of uncertainty in the present: Are there adequate reasons to believe current science can help us to foresee consequences of our actions? And what can scientists, given the uncertainty of many of their research conclusions, contribute to farsighted environmental management? If adaptive management is to be a plausible candidate for a democratic approach to environmental management, and yet we want our policies to be guided by good science, these questions must be answered. Does uncertainty undermine our ability to act decisively to protect people of the future, for example? This question requires in turn a response to the deep epistemological conundrum first articulated by Descartes in the seventeenth century, to the bedevilment of modernists since: If our knowledge is less than certain, how can we act responsibly in an uncertain world?[14]

As a first step in unraveling this complex of problems of uncertainty, we can distinguish between uncertainty as it is experienced within real situations where people face difficult decisions with less information than would be ideal, on the one hand, and the universal skepticism that motivated the Cartesian search for certainty, on the other. Descartes set out to doubt all of his beliefs about the world, limiting himself only to his certain belief that he, a thinking being, existed: "Cogito, ergo Sum." He thus began his heroic attempt to reconstruct knowledge of the world by doubting his senses, relying instead on the "pure light of natural reason." Assuming a general principle, that every event has a cause—which seemed to him to be self-evident—Descartes proceeded from a rationalist base to attempt a deductive reconstruction of certain knowledge, despite the inherent untrustworthiness of the senses. Truths of reason were taken as unassailable and capable of providing unquestionable truth about the world. Today the Cartesian worldview, as enhanced by Newton, has ultimately been unable to deliver the kind of certainty that Descartes demanded. To get on track, we need to somehow get past the demon Descartes left us dealing with: the problem of attaining certain knowledge, coupled with a fear that we cannot act until we are sure of our information. As I set out to show how the pragmatists learned to reject the requirement of certainty and to reject Descartes's dualistic worldview, I will start at the origin of the problem—with Descartes's epistemology.

Many philosophers today would argue that Descartes made at least two very important mistakes. First, as was argued passionately by the pragmatists, especially Peirce, Descartes's starting point of universal and total doubt of all information from the senses misunderstands the nature of doubt as humans experience it. Doubt, Peirce argued, is always a characteristic of a situation in which a protagonist is, or protagonists are, deliberating about what to do. The deliberation arises out of the goals adopted and the actions considered, so that the doubt is directed at particular and relevant disagreements or uncertainties that, in the situation at hand, interfere with choosing a course of action. The value-infused situation—the real choices facing an actor or actors—helps to focus the spotlight of experience on particular, relevant pieces of information about crucial points of fact that appear to be relevant to the choice at hand. This occurs because different participants, coming from different interest groups and adopting different perspectives, will look at supposed facts differently; the disagreements among participants can thus highlight aspects of the situation involving the most doubt or uncertainty. These differences can help to direct the focus of science toward questions that really matter in the management process. At the same time, the situation—again including felt values and a real question of what choice to make—also designates some beliefs and background information as given and unquestioned. For Peirce and the pragmatists, one never questions all of one's beliefs and ideas simultaneously; one always enters into a situation of inquiry, of wanting to reduce uncertainty, with some background assumptions and beliefs that are left unquestioned. According to this understanding, doubt should never be completely general; doubt should be directed by aspects of particular, real situations on particular questions that bear upon proposed actions. It is this feature of pragmatism that explains why Thomas Kuhn's idea of a paradigm is justifiably thought of as an important application of the pragmatist approach—the beliefs, assumptions, definitions, and norms that constitute a scientist's adopted paradigm, the paradigm that makes normal science possible, represent one important example of unquestioned background beliefs.

Second, by choosing reason as his trusted tool, and deductive reasoning appropriate to geometry as his ideal of human thought and intelligence, Descartes set such a high standard for knowledge and certainty that he created for himself and his followers, both rationalists and later empiricists, an impossible requirement for epistemological justification. Descartes set up the problem such that acting reasonably required certain and indubitable knowledge; he formulated the problem of justification so that the mere accumulation of evidence bearing on a question is inadequate to justify action, unless that evidence amounts to unquestionable proof. In essence, Descartes thought that humans, though finite in their intelligence, could apply the methods of rationalism to achieve a Godlike form of knowledge prior to action. Today, many philosophers would suggest a more modest goal of providing a method that, if applied repeatedly, can be expected to reduce error and gradually progress toward the always-receding ideal of perfect knowledge.

The positivist sociologist Otto Neurath provided a useful analogy that has become very popular with pragmatists today. Improving our knowledge and understanding, Neurath said, is most like repairing a ship while on the high seas. Imagine, for example, a ship that is kept in service indefinitely, with no opportunity to be pulled into dry dock for repairs. As particular planks weaken from weather and heavy use, we replace them, standing on the strongest remaining planks while replacing the weakest ones. In principle, it would be possible to replace every plank, resulting in a "new" ship in the physical sense that every plank is new but with the "ship" still the same vessel and having no need to be rechristened. The analogy captures perfectly the pragmatists' approach to epistemological progress. Their pursuit of epistemological renewal doesn't start by tearing down every plank of knowledge, necessitating a completely new construction. Rather, the task of renewal begins for pragmatists by identifying the most problematic planks. Since we must keep the boat afloat as we do our repairs, identifying the problematic planks is not just a matter of identifying what beliefs we have the least evidence for. In the ship analogy, we might fix a moderately rotted plank below the water level before replacing a badly rotted one in a little-used area of the deck; in pragmatist epistemology, this kind of decision involves identifying uncertainties that are particularly relevant to community survival and to other chosen social goals. The assessment is furthermore relative to appropriate values, not simply to a mechanical application of a test of physical strength of the particular "planks" of knowledge. The analogy also illustrates the idea of piecemeal improvement of a belief system in which no belief is ultimately privileged, even though some beliefs are accepted as unquestioned for very long periods of time. So the analogy, by noting that the entire ship is subject to the ravages of time and wear and will eventually be replaced with new materials, illustrates also the pragmatists' idea that every belief is up for reevaluation as necessary. If we can imagine our sailors continuing their back-and-forth passages indefinitely, each and every plank will eventually be tested by use or weather. Similarly, the time will come when each of our beliefs is relevant to an important disagreement regarding what to do.

The ship analogy also suggests in simple terms a way out of the quandaries of subjectivism, antirealism, and relativism. Some readers might think that Neurath's analogy pushes us closer to relativism. They might reason as follows: since the decision of which "planks" to stand on and which to fix is a matter of judgment, surely affected by the goals and values of the crew (survival, at least), it might be argued that different people, including different scientists, will take different things to be "given" and unquestioned at any particular time. Since our society is made up of people with different values and different assumptions—paradigms and worldviews—it could be argued that we must expect different people with different perspectives to express different goals and values and to put different scientific hypotheses up for test. Not only will people cite very different beliefs in support of policy choices, but communication across perspectives may become impossible, because the choice of linguistic categories and meanings depends also on individuals' values and perspectives. This situation undermines appeals to any particular linguistic vocabulary as privileged or valid preexperientially; we must give up, as Dewey saw, appeals to fixed and eternal categories and to fixed and eternal truths. The ways in which we identify, characterize, individuate, and aggregate objects of the senses are deeply affected by our goals and values as well as our perspective and worldview. Through language we construct reality; and different languages reflect, it would seem, different worlds inhabited by different citizens with different perspectives. If we apply Neurath's analogy to language, we find that a kind of linguistic relativity—what we call conventionalism in the appendix—is unavoidable. The choices as to which beliefs to accept as given and which to submit to further test is inevitably affected by what people value. People with different interests and perspectives form different categories, track different variables, and use different descriptive terminology.

Despite this form of linguistic relativism, which plays havoc with our ability to match specific sentences with particular bits of reality, I don't think the ship analogy implies a skeptical and relativist conclusion about beliefs and knowledge. There is still an element of realism in Neurath's story because the decisions made do not escape the test of experience; over time, staying afloat depends on realistic assessments of damage to the ship and realistic models of what will happen if a particular plank fails. If the repair crew constantly repairs the easiest planks to get at, ignoring the ones in awkward places, a disaster will eventually occur. If the crew replaces the bar in the officers' galley every time it is scratched, while ignoring severe rot below the waterline, the whole crew will be selected out of the pool of sailors and a new ship will have to be launched.

Further, although diverse languages may suggest different and relative ontologies for individuals who speak different languages, language is not an individual matter. Just as the repair crew on the ship must communicate with one another to decide what to do next, a seeker after the truth must be a member of an intellectual community. Some form of shared language or linguistic communication is presupposed in the designation of a group as a community, and communication itself becomes a value within a community that faces opportunities and pitfalls. Linguistic choices too are submitted to the test of survival. Those communities that fail to develop an adequate language for describing problems they face and for finding solutions to them will be selected out of the pool of ongoing communities. Similarly, just as the repair crew's decisions will affect the whole crew, giving all crew members an incentive to oversee the repairmen's decisions and work, the progression of the scientific endeavor will bring larger and larger communities—and their collective and individual experiences—into discourse about what to do to protect the environment.

The key point here is that Neurath's analogy shows clearly how we can adopt a position of "limited realism." Limited realism accepts the fact that no substantive knowledge of the external world is knowable a priori; we must construct our system of beliefs without benefit of prior principles to guide us. It also accepts the apparently unavoidable conclusion that our varied linguistic forms, designed to function in many different situations, yield no common underlying structure for all experience. The categories we find in the world are of our own making; they are not given in reality independent of us. For these reasons realism must be limited. But it remains realism in the most important sense. It retains a method of selection based in reality, not in our wishes, dreams, or imaginations. The crew on Neurath's ship either stays afloat or dies. They are held to account for their decisions about which planks are strongest and which the most rotten, which are the most important and which of lower priority, as well as the techniques (however subjectively arrived at) that they develop to execute repairs. Experience, in the end, winnows mistakes.

For the pragmatist it is not so important to separate the hypotheses acted upon from the attitudes and values that bring them to the fore in real situations requiring inquiry; what is important is that we observe and experiment and that there be an external check on claims and counterclaims. There is, after all, an experience that will tell the repair crew they made a mistake. That experience begins with the sensation of sinking into wetness. And so Neurath and the pragmatists, extending the analogy, have available to them a method of truth-seeking, a means to hold beliefs and explanations—no matter how subjectively urged—up for falsification. What we must give up if we accept the pragmatist limitations on realism is not the efficacy of external reality as a check upon our belief system; what we must give up is the idea that any of our beliefs, taken individually, can be compared directly to a chunk of reality. As in natural selection, realism expresses itself in the survival or failure of acting units. In genetic evolution, this realism is applied to individuals. But once we move the Darwinian analogy to the level of community action, survival takes on a new meaning: for the fruits of culture to survive indefinitely, the community that transmits that culture must survive and retain enough continuity so that the stored truths and wisdom of the culture remain alive and meaningful across time. Natural selection can accordingly be applied at the level of communities, as Hadley argued. If the community fails and the individuals who make it up die out, we know they didn't have the truth necessary to survive in the situation they faced; if they survive, they will no doubt continue their practices, perhaps improving upon them in small ways, until environmental changes require further adaptations. As Aldo Leopold said, human communities either survive or fail, and in the process they test the beliefs and associated practices they acted upon.

According to limited realism, then, belief systems, like ships, are tested holistically, not belief by belief (plank by plank). Though the detailed architecture of the world is not given for us, either a priori or by direct experiential apprehension, the entire edifice of our belief system and the actions it prompts us to take must stand the test of more and more experience on the part of more and more observers and their inputs over time. One might object, of course, that the march toward truth described here will be unbelievably chaotic and slow and that the whole process has much randomness. For example, how can we expect that better ideas will appear out of multiple viewpoints and perspectives? How will new and better ideas, institutions, and goals arise out of failures? These questions, however difficult, do not daunt the pragmatist: referring to intellectual roots in Darwin, the pragmatist will answer that, just as in natural selection, there is a need for variety and diversity if evolution is to go forward. In justifying beliefs and proposals for action through conscious experiments, it is important that there be a variety of perspectives, beliefs, and viewpoints. Given that variety, the struggle to survive will winnow out inappropriate practices.

Cultural evolution, which is Lamarckian, is much more rapid than genetic evolution, which is not. Conscious beings can create new ideas and try them out in consciously limited experiments, thereby employing a "safe-fail" strategy to increase knowledge and reduce uncertainty. Safe-fail strategies, in contrast to fail-safe strategies pursued by engineers who use multiple safeguards to eliminate catastrophic failures, are favored by adaptive managers, who undertake pilot projects and other limited and reversible experiments, ensuring that the costs of a failed experiment are acceptable, allowing a reversal of policy when things do not work out. Hopefully, it will be possible to learn that some of our beliefs are wrong without coming to total failure, the equivalent of Neurath's boat sinking. The ability to learn, and to store information in the form of language, makes cultural evolution both faster and less dependent on random events. So the pragmatist search for truth over time, though it can never decisively banish uncertainty forever, can provide adaptive managers some confidence that with many viewpoints and perspectives addressed within the process of deciding what to do, errors and failures of our science—and our values—can be rectified. Even this limited realism recognizes experience of the world as we encounter it in our everyday struggles to survive, both as individuals and as communities, as the ultimate arbiter of truth.

It is possible, I think, for adaptive managers to adopt limited realism as a working epistemology. Since environmental management is clearly a cultural and social endeavor, we can hope that many belief systems, perspectives, and viewpoints will be proposed and tried out but that this initial relativity will be gradually reduced as proposals, the belief systems that support them, and even the perspectives taken are subjected to the ultimate test: do they work in real situations? Hopefully, as time goes on, we can learn new methods and techniques whereby proposals and beliefs can be tested through pilot projects, experiments, and so forth, avoiding the necessity of a cultural collapse to disprove every errant hypothesis.

It is worth emphasizing once again that pragmatism points toward concentrating on improving methods in the search for truth, toward devising techniques and methods that help to separate truth from falsehood without catastrophic cultural failure. If indeed such methods can be devised and gradually improved, it would appear that the optimism of adaptive managers—who, we admitted at the outset, must have some reason to believe that observation and experiment will provide some advantage in an open-ended process of environmental management—may be justified. From our excursion into limited realism and pragmatist epistemology, given many diverse voices engaged in dialogue about what we should do to protect the environment, a general method seems to emerge. This method can, at its best, help us to separate winning from losing strategies by use of case studies, observations, experiments, and a diligent recording of what works and what doesn't work in particular situations. There is thus reason to believe that adaptive management can provide a self-corrective method for pursuing environmental protection.

Pragmatism, we have seen, offers adaptive management a plausible epistemology, capable of justifying patient attempts to learn by doing and capable of justifying the construction of many and sundry scientific models of nature. What pragmatism would change is that these models would be treated as purpose-relative, not as synoptic pictures of reality. Pragmatism also extends the current thinking of adaptive managers by applying the method of experience to purposes stated as well as to specific predictive hypotheses. For example, Leopold, riding on horseback across the Forest Service lands of the Southwest, considered the attempt to impose a European-style grazing culture on those semi-arid plateaus to be an experiment that could only be judged a failure. He created a model from observation and successfully predicted the unfortunate succession whereby grasslands became worthless shrubland that gradually spread over Arizona and New Mexico, virtually useless for grazing or anything else besides photos in coffee-table books. In the end, Leopold did not just reject the implied factual prediction of the ranchers that the lands could support heavy grazing every year; he eventually called into question also the viewpoint and interests that had motivated that hypothesis. He finally concluded that the goal of maximizing meat production on the range itself and even the utilitarian ethic of production and consumption that supported it were inappropriate in the Southwest. He knew this because he had seen the long-term impacts of this grazing on the land. He relied on this experience, for example, when he criticized "boosterism" and economic growth as reflective of unworthy social goals.

Compared to the brash claims of Descartes to have wiped away all doubt and to have placed science on a firm foundation of self-evident truths and deductions therefrom, this little epistemology of adaptive management may seem woefully weak and far too unsubstantial to give much comfort to environmentalists, environmental managers, and the public. What pragmatism offers, in response to the daunting uncertainty faced by environmental managers, is a method that, if applied persistently and in many situations, promises to gradually move the community toward consensus. It does this by creating, through dialogue, sufficient agreement on the nature of the local management problem and on what areas remain in dispute. Managerial choices can then, to the extent possible, be designed to test hypotheses relevant to the remaining disagreements. Compared to either Cartesianism, with its commitment to self-evident truths and pure reason, or logical empiricism, which sought to mirror the world with a logically perspicacious language and with the reduction of all empirical statements to unquestionable building blocks of raw empirical data, the pragmatic method may indeed seem rather weak. And pragmatists cannot make glorious claims for their method, for example that it obtains quick and easy results. On the contrary, the method suggested can at best encourage a slow and rocky road forward, a gradual improvement of understanding and a slow refinement of our goals, as each participant, approaching the shared problems from a distinct perspective, questions some elements of received wisdom and as the community struggles to reduce remaining areas of uncertainty and disagreement in order to find a policy on which its members can, at least provisionally, agree. The pragmatic method, then, is at best far messier and less elegant than either the approach of the rationalists or that of the logical empiricists.

The pragmatic method of experience, though slow, messy, and usually unsatisfying, *works*! Pragmatism works because it simply encourages us to develop methods that have always worked, to seek truth by pooling the community's experience. These methods require us to seek out coalitions and areas of agreement, to struggle to state shared goals and, where disagreement occurs, undertake experiments to reduce uncertainty in these specific areas. Pragmatism—and adaptive management, likewise—works because it creates and builds upon the capital stored in a community of truth-seekers. The strength of the community is precisely in its diversity of opinion and belief systems. What holds the community together, of course, is a shared commitment to act, and to act on the best scientific evidence available at the time when action is required. Rather than seeking an impossible level of certainty, as signified by positing a human-independent reality, pragmatism tries to find objectivity in an ongoing process that is embedded in a community devoted to the method of experience. When there is disagreement within the community—as there will always be—we infer that more evidence must be sought and that actions must be designed to reduce the relevant uncertainty.

Although this modest epistemology falls far short of Cartesian calls for certainty, the pragmatic method, however humble, is adequate to justify confidence in an adaptive management process. If there is enough consensus in a community to act in pursuit of the truth, then the process of adaptive management, which uses science to address specific disagreements relevant to management decisions, can work. Any culture that succeeds both in reproducing individuals on the organismic scale and in transmitting a viable culture into the future will continue to play the game. If a society becomes so hidebound and inflexible that it cannot learn from its environment or change in response to changes in it, that society will disappear, and another "ship of culture" will have to be launched.[15] The learners survive to study and act further; the others get replaced. And perhaps, for adaptive managers, that is epistemology enough.

3.5 Uncertainty, Pragmatism, and Mission-Oriented Science

The shift from a search for correspondence to concern for process in the struggle toward objectivity does not, of course, solve the "problem of uncertainty," which will always be with us. It does, however, shift the epistemological question from a search for an independent and preexperiential reality toward an ongoing attempt to develop and refine methods for separating truth from falsity. Pragmatists stand for—to again borrow a phrase from Dewey—"the supremacy of method." But pragmatists, especially Peirce and Dewey, disagreed somewhat about the functioning of the method because, it turns out, they tended to emphasize different purposes for it. It will be useful to briefly explore this controversy, because it will illustrate both the shared beliefs and the shades of varied opinion that characterize the pragmatist approach to uncertainty, skepticism, and knowledge. It will also illuminate the epistemological possibilities open to adaptive managers today by encouraging a more sophisticated understanding of the interplay of science and values in a policy context. It can be argued that nothing less than a new approach to science and to the epistemology of science—at least in some contexts—will be necessary if uncertain science is to contribute effectively to environmental management.

Before turning to the pragmatists' methods for dealing with uncertainty, I should note that two contemporary European philosophers, Sylvio Funtowicz and Jerome Ravetz, have articulated in broad outlines just such a new approach to science in managerial contexts, an approach that, though they do not emphasize their connection to pragmatism, is compatible with and even evocative of the practical epistemology of pragmatists. Funtowicz and Ravetz divide science into "curiosity-motivated" and "mission-oriented" research.[16] Historically, disciplinary scientists have engaged in curiosity-motivated science, as problems and solutions have been defined and pursued according to standards that are articulated independently of practical applications, "internal," one might say, to the defined intellectual goals of the discipline. Important arbiters of curiosity-motivated science are professional societies reinforced by the peer review process, which are used to control access to journals, government funding for research, and disciplinary status.

In early forays into interdisciplinary work, most scientists have retained the academic bias, attempting to minimize the effect of values on their work and struggling to develop cross-disciplinary science that attains the standards of disciplinary methods. Curiosity-oriented science can be contrasted with "postdisciplinary" or "postnormal" science, science that is undertaken as a part of a mission, science that is explicitly value-laden. It is science of this latter sort that is most likely to be relevant in adaptive management processes.

Mission-oriented science differs from traditional disciplinary and curiosity-motivated science in that the community that reviews scientific results is expanded to include not just scientists from established disciplines, but also affected parties, stakeholders who have varied interests and viewpoints and who express their viewpoints in an open and public process. According to Funtowicz and Ravetz, "Curiosity-motivated and mission-oriented research have complementary properties; the former produces public knowledge but within rigid disciplinary boxes; the latter is transdisciplinary, but bureaucratized and private. The issue-driven research of post-normal science must combine the positive features of these other forms, and develop appropriate institutional arrangements or structures to achieve this."[17] I discuss this useful distinction in section 11.4. Here I will show that Peirce and Dewey anticipated the distinction—and the underlying difference—between curiosity-driven and mission-oriented science and that with some careful adjudication of an unresolved difference between Peirce and Dewey, it can be seen that the pragmatists already had a lot to say about how we should proceed in mission-oriented science.

The story begins with a disagreement—never resolved—between Charles Sanders Peirce and the younger John Dewey regarding the degree of realism that can be wrung out of the pragmatist method of experience. Peirce, as noted above, tried to hold onto a strong form of realism, asserting that the endpoint to which the truth-seeking community tends is both unitary and inevitable; indeed, he even insisted that there was a "correspondence" between that idealized endpoint and "reality." Peirce and Dewey disagreed in the literature regarding the ultimate objective of method, as both struggled to associate the pragmatist method—as a truth-seeking method—with their own deeper concerns. Their difference, it has been pointed out, can be expressed by characterizing Peirce's approach, which attempts to come as close to traditional conceptions of "objective" truth as possible, as a "conform" theory. True utterances in this sense "conform" to reality. But since Peirce took references to a preexperiential reality to be meaningless and of no account, he meant to limit "correspondence" to that between particular beliefs and the idealized outcome of a community that has had an indefinite time to winnow truth from falsehood. In this sense, Peirce's view of objectivity attempted to maintain a similarity in form and purpose between his revolutionary ideas and extant traditional theories of invariant and universal truth. Dewey, for his part, emphasized the particularity and situational nature of intelligence, and he saw truth more as a function of the transformation of an unsettled situation of doubt into a situation allowing action. Subsequent pragmatists argued, then, over two apparently opposed theories: Peirce's "conform" theory, which conceives truth as a relation between current beliefs and an ideal that they must eventually come into conformity with, and Dewey's "transform" theory, which linked truth to a dynamic change from an unsettled to a settled situation.[18]

What is interesting is that Peirce and Dewey, despite a published dialogue on the subject, never resolved their differences over transform and conform theories, even though there existed a fairly obvious "pragmatist" or contextual reconciliation. As has been pointed out by John E. Smith,[19] Dewey and Peirce explicitly stated that they saw different purposes for "inquiry." Peirce often invoked the Scholastic distinction between *logica docens* and *logica utens.*[20] The former represents the precise analyses of logicians and the latter, the unformulated logic, the unspoken standards of reasoning used by individual agents, including individual scientists.[21]

For Peirce, who most respected *logica docens,* inquiry is "primarily a form of logical self-control which focuses on the manner in which beliefs are formed or, rather, should be formed." Dewey emphasized "the motive to control the situation which evokes it and ultimately to reshape the environing conditions of human life."[22] Because Peirce was so concerned to avoid psychologism—the reduction of philosophical concepts to concepts of descriptive psychology—in his philosophical system, he often ignored and sometimes disparaged *logica utens.* A more pluralistic approach to pragmatism, however, might embrace the search for both an improved *logica docens and* an improved *logica utens,* treating these as separable tasks with separable goals and applying different cognitive tools and different "standards of proof" in different contexts in which different goals are dominant. The trick is to design an approach to environmental management that is adaptive by playing *logica utens* off against *logica docens.*

In environmental management, this would mean letting *logica utens* be dominant in that the demands of particular situations may require actions before the scientific hypotheses on which they are based have been adequately verified. Also, in *logica utens,* demands of action may determine which experiments should be undertaken at a given time and could legitimately affect criteria for funding ecological and biological research, for example. But managers must also submit their findings from management-driven decisions to the more stringent rules of *logica docens;* expedient, policy-driven science must eventually pass muster within the more demanding strictures of the disinterested and timeless community of truth-seekers. The resolution is contextual; it depends on the extent to which action is forced in a given context, not on the inherent superiority of the academics' goal of enforcing stringent criteria of scientific verification over the practical goal of acting on the best available evidence to protect social values.

It is important to recognize that although this disagreement seems to align

Peirce with traditionalists *against* his fellow pragmatist Dewey, their agreements are actually more basic than their differences. Peirce put great stock in comparing particular truths with an "ideal" truth, but that ideal remains a construction out of the activities of the community of truth-seekers across time. Both Dewey and Peirce, in other words, treated the truth relation as having only human experience as relata. Peirce, no less than Dewey, rejected any suggestion that truth obtained from a relation between beliefs and a world external to human experience. What this means practically is that although Peirce and Dewey advocate different terminology associated with their different goals for inquiry, they both converge on the central ideas of pragmatist epistemology: (1) when disagreements within the community of inquirers block common action in response to shared problems, experience is the only method that can "settle opinion"; and (2) action and knowledge are internally related, which is to say that significant disagreements about facts or goals are disagreements that make a difference in how we *act*.

Fortunately, we do not have to fully resolve the disagreement between Peirce and Dewey in order to proceed in our search for an epistemology for adaptive management. Peirce, like Dewey, never doubted the usefulness of a "logic" that would separate good from bad day-to-day decisions. He disagreed, however, with Dewey's willingness to leave epistemological matters on the level of the particular. Peirce believed that, ultimately, logic and beliefs must attain the higher standard of surviving indefinite scrutiny and become truths of universal significance. One might say that whereas both pragmatist philosophers recognized the functionality of belief, Dewey was willing to build his theory of truth on many particular functional judgments, while Peirce insisted that if such local and situational judgments are to have the ultimate epistemological force of truth, they must evoke a higher-order, idealized form of truth and knowledge. Since in this book we are mostly concerned with using adaptive management in particular situations that occur in many real, local places, we can follow Dewey and concentrate—for the purpose at hand, which is to come up with a practical epistemology for managerial contexts—on the development of a *logica utens,* a method that will move us toward truth in many local situations. We can set aside Peirce's more expansive notion of truth for future discussion and exploration, because we know, following Dewey, that we at least need a practical logic for managing in particular situations.

If, following Dewey, we have faith in the ability of science and method to address real problems, then *logica utens* is adopted as the logic of environmental management and *logica docens* remains appropriate for the "academic" study of science—for the study of science in a context, that is, where action is not forced. To illustrate how this compromise position would function in practice, it is possible to cite at least three important differences between the operations of *logica docens*, the logic of truth-seeking science, and *logica utens*, the logic of problem-solving and adaptive living.

1. Value neutrality, which remains an ideal in *logica docens*, is not claimed nor required when applying *logica utens* within the policy arena. The application of *logica utens*, in fact, demands the expression of many value viewpoints in the search for policies that fulfill, to the extent possible, the many and competing interests of the community.
2. *Logica docens*, in its application, abhors Type I errors—positive assertions of truth that cannot be fully verified. *Logica utens*, in contrast, must balance the concerns of too quickly asserting a nontruth against the possibility that inaction based on "academic" uncertainty may prove calamitous.[23]
3. Whereas the static, conform theory is the proper ideal of truth within *logica docens*, the more dynamic transform concept of Dewey, with its experimental, problem-solving attitude toward truth-seeking in practical situations, is applicable in the practical disciplines, such as conservation biology and adaptive ecosystem management. If this difference is emphasized, it may be possible to use information gathered in specific management situations to provide the data necessary to continue and enrich academic science, which would still be held to a higher standard of proof.

What adaptive managers need is a *logica utens* for environmental problems and policy. Adaptive managers understand the search for improved environmental policies as one of designing institutions and procedures that are capable of pursuing an experimental approach to policy and to science. In the process of building such institutions and procedures, social learning is expected to improve understanding of the environment through an iterative and ongoing process that will require not just unlimited inquiry but also the encouragement of variation in viewpoints and the continual revisiting of both scientific knowledge and articulated goals of the community.

Adaptive managers doubt that a path to sustainability can be charted by choosing a fixed goal or set rules at the start. We must start where we are; but we do have the ability to engage in experiments to reduce uncertainty and to refine goals through iterative discussions among stakeholders. We also have the ability to develop and reflect upon methods we use for seeking the truth. Environmental management must be a *process* in which managers choose actions that serve as experiments with the capacity to reduce uncertainty and to adjust future goals and choices. In this tradition, the manager tolerates a variety of viewpoints, hypotheses, and proposals for action; this variety of viewpoints and the ensuing experimentation and political discussion are all important parts of the process of selection of more and more "adaptive policies." I am arguing that adaptive management, implicitly, has already operated using the pragmatist method of experience. But I am simultaneously arguing that making this commitment explicit and developing it by incorporating the pragmatists' rich theory of the logic of action and refinements of methodology can provide a coherent starting point for a response to the conditions of uncertainty that pervade environmental problems as communities face them.

The scientists who have developed adaptive management since Leopold have more or less implicitly based their approach on a pragmatist epistemology, but they have not explicitly embraced one important aspect of the program of Dewey and the pragmatists: they have generally emphasized the ability of adaptive management to reduce uncertainty through scientific management experiments, but they have so far said little about Dewey's dynamic approach to value change.[24] For Dewey, philosophy is most vital when it is used to clarify and formulate questions of practical import; and ethics is most alive when it is testing, in practice, goals that have been advanced in pursuit of consensus and social solidarity. Given the pragmatists' relaxation of the correspondence demand on inquiry, which recognizes that there are avenues to truth within subjective experience, attention can be focused on the development of methods to seek the truth and to speed the process of truth-seeking across the board. Within the praxis-oriented tradition of pragmatism, one never—except perhaps as an artificial experiment—separates "fact" from "value." Facts gain their meaning within a value-laden, action-oriented context. By positing a unity of the method of experience based on sociocultural learning, normatively based in the love of truth, the pragmatists can argue for a unified experimental method that can be applied to values and purposes as much as to scientific, causal hypotheses. Pursuit of proposed community goals is no less an experiment and no more subjective than the testing of a hypothesis in a laboratory. For example, when participants in an ecosystem management process articulate tentative goals and revisit these in subsequent discussions, the goals are open to revision in the face of what has been learned and experienced in the meantime.[25]

If the scientific advocates of adaptive management more fully embrace the pragmatic movement and explicitly reject the artificial distinction between facts and values, then they may come to join Peirce and Dewey in declaring the unity of all inquiry and in including values in the purview of their "experimental management." The Darwinian/Deweyan/Leopoldian approach encourages a variety of value hypotheses and enthusiastically embraces a selection process based on results in the pursuit of improved environmental policies. The tradition of pragmatism, in other words, articulates a set of questions sufficiently comprehensive to encompass *both* the epistemological *and* the normative questions that are essential for charting a course toward sustainable living and for justifying environmentalists' goals to the broader population. And if pragmatists, champions of the belief in the normative nature of logic and inquiry, can bring the power of experimental reasoning to bear upon goals and values as well as facts, then environmental ethics may someday be seen as an important subfield of adaptive management science, rather than as an abstract, and sometimes abstruse, subdivision of "the humanities." Time will tell.

I have argued that adaptive managers, being cognizant of the extent of uncertainty that besets any community that is trying to respond to environmental problems, are well advised to explicitly embrace the methods of pragmatists. As a way of furthering this explicit merger of the two approaches, I conclude the discussion by noting some interesting parallels between the idea of sustainability, which I hope to develop into a unifying theory for adaptive environmental policy, and the pragmatists' idea of truth. First, Peirce's conception of truth as eventually emergent from the struggles of the community of inquirers is an interesting model for forward-looking communities that seek to learn to live sustainably. Kai Lee, following Peter M. Haas, emphasizes the importance of the development of "epistemic communities" of scientists who support those managers who patiently undertake to use the scientific method to better tune our conception of sustainable living and sustainable policies.[26] Peirce's respect for a community devoted to the search for truth is consonant with the type of respect that we must develop toward the future and toward the knowledge and wisdom that will be required if we are to live sustainably.

Second, the development of such epistemological communities—communities willing to learn from experience—is an essential aspect of sustainability because, as advocates of adaptive management agree, sustainable outcomes are not definable in advance but must emerge from a program of active social experimentation and learning. Both sustainability-seeking and truth-seeking are best understood as evolving processes, rather than as ideal outcomes. Thus the idea of adaptive management is connected, by virtue of the search for an emergent, temporally sensitive, transformative notion of truth, back through Peirce to Thoreau.

Third, the pragmatic approach to truth rests explicitly on Darwin's idea of natural selection and thus explains, both historically and theoretically, the link between Darwinian/pragmatist thought and Aldo Leopold's land ethic. Leopold, like the pragmatists, clearly sought both truth and right in adaptive behavior and clearly understood both of these in an adaptive, evolutionary sense. The pragmatist interpretation of the land ethic avoids the deep tensions that are introduced into the land ethic by interpreters such as Callicott, who attribute to Leopold a Darwinian ethic and a "modernist" epistemology.[27]

Fourth, the pragmatists' conception of logic and inquiry as a self-sustaining and *normative* process provides a model for normative-descriptive sciences such as medicine, conservation biology, and sustainability studies and points the way around the fact-value dichotomy. That dichotomy, even if sometimes useful in academic science, is inapplicable to the practical problems of management science. Here we need a *logica utens*, and goals as well as scientific hypotheses must be understood and tested as hypotheses. To continue the quest, to ensure the continuation of the community and its truth-seeking ideals, the community must survive. Peirce's normative approach to logic thus points toward a more unified treatment of environmental knowledge, uncertainty, and goals for action.

Finally, and in many ways this is the most basic lesson of pragmatism for adaptive management, pragmatism signals a shift to a functional approach to analyzing beliefs and the associated recognition that the linguistic forms we use are not forced upon us by reality or by authority. If our means of communication do not achieve their goal of communication, we are free to develop new ways of describing and explaining our agreements and disagreements. Whereas the traditionalists saw the subversion of a priori truths and self-evident principles as a great loss, pragmatists saw their demise as liberation from old-think and the fossilized understandings of the authoritarian world of the past. To the extent that the concept of sustainability represents an opportunity to develop a broadly shared ideal of environmentally responsible living, its definition and expanded use can embody a whole new worldview, a worldview appropriate to the future rather than one dictated by the ossified categories of tradition. Pragmatists, that is, made the examination, criticism, and alteration of the concepts we use to communicate a matter of free choice. So the pragmatist epistemology, emphasizing as it does the ability of our present experience to transform our future, provides the key enabling assumption essential to creating, in communities at all levels, a new and sustainable worldview.

3.6 How Adaptive Management Is Adaptive

In the first section of this chapter I showed how adaptive management can be seen as embodying and systematizing ideas expressed by the American pragmatists, especially John Dewey, and by the philosophical forester Aldo Leopold, who first espoused an experimental approach to management within a multiscaled framework of analysis. Since the pragmatists and Leopold were all deeply influenced by Darwin, it is not hard to see how the modifier *adaptive* came to be applied to this approach to management in 1978, by C. S. Holling.[28] We saw in chapter 2 that Hadley and Leopold both applied Darwinian natural selection to societies and cultures as well as at the individual level. Leopold used this idea of group adaptation to develop a model in which long-term survival establishes the fitness of a culture in a place and to establish that anthropocentric activities must not only be judged for instant success in competition; they must also be embedded in institutions and practices that leave the land alive and undamaged. So Leopold and Hadley, at least, understood human survival in a place as a matter of cultural survival over many generations, a process they viewed as driven by Darwinian selection applied to cultural groups. For Leopold managing "ecologically" *was* looking at the "mountain"—the ecosystem—on a scale commensurate with the cycles of nature. This understanding seems to offer more evidence for thinking that Leopold's model of management was a precursor of, and a direct antecedent to, what is today called adaptive management.

This understanding, however, may raise more questions than it answers; we can explore some of them by asking, Did Hadley and Leopold mean to apply Darwin's criterion analogically or literally? We saw in section 2.3, for example, that Hadley hailed Darwin as the deliverer of ethics and social sciences from confusion, because he offered a verifiable criterion for evaluating normative claims about which human societies are preferable. In fact, history is not very helpful in resolving the question whether the *adaptive* in *adaptive management* should be taken literally or analogically, because Hadley and Leopold were writing before the mechanism of biological evolution was understood; it is not surprising, then, that they did not emphasize the distinction between biological selection (which we now know to be genetically guided) and cultural selection (whereby practices are transmitted across generations by the use of language). Once the mechanism of genetic evolution became known, it would be implausible to maintain (as Hadley and Leopold apparently believed) that group selection among highly enculturated social species was simply "Darwinian selectionism," without qualification. Reading what we now know back into the situation, it might be tempting to say that Hadley and Leopold, ignorant of the mechanisms, suggested that the process that sorts adapted cultures over multiple generations is literally selectionist in a Darwinian sense and that they were, we now know, confused.

Rather than speculating about what would have been said early in the twentieth century if Mendel's discoveries in genetics had been recognized for their importance earlier, I suggest instead that, given what we know today, we simply answer, "Both." Adaptive management embodies *both* analogical and literal senses of the idea of adaptation. First, it is useful to recall, as has been pointed out by many biologists and evolutionary theorists, that "selectionism" can be understood as composed of a minimalist model, according to which selection and adaptation are assured by the existence of three simple functions. Provided there is a source of variation in a population (of organisms, societies, or even ideas), and provided there is a means of "coding," by which characteristics are passed from one state to the next, any method of selection will result in selective "survival" and modification of the original variability. This minimal model does not specify the origin of the variability, the exact means of coding, or the mechanisms that sort the population. If we think of selectionism in these stripped-down terms, then it is clear that Hadley and Leopold meant that societies and cultures are quite literally sorted into survivors, who have found the "truth" about their land, and extinct lines, which did not learn the characteristics of their environment and hence disappeared, according to a method of selection. It is also clear that Leopold, at least, saw this as a form of "group selection," which he followed Hadley in regarding as very important in social species, especially species like humans, in which cultural evolution allows rapid adaptation. Once one emphasizes the different mechanisms driving cultural evolution and biological evolution, recognizing that the former is Lamarckian and many times more rapid than genetic change, the key differences in mechanism make us uncomfortable calling Hadley's cultural sorting processes literally Darwinian.

We might summarize this line of reasoning by saying that *with respect to the similarity of mechanisms,* the group selection model that Leopold proposed, based on institutions and practices, is Darwinian only in an analogical sense. When Leopold added ecological meat to the bones of Hadley's Darwinian theory of history, however, the application of the survival criterion was applied literally. Survival over many generations in a place requires sustained success in providing basic needs and also requires the existence of practices and institutions that have been tried in the local environment for many generations. Developing an ethics for the treatment of the land is thus "an evolutionary possibility" because humans have developed consciousness and foresight and "an ecological necessity" based on the rapid change to environing systems made possible by technology.[29] Thus, although consciousness and attendant technological development *created* environmental problems, we need an *ethic,* an act of consciousness, to address these problems. Failure to develop such an ethic has physical consequences that can result in physical collapse of the local ecological system, for example when a culture employs untried technologies in a new place. Leopold believed his observations from horseback justified the judgment that overexploitation of the resources in the semi-arid Southwest was literally destroying the "land community." And of course he turned out to be right. Much of the land in the Southwest has regressed into simplified and largely useless brushland. Survival requires, quite literally, institutions and practices that are adapted to a place, adapted not just in the short-term sense of flourishing for a few years or decades, but in a long-term sense that enables the culture to survive for many generations, through "hundred-year floods," "hundred-year droughts," and many, many cycles of birth and death and birth. So if one thinks of Darwinian selection quite narrowly as individual, genetically coded selection, adaptive management clearly reflects an analogical use of *adaptive*. If, however, one thinks more broadly of group selectionism as occurring among communities, based on learned practices and successful human institutions, then Leopold's model suggests that improved management practices—practices that protect the whole range of human values derived from the environment—would literally increase the likelihood of long-term survival of the community.

Since the actual adaptation, learning to live for many generations in a place, is cultural and Lamarckian and is based on conscious observation and foresight, it may be possible for cultures to develop adaptations to protect the culture over multiple generations and to learn their way out of threatened extinction. They can, based on information such as Leopold's horseback assessments, pay close attention to trends and to changes in the landscape as a result of current practices. Conscious learning, based on observation of the effects of existing practices on the ecological system, makes improved fitness possible without a long series of failed experiments. Long-term survival, for adaptive managers and pragmatists, is thus simply a specialized form of social learning—learning to adapt practices and actions to the opportunities and constraints stored in local ecological systems.

Furthermore, we can also appeal to social learning to answer an objection I have often heard raised against the application of Darwinian selectionism to ethical and social issues. Surely, the objection goes, we should aim for more than "survival." Humans, our objector wants to emphasize, care about a lot more than simple physical survival. And surely we do care about more than physical survival. We are emphasizing not mere physical survival, but survival of individuals as *members of an ongoing culture*. Long-term survival of individuals over many generations requires thriving practices and institutions that hold open valuable options for future generations. Survival, thus, cannot be individual physical survival alone; it must include also the survival and thriving of the culture itself. And we know that if the culture is to survive over many generations, it must be intertwined with the development, use, and protection of the land that represents the habitat of the culture. This point can be taken one step further: interactions between individuals, a culture, and the land they inhabit are not only essential for simple survival; they also give meaning to the experiences that are shared by members of the culture. The institutions that sustain the culture must include practices and institutions that embody the stored wisdom of that culture.

When information from Leopold's horseback "reconnaissance missions" showed declining resources and dying natural systems, he urged his culture to learn to be more sustainable. Such actions can be considered adaptation. Given that Leopold's experimental management was illustrated by his mental model of Darwinian selection, and given that adaptive management involves a literal application of selectionism to groups (cultures), and given that there are also interesting structural analogies between Darwin's survival-by-fitness model and the model Leopold applied to successive cultures that had occupied the Southwest, it seems quite reasonable to conclude that adaptive management comes by its modifier legitimately.

Let us complete this brief discussion of whether adaptive management is properly so called by exploring one interesting and provocative analogy that can be borrowed—and modified as necessary—from genetic evolution. This exploration, I hope, will provide some indication of the richness of the structural analogies that can be generated from the Darwinian idea when it is applied provocatively to problems of environmental management.

Consider an analogy originally introduced and developed by Sewall Wright, a giant of early genetics, in the 1930s as an attempt to explain the puzzling role of the new field of genetics in biological evolution. In his venerable model, Wright suggested that in order to set individual adaptation within a broader context of multigenerational evolution of populations, we could understand biological evolution as a game that is played out on an "evolutionary landscape." The long-term adaptive landscape can be conceived as a topological space in which higher ground represents greater fitness, given an environmental situation. In Wright's model, the best strategy for a lineage or a species will be to find a very high peak and climb as high on that peak as possible. In individual-level genetic evolution, which was Wright's topic, however, we assume that individuals have no foresight: they simply apply the rule of finding the steepest incline available to them. In this game the best evolutionary strategy will be to locate the steepest fitness slope that is reachable immediately. If, however, one recognizes that the steepest-sided peak may take one to a low plateau, going up the steepest slope may in fact cause a given lineage to be stalled on a very low fitness plateau. It remains controversial among evolutionary biologists whether it is possible for a lineage, once trapped on a low peak, to go back down and search for a higher peak, but let us assume for the sake of the analogy that such lineages cannot return down the slope to reengage the search for greater fitness opportunities on higher slopes.[30]

Suppose, now, we consider the problem of choosing an environmental policy as a parallel adaptive game. As in evolution, winning in this game is for members of the group to stay alive to play another generation, and the biggest winners are the ones who keep their markers in the game for the longest time. For individuals, this means surviving long enough to reproduce and make a contribution to the gene pool of a population. But if an individual is a member of a population that goes extinct in a subsequent generation, individual reproductive success will have been futile in the long run. Once evolution is understood as a multiscalar phenomenon, which affects individuals in the short term and communities in the long term, it follows that organisms or individuals can only "win" on both time scales by also leaving a winnable game board for their offspring.

If we were to apply Wright's analogy at the level of cultural survival, two important changes would be required. First, as conscious users of symbols, humans have foresight and, unlike other organisms, humans can, in principle at least, forgo immediate rewards gained by ascending a steep slope to a low peak if they can see much greater benefits in seeking a higher peak. So our cultural application of Darwinism opens the possibility of balancing a choice between steepness of slope (short-term edge in competitiveness) and height of the peak pursued (long-term survival options). If, in order to apply Wright's model à la Hadley and Leopold to human cultures, we alter it to assume that humans are tool-using animals who have developed such powerful technologies that their collective choices can, under some conditions, change the set of choices facing the future, Wright's fiction of an unchanging landscape in which organisms try to survive and adapt no longer holds. Adaptation, because of the introduction of consciousness, culture, and complex technologies, will be adaptation to a moving target, a dynamically changing environment in which traits that are adaptive today may become neutral or maladaptive in the future.

If, in fact, humans can affect the adaptive landscape faced by their successors, and if these anthropogenic changes reduce the fitness of future generations of their society or culture, then we could represent this outcome by saying that the action of an earlier generation has reduced the opportunities of future generations to make a living and develop their culture. Certain choices made by one generation might, on this model, reduce the height of the very fitness peak that past generations have already begun to climb. If the anthropogenic impacts are irreversible, analogous to a culture that violently and pervasively changes its habitat, the actions of the earlier generation could be interpreted as trapping its successors on a relatively low peak. The peak that earlier generations chose to climb could in fact be "lowered" if the collective choices of a given generation alter the environment in irreversible and drastic ways.

Let us recall that the goal of our little game, for individual players, is to perpetuate their contribution for as long as possible. The fitness landscape, then, can illustrate failures of multigenerational sustainability. If a given population proceeds up a fitness peak, one might assume that the population has no effect on the height of the peak. In this case, future fitness will simply be a function of the original choice of a peak to climb, and the "contribution" of earlier generations will be fixed once the population has headed up a given peak. But this is not the case of interest to us. Assume, instead, that the activities (choices) of generations subsequent to the first "climbers" can in fact alter the height of the peak they are already climbing and that they can change the landscape/habitat/place irreversibly. This second assumption simply operationalizes Leopold's belief that today's technologies can alter nature violently and irreversibly, in ways that make it less supportive of human possibilities and choices in the future.

Wright's analogy, I am suggesting, illustrates the sustainability problem when it is applied at the conscious level at which humans foresee the future and develop technologies to pursue their goals; we can interpret the problem of community adaptability as one of developing a multiscale strategy against the background of a changing fitness landscape. If human populations can irreversibly damage an environment so as to eliminate choices that would have been open to future generations, then we can characterize the damage to the future as a reduction in fitness. Such a reduction will result from a reduction of available options to survive and thrive in the place in question. Wright's analogy of the fitness landscape, then, illustrates in Darwinian terms the nature of the intergenerational harm involved in living unsustainably—it involves reducing the options available for future adaptations, by reducing the height of the fitness peak that a culture is climbing. In the worst case, the environment could change so much that future generations are not adapted enough to live; in this case the earlier generation will have directly thwarted its own long-term goals. Or the effect might be mainly cultural and social, affecting the range of opportunities and choices in the future. Some changes to the landscape/environment may terminate certain of our valued practices of today, resulting in a kind of cultural suicide. People of the future will have nothing in common with us because they face such a changed environment—a different set of opportunities—and will feel no sense of community with us.

Provided one generation in fact cares for its "legacy," it follows that choosers must go beyond an analysis of individual well-being to take into account the range of options they leave for subsequent generations. If we add the goal of long-term survival of our successors (sustainability), then we must have (at least) a two-scale decision process with independent criteria of value at each level. If members of a culture act only on a short-term economic basis, as a group they may find the "steepest peak" of rapid growth while ignoring the goal of multigenerational sustainability. That criterion must tell us how to avoid going steeply up a low peak or, even worse, taking actions today that limit future options (lowering the peak we are climbing). A rational chooser who seeks both individual survival and cultural sustainability would at least set minimal criteria to protect the choices of future generations. We can then view decisions as prioritizing at least two criteria. In addition to information about the short-term economic impacts of our choices, we also need to think about the "height" of the peak we are choosing, creating, and sustaining. Good choices increase economic well-being without putting us on a lower fitness peak—one that will leave few options available in the future. We must, as Leopold argued, learn to think—and evaluate—like a mountain, on multiple scales of time.

According to this model, it is the actors who live out the beliefs and values, and their choices write themselves across landscapes. Whether the culture survives indefinitely will be determined by the adequacy of the options that are left open to future generations when they play the adaptive game in their time. Understood in this way, adaptive management is a search for practices that maintain the options important to a culture living in a place. We noted in chapter 2 that Leopold's 1923 comparison of his own management of the land with that of the precursor civilizations of the Southwest represents—in every way except that it does not use the term—a premonition of the idea of sustainable use of the environment. His argument was based on Hadley's Darwinian theory of politics, institutions, and history. Prior cultures survived longer than we will be able to survive, given current trends in land use. His conclusion was based on his observations, described in the preceding sections of the manuscript essay, which he thought showed rapid deterioration of the arid land system of the Southwest and threatened a reduction in opportunities available for future inhabitants of the area. Even if we are thinking anthropocentrically, then, we should admit that the prior cultures were closer to the truth about how to live in the Southwest than were the cattle-grazers following practices imported from Europe.

Interestingly, Leopold concluded that the goal of maximizing meat production on the range—and even the utilitarian ethic of production and consumption that supported this goal—did not justify the long-term impacts of his culture on the arid lands of the Southwest. In this way he did not just bring the activities and institutions of Western culture to account before the tribunals of experience in the service of survival; he also followed pragmatists in submitting social goals and values to the criticism that they cannot be sustained in the face of ecological and other realities. In the true spirit of pragmatism, Leopold apparently rejected the troubling dichotomy between fact and value and adopted the unified "logic" of experience endorsed by pragmatists such as Dewey and Hadley. In this sense, Leopold and his subsequent followers, the adaptive managers, came to recognize that survival—survival, that is, of a thriving human culture—was a matter of community adaptation, community foresight, and social learning, all of which evoke values that transcend individual, consumptive goods of economics and point toward a responsibility to a larger and ongoing culture.

4

INTERLUDE: REMOVING BARRIERS TO INTEGRATIVE SOLUTIONS

4.1 Avoiding Ideology by Rethinking Environmental Problems

Language shapes the world we encounter; language also shapes the environmental problems we encounter. We have seen that ideological environmentalists, who rely upon preexperiential commitments to determine their actions, often frame questions in ways that accentuate differences rather than encourage opportunities for compromise and cooperative behavior. That the traditional ideologies interpret value differences in polarized terms, based on opposed systems of value, does not help, and discussions of environmental policy often sound more like a battle over who owns the words than a search for enlightened and widely supportable policies. My working hypothesis is that deliberation about what to do to protect the environment suffers because the existing vocabularies for discussing and supporting environmental goals and objectives are based on one-sided, impoverished languages. These languages, founded on opposed metaphysical, unempirical assumptions, block progress toward meaningful public dialogue and cooperative action. To be more specific, until the turf wars between economists and moralists are transcended by the creation of a more comprehensive way of discussing and evaluating unfolding environmental changes, ideology and rhetoric rather than reasoning and negotiation will be the order of the day in environmental policy discourse. Further, I have hypothesized that this same lack of a comprehensive and integrated vocabulary for discussing environmental values has also infected agencies charged to protect the environment. It encourages units and cabals within the agencies to join forces with one or another ideological movement; in the process, each side ignores information and ideas generated by opposing factions, and the result is towering—the inability of agencies and parts of agencies with overlapping responsibilities to communicate and to cooperate in seeking solutions. As we begin shifting our attention from problems of management and epistemology to a more explicit treatment of environmental values, I will focus next explicitly on one of the most debilitating aspects of ideology and towering, the impact of linguistic impoverishment on the *formulation and articulation of environmental problems.*

In chapters 2 and 3 I was careful to point out that to the extent that language shapes the world we encounter, it also shapes the problems we face. Muir's theological language encouraged him to see all development as evil, so he and his followers were dismissed as radical romantics, not capable of responding to day-to-day needs and demands of citizens. Pinchot's commitment to individualistic market economics encouraged him to see the importance of resource use and development, because he saw the good of human consumption as overriding other, less consumptive human values. Disagreements, in this context—a context in which two sides have defined the goods they seek in incommensurate terms—are experienced as zero-sum games. Any gain of one side must seem like a loss to the other; to allow the opposition to meet its objectives is to slight one's own.

When the languages we use to discuss environmental values evoke opposed ideologies, compromise and cooperation can become impossible. Perhaps, by building on the insight that ideology and towering do not work, we can reconsider how environmental problems are formulated for public discussion. By looking carefully at many local environmental situations or problems—disagreements about what should be done in a particular situation—one can see that they are not fruitfully expressed either in the all-or-nothing language of ideology or in the exclusive languages used within policy towers. When the various interests are expressed, and one tries to translate all of these statements into either the vocabulary of Muir's moralism or that of Pinchot's individualistic economics, one finds that the monistic versions leave out most of what is interesting in the situation and that preexperiential commitments to one rhetoric over the other virtually destroy the possibility of communication. For this discussion I will dust off a very useful thirty-year-old dichotomy—"benign" as opposed to "wicked" problems—and use it as a starting point in rethinking environmental problems and how to articulate them. The benign-wicked distinction was introduced by Horst Rittel and Melvin Webber in 1973 as a way of explaining the resistance to solution of a large class of complex problems in planning.[1] Their distinction, I believe, can provide insight into a different way of conceiving, posing, and deliberating about environmental problems.

Rittel and Webber noted that a number of social problems—creating sewage and water systems for cities and developing a health infrastructure, for example—had yielded to analysis. Even in situations where resources were inadequate to address these social problems, social planners could help municipalities and governments to set reasonable goals and to undertake technological developments that would predictably lead to social improvements. As the problems in this group were transformed into technical and bureaucratic problems, however, Rittel and Webber found that there remained for municipalities and governments another set of problems, including, for example, crime, urban sprawl, and deteriorating natural systems, which seemed intractable to the usual methods of goal-setting and calculations of efficient means of production of the desired goods. In order to explain the persistence of such problems, Rittel and Webber distinguished "benign problems" from "wicked problems" and hypothesized that some—including many of the ones we would think of as environmental problems—are "wicked" in nature. It will be worth examining this distinction for the light it sheds on the matter of environmental problem formulation.

Benign problems, such as mathematical or scientific problems like providing a chemical analysis of a compound or calculating the most efficient design for removing sewage from a city, have only one solution, and when that solution is found, it can be seen to be a unique solution. Wicked problems seem to yield only better-or-worse solutions; the problems are never solved—they are only "resolved" for some temporary period, until political or social forces demand a different balance among competing values and goals. Rittel and Webber develop their distinction by listing ten characteristics of wicked problems.

1. There is no definitive formulation of a wicked problem.[2]
2. Wicked problems have no stopping rule. (162)
3. Solutions to wicked problems are not true-or-false, but good-or-bad. (162)
4. There is no immediate and no ultimate test of a solution to a wicked problem. (163)
5. Every solution to a wicked problem is a "one-shot operation"; because there is no opportunity to learn by trial-and-error, every attempt counts significantly. (163)
6. Wicked problems do not have an enumerable (or an exhaustively describable) set of potential solutions, nor is there a well-described set of permissible operations that may be incorporated into the plan. (164)
7. Every wicked problem is essentially unique. (164)
8. Every wicked problem can be considered to be a symptom of another problem. (165)
9. The existence of a discrepancy representing a wicked problem can be explained in numerous ways. The choice of explanation determines the nature of the problem's resolution. (166)
10. The planner has no right to be wrong. (166)

Their list is redundant and apparently reveals no internal organization—it is just a list of characteristics. Further, the list identifies a number of rather different contrasts between wicked and benign problems, and the authors invite the reader to accept that all of these differences can be usefully collected under a unifying definition of "wicked problems." To illustrate my point, I will list four somewhat different contrasts that are implied by the listed characteristics of wicked problems and indicate by number related characteristics in Rittel and Webber's list. The first contrast has to do with (1) problem formulation issues (numbers 1 and 9): wicked problems are subject to great controversy over what the problem itself is and how to formulate it. This is related to but logically distinct from (2) the noncomputability of solutions (numbers 4, 6, and 9): solutions to wicked problems are not computable through optimization calculations. A third issue is raised by the observations that wicked problems are unique in their complexity, that there can be no "learning curve," and that, since many interests are involved, the planner has no right to be wrong. We can refer to this cluster of wickednesses as (3) the nonrepeatability problems (numbers 5, 7, and 10). Finally, at least two of the characteristics mentioned by Rittel and Webber have to do with (4) the open-ended temporal frame in which wicked problems are addressed (numbers 2, 4, and 8). This parsing omits only characteristic number 3 from our subcategories, a point to which I will return below. First, I will comment on these four themes running through the other nine characteristics.

Problems of problem formulation represent, in some sense, the key, defining characteristic of wicked problems, because understanding the genesis of formulation issues leads us to realize that value pluralism is present, and that is what makes wicked problems wicked. Wicked problems, that is, defy consensual statement *as problems* because disputants in wicked problems are pursuing different, sometimes conflicting interests. Since social problems that exhibit wicked tendencies are expressions of competing social values and goals, any statement of the problem must state or presuppose some value that is being underproduced. These disagreements are very important because their outcome, clashing visions of what the problem is, determines what data is relevant to resolving the problem. As I understand the structure of Rittel and Webber's argument, the existence of a plurality of complementary, competing, and conflicting values *explains* why wicked problems are wicked. This explanatory hypothesis is thus the heart of the theory, and the theory apparently is committed to a pluralistic understanding of social goods.

The issue of *noncomputability of solutions* is also important, because it explains just how much is at stake in terms of decision theory. This cluster of characteristics represents several ways of stating the conclusion so repugnant to many formalistic decision analysts: many interesting problems, including all wicked problems, have no optimization solutions. There can be no solutions based on algorithms or on computation of a single measure. Rittel and Webber argue that this characteristic calls into question the usefulness, for many purposes, of "the classical systems approach of the military and space programs," which requires that a planning project be organized into distinct phases. Rittel and Webber state that operations research methods become operational "only after the most important decisions have already been made." This characterization seems as relevant and important today as it was in 1973. In other words, single-value optimization models have not provided solutions to problems involving *competing* values. Operations research has become ever more sophisticated in calculating optimums once a problem is formulated as a single-goal, benign problem; further, operations researchers have developed multicriteria, suboptimal dynamic models for many complex problems. The difficult social problems faced by planners, however, remain beyond the reach of such analyses because, in the prophetic words of Rittel and Webber, "setting up and constraining the solution space and conducting the measures of performance is the wicked part of the problem."[3] Such models, no matter how sophisticated, still cannot address this point of Rittel and Webber. Although sophisticated models can *embody* multiple criteria, they cannot tell us how to weight or prioritize the multiple criteria. The system of analysis cannot indicate which system variables are of greatest social interest.

From this simple but elegant theoretical argument, the conclusion apparently follows that wicked problems, especially problems in which multiple interests are competing, cannot be analyzed in a unified computation as would be necessary to apply a single criterion of value. Operations researchers and decision scientists have of course embodied this result in their models by constructing multiple-criteria evaluation models. No algorithm or deterministic model is available, however, to assign the weightings one *should* place on the competing values. Rittel and Webber succeeded in showing that although some aspects of any problem can be formalized and submitted to an algorithmic solution, the problem of how to weight competing value criteria cannot be formalized. No amount of formalization, it follows, can eliminate the need for *judgment* when the target decision addresses a wicked problem.

The fact that wicked problems, including most environmental problems, are *nonrepeatable* is a symptom of the uniqueness of the combination of interests and constraints that bound these problems. Environmental problems, as noted above, do not submit to "one-size-fits-all" solutions; a new balance must be struck among unique demands forwarded by competing interests. These characteristics of wicked problems suggest that there will be a flat learning curve based on repetitive use of methods. In this book I intend to respond to the nonrepeatability problem by first acknowledging that although we cannot create an algorithm for solving such problems, we can nevertheless look at many case studies of successes and failures in cooperative management. Such examinations can help us to develop "heuristics"—guides to asking the right questions and avoiding damaging pitfalls in addressing wicked environmental problems. So abandoning the search for an algorithm for the solution of wicked problems does not imply giving up on using decision-analytic techniques to improve decision processes. It only implies a shift to more concern with processes and less reliance on incomplete analyses of outcomes.

Finally, several of the characteristics, most prominently numbers 4 and 8, emphasize the *open-ended, intertemporal effects* and difficulties affecting wicked problems. These concerns parallel issues already emphasized in this book, issues having to do with the scale on which, and the perspective from which, to model environmental problems in space and time. Wicked problems, because they have no single correct answer and are at best "resolved" by finding a temporary balance among competing considerations and interests, "have no stopping rule." More evidence will always be coming in; at no point can we nonarbitrarily say "we have sufficient evidence" to make a decision.

Environmental complexity manifests itself to decision makers as open-ended and multilayered; environmental action must always be seen as directed at goals in one temporal frame but as also having effects on larger and slower dynamics. Environmental problems are wicked because, given that participants in addressing the problem have many different interests, unintended and delayed consequences of actions undertaken to serve one interest will result in complaints from persons with other interests who count their interests over longer periods of time. Solutions to environmental problems remain, in this sense, open-ended. Rittel and Webber suggest as much when they state, in describing characteristic 4, "The full consequences cannot be appraised until the waves of repercussions have completely run out, and we have no way of tracing the waves through *all* the affected lives ahead of time or within a limited time span."[4]

Temporal open-endedness is endemic to all environmental problems, but fortunately we have a new and evolving approach, hierarchy theory, which offers tools for conceptualizing space-time relations in complex, many-scaled dynamics. To the extent that the wickedness of environmental problems results from problems of modeling multiscaled impacts, hierarchy theory may help to unravel some of the underlying problems of complexity and shed new light on the intractability of wicked problems, at least that aspect associated with open-endedness. However that turns out, an improved conceptualization of space-time relations in assessing impacts would be at least one step toward better problem formulation and perhaps would yield some progress in decision making. It may be possible to use hierarchy theory to better relate model building and decision analysis. Understanding and rating the importance of various dynamics in physical space requires an understanding of goals and of the values that, by not being protected or achieved, define an environmental or planning problem. Regarded physically, the complexity of environmental problems reveals itself in the multiple layers of causation in complex, dynamic systems. But in fact, of course, there is an infinite number of layers and dynamics in every physical system; it is human values, interests, and perspectives that determine which of these dynamics are important and worth monitoring.

The problems with problem formulation are thus all entwined with the problem of open-ended impacts and the need for more and more data about impacts on longer temporal scales. Both of these problems, as Rittel and Webber hypothesize, can be explained by noting that in every wicked environmental problem there will be competing interests and values at stake, including some future values that cannot even be clearly represented in the present. Nonetheless, viewed as a decision-making challenge, environmental problems involve setting reasonable goals and finding the means to protect many, sometimes competing, social values that emerge on many scales of time.

I have resurrected the distinction between benign and wicked problems here because I think it offers several important insights about the nature of environmental problems and their public formulation. Proceeding on the hypothesis that most difficult environmental problems (situations) in which there are disagreements among participants and stakeholders about what to do are wicked problems, we can make several suggestions for anyone who wishes to address such problems through a public, democratic process such as a watershed partnership or an ecosystem management plan.

First, because the formulation of the problem is controversial, and any proposed solution will be relative to the problem formulation favored by its proposer, we can conclude that there can be no solution to wicked problems without moving through multiple and ongoing iterations of model building, problem specification, and articulation of social values and priorities. Solutions to complex environmental problems—on our hypothesis that environmental problems are wicked problems—must never be thought of as final and conclusive. A "solution" represents a temporary stable point in an ongoing negotiation in which today's "solutions" will alter the dynamic and lead to a new problem formulation and new forms of competing claims and interests.

Second, any claim to have a solution must always be viewed tentatively, because the steps we take in pursuing present solutions are almost sure to have unforeseen effects that extend beyond the assumed temporal horizon of today's analyses. Wicked problems thus conform to the expectation of the pragmatists, such as Hadley, who insist that we judge the success of a culture by virtue of its longevity, while recognizing that a judgment of longevity will always be relative to the temporal frame over which observations extend. Rapid success of a policy for developing economic resources may be a sign that the policy is a good one (it apparently addresses one important social need); it may also be a sign that we have not foreseen the longer-term consequences of the policy. On such grounds, Leopold concluded that though wealth was increasing as a result of grazing and exploitation of resources in the Southwest, a careful look revealed that the usually slower-changing arid physical systems that supported that economic growth were deteriorating, undermining the economic success in the long run.

A third consequence directly challenges the assumptions of the ideological environmentalists who, on either extreme, seek to serve or maximize a single value at the expense of others. Rittel and Webber recognized that since wicked problems are wicked precisely because they involve competition among multiple interests, a wicked problem cannot be reduced to a disagreement that can be resolved by measuring a single variable. In fact, most environmental problems involve a competition among multiple goods, all of them legitimate at some level of fulfillment but competing in the sense that the society has neither the resources nor the resolve to fulfill all of the goods to the degree insisted upon by their advocates. Following on this insight, I would suggest that Rittel and Webber should rewrite their third characteristic, going even further. They say, correctly, that solutions to wicked problems do not have the character of being "true" or "false," but I would suggest it is even misleading to refer to such solutions as "good" or "bad." Although I think they mean that a given solution could be a good balance or a bad balance across competing values, the good-bad judgment may encourage either-or formulations and a return to ideological environmentalism. I think it is less misleading to say that wicked environmental problems involve competitions among multiple "goods," that we cannot absolutely rank these goods, and that solutions to such problems will at best involve achieving an acceptable balance among competing goods. The solutions will be acceptable, of course, only for a time, until changing environmental conditions or other changes in the society demand reopening the issues.

Multi-attribute utility theory and related systems for decision evaluation may eventually provide tools for linking multiscalar thinking with multiple distinct criteria for judging complex problems.[5] But decision theory, like operations research, has been slow to give up on the idea of a single, final solution, and multicriteria decision analysis usually involves combining many factors according to an algorithm for computing best policies based on expected outcomes. It may be more useful to develop multiple criteria and, rather than concentrating on computations meant to algorithmically solve problems, concentrate on developing multiple management criteria based on multiple indicators of well-being. This shift would encourage the use of multicriteria analysis throughout the planning and negotiation process. It would then be possible to seek policies that are robust across multiple values and political objectives, rather than insisting—unrealistically—that the analysis yield a single answer and an optimal policy outcome.

To summarize, then, Rittel and Webber examined a set of problems notable for their intransigence that include, apparently, most truly complex—and most "interesting"—environmental problems. We have relied on their analysis as a way of understanding environmental problems as they are faced by modern democratic societies in which multiple interest groups clamor for attention and for resources. Rittel and Webber argue that wicked problems are not correctly formulated until a solution is found. Although this sounds paradoxical on its face, their claim makes perfect sense if one thinks of problem formulation and the proposal of solutions not as a *unidirectional, one-time process* but as a *multidirectional, iterative dialogue* in which incremental agreements on partial solutions can help even contentious partnerships to reformulate their problems in more tractable terms. The argument of Rittel and Webber seems to suggest that wicked problems can be expected to yield only to an iterative and multidirectional dialogue including scientists, policymakers, and the public in a process of social learning. Social learning will be the result of deliberation and discussion and, more basically, of the shared experiences of a negotiating community. If there is a commitment to find a cooperative solution to contentious problems through ongoing discussion, adaptive management and social learning become possible; and with each improvement of our understanding, the barriers to cooperation give a little. Once an adaptive process is under way, we can hope also that learning more about the system being managed may lead to reconsideration and reformulation of the operative interests, goals, and values of the interacting groups. My point, then, is that adaptive management—supplemented with a goal of maximizing information flow and open communication among participants—seems the only reasonable strategy by which to address wicked environmental problems.

4.2 Overcoming the Serial Approach to Environmental Science and Policy

Before we apply the methods of adaptive management to the plural social values exhibited in wicked problems (in part 2), it will be useful to contrast the adaptive management approach to science, ethics, and policy with what is typically offered instead, what I referred to in chapter 1 as the "serial" approach to science and policy deliberation. According to this approach, endorsed since the 1970s at EPA and other agencies, the process of gathering scientific data and building models is supposed to be completed in isolation from policy discussion and formation. This tendency can be illustrated with a couple of examples of what adaptive management *is not*.

First, to pinpoint one multidisciplinary academic study, I reference an otherwise fine essay by Sandra Batie and H. H. Shugart.[6] This paper represents one of the few attempts to bring science and policy analysis together in one piece of work, and yet it is open to criticism because it exemplifies the serial approach to uniting science and policy analysis. The essay provides a method of spatiotemporal scaling as an important part of a system for evaluating changes that may result from global climate change—a move that is consistent with the "hierarchical" approach pursued in this book. The first half of the essay (which one assumes expresses mainly the contribution of the ecologist Shugart) offers general guidelines for describing the scalar aspects of physical systems and gives examples of useful models for interpreting the multiscalar activities of complex natural systems. There is no discussion in the paper of how information about social values might affect the descriptive models developed. Once the descriptive aspect of the problem is treated, the economist Batie, in the second half of the article, brings the evaluative framework of economics to bear on the changing states of the physical system. My point here is that the authors' argument is developed in two distinct stages with distinct languages—ecological description followed by evaluative discourse. This serial tendency may not be inconsistent with an iterative approach to policy choice—it may be possible to revisit the science and then discuss policy options once again—but the serial approach does not facilitate the flow of information from evaluative discourse to scientific discourse. Batie and Shugart attempt to build an evaluation method on top of scientific description—but they do not close the loop in both directions by creating a shared language that allows multidirectional dialogue about social goals and their impacts on scientific models.

The serial view is also exemplified, on a grand scale, in EPA's risk assessment/risk management model, mentioned in chapter 1 as the dominant and official decision model for most of EPA's history. The decision model was laid out in a 1983 National Research Council study, *Risk Assessment in the Federal Government*, which was intended to codify and systematize accepted practices of risk assessors and to clarify the relationship between risk assessors and risk managers in the broader decision process.[7] This document, which was endorsed by several consecutive EPA administrators, came to be called "The Red Book," and it served as the official guidance for the practices of risk assessors and decision makers in the agency. According to the view of risk analysis outlined in the Red Book, the decision process is linear and unidirectional. It starts with surveys of existing scientific studies and proceeds to the development of purely descriptive and quantified models for predicting transport, exposure, and incidence of disease resulting from a given "risk," such as exposure to a chemical suspected of causing cancer or other diseases (see fig. 4.1). This scientific process—which is to be carefully insulated from any infection from values and political consideration—culminates in a "risk characterization," also called a "summary" of scientific data and model outputs. Then, once the science involved is wrapped up in this tidy, descriptive package, it is passed, with minimal commentary or judgment, to the "risk manager," whose job it is to factor in social values and make a "political judgment" on how to manage the risk.

As noted in chapter 1, this holy grail of decision making—the risk assessment/risk management model—became attractive to politicians and scientists and became a foundational, and yet elusive, goal of William Reilly's tenure as EPA administrator. Subsequently it was adopted without question by Carol Browner, Clinton's administrator. The RA/RM model ideally exemplifies what I have called the serial view of science and policy. In the RA/RM model it is suggested that the process of gathering facts and building predictive models begins without an open discussion of values—of what is important. Hardly any emphasis is placed on correct problem formulation or how the problem should be modeled. Worse, the science is thus doomed to go forward based on huge, unquestioned assumptions, with very little sense of context and without guidance that might help to determine what events, activities, and substances are likely to threaten important social values. There is no suggestion that once the science comes in, there will be an inquiry into whether the original problem formulation was adequate. According to the serial view, the system for gathering information and making policy decisions is both linear and irreversible. The flow of information from the science community to the decision maker and the public is a unidirectional flow, with no feedback to the scientists if their data and models fail to address the problems as they are experienced in real communities and in real decision situations.

More examples of serially organized efforts to develop policy could easily be provided, but these efforts—which claim to be based on "good science" (because the science is purportedly untainted by even a peek at social values and goals)—do not in my opinion contribute to a reasonable discourse about policy goals and action. This implicitly assumed "model" of the science and policy process does not iteratively revisit problem formulation, as seems to be required if most environmental problems are wicked problems. This model seems unlikely to establish an iterative and multidirectional discourse that could progressively inform a process of adaptive environmental management.[8] Serial approaches fail because they are based on a false image and an

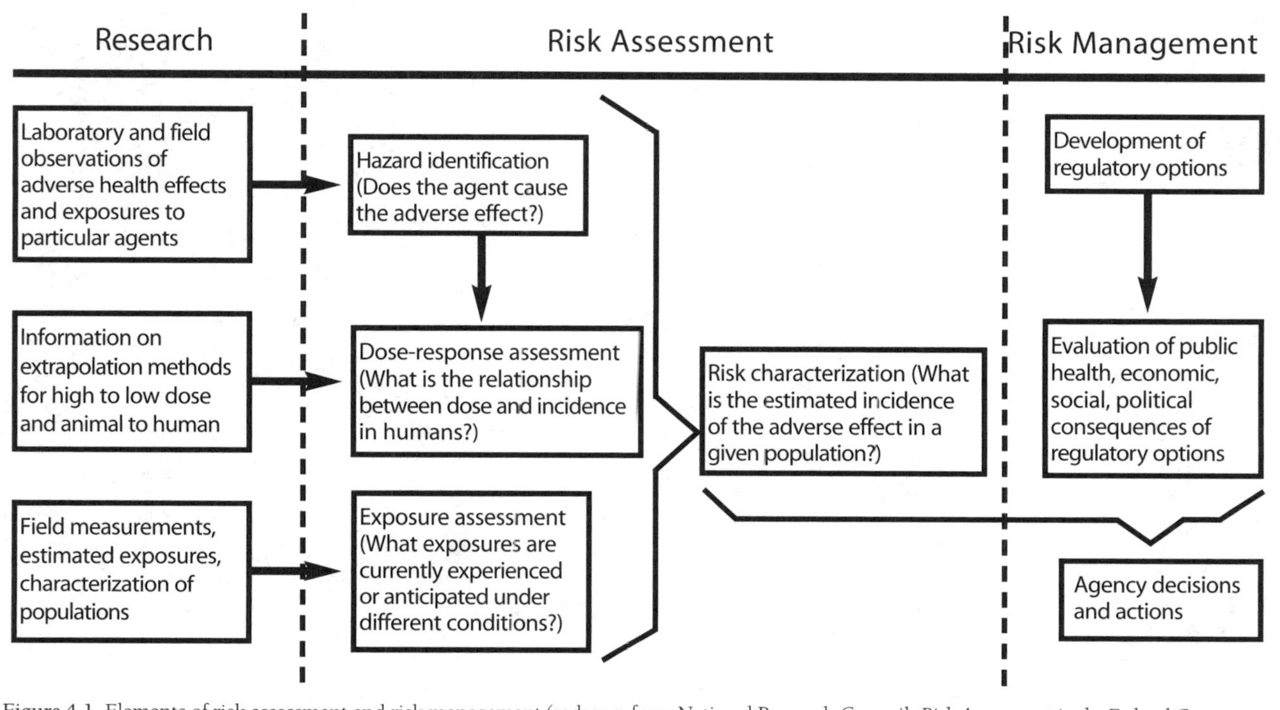

Figure 4.1 Elements of risk assessment and risk management (redrawn from National Research Council, *Risk Assessment in the Federal Government* [Washington, DC: National Academy Press, 1983], 28)

associated myth that is perhaps the greatest barrier to an improved understanding of ecosystem management. The image is that of an ideal environmental decision maker, one who has gathered all the descriptive information regarding the functioning of an ecological system; determined the likely outcomes of further impact from human activities; polled the population to determine the values, goals and preferences in good democratic fashion; and, armed with all the facts, decides what policy to pursue to maximize total welfare. This false image underlies much of our thinking about environmental management today.

We all know that the ideal of a fully knowledgeable manager is impossible and that the idea of a "complete" science is an oxymoron—science is an ongoing process of revision and improvement, not an existing body of fact. But the myth of a complete science nevertheless has a powerful influence on our view of policy process, because we simply substitute "the best available science" for "complete science," and sustain the damaging part of the myth—the assumption that there is a one-way flow of information from scientists toward policymakers and the public.

Once we realize that uncertainty and ignorance are an inevitable part of management—that surprises are to be expected—policies should be designed to survive surprises and to reduce ignorance in specified and useful ways in the process.[9] If we are to improve our management decisions, we must direct future science to reduce uncertainty and to correct our policy course with new and relevant information. Here we see the inadequacy of the serial model, which appears as a mere after-image of a mythologically complete science. The false image of a completed science has been so hard to obliterate because it reinforces and is reinforced by the myth of a "value-neutral science."[10] Again, if pressed, we all admit that the path of scientific research is deeply affected by valuations—by the scientist doing the research, by funding agencies, by foundations and firms, and by other scientists. And yet we want our science, especially the science that affects policy, to be above all bias and reproach, so we pretend that the scientific description comes first and that it is as complete as it will ever get before we begin policy analysis, goal-setting, and practical decisions.

I believe that the serial view is both inaccurate and damaging as a model of science in the real world of policy discourse. It is inaccurate because science in the real world is carried out by scientists, and scientists are persons whose values affect the entire process of scientific study. It is perhaps reasonable, once we recognize this reality, to think of the separation between "facts" and "values" as a useful and temporary convention or ideal, but we must always remember that in the real world science must be understood within a larger social context that includes values. When we view science from the activist perspective of policy and management, it is inevitable that we view it from an evaluative stance. So I suggest that we think of the policy process not as serial, not as like an assembly line with no reverse, but rather as an iterative dialogue in which science and policy are mixed in an ongoing, democratic, policy process–adaptive management.

The serial view is also damaging in that it discourages us from even considering whether and how scientific study and modeling should be guided by policy problems, goals, and objectives. The serial view assumes a one-way flow of information from scientists to the public. But I will argue that action-oriented, adaptive management requires a multidirectional flow of information among scientists, policymakers, and the public. In order for experimental management to reduce uncertainty, it must be able to direct the torchlight of science on questions that are important to management decisions. Communication must flow both directions; and scientists must be willing to accept guidance from policymakers and the public.

Most importantly, however, the serial view fails because it offers no opportunity for progressive integration of information from various disciplines, and it does not allow for the progressive integration of competing values into shared goals. A true commitment to adaptive management requires us to participate in an ongoing procedure for considering and reconsidering management goals and also the values that underlie them. Rather than imagining that the policy formation process is carried out in two realms—the realm of science and the messier realm of policy, goals, and values—we can more usefully think of the policy process (adaptive management) as having two related phases. A phase is here understood as a temporary focus for discussion, experimentation, and learning. Phases, unlike the hard-edged stages of the serial view, are temporary and are thought of more as two ways of looking at a single subject than as two distinct subjects. The subject is environmental policy and management—a normative science—and one can look at it in an action phase and in a reflective phase. In the action phase the focus is on what ought to be done, which includes asking what we do know and what we need to know if we are to achieve stated goals according to specified criteria and measurements. Here descriptive science is used to determine what is possible and what means are likely to achieve the stated goals. The search for rational and democratically acceptable environmental policies, however, requires also a reflective phase, in which policy discourse is focused on choosing goals and the sorts of measures and indicators we will use to keep score of how we are doing in the game of sustainable living. My point is that if we see the context of public discourse about environmental management as one of deciding democratically what to do, then the public must in some way contribute to the choice of goals of management. And since goals are given operational meaning by choosing which variables to track, it seems to follow that there must be democratic input on the choice of indicators and on the standards that are set. This reasoning implies in turn that, in some sense, democratic participation requires that policymakers and at least some of the public can understand models chosen to determine success and failure.

To this end I propose a *process heuristic,* a general guide to thinking through tough problems on a local level. The process heuristic suggests conceiving the process as including two alternating phases, which we call the action phase and the reflective phase. In the action phase, we assume that multiple participants will advance a variety of overlapping and some antithetical goals, and we focus attention on several "action rules," which include a variety of decision rules to be applied in various combinations. The action rules might include very general rules, such as the cost-benefit test, the safe minimum standard of conservation rule (which tells us to protect a productive resource "if the social costs are bearable"), and the precautionary principle. The action rules might also include much more specific, locally applicable criteria designed to track particular important features of local places. These criteria, however, cannot be applied in every situation, nor should they be applied willy-nilly and at random. They arise and are given plausibility within the reflective phase of the process, the phase in which the community discusses which variables to monitor in a particular situation and which goals and values to pursue.

To make decisions such as this democratically, a very open public process is necessary. Scientists must contribute to this process by helping the community to understand key environmental processes and to identify measurable variables that may be important. Interest groups must play a role, because they will want to insist that the indicators chosen reflect the values they support and that the standards chosen are appropriate from their perspective. In the process of disagreement, managers must identify important areas of uncertainty and disagreement and propose safe-fail experiments to reduce this crucial discord. A provisional decision can be made, to proceed with particular, proposed indicators and to apply a proposed set of standards. The reflective phase is then replaced by the action phase, wherein the community chooses particular actions and policies and sets out to judge these according to appropriate criteria—criteria, that is, that were sanctioned in a previous reflective phase. In practice, of course—in active community-based practice—the two phases normally overlap and proceed simultaneously. We accordingly refer to the two-phase policy process as a heuristic—it reminds us to shift our focus sometimes from disagreements about what to do (in the action phase) to the more reflective questions of choosing goals and standards.

Adaptive managers believe in science as a part of the public process, rather than as an input into the process from the "outside." Successful use of science in a public, democratic policy formation process requires a free flow of information in multiple directions. What the idea of sustainability is missing up to now, as argued in chapters 1 and 2, is a multidisciplinary, integrative language capable of supporting multidisciplinary public discourse and deliberations. The main contribution of this book, being philosophical, is to the reflective phase by creating and proposing such a language. In part 2 we will concentrate on developing a pragmatic value theory that will encourage pluralistic, many-sided debate and deliberation. In part 3, armed with a value theory that is appropriate to an adaptive management process, I will be able to address the question that has been left almost completely unanswered in part 1: what would the new science—the interactive, value-driven science I am seeking as a manifestation of adaptive management—look like? That question is the topic of chapter 11.

An improved language cannot be created in ivory-tower isolation. The language must also be constantly tested by reference to examples and case studies taken from the action phase. The result on the ground—in particular communities that undertake such a two-phased process—one hopes, will be a rich form of social learning, in which public discourse contributes to an integration of science and values in the policy process. Now that I have explained what adaptive management is—and is not (it is not serial)—I turn to an examination of possible understandings of environmental values, in the search for an approach to environmental values and valuation that will be equal to the demands of adaptive management of, for example, a watershed partnership or an ecosystem management initiative.

PART II

VALUE PLURALISM AND COOPERATION

5

WHERE WE ARE AND WHERE WE WANT TO BE

5.1 The Practical Problem about Theory

The practical problem we face in this book—and in all discussions of environmental policy—is the problem we symbolized in chapter 1 by the crooked hallways at EPA: you can't get there from here, conceptually. Disjointed conceptual apparatus, like disjointed corridors, leaves too many important social values inaccessible to rational, multidisciplinary discourse. EPA was created as a patchwork of offices corresponding to a patchwork of legislative mandates, and it lacks smooth and transparent pathways from problem to problem and from office to office. The agency is divided by disciplinary schisms and interest-group cabals, and above all, it has failed to establish communication among branches and across disciplines within branches. We were thus forced to conclude that there seem to be many solutions—including rational and proportional policies—that cannot be reached from where EPA is now. As explained in earlier chapters, there is no clear vocabulary that is adequate for open and edifying discussion of environmental problems as problems of social values, rather than as issues separating interest groups. That is the problem, and I prefer to think of it as a *practical* problem, one we can constructively engage, provided we self-consciously seek improved means of communication in real situations.

In part 1 of this book, I present adaptive management as a constructive response to the problems of environmental policy formation and management. Adaptive managers do not claim to know in advance what policies are sustainable or even what the goals of sustainable living are, so they propose an open-ended, experimental approach to environmental management, in the presence of pervasive uncertainty. It was shown that, given expanded human technical power and human knowledge and foresight, communities today face a set of problems that can be analogized as a game of adaptation played out on multiple scales. In order to make a contribution to the future, an individual (or a community) must survive and reproduce; to have one's contribution survive into the indefinite future, it is also necessary for a society or a community to survive as support for that offspring. Adaptive management, thought of as conscious protection of future opportunities, must thus be multiscalar in its planning. And especially, adaptive management must be place-based and community-oriented. Adaptive management is here understood as being undertaken within a democratic society, in which interested citizens, either as representatives of their interest group or simply as individuals, participate in this open-ended, experimental process of management. It is hoped that this strategy will result in social learning, in the emergence of shared goals and policies, and in greater environmental protection.

Adaptive management is a strategy that can both reduce uncertainty regarding particular matters of fact affecting management decisions and reduce disagreement about goals, objectives, and values. Values as well as facts must be an important part of the ongoing discourse; I propose that we make the identification, analysis, and prioritizing of social values endogenous to adaptive management as a public process. In order to make adaptive management so inclusive, we must develop a vocabulary broad enough to support communication from scientists to the larger community and also a dialogue that articulates community values, while using managerial opportunities to support broader social values. A crucial aspect of such a vocabulary must be a way of talking about—and thinking about—environmental values that improves communication among participants in policy debates. In this part, chapters 5 through 9, I introduce a theory of environmental value and, in immediate conjunction with that theory, a normative vocabulary that will encourage interdisciplinary communication about environmental problems in a rich context that includes public discourse about environmental values emergent on multiple scales.

To adopt the task of "straightening the corridors" of communication at EPA may indeed seem a daunting task; all we have at hand is the method suggested in this book, a method that works iteratively back and forth between discussions of actual policy proposals, on the one hand, and discussion of the terms and theoretical constructs that are available in policy discussions, on the other hand. The latter task, especially, is undertaken in a prescriptive and constructive attitude: we are explicitly trying to propose a theoretically grounded language for discussing as many real-world and particular environmental problems as possible. Our strategy of experimental pluralism suggests that we propose multiple vocabularies, use them in particular situations, and select them according to their performance in achieving improved communication and better policy solutions in real-world situations. The goal of the method is to use these terms, at the same time clarify and refine them, and ultimately fit them into a coherent and comprehensive theory of environmental values. This kind of empirical shaping of terms and theories goes on and on as the terms and concepts are used in new and different situations and in endless iterations and revisitings of old problems, even as we develop and clarify the theory of value that supports our purposive linguistic endeavor. This approach, which we call "practical philosophy," begins with a problem-oriented approach, focuses on a few illustrative cases, and then works inductively toward a general theory of environmental values. Theory, though still acknowledged as important, is thus brought to heel by actual, real-world cases, enforcing in a situational way the limits imposed by real-world conditions, as determined by the various special sciences. Because our approach is *practical,* the contexts sought for learning about theory and testing various linguistic formulations is to be sought "in the trenches" of *policy and management,* not within the safety and simplicity of a single academic discipline, philosophical or otherwise.

Pragmatists pay attention to the particularities of unique situations. In action-forcing situations, it is often possible to provide helpful, if context-sensitive, guidance to decide what to accept as certain enough to guide action and what is not so certain and therefore requires further study. These decisions, which occur within a value-laden context, allow us to use agreements about values—however limited and situation-specific—to accept certain goals as consensus goals. Then we can pursue observations and management experiments to reduce debilitating uncertainty regarding techniques to achieve those goals. Shared values and goals can, in this way, sometimes serve as the solid ground on which to stand to undertake experimentation with means to achieve the goals, thereby reducing uncertainty about system functioning. At other times, of course, beliefs about the system and its behavior seem undeniable, and we can stand on these planks to deliberate about realistic and wise goals. The epistemology of adaptive management thus provides for gradual progress and improvement of both our belief system and our preferences and values, by using experience to triangulate between temporarily accepted beliefs and values. The most controversial aspect of this knowledge-seeking strategy, perhaps, is the idea that in concrete situations shared values can sometimes serve as a solid basis upon which to pursue mission-oriented science to reduce uncertainty about outcomes of our choices. To explore this idea, it is essential that we understand environmental values in such a way that through successive applications of our method, values can be improved over time. In this and the remaining chapters in part 2, I provide such a context-sensitive approach that can serve to bootstrap both our values and our factual understanding of management situations simultaneously.

Likening our epistemological problem to a ride on Neurath's boat, which is required to stay afloat indefinitely while repairs are made, we can understand our problem as one of deciding which of our beliefs to accept as strong enough and which should be submitted to immediate and critical review and testing. Sailors on the boat are motivated by their desire to survive, and so they undertake the repairs on the boat with great deliberation and care. They must not only make important technical judgments regarding which planks are becoming weak with age and rot, but they must also make judicious choices regarding which planks must, given the importance of their function, be given priority. Analogously, as adaptive managers, we are driven by the desire to stay afloat and to prosper as a community, and we must similarly decide carefully what beliefs to accept as given, which should be doubted, and which points of uncertainty are of highest priority, given the shared goals of the community. Like Neurath's sailors, we must make such epistemological judgments under pressure; if we guess wrong and stand on a weak board to fix a stronger one, we face danger. If we stand on a strong board and fix a weak one, we could still face danger if, for example, we choose to fix weak boards of no direct importance to the seaworthiness of the vessel and ignore others that might fail catastrophically. We must, like Justice Holmes's judge, act in a way that fulfills several social demands, including the demand that the present decision be both consistent with precedent and legal tradition and also responsive to the new demands of a new situation.

The particular context of a real management dilemma—a context always suffused with value—can be very important for pragmatists in determining which beliefs should be accepted, however provisionally, and which should be submitted to more intense scrutiny by observation and experiment. The necessity of acting—and refraining from action is itself an action—enforces a kind of discipline, a discipline felt in a particular situation with real values at stake. In some situations, for example when the very existence of the community is threatened, decisions can be seen against a backdrop of unquestioned values (community survival); in these situations consensus on values may be far stronger than consensus on science. Epistemological decisions, in situations where decisions are forced and important values are at stake, thus involve judgments of importance as well as truth. We can only examine our whole belief system and try to find some beliefs we can temporarily place beyond doubt. Given the goal of management, we first concentrate on beliefs that are most important to the ongoing voyage, postponing examination of others until later: we keep our ship afloat, gradually transforming it plank by plank. Similarly, adaptive managers sometimes, by hypothesis, help themselves to a platform of beliefs in order to question the goals that should be pursued; and at other times we assume our goals are worthy ones and proceed to test appropriate scientific hypotheses related to the attainment of those goals. Optimistically, the adaptive manager believes that this platform, which shifts over time and in the process of many trials, yields improved understanding and improved goals through an alternation between action and reflection. This may be the only effective way to respond to wicked problems as they arise in a community with diverse and sometimes competing values.

Of course one might object that this whole process is circular and that no "true" justification of goals or actions takes place. We assume facts to support values, and we then stand on the values to support the importance of scientific research to reduce uncertainty and to allow actions to support those values. Now we play our epistemological trump card—the ability of diverse communities, if they operate in an open, democratic mode—to focus attention on weak assumptions and unjustifiable principles. In open public debate and open public processes, when well-informed stakeholders have free access to information and to political institutions, diverse members of a community will have an incentive to identify weaknesses—scientific, economic, and moral—in policies proposed by competing groups. If a process can be created that mimics the process the repairmen on Neurath's boat must develop if they are to survive, then we can give up the dry dock of a priori, self-evident truths and trust science and the observational method, especially if empowered by a strong sense of shared community values, to identify weak planks and keep the boat afloat. So a reasonable way to proceed, in an adaptive management framework, is to inspire stakeholders and participants to challenge and question both the beliefs of science and the proposed goals and values. Democracy, in this sense, can be a powerful engine of truth-seeking. A diverse population, in adaptive management as well as in Darwinian evolution, increases adaptability, by exploring a variety of available options, winnowing out the weak assumptions, and pursuing the most justifiable goals within a particular situation.[1]

Provided Neurath's analogy is apt, we can proceed with our analysis, having established a crucial role for values in our epistemological choices; now we turn our attention to improving our understanding of, and language for describing, environmental values. We want to understand environmental values theoretically. As adaptive managers, however, we are also interested in the way they function in a process of local, community-based experimental management. So far I have emphasized the practical costs of not having at our disposal a coherent and intelligible language, and an associated explanatory theory, for discussing environmental values and policy. These practical difficulties were symbolized by the crooked corridors at EPA; and none of EPA's corridors of communication are more crooked and blocked than those through which information about environmental values and goals should flow.

One important requirement of straightened corridors of communication is the creation of an integrative language that allows cross-disciplinary and cross-interest-group communication. So one task is to develop some clearer ways of talking about environmental values, relating them to the statements of disciplinary and integrative sciences, and—most importantly and most practically—creating an enlightening, integrative discourse about environmental science, values, and policy goals. If we are to go beyond simply improving communication, however, and move toward substantive agreements about what to do to protect resources and live sustainably, we must also provide a theoretical structure that connects the ideal of sustainability to justifiable environmental policy goals that can be operationalized, goals that can be stated and pursued in real-life communities with real-life problems. The purpose of this part of the book is two-fold: to improve our linguistic tools for communication about environmental values and to offer the broad outlines of a positive theory of environmental values.

Pragmatists, from Peirce to Leopold, and adaptive managers are not antitheory; they are, however, very wary of theory cut loose from possible observation. No beliefs are ultimately immune from revision in the face of experience; all theory must sooner or later stand the test of experience, which helps us to separate truth from falsehood and nonsense. This generalization applies to theories of environmental value no less than to empirical hypotheses about causal factors. The goal of such a process is to create theory as a general reflection of experience and to avoid a priori theory invoked to dictate the general shape of any environmental values. By testing proposed theories against their performance in articulating, clarifying, and justifying real environmental goals of real communities, we gradually hone a language that will help communities in the future to ask the right questions and to improve their chances of achieving meaningful improvements in their policies.

5.2 Four Problems of Environmental Values

It is in this democratic, diverse, and contentious world that we must address the problem of environmental values. But what, exactly, *is* the problem of environmental values? On a very general level, the problem is that we have no widely shared language for discussing environmental problems and norms; worse, this linguistic poverty reflects a deeper lack of a unifying theory of environmental values that might make sense of our manifold experience. There are multiple theories of environmental values, some of which invoke values that seem incommensurable with other values. So it might be said that the problem is that there are too many competing and sometimes conflicting values. One strategy—that of theoretically reducing all values to a single principle and measure—has dominated academic and disciplinary discussions of environmental values, including those in environmental ethics. In essence, such reductionistic approaches specify, a priori, that all value must be describable in a single vocabulary. This single, a priori theory then constrains the values that are to count—or at least it constrains the form in which the values that count must be expressed.

An alternative approach, favored here, encourages expression of many values in varied vocabularies as they emerge in real discussions of policy and accepts the challenge to integrate these multiple values into a coherent and acceptable environmental policy. Once we have some experience of this sort, we can start trying to systematize and simplify theory. We might even, someday, succeed in expressing all or most of our environmental values in a single, commensurable framework with a unified vocabulary and thus reduce the chaos in value discussions. But that is only one aspect, really, of "the problem" of environmental values; it would be more accurate to identify several problems about environmental values, all of which emerge in one way or another from our lack of an adequate language and associated theory to make sense of the collection of impulses we call environmental values. Let's begin by separating four possible theoretical questions we might address, and four associated types of theories that might serve as answers to each.

An *ontological theory of value* answers the question, What is the nature of environmental values? Very general theories of this sort are usually the vehicle by which multiple values, as experienced and expressed in actual situations, are sorted and reduced into a small number of ultimate types. Typically, also, an ontological theory of value determines which persons, objects, or processes enjoy "moral standing." One aspect of specifying the nature of value is thus the delimitation of its range of application.

A *theory of measurement of environmental values* answers the question, How should we measure and compare environmental values? Measurement of environmental values, especially as they are understood and projected across time, is a central problem for any theory of sustainability.

An *epistemological theory of environmental value* answers questions such as, How can or should we use environmental values to evaluate and justify particular actions? When so understood, an epistemological theory of environmental values might be very practical and hands-on. More abstractly, we can ask, What counts as an adequate justification for a proposed policy or action?

An epistemological theory of environmental value would thus provide a roadmap toward justifying a given policy proposal, either practically in the particular case or more theoretically when treated more generally.

A *process theory of value* would tell us how to harmonize, or at least balance through a political process, multiple environmental values. Such theories ask the question, How can we choose satisfactory or improved environmental policies, even if participants in a process advocate varied, inconsistent, and sometimes even incommensurable values? Process theories of value are especially important in the (very common) situation in which members of a community have very different environmental values and are still interested in achieving cooperative action.

These problems all have to do with environmental values, so it might reasonably be thought that an ideal theory would simultaneously answer all of these questions in an elegant manner, organizing our conceptualization of environmental valuation in the process. Indeed, some theories of environmental value set out to do exactly that. For example, economic utilitarianism tells us what environmental values are: they are commodity values, understood as reflecting countable units of welfare corresponding to the strength of consumer demand for them. This understanding of the nature of environmental values, in turn, has implications for measurement and justification of environmental values and for the processes of environmental valuation. So in the end we may find that one theory will answer all four questions; but for now it is best to recognize that there are competing theories of environmental values; some of these theories provide answers to only one or two of our questions. In current theorizing, a certain priority is given to ontological theories, because these are considered *dominant* in the sense that answers provided by a given theory to the ontological question entail or at least shape the answers to the other three questions. Thus it may seem as if in current discussions the four questions and types of theories are all mixed together. Let us begin, however, by keeping the questions—and the types of corresponding theories—as separate as possible. Some of the conceptual and empirical relations among them will become clearer as we proceed.

Again, we gain an advantage if we can start any inquiry from a diversity of possibilities, and the mix of problems and theories as listed here offers a variety of choices. Further, by our decision to proceed toward theory within an activist framework of adaptive management, we have provided ourselves a general strategy for choosing among theories of value: we want to choose or develop a theory that will be useful in evaluating the changes in the environment within an adaptive management process. We can ask, then, what types of theories, and which specific theories, are likely to be useful in an action-forced position in which we pursue the goal of adaptive management. More specifically, we can look for theories of environmental value that will be useful in, for example, constructing a theory of sustainability. To this end, let us briefly survey our four general types of theories of value and examine them from the viewpoint of activists who are seeking a theory of environmental values that will be useful in the process of managing adaptively.

Ontological Theories of Value. The word *ontology,* from the Greek *ontologia,* refers to our understanding of being—existence—itself. The Greeks took it that each type of thing has an essence, or true nature—a belief we questioned in chapters 2 and 3. Understanding being in the most general and basic sense would be to grasp the very essence of being. Speaking less abstractly, one important aspect of philosophy, sometimes called ontology, has evolved into the study of the basic types of things that exist. For example, in a materialist ontology, there is only one type of things, material objects: everything that exists, for a materialist, is nothing but a modified and differentiated form of matter. Dualists, taking a different view, insist that nature consists of both body and spirit. We thus apply the term *ontological* to the study of the nature of environmental values, expecting that we will encounter both monists and dualists, as well as pluralists, when philosophers and others theorize about the nature of environmental values. Ontological theories of environmental value thus give an account of the basic types of environmental values, perhaps sorting them into categories or perhaps insisting that all environmental values can be interpreted as being of a single type. Accordingly, we find monistic theories, such as the theory that all environmental values are actually examples of economic, consumer values; and we also find dualistic and pluralistic theories, theories that divide natural values into basic categories and try to relate these categories to each other conceptually.

Even adaptive managers would be pleased if someone provided the "correct" theory of the nature of environmental values. Whatever one's priorities, it is obvious that the development and adoption of a widely accepted ontology of environmental values would be a step forward in our understanding and a huge boon to environmental management. If an ontological theory of environmental value was developed and accepted widely enough, it could be the basis of communal action. Unfortunately, as will become obvious to anyone who examines current discourse about environmental values and policy, such a consensus simply doesn't exist. Many theories are espoused, discussed, ridiculed, and applied willy-nilly. This conceptually and theoretically chaotic situation is not the worst situation in the view of adaptive managers. Sure, they would say, chaos is confusing and uncomfortable, but this very confusion is an incubator for new ideas, compromises, field-testing, and discussion of alternative policies. So although adaptive managers, no less than others, would welcome the simplicity, comfort, and ease of a consensually accepted approach to environmental values and valuation, they tend to prioritize guidance in practical, management controversies and place less emphasis on determining the nature of all environmental value.

Theories of Measurement of Environmental Values. Answers to the measurement question may seem to be just a corollary of answers to the first question, since knowing the nature of environmental values is certainly one major aspect of measuring them, but it is nevertheless best to keep these two questions separate for now. Keeping them separate allows us to explore the possibility that we might measure environmental values while making no or very few ontological assumptions. It would be a great advantage to adaptive managers if we could separate these two questions by devising ways of measuring and quantifying environmental values that are relatively independent of contentious and intractable issues about the ultimate nature of environmental values. But what is undeniable is the importance of any methodological decisions regarding how environmental values are measured and, once measured, aggregated and brought to bear upon policy proposals.

Epistemological Theories of Environmental Values. We can also ask what role might be played by a strong and effective *epistemological* theory that explains how environmental values might be justified and how they might thereby be given importance in political processes. Again, it might seem that the procedures for justifying environmental goals and values would be largely determined by the ontological theory one espouses; but we should not so quickly submerge questions of justification in questions of ontology, because our goal is to leave all beliefs open to disconfirmation by experience at some point in our ongoing, experimental process. When pronouncements about how to justify environmental policies have been derived from general, ontological statements of the nature of environmental values, they have not proved very helpful because of the particularity of each community and its problems, as it faces a unique set of circumstances and attempts to chart a reasonable policy course in unique local situations. What might be useful to environmental activists would be a more systematic, empirically based *procedure* for helping communities to articulate and defend their values in public processes that involve real communication and social learning. The goal of an epistemology of environmental values, then, would be to provide empirically based procedures that encourage the articulation, criticism, and revision of the community's values. Such a procedure might serve, within the broader context of an ongoing community process, as an acceptable *process* of public justification for management and policy goals.

Process Theories of Environmental Value. It may turn out that, speaking from the viewpoint of adaptive managers, the most useful theory of environmental values is not based upon a general theory about the nature of environmental value, but rather upon a theory of process. If we assume that environmental policy is to emerge from a democratic political process, and if we further assume that we live in a diverse society—which for the foreseeable future will continue to be diverse—then a theory about the process of environmental valuation in public contexts, based on observation of many local cases, is likely to be the most useful kind of theory for guiding adaptive management. If we follow Rittel and Webber in considering environmental problems to be wicked problems, problems of balancing multiple legitimate goods, it seems likely that adaptive managers will have to work within a process rather than trying, as Muir tried, to associate one side in such debates with God and the other with Mammon. A process theory could enrich adaptive management by describing how communities, despite diversity of values and interests, can move toward consensus and cooperation in the pursuit of environmental policy goals. For this, one needs a pluralist approach with policy goals and concrete objectives that are robust across multiple values, rather than reduction and aggregation of a single kind of value. The result will be a different dynamic of policy discourse and formation than the one that a reductionistic approach would produce.

At first glance it may seem as if I am proposing something contradictory, almost paradoxical: I argue that diversity of evaluative opinion in a community is conducive to agreement, but what is diversity of values if not disagreement? To see the way out of this paradox, it is useful to rely upon a distinction between environmental values, which include items such as enjoyment of an unspoiled landscape or appreciation of the importance of water quality, on the one hand, and, on the other hand, environmental objectives, which include specific goals and policies such as reducing the nutrients entering a river by 33 percent over a decade or instituting a policy of maintaining buffer zones with native vegetation along streams. This distinction allows us to separate actions that might be taken from the values that are invoked to justify those actions.[2] It can be used to explain the importance of political process in management efforts. Finding better policies is often finding policies and management objectives that support multiple values. We saw a classic case of this phenomenon when Ducks Unlimited and other hunting organizations joined forces with environmental groups such as Audubon to support, and eventually gain passage of, legislation protecting wetlands in the major flyways for migratory waterfowl in North America. Here we have an example of groups with diverse values backing similar policies because habitat protection for waterfowl was essential to the goals of both. If we expect pluralism and diversity and therefore have problems agreeing upon the nature of environmental values, it is often helpful to shift the focus to questions of what to do, trying to push forward the action phase, because there may be policies that will support multiple values in the given situation—what we have called robust policies.

Since we start with the goal of representing the diverse values of many interest groups and stakeholders, we seek a more comprehensive and inclusive conception of environmental values. Is this inclusiveness in direct conflict with the efforts of theorists who start by asserting a particular monism, or dualism, of values? In one sense it is, since we pragmatists eschew a priori judgments that there can be only one, two, or *N* ultimate types of value; but in another sense it is not, because we do not reject, a priori, the hypothesis that values can be reduced to one or more types through a process of clarification, public participation, and social learning. That kind of reduction would be an endpoint of a long process, not an ex cathedra announcement at the outset of the process.

Rather than starting from an ontology of environmental values and deriving guidance to be applied in particular situations, I propose to examine the problem of policy evaluation from the viewpoint of particular real cases. This strategy is not a counsel of confusion; we simply assume that there will be diverse values advocated, and we put our initial energies into a gradual and empirically based attempt at clarification and integration of multiple values, expending less effort on attempts to systematize and simplify value discourse by abstract analysis.

Our embrace of pluralism, based on a reasonable assessment of the current situation regarding discourse about environmental values, does not, then, challenge traditional, reductionistic theories directly by offering at the outset a final alternative to the reductionists' ontological theories. Indeed, one might say that what is proposed involves a shift in the central question to be addressed in environmental value studies, a shift from What is the correct theory of environmental values? to What is the most appropriate *working hypothesis* regarding environmental values? By pushing questions of the single, true nature of all value to the background temporarily, we adopt experimental pluralism as our method. We hypothesize that there exist in the society multiple kinds of environmental values and multiple vocabularies for expressing them.

If we are to understand the current situation in the discourse about environmental values, it is important to learn what we can from extant and popular theories that have already been proposed by philosophers, economists, and others. For one thing, various theories about the nature of value, as they compete in a marketplace of ideas to define, explain, and categorize environmental values, do at least sharpen our concepts and clarify the value claims made by advocates of various values. These discussions, though no longer considered so central to policy decision-making processes, can be useful in the reflective phase, as we attempt to clarify and integrate multiple values. Prior theories are thus not excluded from our discussion. Such theories are viewed as objects of study and as possible outcomes of a long process, rather than as prior commitments that shape our experience of how values can and do affect day-to-day decisions.

In the next two sections we will examine the only two extant theories that claim to be comprehensive and unified theories of environmental value. More will be said about these characteristics of comprehensiveness and unity presently; first we must introduce these two dominant theories. In the remainder of this chapter we shall look carefully at how the two main theories of environmental value have dominated the disciplinary and interdisciplinary discussions of the nature of environmental values over the past few decades. We will call these Economism and Intrinsic Value theory (IV theory, for short). We adopt the convention of capitalizing their names to make it clear that each is, indeed, the name of a type of theory and should not be confused with a particular discipline, even if that discipline has been the home base in which the theory has been developed. Economism can be defined very briefly as the theory that all environmental value is one kind of consumer value among other consumer values, to be compared and balanced against other purchases that might be made with the consumer's presumably limited economic resources. As its name suggests, Economism is espoused by many economists and by others as well. Importantly, however, not all economists accept Economism, the view that all environmental values are measurable as consumptive values.

Only IV theory, the theory that environmental values are to be understood as values intrinsic to nature itself—values that exist independently of human values—represents an alternative theory of competing dominance, capable of informing the answers to the remaining three questions and rivaling Economism for its generality and power. According to this theory, nonhuman elements of nature, no less than human individuals, have intrinsic value. IV theory essentially achieves a monistic and universal theory of value by *extending* human ethics to include the nonhuman world or some elements of it. Such a theory, if it was accepted as the correct theory regarding the nature of environmental value, would, like Economism, resolve many aspects of our three remaining questions, which concern epistemology, measurement, and process.

I have identified these two competing theories as deserving of special consideration because of their importance in the current debate. Speaking technically, we can refer to these theories as claiming both *comprehensiveness,* or *completeness* (in the vernacular of symbolic logicians), and also the highest degree of *connectedness* (also known as *elegance*). To say that a theory of environmental values is comprehensive is to say that it has sufficient semantic power to describe each and every type of environmental value that evaluators legitimately express. The Economists are thus committed to saying that if some proposed value cannot be expressed within the economistic concepts and principles characteristic of their theory, then it is no environmental value at all. The theory of Economism is thus claimed to be complete with respect to the domain of environmental values. To say that a theory is highly connected is to say that all of the types of values can be expressed and asserted by appeal to a very small number of basic principles. The limiting case of connectedness is monism: in monistic theories, the entire subject matter recognized by the theory can be expressed in a single vocabulary and derived from a single unifying principle.

It is interesting that these two theories, Economism and IV theory, espouse both comprehensiveness and connectivity, in that the two goals for theory exist in a certain unavoidable tension. If, for example, Economists insist upon comprehensiveness for their theory of value, then the task of achieving a highly connected theory is made more difficult: the single principle that connects all judgments of environmental value must cover all kinds of legitimate environmental values. If connectivity was reduced, and more principles were introduced to account for various types of values, the task of capturing every type of environmental value within the network of the theory would be that much easier.

Economism and IV theory, as noted, dominate discussions of environmental value. One of the interesting features of this conversation, however, is that these two theories of value have been constructed in differing academic disciplines, with differing vocabularies and articulations of values and ethical concepts. Economism was developed in environmental economics, whereas IV theory was constructed in the emerging field of environmental ethics. So an attempt to describe and evaluate these two theories can double as an introduction to two key theoretical fields that have dominated discourse about environmental values for the past few decades, environmental ethics and environmental economics. Accordingly, the next two sections are devoted, respectively, to recent debates in environmental ethics and in environmental economics.

5.3 Where We Are: A Beginning-of-the-Century Look at Environmental Ethics

Environmental ethics has two histories. If an ethic is understood inclusively as a collection of attitudes and principles that can guide human actions, the concept has a long history that goes back to Genesis 1, and even further back into the oral traditions that shaped that document. In this inclusive sense, environmental ethics has been with us in some form since the beginnings of human consciousness. Chapter 1 of Genesis conveys an environmental ethic in this broad sense by articulating the idea of human dominion, including the view that humans are made in the image of God, and incorporates the ethic of human dominion over nature as an integral part of the Judeo-Christian creation story. Chapter 2 of Genesis adds an obligation to act as stewards of God's creation, an idea that is elaborated elsewhere throughout the Scriptures; so there may be not a single environmental ethic in the tradition, but several. In this broad historical sense, many environmental ethics can be found in literature, myth, and religion.

Here we are concerned mainly with a narrower sense of environmental ethics—of environmental ethics as a practice of professional philosophers—and in this sense, environmental ethics began not so long ago. Its origin is sometimes pinpointed as 1973–74, a short period when there were several articles and one important book published on the subject of environmental values. A prominent event in the narrower and briefer history of environmental ethics as a subdiscipline of philosophy was the initiation of a journal in 1979; subsequently, interest in the field grew and more publications appeared, books and periodicals exploring environmental values.

This fledgling field has concentrated mainly on questions of the ontology of environmental values. Perhaps it is not surprising that ontological theories of value were among the earliest topics addressed by environmental ethicists. What is surprising is that the field continues to be dominated by these questions, despite the existence of a number of alternative approaches to understanding environmental values, and despite the extreme difficulty of, and lack of progress in solving, these deep ontological issues. As a result of their preoccupation with such recalcitrant problems, environmental ethicists have made few contributions to discussions about what to actually do to improve the environment. The specialized nature of philosophers' concerns can be appreciated if we look in detail at the catalyst that generated the subdiscipline of environmental ethics, a 1967 essay by a historian, published in *Science* magazine. This article, by Lynn White Jr., was called "The Historical Roots of Our Ecologic Crisis,"[3] and it has profoundly shaped the philosophical debate about environmental values ever since.

White offered a broad-brush historical account of the ideas and social forces—science, technology, and especially the melding of the two—that have shaped Western culture's view of the human relationship to nature and suggested that these features may be responsible for the degradation of modern environments in the West. White offered several criticisms of Western ideas and culture, including, for example, a brief reference to the conception of time as directional, a Christian idea absorbed from the Hebraic tradition that saw creation as a beginning of history, which also was to have an end. White suggested that this linear conception of time, which differs from the Greek conception of nature as cyclical, with no beginning or ending, has instilled in Western consciousness a directionality and a sense of purpose—and also a form of unjustified optimism that treats all technological change as progress. Westerners, he seemed to be saying, are poor critics of technological proposals because we tend to be technological optimists by default. If developed, this line of reasoning might have led the study of environmental values toward the evaluation of technologies and their long-term and unintended effects; and environmental ethics might today seem more like a specialized form of the philosophy of technology.

This line of reasoning, however, has not been important to most environmental ethicists, who instead picked up on another line of criticism, White's statement that Western Christianity "is the most anthropocentric religion the world has seen." White was of course referring to the creation story in Genesis 1. He notes especially the claim that humans are made in the image of God and in an important sense are separate from nature. It was the charge that Western culture is "anthropocentric" that came to occupy the pages of the journal *Environmental Ethics*. The first dozen or so papers and books that are clearly within the professional philosophical tradition responded in one way or another to White's criticism. Indeed, most environmental ethicists took White's criticism to be valid and compelling and proceeded to propose "nonanthropocentric" ethical positions as an antidote. These proposals proclaimed that nonhuman elements of the environment—elements of many different types, from individual animals to species and ecosystems—had intrinsic value; and hence these elements should have moral considerability in human decision making. Minority forces—in particular, the Australian philosopher John Passmore—argued to the contrary that the Western tradition has adequate intellectual and moral resources to criticize and reform environmental practices, that there are good human reasons to change currently destructive practices, and that introduction of non-Western and nonanthropocentric ethical principles is unnecessary to correct environmentally damaging behaviors. Passmore therefore rejected nonanthropocentric ethical theory as inconsistent with central moral principles of Western social thought and unnecessary to support improved environmental policies.[4] Much of the writing on environmental ethics since those early days has addressed this issue of anthropocentrism in one way or another.

The focus of environmental ethicists on anthropocentrism versus nonanthropocentrism has interacted with disciplinary boundaries to create an unfortunate impasse. Since economics, especially as represented by its mainstream interpretation as a science concerned with human welfare, offers a ready-at-hand example of an unapologetically anthropocentric discipline, environmental ethics took shape as a discipline defined by its opposition to economics as a means of evaluating nature. The result was polarization along disciplinary lines, with environmental ethicists developing arguments showing the inadequacy of environmental economics as an ontology of environmental values, but stating these arguments in a vocabulary that made it impossible for the arguments to have any effect on economists. Meanwhile, environmental economists either dismiss environmental ethicists as talking nonsense, or they incorporate various of environmental ethicists' ideas into their own valuations, treating environmental ethicists' assertions of intrinsic value, offered as assertions of a new ontology of natural value, as dollars-worth of "willingness to pay" to protect "existence" values. The two disciplines most likely to clarify how we value nature are thus trapped in a bifurcated discourse, each rejecting the other's ontological position but doing so in a language that is not understandable by—and has no standing for—their real opponents. The result is that each of the two disciplines can often present a united front against the other, but this united front simply reflects the fact that their real differences are ossified into different languages, within which they can communicate with their allies but cannot even understand the arguments of their opponents. They exist within disciplinary towers.

Was this polarization inevitable? No. Environmental ethicists, as noted above, could have developed White's concerns about our technological optimism. Was the polarization inevitable once White had introduced the label "anthropocentric" and made his charge of anthropocentrism against Western culture? No again. With hindsight, we can see that White's charge of anthropocentrism can be given two rather different interpretations. White might have meant, as environmental ethicists have taken him to mean, that anthropocentrism is an *ontological theory of environmental value:* the theory that all and only human beings have intrinsic value and are morally considerable. But White never says this; in fact, the only explanation he gives of the term is by reference to the doctrine that humans are made in the image of God. White could as well be interpreted not as criticizing a *theory* of environmental value but rather as criticizing an *attitude* of human-centeredness, a kind of hubris about the importance of humans in the larger scheme of things. The charge of hubris ties nicely to White's other concerns about our optimism regarding technology and cultural progress and requires no positing of intrinsic values in nature. Although clearly anthropocentric in their beliefs, the ancient Greeks easily found the moral resources—in epic poetry, theatrical tragedies, and Aristotle's ethics—to criticize hubris, understood as the temptation of humans to usurp the role of the gods. One need not extend moral citizenship beyond the bounds of humanity to criticize hubris: it is a human failing with human consequences. If one interprets White in the second way, no particular antidote for hubris, theoretical or otherwise, is predetermined; and under this reading, agreement with him requires acceptance of no particular ontological theory of environmental value. Anthropocentrism, in this attitudinal sense, can be rejected without embracing a theory that opposes it.

Early environmental ethicists interpreted White's argument as an attack on a *false theory,* the theory that (all and) only human beings have intrinsic value. It is not surprising that once the question was posed in this unfortunate manner, environmental ethicists responded with an alternative, competing moral theory, claiming that some (or many) nonhuman entities have intrinsic value. Unfortunately, this formulation of the problem of environmental valuation leads to polarized rhetoric and plays into the hands of the ideologists. Communication broke down as Economists and IV theorists, both hung up on the ontological question, could not agree about where to draw the line between entities that are intrinsically valuable and those that are merely instrumentally valuable. Polarization of the public discourse about environmental values was thus engendered at the very inception of the field of environmental ethics as a "distinct" discipline.[5] From that starting point, the antipathy of environmental philosophers for economic valuation was inevitable.

5.4 Economism as an Ontological Theory

Let us next examine the main tenets of Economism, which here refers to a theory held by many economists, but by no means all, that all important environmental values can be understood as economic values. These are the values expressed by human consumers as they participate in economic exchanges. According to this theory, environmental values and goals must compete with other good causes for effort and investment, and environmental goods and bads should be understood in market terms—the aggregated "willingness to pay" of consumers for discrete goods and ecosystem services. I do not follow most environmental ethicists in rejecting economics as a basis for environmental valuation; rather, I believe economics offers one important perspective on environmental values. Indeed, I believe the economists' emphasis on trade-offs and the importance of making good investments in environmental protection—and making these by comparison to other investments the society might make in education, infrastructure, and so forth—is an important part of environmental policy analysis and decision making. My target, rather, is the general claim that economic valuation can be *monistic* (interpreting all environmental goods as economic consumer goods) and at the same time provide a *comprehensive* treatment of environmental values.

One area of special concern is whether economic valuation can accurately characterize future values in terms of current consumer behavior; this problem of assessing long-term, multigenerational impacts of our actions, especially impacts that will mainly be felt by future generations, is so important that all of chapter 8 is devoted to better understanding it. Long-term impacts are only one example of values that are difficult to characterize in purely economic terms; in this section I will emphasize other important environmental values that apparently defy economic interpretation.

A central battleground in environmental economics is the status of contingent valuation (CV), the methodology that uses questionnaires, bidding games, and other interactions with respondents to assign economic value to increments and decrements of various environmental goods. Much has been written about the reliability and theoretical viability of quantifications of individual willingness to pay (WTP) and about our ability to assign reliable dollar values to environmental goods, especially to nonmarket environmental goods. Our discussion in this section, however, will not turn on the reliability of CV studies, but rather on their "scope," or "power." We are specifically examining the question of whether monistic Economism can plausibly claim to offer a *comprehensive* theory of environmental value. It turns out that any claim of comprehensiveness for economic values puts a huge weight on the practice, and the theory, of contingent valuation. To evaluate the comprehensiveness of Economism, then, it is crucial to understand the strategic role CV plays in discussions of how to evaluate policies to improve the environment.

Here's the nub of the matter. If one adopts Economism, and argues that economic valuation provides a comprehensive approach to environmental valuation, one must believe that there exists, in principle, an adequate economic marker for each significant change in social value that results from a change in the human environment. Even if one were to take lavish refuge in the "in principle" caveat, one would apparently be obliged to provide some account of how unquestioned social values will be given economic expression in terms of market values—that is, in terms of the willingness of individual consumers to pay to effect or to avoid a given change. For the doctrinaire Economist, this means either one can provide a means to measure such values in terms of WTP, or one can claim that the value is simply not a value worth measuring. The plausibility of Economism depends, then, on being able at least to perform hand-waving ceremonies with respect to how an economist might proceed to "measure" WTP for any unchallenged environmental value.

The dilemma derives from special aspects of the problem of measuring environmental values. Speaking strictly, value, in the economic perspective, is evaluated at that point at which a transaction or exchange takes place—when consumers exchange something of value that they own for something they would like to own—because it is at this point that the consumer's actual behavioral tendency to spend scarce resources can be measured. The problem is that most environmental goods are public goods, and hence they are not, strictly speaking, "owned" or routinely traded in markets. True ownership, by definition, includes the right of an owner to exclude others from the use of an owned good. But public goods, such as enjoyment of a state park or clean air to breathe, fail this requirement of exclusion and so are not tradable in real markets. The dilemma, then, is whether to be scientific, behavioristic, economic purists, measuring actual *observed* market behavior, or whether—in the face of so many intuitively perceived nonmarket values—to relax the behavioral criteria and allow supplemental empirical information gathered from questionnaires and other such means to measure shadow prices and allow these to count as measurable economic behavior.

CV was first suggested in the early 1960s by Robert K. Davis, who used questionnaires to estimate some nonuse values of outdoor recreation.[6] Several initial efforts to identify shadow prices by using hypothetical markets were tested in the 1970s. Since then the practice has flourished, and literally thousands of such studies have been undertaken. According to Robert C. Mitchell and Richard T. Carson, "the CV method uses survey questions to elicit people's preferences for public goods by finding out what they would be willing to pay for specified improvements in them. The method is thus aimed at eliciting their willingness to pay in dollar amounts."[7] Applications of the method typically involve three elements: First, researchers provide background information and a careful description of the good or goods to be valued. They then construct a hypothetical market in the form of a plausible scenario in which respondents face a choice regarding a well-defined "commodity"—such as reduced risk of cancer from exposure to a chemical in drinking water or an increase in visibility in a national park. Second, direct questions (such as how much they would agree to pay as additional charges on their water bill) are asked, or bidding games are developed, in order to ascertain how respondents assess the dollar value of the hypothetical good. Finally, respondents are asked about personal characteristics such as age, income level, number of years in the community, and so forth, as well as more pertinent questions about how the good in question affects them in their own lives. Survey researchers believe that if sampling techniques are adequate, information from such studies can be generalized to the population in question, providing a measurement of the willingness of a population to pay for a good—a public good—that cannot be "owned" and for which there are, accordingly, no markets to observe directly.[8]

For the purist, shadow prices—these hypothetical representations of the willingness of consumers to pay—are at best pale images of true economic exchange behavior. For advocates of contingent valuation, shadow prices—provided the research methods of the contingent evaluator fulfill requirements of good social science methodology—are legitimate estimates of economic value; they measure what consumers would be willing to pay *if* there were a market for the good in question.

This dilemma, which might be considered simply an interesting methodological question for those who see economics as one of many approaches to assigning value, becomes absolutely crucial for those (Economists) who claim that economics is, even in principle, a *comprehensive* means of characterizing and measuring environmental values. Contingent valuation is the only known method for assigning market values to nonuse values and to many public goods; and, as noted, most environmental goods have a nonexclusive, public aspect. Therefore, if Economists wish to maintain *both* monism (the view that environmental values can be understood as economic values) *and* comprehensiveness (the view that economics can capture *all* environmental values that exist), then they must provide some economic method for capturing each and every type of environmental value. They must, for every environmental value cited, either provide a method for measuring that value or deny that the value proposed is in fact a "true" value. The monistic Economist must either use contingent valuation techniques to place a value on environmental goods that are public goods or "reduce" environmental goods to those few environmental values that can be exclusively owned. This would mean, for example, that if I own a patch of old-growth forest that happens to protect the headwaters of an important river, the standing trees (which can be cut down and sold) will be given a value; but the value, in improving river quality, of those trees and the undergrowth they protect must be treated as worthless (in narrow, economic terms). Again, this dilemma arises *not* from placing economic values on *some* environmental goods, but from the insistence of Economists on *both* the monism *and* the comprehensiveness of their value measures. Monistic theories of environmental value can achieve comprehensiveness only if their basic value categories are exhaustive of all legitimate social values, and it seems undeniable that there are nonuse environmental values that are not traded in markets—so the Economist is apparently driven to embrace CV as essential to the claim of comprehensiveness.

Herein lies another story about my adventures at EPA. In chapter 1 I described how I needed to be bilingual—fluent in both economese and toxicologese—as I shuffled back and forth between different boards and panels at EPA. We can now use some of my painful lessons in the conceptual corridors of EPA as means to illustrate the current poverty of interdisciplinary discourse about environmental values. The setting was the early 1990s, and I was serving on the first-ever Environmental Economics Advisory Committee of the Science Advisory Board. We met two or three times per year to examine the use of economic science at EPA; topics covered were sometimes raised by us, sometimes posed to us by administrators of the Science Advisory Board, and sometimes formulated by individuals in various program offices.

It is an experience for which I will always be grateful—especially to Allen Kneese and V. Kerry Smith, giants in the field of environmental economics, who were also interested in broader issues of value. I am grateful to them because they were very nice to me as an "outsider" to their discipline and because I received an education in environmental policy formation that was not available at even the best university. Since I worked with them as an EPA consultant, I was even paid (a little) to boot! As the only noneconomist regular member of the committee, I tried to listen and learn and also comment as an outsider on various issues. Most of the time the proceedings were amicable and relaxed; and discussion of policy was at a high level, often higher than my head, but I tried to contribute as best I could, and my fellow committee members were unfailingly gracious and supportive. In short, it was a great atmosphere for learning about environmental economics.

But a few meetings were different. As I entered the meeting room one day, I noticed the atmosphere was tense, and it turned out that the stakes were high, very high, and the top industry lobbyists and staffers were all there. The subject hardly seemed earthshaking: a contingent valuation study, commissioned by the Office of Solid Waste, to determine the nonuse value of a proposed groundwater cleanup.[9] I was impressed, however, when I heard that Exxon's top environmental attorney was there; he sat quietly almost all day in order to speak for five minutes or so in the late afternoon. I was also impressed that he provided us with a manuscript, under his authorship, impeccably referenced, on the subject of the unreliability of measurements of nonuse values. It turned out that it was more the context than the content of EPA's study that mattered—timing was of the essence. At that very moment, the Exxon Corporation was fighting a case in the courts in which they subsequently were assessed over $5 billion in damages as a result of the Exxon Valdez oil spill, and many of these dollars were tallied on the basis of contingent value study estimates.[10] Even a philosopher knows that's a lot of money, and I noticed my palms were a little sweaty. I had a front-row seat at the prizefight. The subject was science—economic science; but it was science that was politically hot, because it was science that mattered outside the academy, to the tune of billions.

We spent several days discussing and critiquing the study, with the top experts in contingent valuation, environmental economics, and policy allowed to make presentations and answer questions. It was all right here. The nexus of science and policy at EPA. What do economists do when stakes are in the billions? They argue, just like philosophers. So this study was taken apart, criticized, and evaluated, as we heard "testimony" by top economists and survey researchers; the intense examination went on for several daylong sessions, interspersed with revisions, and then more hearings were conducted.

The study itself had a modest beginning. A midlevel manager in EPA's Office of Solid Waste, who had the responsibility to assign values to the costs and benefits of all agency projects as a part of a cost-benefit analysis, felt it was important to assign some dollar value to nonuse as well as use values if the EPA Office of Solid Waste was to make intelligent decisions regarding when to clean up Superfund sites in various parts of the country. Under pressure from Congress to conduct cost-benefit analyses of both their activities and their regulatory policies, bureaucrats suffered from what might be called missing-number angst. Having been told that it was their job to provide complete accountings of the costs and benefits of a policy, but having no obvious means to measure the actual benefits perceived by citizens, bureaucrats reasonably concluded that they should learn how to measure the various kinds of values economists list as candidate values for a cost-benefit analysis. The question that apparently motivated this study was whether, once all of the use values of cleaning up a contaminated groundwater source were counted, there might also be nonuse values that should be added to the benefits side of the ledger when EPA contemplated cleaning up a Superfund site that involved groundwater contamination. Although the subject seemed both a little contrived and unlikely to seem interesting to anyone but an economist or a bureaucrat, it apparently was not considered uninteresting to Exxon. So, during lulls, I amused myself by trying to figure out how much it cost Exxon to have their top environmental attorney, and his entourage of henchpeople, there for the day.

Despite the charged atmosphere and the high stakes, the experts chosen by the EEAC told it like it was, as far as I could determine. They were sympathetic to the effort, but most of them complained that the "commodity" that was presented to the respondents was not sufficiently clear to allow unambiguous interpretation of the responses to the questionnaire; therefore, it was argued by the leaders in the practice of contingent valuation, the dollar amounts representing the WTP could not be accurately attributed to nonuse values. I guess the clear implication was that the manager should not put those dollar values on a line in the cost-benefit analyses (CBAs) for projects he hoped to fund. The scientific process of peer review proceeded, and the work that had been commissioned for a quite practical purpose by one of the program offices of the agency was systematically shredded by the leading experts in the field. Despite a large price tag and a cast of thousands, including some of the most respected practitioners of the art of contingent valuation, the investigators could not defend the accuracy, or even the meaning, of their data to their peers. The study, I'm afraid, was more or less pronounced dead-on-arrival as a scientific study, but I learned a lot at the three-day autopsy. One thing I learned was that WTP estimates of nonuse benefits can be highly controversial, with experts deeply divided about how to count them and how to interpret them once they are counted, even when they have been done with adequate funding according to state-of-the-art methodology by highly respected researchers. I'm afraid that if I had been a judge of the prizefight, I'd have had to rule that Exxon's attorney had won, having scoring a TKO when, in his brief statement during the public comment period, he claimed that the very idea of "pricing" nonuse values was impossible to operationalize and meaningless. He didn't even have to state his real conclusion aloud: CV studies never should be used to assess damages against unfortunate oil companies whose tankers run aground.

But let me quickly balance this picture. If we simply declare Exxon and its high-priced attorneys the winners and thereby banish CV from courtrooms and policy evaluation processes—while holding to the beliefs of Economists—we would in essence eliminate all damage assessments for dead seabirds, for reduced ecological functioning, and for many of the other damages to Prince William Sound. Exxon would be held responsible only for measurable damages to economic productivity, as measured by drops in the take of fisheries and for impacts on tourism. Critics of economics should be very reluctant to concede this much. The bad showing of CV in this particular case should not be taken as sufficient reason to give up on the technique, but the example highlights the general contours of the argument. Commercial interests will often wish to minimize or reduce the value of environmental goods (especially when they expect to have to pay for having destroyed them). So it is not surprising that Exxon sent its best legal team and that the team attacked CV. The problem is that if impartial Economists hope to stand up against Exxon, insisting that there are nonmarket values, then they are stuck with this poorly understood methodology of CV as their only means to express nonuse values as economic values. Again, context affects the quality of science, and even the experts agreed that the way the question had been posed by EPA left the researchers with a virtually impossible task.

What I found most revealing, in listening to the several days of expert testimony, was the disparity between two explanations of the goals for the research: the one given by the EPA manager who had commissioned the study and the one given by the principal investigator who led the team that had performed it. The manager, who impressed me as conscientious and feeling somewhat persecuted, explained that he felt pressure to support Superfund and other cleanup activities with CBAs that were complete as well as accurate. They had done pretty well, he thought, in detailing the various reductions in health risks that would obtain by cleaning up the toxins in groundwater; and he felt, since nonuse benefits were one possible kind of benefit, that he needed some research to determine whether such benefits existed and how large a figure should be recorded for them in the case of one particular cleanup. He said he thought that commissioning such a study would help him in other cases where he had to make decisions about where to invest cleanup dollars. He had commissioned a study to determine what number to put on a line in a CBA, and now he was caught in a cross fire between scientists and Exxon's top legal gun, feeling very much—I suspect—like a perforated bureaucrat. I think he was justified in his attitude. He could say, unapologetically, "It's in my job description; I was just doing my job, fulfilling a presidentially mandated requirement that a complete cost-benefit accounting be done prior to the promulgation of regulations and undertaking cleanup efforts."

The principal investigator described the research and its goals quite differently. He said that when he learned of the possibility of doing CV research on the topic, he immediately realized the difficulty of the task involved and seriously considered not to pursue it. It turns out that Robert Mitchell and Richard Carson, authors of perhaps the most respected book on the methods and limits of CV, had already attempted a study of nonuse values attributed by respondents to groundwater cleanups, only to suffer "scenario rejection" whenever they tried to instruct their respondents to state only WTP values that were in no way connected to present or future use. The respondents could not accept the assumption that the water, once cleaned, would not be used. The problem, in other words, was that ordinary respondents did not have a commonsense grasp of the theoretical distinctions between use and nonuse values in cases like this, even though these theoretical distinctions had been embodied in the bureaucratic template of cost-benefit analyses. If bureaucrats and economists cannot manipulate their respondents from the public into parsing their view of benefits as economic theory requires, there will be—God forbid—an empty box in a supposedly comprehensive cost-benefit analysis, resulting in missing-number angst.

The framework used for cost-benefit accounting presupposes that WTP can be understood as an aggregation of individual willingness to pay for multiple packaged commodities that might be derived from the particular good. But in this case, as Mitchell and Carson had already established, the public did not have ready-made preferences that corresponded to the theoretical categories of economists. This was crucial because if the respondents included willingness to pay for such use benefits as reducing health risk to drinkers of the water, the data would involve double counting and would be misleading.

To their credit, the researchers foresaw the difficulty and, based on the failure of the Mitchell and Carson attempt, chose another approach. Rather than try to create a scenario in which groundwater would be cleaned up but never used—which made no sense to respondents—they asked their respondents to first report their total WTP for the cleanup; then they were asked to partition this total value into use and nonuse values, both by responding to direct questions and by a method of "scenario differences," which required that respondents rank-order various outcomes. Unfortunately, the researchers found that once respondents had chosen a favored outcome, they rejected all others as infeasible, thus making the scenario-differences approach useless. As noted, the researchers recognized the difficulty with the project from the start; in a supplemental memorandum written and submitted to the EEAC between the first and second hearings on the CV study, they stated that they had concluded during pretesting that "the 'commodity' [of nonuse value] as economists would term it was, of course, nothing of the sort in the minds of respondents," because "respondents [had a tendency] to override assumptions made in the design of the survey instrument with their own beliefs if those beliefs contradicted those implicit in the survey instrument."[11] In the end, the researchers resorted to "verbal protocols," which are informal comments and explanations added by respondents to their WTP specifications. The researchers had found it necessary to adjust WTP figures in light of these verbal cues; it is not surprising that expert readers of the report showed considerable skepticism that the study provided a clear indication of nonuse values, since interpretation of verbal protocols clearly involved subjective judgment by the researchers as they collated data.

The difference in perspective between the bureaucrat who commissioned the study and the researchers who carried it out was that the latter apparently saw the difficulty of the task as an intellectual challenge. The principal investigator saw the contract as a chance, he said, to test the limits of the CV methodology. I could see his point, and as a professor in a large university that expects professors to do funded research, I could also assume he was legitimately concerned to raise funds for his university, support graduate students, and teach them by involving them in cutting-edge research.

Notice the difference this makes in the attitude of the investigator compared with that of the bureaucrat, however. For the investigator, failure could still be interpreted as success. If one's goal is to learn both the possibilities and the limitations of a CV methodology, then attempting daring experiments at the edge of current methods and techniques is a sign of good scholarship. If the study fails because of bad design, this fact will be pointed out in the literature, warning future users of CV techniques to avoid these practices; or if the scientific community accepts the methodology used as state-of-the-art and still finds the numbers generated unreliable, the study will have shown the limitations of the central distinction between "use" and "nonuse" values. These outcomes would both contribute to the advance of CV science. In contrast, the bureaucrat was left, after the study had been executed and autopsied by experts, in a charnel house. If the numbers have been publicly argued to be unreliable by the top experts in the field, only a brave, or brazenly foolish, bureaucrat will enter them on a line in a CBA. Whatever comfort there is in being able to cite numbers is lost once the numbers are suspect. A bureaucrat will hardly see it as consolation that his office has contributed to the development of the CV method by paying top researchers to undertake a nearly impossible task that he understood to be required by his job and by a combination of governmental mandates and entrenched economic theory at EPA.

We can understand the differences in the attitudes of the investigator and the bureaucrat a little better, I think, if we return to the distinction between curiosity-driven, supply science, and postnormal, or demand science (see section 3.5). On the one hand, the bureaucrat was asking a particular question and needed an answer tailored to a very particular policy context: he had a need (demand) for a study that would place a dollar value on a particular "commodity." The investigator, on the other hand, could supply a model and numbers, according to the methods and concepts that existed in his field. The result in this case was, arguably, acceptable to the supplier: the university was paid for the study and publications could result. But from the point of view of the bureaucrat, and in terms of protecting and enhancing public values—the demand side—it was difficult to identify any value derived from the costly study.

There is one other contextual aspect that affected the situation at these hearings, but how greatly, and what exactly was its effect, is difficult to say. Because we were working subsequent to the Exxon oil spill, a huge amount of research was supported by all sides in the conflict. Much of this research attempted to assign particular dollar figures to losses of nonuse values, and some was also directed at assessing the reliability of figures representing losses in nonuse values by residents of Prince William Sound. But this research—which had drawn most of the top CV practitioners into consultation with either Exxon or one or another of the litigants who sought compensation for damages, such as fishermen, state government, and Native American tribes—was generated in the process of litigation. At that time, and to this day, most of these studies have not been published because of their proprietary nature. The owners of the studies are thus free to use favorable findings, and presumably they are also free to hide data unfavorable to their case in court proceedings. With the best practitioners drawn into secretive research, and attention turned to damage assessments in a court of law, our committee meetings represented an opportunity for litigants on all sides to make their points about CV in a public forum. In effect, our public committee meeting and the sparring about this particular CV were only the visible tip of a litigious iceberg.

Again, it wasn't that difficult to see where the players were coming from; and in fact the process of real science was almost palpable in the room. Even though the SAB seldom finds itself in quite this role, that of dissecting a particular study for a particular purpose, I believe its members honestly sought the truth both about this study and about CV studies in general. But I still feel a twinge of sympathy for both the bureaucrat and the principal investigator. What neither of them could have foreseen was that the timing of the study placed it in the middle of a regulatory and legal firestorm; the success of a speculative, less-than-ideally-designed study was suddenly seen as representative of the usefulness and accuracy of CV studies in determining nonuse values and against the backdrop of a multibillion-dollar lawsuit. Neither of them could have known that, whatever their motives, their study of nonuse values would be seen as a threat to the economic well-being of the Exxon Corporation!

For us, several points can be gleaned from the proceedings. First, if CV studies are to gain respect and be given weight in policy discussions, they must be carefully designed; especially, they must define a clearly understandable "commodity" that the respondent is paying for or bidding on. This is essential if CV-described values are to be made commensurate with the real exchange values that characterize economic measurement of values. Second, the distinction between use and nonuse values is obscure, difficult even for experts to locate precisely; it is nevertheless important to economists as scientists, because if nonuse values cannot be separated from use values, there are dangers that, in aggregation, double-counting will occur. Third, the contingent valuation method, in practice, does not work well for some tasks. The methodology of CV works smoothly only if the benefit to be evaluated turns out to be describable as a discrete commodity; and sometimes the values served by environmental policies are impossible to separate from other, related preferences.

For our purposes, as we examine the claim that economic valuation is both monistic and comprehensive with respect to environmental values, the third point is especially pertinent. It is clear that we can define what, at first glance, appear to be reasonable "commodities" and describe them to respondents in CV studies but receive in response no repeatable or stable "value" because the respondents do not understand the commodity in anything like the same way. Aggregations that include dollar values representing these respondents' WTP with other economic behaviors are inaccurate or even nonsensical—the mixing of the proverbial apples and oranges. And now the dilemma facing the monistic Economist in policy-relevant science is unavoidable. For each such definable "commodity," the Economist must either find a way to design CV studies that will register the value in question or deny that the commodity description refers to a "true" environmental value. The more inclusive one is, the more environmental values one considers to be legitimate and worthy of measurement, the more creative—and speculative—must be the methods used to accomplish the purpose at hand. But if the hard-nosed, professional CV analyst is too critical—if he or she rejects all "borderline" values as untestable—then it will begin to look as if economic analysis is not really comprehensive: there are environmental values that cannot be captured by "packaging" commodities and querying respondents about their willingness to pay for those commodities. This is where the rubber meets the road with respect to contingent valuation of environmental goods: the success of comprehensive evaluation of environmental goods rests on the ability of CV researchers to devise studies to make sense of nonuse values. Otherwise, a "complete" CBA will include use values only, Exxon's attorney will get a raise, and a lot of CV researchers will be out of work.

Despite the stakes, honest and respected economists have been driven by arguments to the conclusion that comprehensiveness of economic valuation is not a realistic goal. First, consider the work of Richard Bishop, an important resource economist at the University of Wisconsin. He argued, way back in 1978, that "benefits" analysis in the valuation of endangered species is haphazard at best. He recommended applying the safe minimum standard of conservation (SMS) rule, which tells us always to save a resource from irreversible loss, provided the social costs are bearable.[12]

Myrick Freeman III, who wrote *the* book on the subject of evaluating benefits of environmental and natural resource programs in 1979 (and who was cochair of the EEAC following Kneese and Smith), published an updated and expanded version of this book in 1993. The new book was intended to "integrate in a comprehensive, unified manner all that was known about the theory of environmental benefit estimation."[13] Freeman's helpful, even exhaustive, book was described by his peers as "the definitive reference"; it stops short, however, of asserting comprehensiveness for the approach. Near the end of his updated survey of the field of economic valuation, Freeman adds a "qualification":

> The economic framework, with its focus on the welfare of humans, is inadequate to the task of valuing such things as biodiversity, the reduction of ecological risks, and the production of basic ecosystem functions. When policies to protect biodiversity or ecosystems are proposed, economists may be able to say something sensible about the costs of the policies; but except where nonuse

> values are involved and where people use ecosystems (for example, for commercial harvesting of fish or for recreation), economists will not be able to contribute comparable welfare measures on the benefit side of the equation.[14]

Note that Freeman includes nonuse values as types of values that can be explored using the economic framework (a point he consistently defends throughout the book), but that despite this he apparently acknowledges that there are no doubt important values—especially values that express themselves as "ecological" values or values of "biodiversity"—that cannot be measured by existing economic methods. This is a major concession, since "ecological" values and "biodiversity" values are among those most cherished by environmentalists. One might discuss the exact import of the assertion that economists "*will not be able* to contribute . . . welfare measures." Does Freeman mean to say that *given current methods,* and methods available in the foreseeable future, economists will be unable to contribute welfare measures, or did he mean to make the stronger statement that it is *in principle impossible* for methods based on aggregating individual welfare to comprehend these important environmentalists' values? From a theoretical viewpoint, this question is important; but we are discussing the current and near-future situation in environmental valuation; within this horizon, Freeman is unambiguous. He says that to limit environmental values to those that can be expressed in the economic rubric would be to considerably narrow the range of values environmentalists act to protect. In short, for the foreseeable future at least, there will be important environmental values that will be inaccessible to economists' methods. This argument, in my view, renders Economism untenable; there are some human values that will always go unexpressed in a system of cost-benefit accounting.

A much more sensible position emerges from the discussion of Arild Vatn and Daniel Bromley, two economists who argue that among environmental goods, some are fairly easily and plausibly turned into "commodities."[15] These are the "goods" that are easily discretizable (separable from other elements of value in nature) and the goods that are sufficiently similar to market goods that respondents to questionnaires and participants recognize them as something they might consider purchasing. But for many environmental values, those that are not easily discretizable and those for which no reasonable market analogues exist to inform respondents' choices, the Economist will have to choose the other horn of the dilemma. Unable to include these values within her or his monistic framework of valuation, the Economist will have no choice but to argue that they are bogus values or somehow unimportant. Maintaining a claim of comprehensiveness for economic valuation techniques therefore dooms Economists to implausibility in the eyes of any committed environmentalist and to an indefinite number of rear-guard actions in which they must argue that some value treasured by environmentalists is no legitimate value at all. Since they are left in this implausible position by their joint commitment to monism and comprehensiveness, environmentalists less sympathetic to the demands of economic rigor are likely to dismiss the Economist out of hand.

Remember that the argument I have just given, concluding that economics cannot be considered a comprehensive approach to environmental values, should not be taken to deny the value of the economic approach or of contingent valuation studies as one important element in a comprehensive approach. Indeed, the failure of CV researchers to properly capture the particular value of nonuse values associated with groundwater cleanup may have been due more to the subject matter than to failures of technique. Remember, the objection to the EPA study of nonuse values of groundwater cleanup was not that the technique of CV is useless when it is well applied, but that—for a variety of reasons and motives—the technique was badly applied in this particular case. In any case, since there seem to be quite a few values for which constructing an associated commodity is impossible, even Vatn and Bromley's compromise position entails the rejection of claims of comprehensiveness for economic valuation of environmental goods.

In my opinion, it would be liberating for economists themselves if they would dispel the myth that their approach can be *both* monistic and comprehensive.[16] If they gave up on that myth, two broad strategies would open up to them. These are not exclusive options, except perhaps in terms of opportunity costs. One option would be for economists to relax their requirement of monism and experiment with new, nonmonetary, ways to measure values, as Bishop did when experimenting with SMS; economists might then find opportunities to work with other social scientists to develop more comprehensive measures of value. Or economists could give up their claim of comprehensiveness, admitting that the range of values they are prepared to measure does not exhaust the larger range of environmental values. Some, but not all, environmental values are economic values. This would be a strategy of specialization, of honing the method of CV and applying it to a subset of valuation problems, those that allow discretization of goods, for example; here, economists would be leaving a part of the field of environmental valuation to cognitive psychologists, philosophers, anthropologists, sociologists, and so forth. A mixed strategy seems better than either pursued exclusively, I think, because a mix of specialization and technique development in a narrower range of cases would ideally prepare economics as a discipline to contribute to a more interdisciplinary and more comprehensive approach to characterizing environmental values. Either way—or both ways—economists would be free to improve their understanding of economic valuation of environmental goods as an internal matter and also contribute as possible to a larger, multidisciplinary and pluralistic discourse about environmental values and valuation.

5.5 Breaking the Spell of Economism and IV Theory

Where are we, then? What is the current state of our understanding of what we call environmental values? To summarize, there exist two general theories of environmental values, Economism and IV theory. These theories are general in the sense that they claim to have, or at least seek, accounts of environmental values that are both comprehensive (capable of expressing all legitimate environmental values) and tightly connected (capable of expressing all such values as special cases of a single, ontological type of value). There are, of course, many variants of the two theories, and even attempts to combine them, and so the situation is surely more complex than I have intimated. But I think it is not too far wrong to say these two theories dominate academic and disciplinary research and discussions of environmental values today. A key aspect of this debate—if one can call it that when competitors so consistently speak past one another—is that the two theories are developed and expressed within two distinct disciplines, the practitioners of which very seldom speak to each other. Philosophers and economists are apparently hell-bent upon occupying the full territory of environmental values discourse; and so far they have each succeeded in the dubious sense that they have developed, within their own disciplines, languages capable of expressing all environmental values that are important according to their own general theories of value. The two fields have, in other words, each used their own linguistic frameworks in conjunction with their own theoretical and ontological assumptions. Not surprisingly, the specialist languages they develop are capable of expressing the values their own theories recognize, but the same languages blind their users to values central to their opponents' theory. The really interesting question is the one that cannot be framed in either of the two self-limiting languages: which of the general theories is closest to capturing the actual range of values human beings feel toward the environment?

And here, in this more public and more practical context, we hear real people expressing many and varied environmental values. One hears appeals to the rights of other species, economic/prudential arguments, justice and fairness arguments, arguments based on obligations to future generations or to God regarding creation, and on and on. One can understand how scholars faced with the problem of giving an account of such a diverse and unruly subject matter would be anxious to introduce some order into the fray. Both environmental philosophers and environmental economists have deployed their theories to limit the range of important values and to bring order to bear on the subject of environmental values by oversimplifying it. Pragmatists will not fault traditionalists for wishing to introduce order into our public discourse; pragmatists simply demand that the order not be gained by arbitrary definition or self-evident fiat.

Despite the fact that public discourse about environmental values is highly pluralistic—many voices express diverse values in the course of public debate about what to do—Economists and IV theorists have insisted on interpreting these many voices within the vernacular of their own theories, respectively. The result has been to polarize the debate about environmental values and how to measure them. Because they speak different disciplinary languages, environmental economists and environmental ethicists have hardly engaged over their differences. They simply speak past each other, making contrary ontological claims about the nature of environmental value and what has it, but never engaging the practical problem of how best to measure environmental values as they are expressed in real situations by engaged citizens. Why, then, have both environmental ethicists and environmental economists, especially those who have bought into IV theory and Economism, looked at a multisided value discourse about what to do to protect the environment and seen polarized options? Why is the academic study of environmental values seen as a war between opposites, when public discourse about environmental values is usually a matter of more-or-less, trade-offs, and balancing competing values? The answer, I think, is that these scholars have placed so much emphasis on having ultimate theoretical connectedness—monism—that they have dismissed out of hand the possibility of a conceptually pluralistic method for the study of environmental values. The polarization of the discussion of environmental values between economists and environmental ethicists results from four shared assumptions that cluster around this commitment to monism. One way to break the spell of Economism and IV theory is to make explicit the extent to which their polar opposition emerges as much from their shared assumptions as from their disagreements. For they would not clash so directly were it not for the remarkable similarity of the assumptions they share in formulating the problem of environmental values.

1. Both Economists and Intrinsic Value theorists accept a sharp dichotomy between values that are "intrinsic" and those that are "instrumental"; further, both groups proceed to use this sharp dichotomy to separate nature into beings or objects that have "moral considerability" and those that lack it. In a particularly strong version of Economism, for example, Gifford Pinchot (first head of the U.S. Forest Service) said, "There are just two things on this material earth—people and natural resources."[17] Pinchot was thus enforcing a clear

distinction between persons and other things, living or nonliving, and saying that the well-being of the former, but not the latter, should be taken into account in our calculations regarding what is acceptable behavior. Interestingly, the position of Pinchot and his Economist allies coincides on a more basic level with that of Immanuel Kant, who is typically cited as opposed to the consequentialist emphasis of utilitarians. For Kant, only rational beings could be "ends-in-themselves" and rights-holders. Both Pinchot's utilitarianism and Kant's rights theory are thus based on a sharp distinction among entities—those that are to be regarded as being "ends in themselves" and those objects that can be used, without restriction, in service of the beings who are ends in themselves.

Economists and Intrinsic Value theorists, then, agree that there must be some special status for those beings which have noninstrumental value. They simply disagree regarding which objects in nature actually have this special status. For Economists like Pinchot, the special status is coextensive with humanness; for the IV theorists, moral considerability is coextensive with a much larger subset of nature's components. Either way, the sharp distinction between instrumental and inherent values reduces to a typology of ins versus outs based on a falsely dichotomous classification system.

2. Another, related aspect of the bipolar formulation of environmental valuations is the apparent bias of both sides in favor of evaluating *objects* or *entities* rather than evaluating *dynamic processes* and *changes in processes*. Protection is assumed to be protection of items in an inventory: should we try hardest to save genes? Individuals? Populations? Species? Ecosystems? This object bias is of course endemic to all of Western culture at least from the classical period on; it represents the triumph of Plato's belief that constancy of forms constitutes reality over the ideas of Heraclitus, who declared around 500 BC that "all is in flux."[18] This ideological triumph also led to modern scientific reductionism, which seeks explanation in the motion of elementary particles.

The "atomistic" idea, which emphasizes elements and entities, is so deeply engrained in Western thinking that alternative conceptualizations of nature such as that of Heraclitus have been revived only relatively recently. Since the publication of Charles Darwin's evolutionary theory, however, the importance of systemic change and irreversible developments—of complex, dynamic processes—has asserted itself. This revolution has extended to physics, and physicists are now leaders in an interdisciplinary effort to develop a more dynamic worldview, as is evidenced by ever-increasing emphasis placed on nonequilibrium dynamics. The full implications of a dynamic worldview are just now being felt. It may be decades before these concepts are well understood, but creative work in nonequilibrium system dynamics is already leading to new insights in ecology, and this direction holds promise for applications to environmental policy.[19] This much we know for sure: full absorption of systems thinking, as it evolves into environmental management, will have far-reaching impacts on the policies we advocate and will almost certainly require more attention to interspecific relationships and system-level characteristics. It will, accordingly, downplay the role of entities and objects.

3. Another similarity between Intrinsic Value theorists and Economists is that they both defend monistic theories, speaking ontologically. They have as their most central commitment a belief that there is ultimately only one kind of value worth counting in decisions regarding what to do to protect the environment. This commitment, in turn, affects their answer to other questions about environmental values, such as how to measure them and how to justify them. An ontological theory of environmental value is thus monistic if every type of value it includes can be interpreted as a special case of a type of one single "ultimate" value, with ultimate values being ones that cannot be included as a special case of some more basic type of value. Scholars from both economics and philosophy have thus reduced multifarious values to the categories of their theoretical commitments.

4. Finally, both theories treat environmental value as placeless—neither dollars nor "intrinsic values" are contextualized or sensitized to the particularities and idiosyncrasies of local places; by insisting that all value be expressed in a common coin—dollars or units of intrinsic value—advocates of these monistic theories tend to "wash out" out the particular, the special, and with it any human attachments to places that are special to people, including "home" places and places of great beauty and other human significance. In economics, therefore, place locations disappear as values are turned to dollars, which are fully fungible, transferable, and mobile across international boundaries and participate in worldwide markets. These, in turn, exert homogenizing influences on behavior and policy of particular communities.[20] Intrinsic value, similarly, directs attention to the characteristics an object has "in itself," as distinct from relational characteristics; and place identification is all about relationships between a culture, its wildlife, and the physical context. To wash these commitments out by expressing all value in terms of dollars or units of intrinsic value leaves environmentalists with the necessity to defend their policies by expressing values in one of these two ways. Neither of these vocabularies is able to convey the importance of valuing the particular, the local, and the place-based. The drive toward monism, shared by both Intrinsic Value theorists and Economists, inevitably makes for an abstracted, contextless, and placeless sense of value.

What is interesting, then, is that—for all their polarization—the two most often-cited theories of environmental value rest on four shared assumptions.

What is even more interesting is that these shared assumptions about the nature of environmental values are all questionable if treated individually and separately and also questionable taken together, as a general understanding of what we need in the way of a theory of environmental value. The advocates of these theories understand the problem of environmental valuation as a problem of classifying entities as *either* instrumentally *or* intrinsically valuable on the very general, ontological scale; they then seek to maximize a very abstract and placeless value, a value that derives none of its efficacy from the distinctiveness and unique charms of a place. Indeed, it is these four shared assumptions about how to value the environment that create the tendency toward polarization in environmental ethics and environmental policy discussions, and toward ideological environmentalism more generally.

This polarization shows up in many kinds of practical contexts. For example, on the most general and global scale, it is seen in the tension between economic/development values and intrinsic values; on smaller scales, it tears forest communities apart as they disagree whether to cut the last old growth to sustain jobs or save the old growth and set it aside. And in the most excruciating and difficult management situations, neither of the theories of value provides a useful framework for comprehensively articulating the values at stake in hard decisions.

Take the fairly common example, repeating itself in many places today, of whether to cull herds of "wild" ungulates on ranges they have "outgrown." Is this problem any easier to solve once it is translated into one of our monistic theories of value? Assume you are the manager of a reserve where no hunting is allowed and the elk population you manage is well above its historical size; significant changes in vegetative patterns are being reported by consulting botanists, and there is a measurable decline in the weight of yearling elk. So you call two consultants—one an Economist and the other an Intrinsic Value theorist. The Economist, who claims to offer a comprehensive account of environmental values, recommends that we try to identify all the economic impacts of culling and to choose the course of action that will lead to the greatest aggregated willingness to pay. The consultant points out that we must look for nonuse values as well as use values. Thus, to make a decision regarding whether to cull the animals, to drive some animals off the reserve where they will be accessible to willing hunters, or to do nothing, you will need to gather data on the economic value of every plant that is being eaten, every example of someone paying admission to view plentiful elk, how much hunters would be willing to pay to shoot elk as they enter private lands, and countless other questions about actual economic values. But that's only the beginning; after gathering the actual market impacts of the three policies, you must then undertake to assess the nonuse values, such as the willingness to pay for "existence" values—the value people profess to be willing to pay "just to know something exists." The consultant, before leaving, says he can think of several important valuation studies that must be done to make a reasonable decision. His consulting firm, for only a few million dollars, could put together a package of valuation studies to determine (1) the willingness of the public to pay for not having the antelope culled, (2) the willingness of the public to pay for plant protection on the reserve, and (3) whether the public is more likely to attend the preserve (and pay the entry fee) if the herd is at a large, unhealthy size or at a smaller and healthy level. And he will soon think of a few more studies that will be needed.

Since your annual budget for "research on social values" was zeroed out in the Reagan administration and never restored, you politely demur, see the Economist out, and invite in the Intrinsic Value theorist. "At least the going wage for philosophers is a lot lower than for Economists," you grumble to yourself, as the philosophical consultant sinks into the chair across from your desk. After you briefly explain the quandary, the philosopher says, "Well, I don't know anything about elk, or about the plants and animals you have here in the reserve, but I have an excellent theory that allows you to make a decision. First, make a list of all the things, both inside and outside the park, that will be affected by your decision. Second, for each thing, and each kind of thing (species, for example), that will be affected, you must determine whether that thing has 'intrinsic' value or not. For this task, I can offer you a bibliography of readings by environmental ethicists. Unfortunately, you will find that they differ with regard to which objects in nature do, in fact, have intrinsic value. It's also a problem that some intrinsic value theorists think only individuals can have intrinsic value, while others think ecosystems and species can have it. Worse, there's quite a disagreement among environmental ethicists whether things have grades, or degrees, of intrinsic value, or whether everything that has intrinsic value has it equally. But these are important problems, and once you answer them, and once we resolve the outstanding philosophical issues, the decision will be simple. Simply choose the policy that will protect everything, inside the reserve and outside the reserve, that has intrinsic value." You point out that you're trained as a biologist and the classification system you learned was the phylogenetic one, but you don't know how that will help with this normative taxonomic task. At first the philosopher is downcast, sorry you can't see the value of his work. But then a bright idea spreads across his face; "Perhaps the two of us can go to the government and get funds allocated to set up a Center for Environmental Ethics." The center would be of great value to managers, the philosopher says. "Managers like you cannot do your job without someone helping you to sort entities affected by your decision; otherwise, you'll unknowingly commit countless immoral acts.

If we had a Center, however, we could hire several philosophers who represent all the different viewpoints on what intrinsic value is, how it is discovered, and what has it, and we could undertake in-depth studies. Once we have reached decisions on all these questions about intrinsic value, we'll have your answer. Then you'll know how to do the moral thing."

Because of the sincerity and the rapt gaze of the philosopher, you are almost drawn into a discussion about what has intrinsic value and how to recognize such things; but you glance at the clock. Your next appointment is waiting, and a deadline has been set for you to propose a decision that satisfies both the governor, who is an avid hunter, and the head of the local Citizens Wildlife Committee, who routinely refers to hunters (and professional cullers) as murderers in her weekly column in the local paper. You thank the philosopher for the stimulating ideas, saying, "Well, you've certainly given me food for thought," and show the philosopher politely out the door.

What managers quickly learn, as dramatized in this imaginary anecdote, is that there is not the data, the time, nor the money to "apply" monistic theories of value in particular situations to achieve decisions that can be judged "rational" according to either of the theories of environmental value. Because such tasks are essentially additive—they recognize items that are of value and we are told to maximize that value—it is not obvious how to use a radically incomplete sum. Worse, given the ways that shared assumptions shape policy questions, the manager is caught in the middle, because these assumptions apparently preordain that any evaluative decisions must be all-or-nothing, either-or sortings of objects into types. We have seen that if the manager tries to apply either of the standard theories, the information demands are overwhelming; and even if these problems were overcome and the necessary information was available, the question of environmental value is posed in a way that encourages polarized answers and discourages the search for acceptable compromises.

We have found that theoretical discussions of environmental values have fallen into polarized rhetoric, with Economists and IV theorists sharing key assumptions that greatly constrain approaches to developing a comprehensive theory of environmental value. What, we can now ask, would be the consequences of denying the crucial, shared assumptions of the Economists and the IV theorists? If we can free ourselves from the shared assumptions that bind *both* Economists and IV theorists in polar opposition over classifying objects of value, it will be possible to look with fresh eyes at questions of value and policy. In the paragraphs below I consider the effects of rejecting the central assumptions shared by these opposed groups.

1. Denying a sharp dichotomy between instrumental and intrinsic valuing. Economists and IV theorists share a complex web of assumptions about

the nature of environmental value; the defining feature of that web is the belief that a sharp distinction must be drawn between two kinds of value, intrinsic and instrumental, and between two associated types of things, those that are valued for their own sake and those that are not. Following J. Baird Callicott, however, we can argue that *value* is more basically a verb, not a noun, and that valuings are always acts or dispositions of conscious beings.[21] On this interpretation it is possible to consider a range or continuum of *ways of valuing*, each of which constitutes a relationship between a conscious valuer and the element of nature in question, experienced in specific situations. But given this conception of environmental valuation, and once we give up the entity orientation, there is no reason to separate objects into those that must be intrinsically valued and those that are instrumentally valued. Those separations are simply confused attempts to reify one element of countless irreducible relational acts of individual valuers who experience value in many particular cultural or natural contexts. The assertion that nature, or one of its elements, has intrinsic value, on this view, is a confusion. Natural objects—and processes—can be valued intrinsically, perhaps, but that value emerges from a relationship and does not in any sense "belong" to the object.

According to the view explored here, the task is not to determine which objects "have" some reified type of value, but rather to determine whether good reasons can be given for invoking a particular value in a particular situation. This line of reasoning apparently opens up the possibility of reconciling the two sides in the debate over intrinsic versus instrumental value in nature: it is possible to include both instrumental and noninstrumental reasons for preferring one set of policies over another, without asserting that "intrinsic" values exist independently of human valuing agents and their actions. If we reject this sharp dichotomy between instrumental and intrinsic values and the associated classification of natural objects as instruments or as moral beings, a pluralist and integrative position emerges as a possibility: there are many ways in which humans value nature, and these ways range along a continuum from entirely self-directed and consumptive uses and include also human spiritual and aesthetic values and other noninstrumental valuations. If one forgoes a sharp, definitional distinction between two types of value, the moral task of sorting entities into those that have, and those that lack, this special feature of noninstrumental value becomes a nonproblem. The sorting question, it turns out, has interest only after one enters the polarized conceptualization of environmental values that comes with the web of assumptions shared by Economists and Intrinsic Value theorists.

2. Rejecting the entity orientation. Suppose that, following a more pluralistic approach, we stop thinking of environmental evaluation as an exercise in categorizing objects at all and accept instead the goal of choosing indicators of

the adaptability of various technologies and policies. Attention would then turn to impacts of existing and proposed technologies and policies on physical and social processes and the ways they contribute value to humans. The task would be to develop an indicator, or a suite of indicators, that would allow the ranking of "development paths." A development path would be thought of as a scenario that could be projected to unfold under a given policy or set of policies. Evaluation would then be a matter of ranking various development processes that might occur from the present into the future. We hope, in the end, to be able to say, "Development Path A is more (less) likely to protect and enhance social values *V1, V2, V3, . . .* than Development Path B." The process approach advocated here simply ignores the problems and possibilities of entification and sets out to evaluate processes of development and change as they play out on a landscape at a particular place.

Rejection of the entity bias has an even more profound implication for the theory of environmental value. If we reject the assumption that environmental evaluation is basically a matter of sorting entities, and focus instead on evaluating processes and paths of change and the values experienced by people and cultures within these processes, it is possible to recognize a deeper source of value in nature, what might be called "nature's creativity." Ilya Prigogine and his coauthor Isabelle Stengers have argued persuasively that Western thought has for too long emphasized "Being" at the expense of "Becoming" and entities at the expense of processes.[22] Prigogine and other leaders of the emerging science of chaos and complexity have set out to repair this imbalance, arguing that change, process, and becoming are more basic than being and that the world of objects we see is simply our stilted perception of a rich, multiscalar, evolving system.[23] The more dynamic models suggested by the sciences of chaos and complexity place humans and their societies within a larger evolving ecological and physical system. All description *and all evaluation* occurs from within a dynamic system, so humans value nature *from within nature*. As parts of nature, they are also conscious and autonomous agents who can, inadvertently or purposefully, affect the larger systems of which they are a part. Environmental values emerge within this key natural-cultural dialectic. If we were to apply this kind of thinking to biodiversity policy, we would focus on the processes that have created and sustained the species or elements that currently exist and populate the world rather than on the species or elements themselves. Indeed, emphasis on the value of creative processes in nature may go a long way toward expressing the common factor in most people's valuing of nature. By attempting to evaluate or rank various development paths from where we are into the future, by identifying and evaluating changes that emerge on multiple scales, and by treating changes in the human environment as complex dynamic processes that can go well or badly for us and for other species, we can rank proposed policies according to a variety of criteria. In this way we can avoid evaluating objects, or "atoms," of nature and focus our attention on processes instead.

3. Not assuming reductionism and monism. We have seen that monism, if adopted as a prior constraint on all acceptable systems of evaluation, needlessly constrains experimentation with multiple methods of evaluating environmental changes. If we reject the task of sorting entities into those that are instrumentally and those that are inherently valued, we can concentrate on identifying and classifying a variety of environmental values and types of environmental values. We can start from the pluralistic viewpoint that all cultures value nature and natural processes in many ways. We should, as a first step, develop a vocabulary and operational measurements that are rich enough to express these multiple values. We thus embrace pluralism as a working hypothesis, setting out to characterize and operationalize as many values and types of values as possible. Adopting such a stance leaves for subsequent discussion the question whether some of these types of values can be usefully "reduced" to other types, assuming that some level of consolidation of multiple frameworks will eventually emerge. Our evaluations are no longer constrained by the *requirement* that environmental values must be commensurable and measurable within a unified system of evaluation; instead they are expressed in rankings of possible development paths, and we accept the challenge of choosing the best possible criteria and indicators for establishing such rankings.

4. Rejecting the assumption of placeless evaluation. Evaluation models like Economism and the IV theory are constrained by their monism to express all value in a common currency, so their accounts of value tend to lose, in the process of aggregation, the place-relative knowledge and value that arise within a specific dialectic between a human culture and its physical and ecological setting, or context.[24] One implication of the adaptational model for understanding environmental problems is to emphasize the importance of localism. As we relax the requirement for a single, universally aggregable accounting system for all environmental values, it becomes more possible to hear, and register, the very real concerns of local cultures trapped between the hard realities of international economic forces beyond their control and the limits and constraints that manifest themselves at the local and regional level. Localism, as a replacement for universalism, leads to an emphasis on local variation, on diversity from locale to locale and from region to region, and to many local "senses of place," each of which expresses a unique outcome, at each particular place, of the infinitely variable dialectics between local cultures and their habitats. Development and various development paths can therefore represent differing trajectories created by the nature-culture dialec-

tic in specific, culturally evolved places. This trajectory, given the above conceptualizations, can be measured both in the shorter, economic frame of time and on a multigenerational scale that counts impacts on future inhabitants of the earth.

Once one recognizes multiple types of values and attempts to use this pluralistic approach in actual situations, it is possible to work toward a more systematic and integrated account. If useful systematizations and reductions among value terms are fruitful, we may move toward a system of fewer and fewer basic types of value—that is, toward more connected theories. However, since we are far from being able to accomplish the reductions necessary to express the many environmental values we see as a single kind of value, monism in this sense remains a distant and not-very-relevant objective. In the meantime, we act as experimental pluralists. Once freed of the constraining assumptions of monistic theories, we can account for environmental values pluralistically and in an experimental spirit.

An important lesson we can draw from our analysis is that if we look at theories of value and approaches to evaluating environmental change from the viewpoint of an active, adaptive manager, the goal of achieving a *comprehensive accounting* of values affected by a decision should—for the foreseeable future—be ranked much higher than the goal of achieving *connectivity of values*. Economists and IV theorists have unfortunately reversed this priority, allowing their preexperiential commitments to dominate the goal of achieving a more comprehensive accounting of environmental values. It is the latter that we need, however, if we are to make good choices in managing complex problems in complex situations.

5.6 Pluralism and Adaptive Management: What the Study of Environmental Values Could Be

Both environmental ethicists and environmental economists embrace a general theory of environmental values that purports to be both complete and monistic; preexperiential ontological commitments have led them to emphasize monism over comprehensiveness. The central goal of both disciplines has thus been to represent all types of environmental value as a single ontological type. Within disciplines, this focus has led to articulation of disciplinary orthodoxies rather than to a diversity of understandings or an experimental approach to describing and analyzing environmental values. Despite very similar conceptual assumptions about the nature of environmental value, these two disciplines have largely gone their own ways, failing to establish an effective cross-disciplinary dialogue about environmental values. Worse, the extradisciplinary discussion between economists and environmental ethicists has not been engaged, as each discipline remains trapped within its own theoretical structures and interpretations.[25] I have argued that it does not need to be this way if we (at least temporarily) relax the expectation that all environmental values must fit into previously designated categories and encourage the development of a variety of theories and approaches to understanding environmental values. If these approaches are tried in many local situations, especially as an important part of an adaptive management process, it may be possible to start with pluralistic conceptions and pluralistic measures for expressing evaluations and yet work toward a more integrated set of environmental indicators.

What I think is needed is a more empirical approach to environmental values and goals; by an empirical approach, I mean something like John Dewey's view of science and social action, and of social learning as a real possibility for democratically organized communities. Two of Dewey's ideas are particularly pertinent to adaptive management. The first, which is discussed above, is Dewey's treatment of science and moral inquiry as subject to the same rule of logic: the rule of experience. He simply denies that there is a problem about the "fact-value gulf" and embraces the idea, often decried as "naturalism," that we resolve ethical dilemmas through experience, especially through community-based and community-sensitive, participatory experience. Because Dewey and the pragmatists engage values in real situations, with important stakes for communities, values are as open to revision in the face of new experiences as are uncertain beliefs. In real situations, values can be so compelling that they provide solid planks for undertaking experiments and reducing uncertainty by the judicious use of science within a management process. As we noted in chapter 4, when we are engaged in active, adaptive management, Neurath's analogy encourages us to find refuge in consensus wherever possible; an epistemic community can be built upon shared social values. At other times, experience leads us to reconsider some of our values; at these times, values, like other beliefs, can be undermined by experience. For example, many U.S. cities—especially my city of residence, Atlanta—have over the past few decades carried out an experiment in unlimited suburban growth. This experiment was animated by a value, the value individuals place on freedom and unlimited mobility, which is observable in the pervasive choice to travel in single-occupant vehicles. Dewey would say that despite the *fact* that many residents *desire* the mobility associated with the automobile, these desires should be seen not as unalterable "preferences" but as a *hypothesis* that an automobile culture is *desirable*. Time and future observation will test this hypothesis.

A second idea of Dewey, mentioned in chapter 2, is "social learning," an idea that, once refurbished and modernized, can serve us well as a guide to adaptive management. Social learning has important implications for the environmental policy process, and it also has implications for the way we think of stakeholders, the general public, and professional policymakers and bureaucratic managers.[26] Most centrally, Dewey's idea of social learning provides a broad framework for understanding the goals of adaptive management within a diverse and democratic society. Here, Dewey's work on democracy in an increasingly technological society closely parallels and supports Aldo Leopold's idea of involving the public in broad management decisions. This commitment, however, apparently demands a faith that through the educative function of public agencies and their representatives, the public can be informed and made capable of addressing complex environmental dilemmas. Dewey's idea of social learning and Leopold's idea of public involvement in management therefore require an important shift in the role of environmental managers and experts involved in managing resources. Dewey perceived that the growth in complexity of modern industrial societies was such as to bewilder the layperson; so he conceived of communities as capable of strategically organizing themselves to incorporate directed, mission-oriented science at specific areas of uncertainty and disagreement. Dewey also saw that under the best conditions, such an organization might lead to a cooperative interaction of the public and experts, unified by a shared goal of protecting an important environmental feature of their place. Such an organization would encourage individuals to work together to learn about and discuss openly both the goals and the methods of environmental management. If this ideal could be achieved in an open, iterative process of community-based adaptive management—through the use of citizens' advisory committees in conjunction with blue-ribbon scientific panels, for example—it might be possible for the community to cooperate to reduce uncertainty, to adjust environmental goals, and to engage in management activities that improve local conditions and at the same time contribute to the learning curve with regard to environmental values more generally. What we need, in other words, is an approach to environmental values and valuation that fits comfortably into the experimentalist framework of an adaptive management process.

6

RE-MODELING NATURE AS VALUED

6.1 Radical, but How New?

One of the advantages of accepting lots of speaking engagements around the country and in Europe is that one meets many fine and charming people who share a deep concern for the environment. I have greatly enjoyed meeting leading scholars in other fields of import in addressing environmental problems. This is a benefit of doing philosophy in a public policy context; had I remained a narrow disciplinarian, limiting my studies and writing to philosophy, I would probably not have this opportunity very often. Whenever I am among the giants of another field, I try to learn something about the way they think; and I cannot begin to describe how much I have learned in such discussions, formal and informal. When I had a chance to engage Donald Worster, a pioneer among environmental historians and a wide-ranging and unconventional thinker, I expected provocation. About a decade ago, I had a conversation with Worster at a Prairie Festival, and though I've forgotten most of what we conversed about—as I'm sure he has—I remember one brief exchange about environmental ethics that left me thinking. I finally have something to say in response to Worster's thought-provoking comment on the enterprise of environmental ethicists.

Somehow our topic drifted toward radical environmental philosophy and the philosophical discussions of Intrinsic Value theory; I suppose I said something brashly critical of the whole idea; as I was wont to do in those days (and sometimes today), I may have decried the concept of intrinsic value as unclear and useless, and on and on. Worster's response, which brought me up short, was "I see these people as the 'deep divers' of environmental thought; it's important that someone ask the really deep questions about environmental value, and even if they don't get it right, it's important to have people like them stirring things up." Two things caught my attention, as I thought about the matter: first, Worster did not address the specific conceptual issues I raised, but rather recontextualized what radical environmentalists say, making it a part of a larger, community-based and historically dynamic situation. He did not evaluate the movement according to the usual critical standards a philosopher would apply to proposed moral theories. Second, I did agree strongly with Worster's recognition of the important function played by deep divers; and yet I wanted to question—though I could not find the words at the time—whether current philosophical speculation about intrinsic value in nature represents truly deep thinking. Let us then begin our positive discussion of environmental values, which will occupy this and the next two chapters, by thinking a bit about what really counts as "deep thinking" in response to the modern environmental "crisis" and the role of environmental values in it.

Both environmental economics and environmental ethics, as professionally staffed, distinct subdisciplines located in university departments, represent relatively new fields of study. In the United States the development of these disciplines responded to increased environmental awareness, fueled by underlying shifts in social demographics, which began after World War II and accelerated in the 1950s and 1960s. These trends saw population shifting from the countryside to urban areas; increases in standards of living, which provided people more leisure time; and changes in people's priorities, which shifted from subsistence exploitation toward amenities and recreational use of natural areas. Eventually, these trends, and associated increases in the popularity of outdoor recreation, resulted in vastly changed attitudes and norms regarding nature, natural systems, and wildlife.[1] Corresponding to these trends, environmental economics took on recognizable form in the 1950s and early 1960s,[2] and it became an important policy force with the creation of Resources for the Future (RFF) in Washington, DC. Begun as a unit of the Brookings Institute, RFF split off and became an independent research institute; it remains an important player in environmental economics and policy formation today. Environmental ethics, as noted in chapter 5, developed as a distinct subdiscipline of philosophy in the 1970s.

Both new fields were immediately recognized as involving important, and controversial, departures from the theoretical commitments and viewpoints of the core disciplines of which they were modifications, economics and philosophy. In economics, the development of theory to support monetary valuation of nonuse goods, for example, was strongly criticized and resisted by mainstream economists. These values, and the contingent valuation methodology they require, are still looked upon with suspicion by many purists among professional economists. From a disciplinary viewpoint, then, pioneering environmental economists' innovations seemed new and jarring. The theoretical applications of environmental ethicists, especially nonanthropocentrists, were likewise considered strange, even unintelligible; practitioners of the new field are thus seen by their more mainstream philosophical colleagues as challenging central beliefs and principles of traditional ethics; both practitioners and commentators describe this approach as radical. Nonanthropocentrism lays claim to its label because it calls into question the moral significance of the human-nonhuman divide, central to all ethical theorizing in the Western tradition at least since Socrates. Other environmental ethicists are considered radical because they question moral individualism, the view that moral obligations must always be obligations to individual human persons, which has been pervasive since the Enlightenment and has gained strength throughout the modern period in Western thought. According to these lines of thought, environmental ethics must depart from traditional ethics by considering the possibility that nonindividuals, such as species and ecosystems, could be morally considerable in themselves. The rise of environmentalism and the loud, if sometimes inarticulate, advocacy of new classes of values, it can be argued, has indeed made a convincing call for new and creative thinking about environmental values.

I heartily agree that we need more creative thinking. For all the claims of novelty, however, a careful look at the new disciplines of environmental economics and environmental ethics reveals no creative leaps in conceptualization of environmental values. Their practitioners build upon the basic concepts of their parent disciplines, and they have not provided any new and useful vocabulary for thinking and talking about environmental problems or environmental values. In fact, the basic vocabularies of the two disciplines involve fairly straightforward *extensions* of existing disciplinary vocabularies with new applications to the changed situation and to changed sensibilities of the populace. Thus, although both new disciplines apparently embodied important leaps in theory, they merely extended the application of existing theory and did nothing to enrich the normative or theoretical vocabularies available for conceptualizing and evaluating actions. Environmental economics expanded the realm of application of the vocabulary of commodity and exchange values, which was developed for describing and measuring privately owned goods, to apply to public goods. Since public goods lack the key aspect of exclusivity, which anchors the idea of private ownership, the application of measures of exchange value to nonuse values—by definition not exchanged in markets—was indeed conceptually jarring. Decisions affecting environmental public goods were thus assimilated, through the fiction of hypothetical or "shadow" markets, into the system of price values, creating a new "kind" of economic value, value that is represented not by a real price, but by a "shadow" price, as explained in section 5.4.

For environmental ethicists, the so-called radical innovations have involved a similar expansion of the existing theory without improvement in conceptualization or the addition of new terms with which to discuss environmental values. In ethics, the theoretical extension proposed was to expand the class of morally considerable individuals or entities and to apply existing normative terms and concepts of human-based ethics: rights, interests, and especially the idea that some beings can be said to have intrinsic value, making them objects of moral concern. Elements of nature, by virtue of this expansion, become moral subjects, or at least moral "patients" (as they are sometimes called).

What we have here, then, are two *theoretical* leaps that were not accompanied by associated advances in conceptualization and vocabulary; and it is usually new vocabulary and new conceptualizations—new "mental models" of nature, one might say—that create true revolutions in human thought. In both cases, basic evaluative approaches that were developed for carefully delimited domains were simply expanded in application to new values normally untouched by the original theory. In both of these cases, the proposed expansion flies in the face of original, key distinctions that virtually defined the boundaries of the parent disciplines. Environmental economics came into existence by flaunting the original restriction of economics to the study of *economic* value (value that arises in exchanges in the processes of production and consumption), a restriction that originally defined economic value, *as distinct from other kinds of value*. Shadow pricing ignores this foundational distinction and applies the idea of price far beyond its original application. Similarly, environmental ethics flaunts the original distinction between human *persons*, who were traditionally thought to demand a special kind of respect and standing as free and rational beings, and nonhuman "objects," which were thought to fall outside the realm of ethical behavior.

Although one might consider these innovations brave, one might just as well describe them as obstinately foolhardy, since they flaunt the very distinctions that gave rise to the parent theory's specialized evaluative vocabulary. These theoretical departures may seem radical, deep-diving, because they attack what has been considered "common sense" for decades or even centuries. I would argue that they are radical only in the limited sense that they create conceptual chaos, ignoring as they do differences crucial to the very application of the key normative and theoretical terms central to their parent disciplines. In this sense they provide no new terms or distinctions that might reconstitute the normative landscape and truly replace—rather than merely extend—their disciplinary orthodoxies. One might say, then, that both environmental economics and environmental ethics have labored to force the new wine of expanded environmental concern into old skins of disciplinary conceptualization. In my view, neither environmental ethics nor environmental economics is *radical enough* in its departure from the traditional theories and conceptualizations, including the operative vocabularies, of its parent discipline.

We need to do more than simply broaden the range of application of old, narrow, discipline-based terms and theories. As we have seen, terms and vocabulary are constrained by theories, which identify the types of objects recognized, and these theoretical constraints infect the very meaning of the terms used. As a result, trying to pour new, breakthrough ideas into old terminology is fraught with confusion and often with unnecessary conflict, as new and creative ideas, rather than replacing older and more limited ones, appear as direct assaults on widely held and comfortable viewpoints. If we are truly facing a new kind of crisis, a global pressing-against-limits that is unprecedented in human history, why should we expect simple transfer of the normative vocabularies of old-think to somehow transform consciousness and get us thinking aright? Maybe stretching the old wineskins is too limited a response; maybe we need to be *really* radical, and develop new conceptualizations and terms. What we need is a whole new way of looking at, and speaking about, nature and the human place in it.

Now after a rather theoretical examination (here and in chapter 5) of the failures of environmental economics and environmental ethics to provide a comprehensive and effective theory of environmental values, we have arrived once again at the beginning point of this book—the realization that we do not have a vocabulary that is adequate to the task of communicating human hopes and fears about the environment. Failure of communication about environmental goals and values is ultimately the most important intellectual problem in the search for more acceptable environmental policies. The theoretical analysis of environmental values, begun in chapter 5 and unfolding throughout part 2, explains the theoretical basis of the practical problem of communication that so confounds environmental policy discussions. Our examination of these theoretical problems has thus led us back to the central working hypothesis of this book: there are serious problems in communicating about—and even thinking about—environmental values because of what I have described as towering at the U.S. EPA and because of the more general poverty of vocabularies available to evaluate and integrate environmental policies.

We have reinforced, from a theoretical direction, the hypothesis that environmental problems have resisted solution because of failures of communication, which stem from the lack of an integrative vocabulary for discussing environmental goals and values. These failures of communication are public symptoms of the lack of adequate normative theories in economics and philosophy. The response of those disciplines to growing environmental problems has been to extend inadequate terms and concepts to apply to more instances—to extend existing value theory to apply more broadly; it is a too limited response to the novelty of our situation. That, I think, is what I wanted to say to Worster: Yes, it is important that we dive deeply in our search for better ways to describe the value of nature, but the simple extension of traditional disciplinary concepts may well be diving to the bottom of the shallow end of a very deep pool.

If we think of Economism and IV theory as practice dives at the shallow end of the pool, our critical attitude shifts from evaluating such ideas as the "one and only correct theory of environmental value" to welcoming them as contributions to an ongoing chaotic but creative discussion. Better ideas emerge from more diverse mixes, according to pragmatist epistemology, so we see value in efforts to extend traditional concepts. But these methods are not adequate, in and of themselves, to the situations we face. It's time to move to the deep end of the pool.

But how do we know, surveying from above, before we dive, which is the truly deep end? How will we know when we are even swimming in the right direction? We have just seen an important part of the answer: expanding the range of theories and approaches—putting lots of swimmers and divers in the pool—allows us to take many more soundings into the depths of environmental values and valuation. Provided we have means of communication, multiple divers can expand our experience base. Hence the tolerant attitude, within an adaptive management process, toward multiple expressions of value (pluralism) and toward having many local experiments and pilot projects as tools of management and as opportunities to learn and reduce uncertainty. Disciplinary "models" should never be considered a priori constraints on the variety and diversity of value expression in public discourse. A free and open competition of ideas and vocabularies is a resource and can encourage social learning.

Another clue is what we have learned by our examination of unsatisfactory *communication* regarding environmental values, as in the case of wetlands mitigation and countless other values. Most environmental policy discussions are framed as dichotomies (e.g., conservation versus preservation, human versus nonhuman interests) or as extremes (e.g., exploiters versus tree-huggers). Again we are driven back to the conclusion that there exists no useful, neutral vocabulary for talking about environmental problems, and especially for talking about values and environmental management goals. Early efforts at theories of environmental values, we have just seen, did not dive deep enough. Merely extending the application of old concepts, limited to individualistic, human-based ethics, does not move us past all-or-nothing classifications of objects as mere instruments or as goods in themselves. Extensionist theories understood monistically result mainly in academic turf wars and in failures of communication. If we think of them as adding diversity and new ideas to the intellectual mix, however, they can be seen as practice dives for the real intellectual deep diving that must take place if we are to improve communication across disciplines and among interest groups.

6.2 A Naturalistic Method and a Procedure

In section 6.1 I argued that what has passed for deep diving—radical environmental philosophies in environmental ethics and pricing "non-market" values in environmental economics—has not gone deep enough; these approaches do not, I concluded, provide a vocabulary well designed to communicate the dangers and opportunities we face today. What has changed in recent history, long after our moral codes were developed, is the human ability to employ pervasive and powerful technologies, as humans exert more and more dominance over natural systems. The effect of these changes on human morality is that we live in a hugely expanded moral universe of human responsibility. The changes in natural processes attendant upon human population and technological expansion represent a series of more and more irreversible experiments in reducing the complexity and diversity of the planet's life and cultural practices. Natural systems, as well as conventional cultural practices, are undergoing constant "disturbance" at every level and on every scale. What we need is a new way of talking about and evaluating rapid, often irreversible changes that will result from continued economic and technological growth. Here, I think, we are getting near the deep end of the pool. Here we may finally figure out a proper role for environmental ethics, or environmental value studies, in a process of adaptive management.

I wish I knew how to take accurate soundings, to tell when we have found the deep end, but I don't. I do, however, have a hunch. My hunch is that if we're going to find a conceptual breakthrough that will help us to reconfigure our thinking about environmental problems, it will result from reexamining deeply held assumptions about the relationship between facts and values. We are thus diving in a quite different part of the conceptual pool, far from environmental ethicists, who have hardly questioned the dichotomy between facts and values. They have instead sought to reconfigure our thinking about nature by disturbing the traditional application of the categories of standard ethical theory, working mainly within the theory of value—ethics—and hardly questioning the sharp separation between descriptive and prescriptive discourse. Similarly, we dive far from economists, who simply broadened their concept of commercial value—commodities—to cover nonmarket values. My hunch, however, leads me to believe that before we can get the moral and normative concepts right—before we can develop a new ethic for the new, dominant situation in which humans find themselves—we will have to reconceptualize our place in the world. And such a reconceptualization may have more to do with the way we *understand* nature than it does with how we *value* nature.

In order to explore the implications of a reexamination of the fact-value dichotomy in the next section, I here propose an alternative to the usual ways of thinking about the interaction of these "realms." What I propose, if it needs a label, is a form of "methodological naturalism." I first explain the need for methodological naturalism by showing how it is an essential element of a philosophy of adaptive environmental management. Once the context is set, I show how a naturalistic method, based in experience—experiment and observation—can serve as a self-corrective device capable of gradually pointing communities toward improved goals and values.

Let us then begin our fresh look at environmental valuation with an examination of the relationship between facts and values in public political discourse. We abandon, at least for now, the search for a monistic theory of environmental value and adopt moral pluralism as our theoretical starting point. We embrace pluralism in two senses. First, we believe it represents a true empirical statement about the values expressed by citizens in diverse, democratic modern societies.[3] It is a simple fact that when citizens are asked to articulate their environmental values, they express those values in many ways; and even the most ardent advocate of monism is hard pressed to encompass even most of these values in a single, monistic vernacular. Monists inevitably end in intellectual hand-waving at best and in marginalizing important values at worst.

Second, we embrace pluralism as the best *starting point* in the search for improved theories and expressions. Pluralism encourages us to think of environmental conflicts as problems of choosing among multiple goods, not all of which can be fully supported with available resources, rather than as problems of maximizing a single kind of good such as intrinsic value or economic efficiency. This formulation encourages a search for creative, win-win situations; and sometimes it is possible to form coalitions of citizens and groups who support common *objectives* on the basis of very different *values*—as when, for example, Ducks Unlimited and the Audubon Society collaborated to increase protection of wetlands in the major flyways of migratory birds, the classic case mentioned in chapter 5.

What we need, if we are to live with pluralism, is a process that can create at least temporary agreements to act together, even in the face of disagreements regarding some elements of the science base and in the face of pluralism and disagreement about values. For this process to work, given pluralism and a need to act, we want a method that can do three things. First, it must be capable of supporting action, even when there is uncertainty and disagreement about facts and values, on a day-to-day basis (the action phase). Second, it must provide a method—careful description and experiment—that is iterative and self-corrective and applies to value disagreements as well as to factual ones (the reflective phase). Third, this method must be embedded in a process that is inclusive, open, and democratic. The method will be inseparable from the process in the sense that the process has to be stable enough to allow both action *and* reflection; effective reflection provides feedback about the process, encouraging social learning that will continually improve the process over time. The method we embrace must thus be compatible with, and effective within, a diverse community committed to discourse, deliberation, and cooperative action. We need, in short, an evaluative method that will complement an activist agenda such as adaptive management.

This sounds like a daunting task, but fortunately we are not starting from scratch. Justification of actions, like justification of beliefs, is illuminated by the analogy of Neurath's boat. Pragmatism is committed to a community-level application of a method that expands collective experience through observation and experiment. Pragmatic ethics, like pragmatic epistemology, welcomes diversity and pluralism; faith in the method to sort out errors, however slowly, provides for the rudiments of a self-correcting system for ethics as well as for science. Pragmatism, then, by asserting the unity of the method of inquiry, with both ethics and science based on experience and sharing a common logic of discourse and reason-giving, recognizes no sharp divide between facts and values. The pragmatic method sets out to avoid, rather than to answer, that most puzzling question in moral epistemology: can facts—the collection of data and information—fully support moral judgments? To see how to avoid this question, we need a brief history of the fact-value conundrum.

This question has been the bane of naturalists' existence since at least the middle of the eighteenth century, and it is worth taking a detour into the history of ethics to see how and why a "methodological" form of naturalism is an improvement over the old-fashioned variety. Naturalism is generally reviled, like some shady and persistent loiterer on the edge of ethics, at least partly because of a rather cavalier reference by G. E. Moore to the "naturalistic fallacy," which Moore took to be an error of reasoning. Moore, you see, was defending his own theory of value by which goodness is a nonnatural quality of objects, a quality not reducible to observable qualities. So he described proposed reductions as a fallacy, the naturalistic fallacy. Many philosophers have noted that Moore's attribution of fallacy is largely an unfounded marketing tool for Moore's own, rather eccentric theory of nonnatural value.[4] Surely there is no reason to tar all forms of naturalism with Moore's intellectual slur.

But I get ahead of the story, which actually starts with David Hume, the Scottish philosopher and skeptic who observed in 1739 that speakers often proceed from sentences with "the usual copulations of propositions, *is* and *is not,*" to sentences "connected with an *ought* or *ought not.*"[5] Hume proceeded: "For as this *ought* or *ought not,* expresses some new relation or affirmation, 'tis necessary that it shou'd be observ'd and explain'd; and at the same time that a reason should be given, for what seems altogether inconceivable, how this new relation can be a deduction from others, which are entirely different from it." If we wade through some of Hume's terminology about relations and concentrate on his view of the "copulation" of propositions, we can see that he is making quite a narrow point: that one cannot move by deduction from *is* statements to *ought* statements.

Hume's formulation, however, is unfortunate in another way; Hume's criticism of the move from "is" to "ought" presupposes that *is* statements and *ought* statements are sharply separable, a view that is based on Hume's metaphysical assumptions. It turns out that Hume's "insight" about what kind of deductions are legitimate, though not undermining naturalism as such, did have the more unfortunate effect of setting a precedent for empiricists who, since Hume, have generally sharpened, not softened, the linguistic divide between descriptive and prescriptive discourse.[6]

Hume, who may have intended a rather narrow point about logical relations among certain kinds of sentences, was understood by readers to have established that there exists a "gulf" between facts and values, between descriptive discourse and prescriptive discourse. Given the modern faith in science, and once religious bases for ethics began to fade with the secularization of modern society, it is not surprising which side of Hume's gulf has been considered the slums of the intellect. Whereas science was considered determinate, determined, predictable (at least in principle), and "objective," values, as they were severed from their traditional bases in religious faith and dogma, descended into epistemological purgatory and became expressions of, first, Hume's "sentiments" and later, according to the positivists, subjectively felt emotion. So the separation of descriptive discourse from prescriptive discourse, noted by Hume and accepted as some sort of philosophical gospel since, has gone badly for prescriptive discourse, resulting in a sort of epistemological apartheid, enforced by Hume's uncrossable gulf. The practice of normative ethics in the gulf region has become, as a result, terribly dangerous. Perhaps it's time to reconnoiter.

To use the categories of theories I introduced in section 5.2, naturalisms are epistemological theories. They are theories about how we justify and support claims with moral content. Speaking broadly and historically, theories of epistemological justification of moral norms can be placed in three categories, corresponding to a commonsense separation of possible sources of evidence or relevant knowledge. Some positions, such as Immanuel Kant's theory of rights and individual obligations, seek to base normative judgment, including moral judgment, on reason, "practical" reason, as he said. A second position is sometimes called intuitionism. Moore—just noted because he so colorfully expressed his distaste for naturalism—is often cited as an example of this view, since his belief that the good refers to nonobservable properties apparently left him with little but "intuition" by which to apprehend it. Intuition, one might suggest, is close to what is sometimes called conscience in ordinary speech, the "still, small voice" inside, which urges one to do good. Intuitionist theories, however, do not fare well epistemologically. To what does one appeal when different individuals' consciences demand contrary actions or policies? The third possible source of evidence for moral judgment is experience, observation, science, and descriptive knowledge—the very source of information that was considered irrelevant by Moore to discovering "the Good." Let us call these three epistemological theories about the source of moral knowledge a priorism, intuitionism, and naturalism, respectively, Now in reality, all of these views exist in endless variations and combinations, but here we are interested in the big picture.

In the big picture, methodological naturalism is a form of naturalism. It asserts that the only basis for correction of inaccurate beliefs *and evaluations* is more experience, experiments with controls when possible, careful observation otherwise. Methodological naturalism advocates developing self-corrective processes in public discourse, whether scientific or evaluative. Methodological naturalism does not, however, set out on the basis of purely factual assertions to *deduce* value conclusions. By concentrating on fair political processes, rather than deductions from general principles, we believe the boat of deliberation can be both kept afloat and renewed over time. What is ironic is that contemporary empiricists have accepted Hume's caution concerning derivations of *ought* from *is* statements—and then they have taken this caution to imply that there is an unbridgeable gap between the two kinds of statements. In doing so they have overgeneralized Humes's insight and failed to observe how factual and value assertions do in fact occur within their natural habitat of public debate and discussion about what to *do*. In that habitat it is undeniable that facts and values are all jumbled together and that reasoning about what to do crisscrosses back and forth over Hume's putative gulf, sailing Neurath's boat unceasingly and gradually improving both day-to-day practice and the methods that support more rapid improvement in the future. To separate fact from value, description from prescription, is to do violence to the context in which language gains meaning. As Bernard A. O. Williams has argued, facts and values are not "separate" in ordinary discourse.

Philosophers who articulate theories about the nature of ethics and value have *theorized* that values are separable from facts and that factual assertions and assertions of value have a very different "logic."[7] But in so asserting they refer not to the real world but to a philosopher's idealization, which very likely gains meaning only in conjunction with that philosopher's theoretical stipulations. Such theorizing is sterile and circular; worse, it can dull our ear to the real resonances between facts and values in our actual, day-to-day work of deliberating about and justifying environmental policies within a pluralistic but open and experimental process. Thus it is pragmatists' aversion to sharp dichotomies and their preference to explore a theoretical question in its natural habitat of public, ordinary discourse that allow them to avoid the question, Can values be derived from facts? A better question is, What processes of deliberation are more likely to achieve a proper integration of facts and values in a community's struggle for improved environmental policies?

Pragmatists, then, avoid the issue of crossing the "fact-value gulf" by engaging in ordinary discourse, where the gulf does not exist. We need not deny that it is possible, in some idealized philosophers' language, to separate fact sentences from value sentences or to approach such a separation in laboratory contexts by conscious and artificial efforts. The relevant discourse for adaptive management is ordinary language, however, the language that communities use to deliberate and decide how to use their resources. Individuals and communities express aspirations and face problems and uncertainties in ordinary discourse. Every factual statement in that discourse also expresses values of the speaker to some degree. Our form of methodological naturalism thus avoids the question whether pure facts can imply pure values by avoiding their unnatural isolation from each other.

Following Peirce and Dewey, we declare there to be only one "logic"; it is a logic whereby assertions are challenged on the basis of broader or contrary experience. This logic functions within a diverse community of truth-seekers, who may adopt very different viewpoints about some values and some facts. They act as a community, nonetheless, when they agree to enter a process of deliberation about what to do collectively to solve social problems such as resource degradation and species extinction. We deal with pluralism, then, by adopting an open process of deliberation about action. The process is open in the sense that everyone can participate and also in the sense that no assertions, if offered as justification for what to do, can be considered immune to criticism; all assertions, whether apparently factual or reeking of values, must be open to challenge on the basis of contrary or new experience.

Pragmatists do not expect all reasons, either in science or in ethics, to come in the form of "deductions" from unquestionable facts or principles. All that is claimed by the pragmatist is that, given the right conditions and public attitudes, the method of challenging reasons in an open, deliberative, pluralistic community will, over time, contribute to consensus. In the long run, this same method can ferret out error and decrease uncertainty, but this outcome requires a community capable of social learning. Adaptive management processes must therefore encourage open forums and community advisory committees. The epistemological emphasis, once one adopts a process orientation, shifts from deduction based on prior knowledge to creating fruitful and provocative models and explanations that work in more and more situations. These models and explanations are then, in the process advocated by pragmatists, subjected to more and more experience of more and more truth-seekers, to see if they hold up over time. In ethics, as in science, the search for justifications is always provisional, based on all kinds of current assumptions and possible biases in our viewpoints.

The question arises, of course, whether this approach does not make ethics subjective, relative, or nonrational. But this issue can be sidestepped here, with a brief explanation. In the appendix I explain in some detail the importance of an argument (originating in the writings of Pierre Duhem and restated in more modern form by W. V. O. Quine) that there can be no crucial experiments in science. Every experiment—every experience—is understood and interpreted based on background knowledge. To put the point slightly differently, we never can compare single sentences with a single experience. Language maps onto the world in clumps much larger than sentences, and indeed, once that is admitted—in a context of action—it can be shown that the relevant clump is our entire belief system. Once we reject the sharp separation of facts from values, and once we follow the pragmatists in insisting on a single process of inquiry, it is obvious that Quine's insight applies also to value assertions. Value assertions must be understood as part of a larger pattern of beliefs; they are vulnerable to experience, since individuals often reconsider their values in the face of new or unexpected experience. Value assertions, then, have the same epistemological status as do factual ones. They are both vulnerable to experience, but one cannot model the relationship between experience and changes in either factual beliefs or values as a deductive relationship. So ethics and science are on exactly the same footing. In a process of inquiry, individuals will change their beliefs or values in reaction to new experiences; understanding this reaction is ultimately a matter of psychology and sociology, not of logic. Since this reaction is not strictly speaking "logical" either in science or in ethics, we can only describe and assess the choices individuals make. Ethics, in other words, is no more subjectivist or relativist than science. All assertions must stand the test of new experience. And the best way to submit a controversial assertion to wider experience is to inject it into public debate in which people have a real stake. Public deliberation about what is of value finds its natural habitat in discussions about what to do. Values, as they are encountered in everyday situations—such as citizens or stakeholders discussing what to do to protect an environmental asset—are expressed in ordinary language, where facts and values are never segregated.

Adaptive management and pragmatism thus suggest an alternative empiricist approach to thinking about values in pluralistic, democratic societies. By interpreting naturalism methodologically we make no claims about what "derivations," "copulations," or "deductions" are possible among types of sentences. A methodological naturalist focuses not on deductions of "truth," but rather on encouraging open discussion and deliberation, on ensuring a political process that promotes public discussion and deliberation as a necessary basis for choices. Along the way, the methodological naturalist pays attention to how people learn and how consensuses are formed; the methodological naturalist also pays attention to which vocabularies resonate in public discussions and which goals can gather support in the form of broad coalitions.

For adaptive managers, the epistemological process of justifying values is very similar to Neurath's strategy. In any given situation in which action is forced, there will be some beliefs or norms that gain more acceptance than others. The pragmatic method uses the stronger "planks" as support while testing other planks and as a place to "stand" while replacing planks determined to be weak. In a situation in which action is forced and knowledge is limited, the best strategy is to learn by doing. For example, Kai Lee describes how the Northwest Planning Council, united by the objective of increasing the percentage of juvenile salmon that reach the open sea and by an agreement to achieve this goal with minimal economic disruption, undertook pilot projects and experiments to compare trucking salmon smolts, building spillways, and other methods as means to increase survival of juveniles.[8]

Again, Neurath's analogy helps. His epistemology does not encourage doubting every shared belief and value of the community but, rather, bringing the best science available to bear upon areas of disagreement. Public deliberation that includes expressions of values as well as discussion of science is essential to identifying disagreements that separate community members and lead them to advocate differing policies. The problem, for the adaptive manager who is advocating new policies, is not to justify all of our beliefs at once, or some of our beliefs in isolation, but to focus experience and science upon crucial disagreements in the situation at hand. Discourse does not occur discretely, in units of fact and units of value; and any attempt to represent it as such will falsify the role of language in real life. Having deserted the attempt to derive one kind of "truth," ethical judgments, from other kinds, pure facts, we can once again resort to Neurath's strategy and apply that strategy evenhandedly to descriptive and prescriptive discourse. The strategy is to develop a sufficient consensus regarding some unquestioned values, some solid and uncontroversial information, and to put these agreements to work in experimenting our way through areas of uncertainty and compromise.

Even if one abandons the search for deductive relations among atomic fact-sentences and atomic "oughts," however, it still seems reasonable to ask about the impact of experience, observation, and sensory information on our evaluations. Pragmatism accepts that our language embodies both fact and value; in any argument one can appeal to facts and to values; but if one appeals to a value, no less than if one appeals to a fact, one must stand open to challenge on the basis of new or different experiences. No principles are rated as unchallengeable within the deliberative process.

Naturalism of the methodological variety suggests new relationships between social sciences and environmental policy. Since we are setting aside ontological questions about the nature of value and right action, we can turn our attention away from trying to deduce applications of general principles, to a process in which diverse participants make claims and counterclaims. The social sciences can now become important participants in the general deliberations about what to do, by providing evidence through questionnaires, surveys, and inference from behaviors about the interests, goals, and values of the general public, as well as the more restricted group of active participants in the management process.

As noted in section 5.4, we reject the assumption that only evidence about individual, "sovereign" preferences is relevant in deciding what to do. Methodological naturalism is dynamically applied within an ongoing process of social learning. Rather than simply eliciting individual preferences in isolation from ongoing deliberation, social scientists should be actively involved in the ongoing search for coalitions and consensuses by studying actual citizens and stakeholder groups that participate in actual processes in actual situations.[9] Although most citizens are unwilling to invest the time and effort necessary to participate fully in processes of deliberation, model-building, criticism, revision of beliefs, and reconsiderations of values, the good news is that the decisions of participants who *do in fact* participate actively in an iterative process can be taken as an important approximation of what the whole community would decide if each member were willing to make the same investment. The stakeholder-representational process may be imperfect, but it might nevertheless—if it represents all the stakeholder groups—provide a useful proxy for the counterfactual outcome of a full-society deliberative process. Much more is said about this use of this method in sections 7.5 and 11.5, but the point is that information about the opinions of active participants should be relevant information in choosing what to do. The outcome of a complex and ongoing process of monitoring, negotiation, pilot projects, focus groups, and revisions based on such is surely as important a piece of information for decision makers as would be the "isolated" opinions of self-interested individuals.[10]

This epistemological method must help us to do two things. First, it must help us, in situations where there is some agreement about the factual situation and the goals the community wants to pursue, to use these agreements to find a viable policy. Doing so will often mean seeking a policy direction that is "robust" over a wide array of values and over considerable variation in factual beliefs of participants. This task of our epistemological method is to help us to move from agreement to action, by providing a means to identify policies that are supportable by many groups with differing viewpoints and interests, and to provide guidance as to when action is justified on the basis of given agreements and compromises. If there is adequate trust among parties to allow experimentation, the promise of such probes to reduce uncertainty and inform future actions should encourage support for pilot projects and other limited actions. This epistemological task, which we can call the problem of legitimating action, requires a method that leads to robust policies, allowing us, when we are operating in the action phase, to progress from real agreements about the situation and preferred goals to actions that are justified in a particular situation. They are justified in the sense that a group of participants with diverse backgrounds, with some shared beliefs and goals, and with some disagreements can reasonably support a particular policy as an improvement over the status quo.[11]

Our epistemological method must also allow us, especially when we move into the reflective phase, to survey a larger range of experience and to criticize, revise, and reform our preferences and value positions. We need to be able to sustain an ongoing public dialogue that allows us to question values, even ones that are widely held, and to support a process of social learning in which members of communities, upon seeing the consequences of acting in pursuit of particular values, may come to question and revise some of the values they have been acting upon.

Both of these tasks, as noted, are in one sense more epistemological than moral. The shift to an active, experimental science of management encourages us to *justify policies,* not to argue about the correctness of general theories of value. This does not, of course, preclude hard debate about values, because citizens' values will be a major part of any justification for a proposed or opposed policy. Any value that is cited in favor of a proposed policy is open to challenge, so values will often be at the center of disagreement and deliberation. But what we have learned is that there is no point in discussing whether "pure values" can be supported by "pure facts"; there are neither. The question is whether this policy can be supported given the mix of beliefs and values exemplified in the community. What we seek is a method-guided process that will allow us both to act reasonably in the face of day-to-day challenges and to engage, in the longer run, in an ongoing process of identifying weak elements—whether factual or normative—in our current belief structure, undertaking steps to repair or replace those weak elements.

To apply this method in an adaptive management process, one would ensure that all stakeholders in decisions have a voice and that they are free to challenge any of the beliefs and evaluative statements that are given to explain and justify environmental policies. Only in such an inclusive community of inquirers are we justified in accepting unquestioned assumptions as even a temporary basis for cooperation and community action. If one hopes to create a practical epistemology of environmental values, one would also empower these stakeholders, either by providing them public funding or by giving them a voice in the choice of research topics undertaken with public funds, to seek evidence that might contradict assertions made in support of policies. If there is a lot at stake in a decision—if important social and community values are affected—and if participants have the means, they will root out errors and expose short-sighted and selfish goals as a part of an ongoing and iterative public discourse. Of course this result will not emerge from just *any* process: the process must treat all participants and their opinions with respect, and the process must be directed at truth-seeking, not be subverted to serve the interest of power or of any exclusive subset of the community.

Our methodological form of naturalism has the advantage that it focuses, at least initially, on *political procedures,* procedures designed to arrive at acceptable policies even in the face of diverse belief systems, perspectives, interests, and values distributed in various combinations in the population. The procedural approach tries to encourage wise action, based on current agreements and coalitions, even as the process of reason-giving, deliberation, and justification proceeds. That the process of seeking better policies is ongoing takes the pressure off getting "the right answer" or "the right evaluation" and allows discourse to proceed at the reflective level while agreed-upon actions probe possible system behaviors, with monitoring to reduce uncertainty and in some cases to provide new evidence that will support changes in policy direction.

The point, in my view, is not to seek as many facts and generalizations as possible and try to *derive* from them conclusions about what to do; instead, scientific models and understandings shift and reset the context in which environmental problems are conceptualized and discussed. Facts, I will argue in the next section, can cause us to reframe value questions in crucial ways. This does not mean that value questions get resolved by accumulating facts—there is no "entailment" of values by facts—but facts can alter the way we see a situation, changing the rational, factual, and moral considerations that are considered relevant, revealing new areas of moral responsibility. Facts, I will argue, have a very special role in place-based, committed, democratic deliberation; facts can disturb comfortable assumptions and cause us to ask new and disturbing questions.

So it is useful to recognize that the relationship between factual information and evaluations shifts according to the context. In a settled situation in which management goals are well agreed upon, facts and science tend to dominate, because there is apparently a working consensus about goals, and science is employed to determine what is possible and necessary given certain goals. But when values become controversial, facts come under scrutiny not only for their truth but also for their applicability and importance in a given situation. If, for example, a forested community agrees that it should pursue maximum sustainable timber harvest from the forests in the region, there is a whole set of scientific information and research that will help them do this. But if the community becomes divided about values and factions come out in support of competing and conflicting values such as managing the forests to protect biodiversity, or traditional landscape features, or whatever, then we can expect very different science to become relevant, as well as very different forms of uncertainty. In such situations of ferment, a new perspective, a new way of looking at the world, sometimes emerges. These shifts in perspective can lead to a process in which values and evaluations are recontextualized and reconsidered.

6.3 Re-modeling Nature: Learning to Think like a Mountain

According to the strategy of methodological naturalism just advocated, we ought not to expect science to provide a "justification" of our moral inclinations in terms of "hard facts"; indeed, the idea that science provides value-neutral, "pure" facts that should be privileged as the basis for value assertions or policy recommendations is at best an artificiality. In the rough-and-tumble of policy discussion, facts and values are all jumbled together and the lines of justification in public deliberations constantly crisscross the artificial divide between facts and values.

Nevertheless, as students of the policy process, especially in the reflective phase of our discussions, we may find it both useful and important to temporarily separate out bodies of information generated by the various special sciences and consider this information as distinct as possible from the push-and-pull of public values and interests. In this context we can recognize a difference between a careful presentation of data gathered by a limnologist on the response of a lake to surges of excess nutrification and the testimonial of a long-time resident of the lake area who, with tears in her eyes, expresses her dismay at the losses of water clarity that have occurred over the years. Our ability to see this difference—and a universal recognition among adaptive managers that it is important to have "objective" science as a basis for policy—leaves us with an important question not resolved by our rejection of the doctrine that there is a logical gulf between descriptions and prescriptions. We still want to know whether and how breakthroughs and advances in science affect human values and evaluations. We cannot sufficiently separate factual assertions from the values they implicitly embody to study logical relations among pure facts and pure values. We have just learned that this interaction is not well captured as a deduction or any other "logical" operation. It is nevertheless undeniable that exposure to new facts and information, when integrated into ordinary discourse and animated by values and normative concerns, can lead to a revision of our moral beliefs. To understand this phenomenon, it turns out that we must look at experiences of individuals who have reacted creatively to factual information.

One way to explore the phenomenon is to recount a few anecdotes regarding conservationists of yore. Let us start with John Muir, whose life was changed—according to his own account—by an encounter with a rare orchid in the remote forests of Canada. It was the height of the Civil War, and Muir (a pacifist) was sojourning in the Canadian woods, ostensibly gathering botanical samples, because he (mistakenly) thought he was being sought by the authorities for draft evasion. Muir came upon a specimen of the rare and beautiful orchid *Calypso borealis* in the trackless and remote woods and sat down beside the plant and cried with joy.[12] Decades later, Muir described this experience (along with meeting Emerson much later) as one of the two most important experiences of his life. Muir saw a rare orchid but interpreted its location in the wilderness as establishing that God's beauty was not distributed for maximal enjoyment of humans. This recognition dislodged just enough religious orthodoxy to allow Muir to see the orchid as a fellow traveler and to begin a process that led to Muir's adopting pantheism and, at least in his private thoughts, a nonanthropocentric ethic. A botanical observation caused Muir to reframe his thoughts about humans and nature and eventually caused him to significantly adjust his moral beliefs.

What is important here is that the observation had more of a psychologically catalytic effect than a "logical" effect. Sure, we could split hairs and try to separate the "observation" from the interpretation and the act of creative transformation it catalyzed in Muir's understanding. But this would surely do violence to the holistic experience Muir described. The experience was more than the imaging of the flower; Muir's lifelong passion for collecting botanical samples, mixed together with his feelings of isolation and alienation from human society, created a context ripe for a creative restructuring of Muir's theoretical *and perceptual* experience that day. All of this was part of the experience; further, the experience continued as an important determinant of Muir's future. The encounter with the flower was enriched and given new meaning throughout Muir's life as he rearranged his thinking about creation. By the time he was jotting notes in his journal while walking from Kentucky to Florida several years later, Muir was attributing rights to alligators, bears, and rattlesnakes. Muir's rich and creative voyage was neither a matter of "pure" observation nor a moral conversion in any simple sense. An observation, in a particular context, caused Muir to place that momentary observation within a new intellectual firmament; it was more of a transformation of experience than of changing value theories. Most importantly, it was not a deduction or a conscious reevaluation; it was simply an experience that took on deeper and deeper meaning as Muir's beliefs and evaluations changed around it. This basic perceptual shift of course opened new paths of thought for Muir's scientific interest in ecology and also led to a deep moral reevaluation, an embrace of pantheism, and a robust nonanthropocentrism. My point is that the "key log" that led to these subsequent changes was a shift in perception—Muir *saw* the orchid in a new way, as existing independently of his motives and motives of his kind. Once he had seen the orchid in this new way, he was able, over the next few years, to root out more dogma and eventually articulate an inclusive philosophy of pantheism, complete with batteries of rights for all of God's creatures, and especially plants.

Let us look at another case, briefly. Thoreau, in the puzzling and sometimes jarring chapter of *Walden* called "Higher Laws," lays out his philosophy of humans and their place in nature. After stating that he loves "the wild not less than the good," and endorsing hunting as an experience necessary for young men to thoughtfully enter the realm of manhood, Thoreau begins a paragraph with the phrase "It is a significant fact that . . ." Thoreau's reasoning in that paragraph meshes nicely with Muir's in his botanical experience, although the style is far different and the experience is zoological. Thoreau reports the "significant fact," drawn from an entomological textbook, that butterflies and other species of insects, when they pass from a larval stage to a later winged stage of life, eat much less in the latter, "perfect" state. Thoreau goes on to explicitly discuss how the maturation of a society, maturation beyond what he called a "savage" state, is tied essentially to a change in perception, a change in which the individual comes to experience nature not as mere material for consumption and use, but as emblematic of a higher life of freedom from physical demands. Thoreau treats the fact that the butterfly gains flight and freedom by transcending its heavy animal body as an inspiring metaphor for human transcendence of a consumerist lifestyle: "The gross feeder is a man in the larva state; and there are whole nations . . . , without fancy or imagination, whose vast abdomens betray them."[13]

Again, as in Muir's case, we encounter a normative insight, one that captures the essence of the self-transformation that Thoreau set out to describe in his famous metaphorical "year" at Walden Pond. Here Thoreau uses a generalization from a textbook as his starting point; but he often compared direct experiences of animals with human action and behavior. Again, Thoreau's insight is presented not as a deduction from some general or self-evident truth, but as an insight suggested by an analogy from nature, an insight that is given life by drawing out the elements of the analogy. Thoreau is trying to *encourage* moral change by describing experiences and making generalizations from experience; he does not operate in terms of premises deduced from foundational moral principles. He rather redirects moral attention by citing facts that suggest analogies to human experience and behavior. In this way he uses information and analogy to recontextualize moral decisions.

Finally, in perhaps the most important recontextualization in the history of environmental thought, already cited in this book, we return again to Aldo Leopold's simile of thinking like a mountain—the conceptual leap by which the observant Leopold reconfigured the spatiotemporal landscape on which he had experimented with his wolf removal policy. As usual, Leopold was onto something profound, and we should pay attention.

Many readers of the essay "Thinking like a Mountain" have concentrated on Leopold's poetic and moving description of the "green fire" dying in the eyes of the old she-wolf: "I realized then, and have known ever since, that there was something new to me in those eyes—something known only to her and to the mountain."[14] In the face of such poetic power, it is natural for the reader to pick up on the empathy Leopold felt for the dying animal. A careful reading, however, carries one far beyond the sentiment in Leopold's poetic expression. Near the beginning of the essay, its governing idea is stated: "Only the mountain has lived long enough to listen objectively to the howl of a wolf." The wisdom Leopold saw in the old wolf's eyes is given operational meaning in the next paragraph when Leopold describes his observations, apparently years later, of the scientifically evident effects of his wolf eradication program. Leopold does not respond to the death of the wolf by exploring the animal rights or animal welfare implications of hunting. No, he interprets the death of the individual wolf as symbolic of the degradation of the ecological systems that covered the mountains of the southwest territories. However moved he was by the green fire dying, Leopold's deepest concerns are directed at the observable changes he sees on the landscape. Leopold is using the death of an old wolf to symbolize the disappearance of a species from the ecological systems of the Southwest, and he begins to see that those changes bring anything but a "hunter's paradise." The loss of wolves from the mountain is important because, in Leopold's personification of the mountain without wolves, the mountain must "live in mortal fear of its deer."[15] His analysis drives him to the conclusion that, as a conservation manager, he must manage the mountain, not single species, and that although the system has considerable resilience, violence of increasing pace and scale threaten that resilience. Having moved from wolf to mountain by analogy, Leopold expresses concern for the "health" and the "integrity" of ecological communities that change at the pace and on the scale of ecological change. As noted, the theme of the little essay is time.

In mentioning time and duration, Leopold was referring to "Marshland Elegy," another essay in *A Sand County Almanac,* where he carefully identifies three scales of time.[16] He differentiates human, experiential time from what can be called "ecological" time and "geologic" time. Human, experiential time is illustrated at the beginning of the essay as the narrator tells of the arrival of cranes at a crane marsh; Leopold accentuates the "slowness" with which we experience time when we are waiting for something—pots to boil or cranes to land in the marsh. Ecological time, the second temporal scale of interest, is determined by the pace of colonizations of plants and animals, by the forces of competition among species, and by the development of symbioses and predatory relationships. Changes on these scales, which range from decades to a millennium or so, modify landscapes and form the habitats that animals, including humans, live within. Geologic time, sometimes called "deep time" today, refers to the slow processes by which the physical features of a landscape are shaped. Leopold's point in the essay on the crane marsh is that humans must recognize that they have now gained the power to be the dominant force in changes on larger scales, that we ignore long-term consequences of our actions at great risk, and that the use of our technological power greatly expands our responsibilities as actors in a multigenerational drama. That is the lesson to be drawn from Leopold's famous insight that he had erred in removing wolves from the wilderness—he failed to think about the long-term ecological impacts of his action. He wanted a large deer herd for hunters, but when evaluated over a longer period of time, his actions failed. Learning to think like a mountain is to expand one's temporal consciousness to see humans as actors not just on a short-term economic stage, but also as increasingly dominant actors on the ecological scale, capable of changing not just the actors on the stage but also the very stage itself. Thus Leopold recognized that in accelerating ecological change by removing species, for example, we can also impact geologic formations—such as mountains—by increasing erosion. He was warning that with new and powerful technologies, and with ever-grander dreams of human dominion, we are in danger of destroying (or at least diminishing and uglifying) the very habitat that, having slowly evolved, supports us and is, to us, a beautiful home. The most central insight of Leopold's simile, then, is an insight about the importance of scale, both temporal and spatial, in our thinking about environmental management.

What these three examples—from Muir, Thoreau, and Leopold—have in common is a mental transformation that is triggered by a meaningful experience. For each man, an encounter with new information—or, one might also say, a newly meaningful encounter with available but hitherto overlooked information—reconfigures the architecture of the mental model by which he relates humans and nature. Factual information, in these cases, creates a new perspective, reframes a question, causes a reconceptualization on such a basic level that worldviews are altered and a new perspective emerges. Such a reframing can lead, as in Muir's case, to an associated change in values and attitudes, but the path from perceptual reconfiguration to moral implications may be quite indirect and may not resemble a deductive relationship between factual and evaluative sentences. Indeed, the relationship may appear almost magical; its mysterious nature, however, does not detract from its importance in understanding how facts and values affect each other in real-world situations.

The importance of these experiences, both to the individuals who had and recounted them and as meaningful symbols of environmentalism, lies mainly in the way they encourage a change in perception or focus. Most environmental ethicists, I think, have been too quick to emphasize the moral aspects of such experiences, especially as they bear upon moral theory. In each of these cases, although moral concern is evoked, the intellectual and emotional transformation involved has more to do with rethinking the human place in things than with a change of values. What does this generalization tell us about how to better understand environmental values and evaluation? One thing it does is to call into question the general strategy of extensionism, as pursued by environmental ethicists and as described in chapter 5.

Much of the writing of environmental ethicists is premised on the assumption that modern societies damage nature because such societies believe a false theory of values, the theory that all and only humans have intrinsic value. In challenging this false theory, environmental ethicists, at first, formed intellectual alliances with animal liberationists, arguing against anthropocentrism and in favor of a countertheory, that at least some parts of nature, such as animals, have intrinsic value. In an important 1980 essay, my friend and philosophical raconteur J. Baird Callicott pointed out that since animal liberationists had concentrated on individual domestic animals, such an assimilation of environmental goals to those of animal welfarists or animal rightists would not support many of the key goals of environmentalists, who seemed more concerned with ecosystems and wild species.[17] He proposed as an alternative that one could follow Aldo Leopold and become a "holist" by attributing intrinsic value to ecological systems ("the land community"), concluding that the struggle to settle upon a new theory of value for environmentalists is "a triangular affair." Animal liberationists occupy a corner of the triangle, as do of course traditional anthropocentrists; but one must also consider holism to be a possible basis for a new environmental ethic.

Perhaps most readers, like myself, were so impressed with the plausibility of the argument that the affair was indeed triangular, that they were slow to recognize that Callicott had described not a triangle but a truncated rectangle. To see this, consider a little diagram (fig. 6.1) that I put on the blackboard on the first day of every course I teach in environmental ethics.

For those beginning to study the field of environmental ethics, my diagram helps to sort authors they are reading into categories according to the moral basis on which they support their views, and it helps to define the major fault lines dividing the ethical terrain. It turns out that as we progress through a standard anthology, we read a number of authors who can be placed in categories IA, IB, and IIB. And of course these three categories fit perfectly with the three corners of Callicott's triangle. Category IA corresponds to the "standard ethical position" (what Callicott called "moral humanism"), un-

	A Anthropocentrism	**B Nonanthropocentrism**
I Individualism	anthropocentric individualists *traditional ethics and mainstream economics*	nonanthropocentric individualists *animal liberationists*
II Holism	anthropocentric holists *E.g., Leopold and Norton*	nonanthropocentric holists *E.g., Callicott*

Figure 6.1 Two intersecting distinctions in environmental ethics

questioned at least since the Enlightenment, that all and only human individuals have moral standing. Category IB is represented by animal liberationists (Callicott called them "humane moralists"), who extend moral standing to individual animals, attributing rights or legitimate moral interests to them, or at least to those animals above some level of complexity. The ethically holistic position, IIB, which considers species and ecosystems to have moral standing, was attributed to Leopold and unabashedly advocated by Callicott himself. Position IIA seemed impossible to Callicott because of his commitment to monism: if wholes are valued, then they must be valued *in opposition* to human individual goods (IIB) or else one must choose an individualist position. But Leopold, by identifying a plurality of values distributed over different landscape scales, could embrace IIA, the view that we should value *both* humans *and* ecological wholes.

Callicott, assuming that the antidote to environmental destruction must be found in a new moral theory that spreads intrinsic value more widely through the community (extensionism), says, "the good of the community as a whole, serves as a standard for the assessment of the relative value and relative ordering of its constitutive parts and therefore provides a means of adjudicating the often mutually contradictory demands of the parts considered separately for *equal* consideration." The holistic nonanthropocentrist, such as Leopold and himself, Callicott said, "locates ultimate value in the biotic community and assigns differential moral value to the constitutive individuals relatively to that standard." According to this view, the worth of individuals of all species is a function of their rarity and their contribution to the system. Callicott recognized that this ethic would be "somewhat foreign to modern systems of ethical theory" but did not shrink from applying the holistic value scheme to humans, who, he allows, have become ecological liabilities because of their large populations. Applying the holistic ethic, Callicott concludes that, correcting for the weight of the organisms, the human population should be "roughly twice that of bears."[18]

Because he had framed the question as one of determining who or what has "moral considerability," Callicott formulated the moral problem as one in which humans, being so numerous and so resource-hungry, neglect other moral entities in nature that have not been afforded fair access to resources according to the traditional, anthropocentric theories of ethics of the Enlightenment. To balance this situation, and to operationalize the idea of moral considerability for nature, Callicott simply attributed moral standing to natural communities and wild species, elevating them to the role of justified competitors for the bounties of nature. Thus, he implied that, once Western civilization had followed Leopold and adopted a holistic nonanthropocentric ethic—an act he described as similar to attributing intrinsic value to one's child—ecological wholes would be treated as legitimate claimants on available resources.[19] This view seems to place humans in competition with nature itself. If humans are using more than their share of resources and ecosystems are suffering for this excess, then human populations must be curtailed and human consumption minimized to reestablish a balance among individual and holistic competitors for nature's resources.

Since this essay was published in the early 1980s, Callicott has been forced to back off considerably from this strongly holistic viewpoint, because it was quickly pointed out that in such a value scheme, interests of individuals, whether human or nonhuman, would regularly be sacrificed for the good of systems and species. Given that humans are neither rare biologically nor, apparently, necessary ecologically, Callicott's reasoning seemed to point toward grimly misanthropic policies. Tom Regan, for example, accused Callicott and (assuming Callicott had correctly interpreted Leopold's holistic tendencies) Leopold of "environmental fascism."[20] Callicott quickly retreated to a weaker form of holism whereby ecosystems and species could have legitimate claims and interests, but since the obligations to the land community were "layered upon" obligations to family and other humans, those obligations could not override the more intimate and prior obligations to family, clan, or humanity generally.[21] In this way Callicott backed off from ever favoring whole systems over individuals when conflicts arise among them, by subordinating holistic ecological entities to human individuals; his revised position amounts to a form of human favoritism under the label of nonanthropocentrism. What has gone wrong here?

First, notice that a central dilemma is created by the way Callicott formulates the moral issue. Because he assumes that human destruction of nature is caused by humans living according to a false belief that all nonhumans lack intrinsic value, he challenges this theoretical belief, thereby leveling the playing field by attributing such value to nonhumans. Correctly perceiving that Leopold, a hunter and a naturalist, was more interested in the perpetuation of ecosystems and species than of individuals, Callicott thus found a third corner of the theoretical rectangle, distinct from the moral individualism of both moral humanism and humane moralism, by attributing such value to ecological wholes, including the "land community." But this problem formulation, and Callicott's response to it, confronted him with a nasty dilemma. Either he could treat the intrinsic value of ecosystems and species as similar to, and commensurate with, the good of human individuals—as would be suggested by his analogical use of familial love and caring as his interpretation of intrinsic value attributions—or he could treat the intrinsic value of ecosystems and species as representing a new kind of intrinsic value that is not commensurate with the intrinsic value we regularly attribute to human individuals. Neither horn of the dilemma, it turns out, is tenable. If Callicott assumes parity among individual and holistic intrinsic values, he seems committed to adjudicating the competition between humans and ecosystems by noting that humans, with exponentially growing populations, must be denied their previously unchallenged rights to consume resources. Individual interests, whether human or nonhuman, will thus be vulnerable to being overridden by fair demands of ecosystems for resources. Once the fascistic tendencies of this solution to the moral dilemma were pointed out, however, Callicott quickly retreated to the other horn of the dilemma, suggesting that the claims of nature and the claims of individuals exist in different realms of concentric obligations. Familial and tribal responsibilities, as well as later commitments to the intrinsic value of all humanity, are thus rendered incommensurable; the needs of ecosystems, no matter how much rarer they are than humans, cannot trump obligations based on human relationships. But this resolution of the problem either provides us no basis for decision when wholes are threatened (because the decision requires comparing incommensurables), or else it simply reinstates human-centered ethics disguised behind a myth that prior human recognition of familial and personal obligations establishes their moral priority.

What has gone wrong is that Callicott undertook to extend our moral commitments and obligations before he had reconfigured our perceptual world. Neglecting the central insight of Leopold's simile of thinking like a mountain—the necessary reconfiguration of the world into multiple scales—Callicott tried to solve the moral dilemma of humans versus nature on a single scale. He did not begin by correcting the short-term bias of most human thought and concern and then reconceive the problem of valuing humans and nature within a complex, multiscalar system. Learning to think like a mountain is, first of all, a perceptual shift that may prepare the way for new moral thinking, but only after the moral question has been reinterpreted within a multiscalar world. In this new world, value exists on multiple levels and unfolds over different horizons because it is enmeshed in different dynamics. Many of these dynamics may be of interest to humans, but since the dynamics are independent of each other—because they exist on different scales—we should not be tempted to see their demands as in direct competition with each other. We thereby avoid Callicott's dilemma. The consumptive values of human individuals exist on a short-term economic scale and are associated with a relatively rapid, individualistic, economically organized dynamic, whereas human concern for ecosystems and species ("the mountain") unfolds in the multigenerational frame of ecological change.

For more than a decade Callicott has championed monism, arguing that a pluralistic theory of environmental values will necessarily lead to unacceptable relativism with respect to moral claims regarding others. If he were to give up monism and his insistence that a single theory must encompass all human value, he would see that there are multiple types of human value, that these values emerge on different scales of the system of interactions between humans and nature, and that individual values exist on different scales than do multigenerational, communal values. This separation of values onto different scales reduces direct competition between them, since different dynamics will be associated with their production. More importantly, locating individual and multigenerational values on different scales of the system would allow Callicott to avoid the debilitating dilemma he faces in his account of environmental values. He could avoid the apparent implication of a monistic holism, that one must choose between environmental fascism (if values of wholes override values of individuals) or a collapse into individualism and anthropocentrism (if obligations to individuals, especially to human individuals, override the values of wholes). Such an approach responds to the apparently unavoidable pluralism of any "model" for facing wicked problems, and it would embody Leopold's call for thinking like a mountain within the scale-sensitive, multiscalar, open, self-organizing systems of nature. Next I explore how such an operationalization might be accomplished and reflect on the possibility of linking human, scaled values with the dynamics of a describable physical system.

6.4 Hierarchy Theory and Multiscalar Management

We can see the more positive implications of Leopold's important simile by considering again the three central principles of adaptive management, as introduced in chapter 3. These axioms are the endorsement of an experiential, experimental approach to management; a statement that natural systems, as managed, are multiscalar; and a statement that all observation, measurement, and activity are experienced from some identifiable point in a larger, dynamic system. We noted that the first principle stands as the defining characteristic of adaptive management; and we have just now noted that Leopold, more than today's adaptive managers, realized that evaluations as well as descriptions must be tested experimentally. The second two principles, of multidimensionality and internal location of observers, embody an ecological theory called hierarchy theory (HT), which has been introduced into ecology as a way of relating processes that occur on different scales.[22] Adaptive managers have incorporated the basic axioms of HT into their principles, providing themselves with a schematic treatment of space-time relations. We might say, then, that the first principle provides the motivation for modeling natural interactions, whereas the other two principles (HT) provide the "architecture" of a set of models that we expect to be useful in management situations. Holling, for example, expresses HT as follows: "The landscape is structured hierarchically by a small number of structuring processes into a small number of levels, each characterized by a distinct scale of 'architectural' texture and of temporal speed of variables."[23]

Throughout the last half of the twentieth century, ecologists—who were often perplexed by stability and change and how to measure it—became increasingly aware of and vocal about the important role that scale plays in ecological science.[24] Whether one describes a system as stable, cyclical, or chaotic has a lot to do with the scale on which one probes the system. Even more basically, ecologists were constantly reminded that when different ecologists study similar systems on different scales, they report different dynamics and emphasize different relationships. So the field of ecology desperately needed a way to organize complex space-time relationships, and hierarchy theory, the structure of which was borrowed from general systems theory, was developed to respond to these problems. Speaking technically, hierarchical models are specifically those models that are structured by two defining assumptions: (1) all observation and measurement must be taken from somewhere within the complex, dynamic system that forms an "environment," and (2) smaller subsystems change at a faster rate than do the larger, encompassing systems that form their environment. HT can thus be understood as providing a collection of models, designed to organize space-time relationships in a multiscalar, complex ecological system. In these models, objects appear on one level as agents, but when viewed at a higher level of the hierarchy, they join with other objects, constituting a collective "agent" that acts in a larger dynamic.[25]

Functionally, the axioms limit the types of models we consider and thus give shape to the world we experience. Methodologically, we can use hierarchical models to organize information about the physical world onto three scales. The "focal" scale is the level of system that is being studied. From any given focal scale, it is possible to move down to level −1, where one observes component elements of the focal level. Observations and explanations on the −1 scale track the mechanisms and dynamics driving activities on the focal level. If, however, one wishes to understand the context in which the focal-level dynamic functions—the constraints placed on the focal dynamic by the larger, slower-changing environment—one looks at the focal dynamic as one element in the larger dynamic at level +1. Axiom 2 should be interpreted as saying that if two dynamics exist on neighboring levels or scales of a system, then they differ by an order of magnitude. For example, the dynamic driving the birth and death of individual animals must be such that multiple generations of births and deaths will occur in the process driving the larger, contextual environment—the level +1.

Ideas similar to HT had been kicked around in graduate seminars and in discussions of ecological structure for decades, but the first fairly formal treatment of the subject was published by T. F. H. Allen and Thomas B. Starr in 1982; a subsequent book was published in 1986.[26] The somewhat abstract speculations of theorists were then given more empirical content with the development of hierarchical patch dynamics as a working hypothesis about landscape constitution and change over multiple scales of time and space, and now a considerable bibliography exists on the subject of hierarchical organization in ecosystemic studies.[27]

What strikes me especially is how similar HT is, in its basic insights and supporting intuitions, to Leopold's simile of thinking like a mountain.[28] Leopold's analysis of his actions (the campaign he waged to destroy wolf and mountain lion populations) and their outcome (starving deer and degraded vegetation) was a practically useful exemplar of the kind of model HT proposes. To think like a mountain, he had, first, to locate himself on the mountain, observing the mountain from within its context, a context that is constrained on all sides by dynamics unfolding on multiple scales. Human management choices and affected dynamics were taken to set the focal level of Leopold's observations.

From that focal level, Leopold could either look downward in the hierarchy, analyzing the mechanisms and motives of individual interactions of deer, wolves, and hunters, or he could look upward and explore the constraints affecting the viability of his management choices. Within this framework of analysis, Leopold recognized that his original evaluation of the situation, which presupposed a relatively stable system that would not be permanently altered by the severe action of eradicating all wolves and most mountain lions, focused only on the short-term relationship that created a temporary increase in deer populations and greater opportunities for hunters. For Leopold, learning to think like a mountain was to recognize the importance of multiple temporal scales and the associated hidden dynamics that drive them. These normally slow-scale ecological dynamics, if accelerated by violent and pervasive changes to the landscape, can create havoc with established evolutionary opportunities and constraints and threaten the society with extinction. A society, in Hadley's words, has found the "truth" if its cultures are adapted to the physical system in which its members live. But Leopold discovered that success on short-term economic scales, especially when associated with "violent," technologically supported tactics, can set in motion destabilizing ecological processes and threaten the contextual processes that support individual uses of the environment.

Hierarchy theory, if built into an activist, adaptive management system, then, provides a Darwinian unification of Hadley's model for cultural/societal competition and evolution with Leopold's ecological insight about wolf management. He uses similar reasoning to determine that the grazing culture in the Southwest was, despite economic growth, failing as a long-term adaptation. The people who transplanted the European grazing culture did not have the "truth" about the new context in which they were acting so confidently. They did not have the cultural practices and adaptations that had been honed by generations in that place by previous, indigenous cultures. Whereas Hadley had assumed that capitalism and boosterism, combined with technological growth and development, would work everywhere to the same advantage they had in Europe, Leopold relied on his horseback observations to show that the transplanted European culture—perhaps because its *values* were inappropriate in a new ecological context—was racing toward its destruction. Most people in the transplanted culture had not lived through a hundred-year drought. Leopold's careful ecological observations convinced him that in the longer run, over multiple generations, the imported, growth-oriented system was not well adapted to the arid Southwest, with its periodic droughts. And Leopold realized that wolves are key players in a slower dynamic, a dynamic that trims the deer herd to fit the vegetative cover of the mountain.

So thinking like a mountain can be seen as an exemplary model of multiscalar relationships and how they affect—and are affected by—our decisions. This model involved viewing a multiscalar system from the perspective of an actor inside it and anticipated the spatiotemporal insights of the HT axioms. Further, Leopold recognized that his original evaluation of the policy of wolf eradication took into account only its impacts on hunters' short-term welfare, a too-short scale of time to see important impacts. He learned that total removal of wolves exerted too violent an impact on the system; in the long run, the result was starving deer, stunted vegetation, and erosion of the mountainsides. Surely, it was a satisfying intellectual bonus to Leopold that this physical model dovetailed perfectly with the Hadleian model of Darwinian historical selection of cultures that are well adapted to their habitat.

Leopold had, then, a fairly complete—if metaphorical—characterization of the set of models we today call hierarchical models. He clearly saw that such models were important for monitoring and management because they could help him to organize data and analyze what was going on at different temporal horizons, in more subtle and slower-changing dynamics.

As noted, Leopold was considerably more advanced than today's ecologists and adaptive managers in one important respect. He saw that the reasoning that compelled him to pay attention to multiple dynamics operating on different scales if he wanted to predict impacts of his activities applied also to any attempts he might make to intelligently evaluate those impacts. Evaluations, no less than models of the physical relationships, had to be made from within the system, including an accounting of unintended consequences of our probings and our actions on larger-scale systems that support further, and future, productivity.

When Callicott interpreted Leopold as an extensionist and a moral holist (square IIB in fig. 6.1), he missed Leopold's most important discovery, the recognition that we act, learn, and evaluate within a multiscalar world. Leopold's greatest contribution is not in musings about extending moral considerability but in reconstituting the perceptual field of environmental managers. He transformed that world into the world of an adaptive manager, measuring, testing, and evaluating from within a complex, dynamic system as a guide to adjusting our behavior. Leopold first recognized that managers cannot view species and ecosystems as simple, "external" objects because the manager is also an actor on the same stage with individuals, species, and ecosystems; the manager necessarily participates on all these levels and scales, having impacts on multiple dynamics that will play out over different time horizons. Since he was, first of all, a manager, Leopold's choices—as well as the observations and evaluations that drive those choices—are necessarily human-centered. In this perspectival sense, Leopold's new model was most basically anthropocentric. It models human decisions based on human perceptions and evaluations. Once one's perceptual world is thus reconfigured, it would be a simple confusion to extend this type of value to ecosystems as objects in competition with humans for resources, as Callicott does.

So we should, I conclude, place Leopold in box IIA, outside Callicott's triangle, and consider Leopold to be an anthropocentric holist. This category is open to him because, contra Callicott, he was a pluralist, recognizing different values emerging at different scales of a complex system. By reconfiguring the world we humans live and act within, Leopold was able to appreciate the full implications, including value implications, of living within a complex, dynamic system spread outward from our own experiential perspective, but spread out also across multiple scales of space and time. Within this newly configured, experientially constituted universe, it is possible to ask how humans do or can value ecosystems and communities and perhaps to conclude that humans value them noninstrumentally. But this is a statement about human values and does not entail human-independent values "discovered" in nature. Ecosystems, in Leopold's new universe of experience, cannot be independent centers of values that compete with human values; ecosystems rather are experienced, and evaluated, from within a multiscaled and open system in which humans are embedded. Humans, therefore, observe and evaluate changes on multiple scales and thus human values cannot be treated "monistically"—as expressions of a single type of value that is scale-independent. If Callicott had seen that Leopold's reconsideration of how to evaluate human-induced change was logically subsequent to a reconfiguration of his perceptual world, he would never have been tempted to put the interests of ecosystems in competition, on the same level, with individual, human goods; and he could have thus avoided his dilemma forcing a choice between environmental fascism and a green light to degrade nature.

By reconstituting our perception of natural systems as multiscalar, Leopold encouraged a pluralistic approach to evaluation, an approach according to which humans may, and eventually must, evaluate changes that emerge on multiple scales, including both the scale of individual economic activity and community-wide, multigenerational scales. By placing all of these evaluations within a reconfigured perceptual field, however, Leopold also eliminated appeal to values that transcend human experience. He was, in this sense, unquestionably an anthropocentrist. Evaluation, as well as our observations, must be made within our experience, once it is reconfigured to allow us to think (including to evaluate) like a mountain. As such, values, no less than our factual conjectures, must stand before the tribunal of experience. Leopold, like Dewey, recognized that there could not be separate logics for scientific predictions on the one hand and evaluation on the other. Choices of what to monitor, on what scale, with what instruments simply cannot be separated from questions of what is valuable and which dynamics are important because of their impacts on social values. Scale thus becomes a crucial aspect of this complex of understanding and evaluation.

Leopold began to restudy deer-wolf relationships because of a double failure. First, his implicit prediction that the deer population could be held in check by hunters proved false. Second, however, there was a catastrophic failure in achieving his social goal; that is, a social disvalue was caused by the interruption of the traditional deer-wolf-hunter-vegetation complex. By the late 1920s Leopold was recommending reduction of wolf populations but maintenance of smaller populations in wilderness areas. Manipulations of hunting regulations, he said, "are a crude instrument and usually kill either too many deer or too few. The wolf is by comparison a precision instrument; he regulates not only the number, but the distribution, of deer."[29] By the mid-1930s Leopold had become an advocate for predator protection, especially in wilderness areas. What is interesting is that way back in the teen years he had formed a simplistic hypothesis, that wolves and mountain lions limited deer availability for hunters. He formed an objective, to extirpate the large predators from the Southwest. He succeeded in his goal, at least with respect to wolves, while driving mountain lion populations to ecologically insignificant levels.

Reflecting years later, Leopold did not like the outcome of his "experiment." One thing he learned was how values and facts become all jumbled together in environmental management decisions. Hierarchy theory provides a useful framework to encourage consideration of management actions on the larger-scaled and slower-changing systems. But one can never eliminate the crucial role that values play in directing our attention toward certain dynamics and away from other dynamics.

Leopold came to believe that human values themselves are scaled, that human values unfold over different horizons and that a reasonable assessment of any policy must first identify values and goals that may emerge over longer periods of time, according to slower and more subtle dynamics. Leopold's emphasis on economic values caused him to pay attention to shorter-scaled physical dynamics. If one were to emphasize long-term values, such emphasis might encourage study of longer-term dynamics, for example erosion and siltation of streams. Choosing important dynamics to study and choosing which values to protect stand in a chicken-and-egg relationship. Importance cannot be judged on purely scientific grounds, so if one has no idea what values to protect, one cannot determine which dynamics to monitor, what to study, and what indicators to emphasize in setting management goals. There are so many natural dynamics and so many possible ways to model them, that failure to focus on a few key dynamics will create a situation of such great uncertainty that management decisions will be impossible. Conversely, to talk about environmental values in universal terms, not based on any specific, local models of actual natural dynamics, will not result in progress toward locally chosen indicators and management goals. Discussions of social values must inform decisions regarding what to monitor and what models should be constructed; meanwhile, information about natural dynamics and likely impacts on them by human activities must inform and shape our understanding of what we value.

Before closing this discussion and moving forward to consider how members of a diverse society, expressing multiple values associated with multiple scales, can move toward cooperative behavior, we must clear up two important ambiguities that result if we follow Leopold in thinking like a mountain. First, we must inquire as to the logical and epistemological status of the hierarchies we posit and use to organize our perception and understanding of the world we encounter and act within. This is an issue that Leopold (who, after all, simply offered a provocative simile) never answered, or even addressed. Second, we must ask whether, after adopting Leopold's reconfigured world of managerial perception, we can make any sense at all of claims that nonhuman objects in nature can "have" intrinsic (inherent) value.

What then is the logical and epistemological status of the scalar specifications for a particular ecological hierarchy? We can mention three possible interpretations. (1) The hierarchical organizational structures are an essential element of human understanding; all knowledge of nature must necessarily be expressed in hierarchically ordered systems. (2) Hierarchical organization is inherent in all ecological systems; every correct empirically descriptive model of ecological systems must exhibit hierarchical structure. (3) Hierarchical models, because they provide a series of useful conventions for organizing complex space-time relations in complex, dynamic ecological systems, are useful for a variety of descriptive, explanatory, and management purposes.

Of these explanations, we can eliminate the first interpretation out of hand. On this interpretation, we would be allowing a priori knowledge to place restrictions on what can be observed, a position we have rejected in favor of a more pragmatic, experience-based approach. Worse, the explanation seems false on its face—there are many alternative models that are not hierarchical and that are clearly useful in some circumstances. The second interpretation, however, may well be true. For example, C. S. Holling advances evidence that the size distribution of organisms in an environment does not vary continuously, but rather organisms exist in "clumps" of possible body sizes. Holling suggests the following hypothesis, which he calls the "Textural-Discontinuity Hypothesis": "Animals should demonstrate the existence of a hierarchical structure and of the discontinuous texture of the landscape they inhabit by having a discontinuous distribution of their sizes, searching scales, and behavioral choices. Landscapes with different hierarchical structures should have corresponding differences in the clumps identified by such a bioassay."[30] In other words, the body size of animals is expressive of the organizational structure of their habitats. Since humans are animals, and Holling's hypothesis suggests that body size is constrained by system structure, one might infer that there is an important empirical relationship between humans and the structure of the environments in which they survive. Holling builds on this idea, arguing that for each species, and given the body size of organisms of the species, there will be a "natural" microscale, mesoscale, and macroscale built into our landscapes. The microscale refers to the mechanisms on which individuals depend within the dynamic of a particular environment, the mesoscale is the scale on which populations must successfully negotiate a niche that allows them to reproduce and maintain relatively stable populations, and the natural macroscale refers to the dynamics of geomorphology, affecting topography, hydrology, and so on. According to Holling, then, the three-scale structure embodied in HT can be applied to environmental structure encountered by all animal species, and the body sizes of species are adapted to structural aspects of the environment on these three levels. This hypothesis is really just a more technical statement of the temporal-scale model of individual time, ecological time, and evolutionary or geologic time that Leopold sketched in "Marshland Elegy," as noted above.

Whether Holling's empirically based speculations are true or not, I do not think it is necessary to rest our admonitions to think like a mountain, a stream, or a wetland on such an empirical hypothesis. Leopold's admonition, we must always remember, was offered in a *management* context; so even if there are natural gaps and "lumps" in the systems we observe, our choices of models will be affected as much by our goals, values, and concerns as by purely physical structures in our environment. We therefore follow Allen and his colleagues, who describe the choice of hierarchies as "utilitarian" in nature, treating these hierarchies as pragmatic choices about how best to make sense of our experiences.[31] But we differ from these authors, who apparently think of the choice of hierarchies as a matter of choosing the best constructs for the purpose of *pure physical descriptions* of the system in question. Since I doubt that there is such a thing as pure description, and since, following Leopold, we are addressing the problem within an activist, management context, it is clear that our values as well as our observations will affect our choices regarding the constitution of hierarchies by which to organize and understand our perceptions. This viewpoint, then, brings us back to conventionalism and linguistic tolerance as advocated by Rudolf Carnap, by pragmatists, and by students of the pragmatics of language and communication, as discussed in the appendix. According to this view, we develop physical models and theoretical languages as useful tools for understanding experience, and we try to improve these models even as we use them to inform our interactions with nature. Carnap's principle of tolerance simply applies the experimental method to the use of models and languages in active situations.

If this strategy seems to point us toward subjectivism and the development of models that serve special interests, we turn to Peirce and the constraints set by the needs of community, by the need to communicate even with those who have opposed interests. A community committed to acting cooperatively will, in its deliberations and struggles to communicate, test various languages and vocabularies. If one purpose of models and languages is to encourage cooperative behaviors, then our linguistic conventions must also be tested against experience in public discourse, and they must be refined and improved through the processes of communication and deliberation. On Neurath's boat, *all* beliefs, including beliefs about how to configure our experience into useful hierarchical structures for understanding and managing the environment, are open to revision in the face of new, especially practical experience. The necessity of communication for cooperative behavior can thus be the engine that drives new conceptualizations and new and more useful models of environmental problems.

With this clarification made, we can now address the question that has been central to so much writing in environmental ethics: do any nonhuman elements of nature have intrinsic value? Our interpretation of Leopold's thinking-like-a-mountain simile, when coupled with our rejection of all a priori sources of knowledge, leads us to reject any claims of intrinsic value that exists *independent* of human perception. Such values would be, essentially, "intuitions"; they would, to use Moore's phrase, refer to "non-natural qualities." But such values violate the pragmatists' insistence that all of our beliefs be open to challenge on the basis of experience. However, if we next ask, Do human individuals ever value nature intrinsically? we can quickly answer that they do. Placing value on things independently of their immediate, selfish uses is certainly one of the several ways humans evaluate changes in the natural systems that form their habitats. Such experiences are part of the raw material from which we must construct our holistic approach to understanding and evaluating changes in natural systems. But such assertions of value should not be treated, following Callicott, as values that may conflict with, and must be balanced against, opposed human values. They *are* human values. If there is a conflict among human individual interests, human communal interests, and human sentiments of altruism—as there always is—the conflict occurs within human understanding and evaluation. Failures to protect things we value intrinsically is a human failure to be measured in human terms, and if such failures leave us perplexed, the place to look for enlightenment is in improved ways to scale and integrate human values, rather than in an attempt to adjudicate conflicts between one set of values defined as humans claims on nature and another defined as nature's claims on humans. If we accept re-modeling of nature to insert humans within a multiscaled system, the conflict disappears. Individuals suffer economic losses and gains; but damages to the "mountain"—the ecological context of the human activity in question—emerge on a multigenerational scale and have effects at the community level. Here, the "good" is multigenerational and not commensurate with individual human economic "good."

As a pluralist, I believe that assertions of noninstrumental value in nature should be taken seriously in our search for a coherent and comprehensive theory of environmental value. If such values are included as an important type of value in our comprehensive theory of evaluation, we should not forget that such values are highly theory-dependent; they have no theory-independent existence as trumps or counterweights to human values. We recognize multiple types and expressions of human values, but since these are associated with different scales and different dynamics, one need not choose between honoring other humans and honoring ecosystems. An adequate theory of environmental value must make a place for all human values, instrumental and noninstrumental; it must also—and this is the hard part—provide a way to reconcile or fairly balance competing human interests. Although we are far from having such a theory, Leopold taught us that such a theory will be possible only if we reconfigure our world and reorganize our experience. If we can begin to judge proposed actions and policies for their economic impacts and also for their ecological and evolutionary impacts, we will, in effect, be associating different *human* values with multiple dynamics of natural systems. A good policy is one that has positive impacts on all levels. It is *not* the one that does the best when measured against a short-term economic criterion. Leopold's insight, then, shows us how to array human values on multiple scales and to avoid Callicott's false dilemma, which puts human interests in conflict with nature's interests on a single scale.

We can now summarize the key idea suggested by Leopold's analysis of his wolf-eradication policy: It is useful to perceive human choosers as embedded in a hierarchical system and to see that the human values delivered by that system emerge on different scales of space and time. Once values are so sorted, it may be possible to associate these variables with natural dynamics essential to their continuation. Viewed in this way, thinking like a mountain is thinking about human values as time-sensitive and as produced by specific processes and dynamics that unfold on identifiable scales. These relationships are shown in figure 6.2. The first level of the hierarchy, with the shortest temporal horizon, corresponds to economically based decisions, the horizons of which are bounded by the horizon of individuals' economic concerns. The second level of the spatiotemporal hierarchy is especially important because it is the level at which humans shape their own culture and multigenerational community through individual and cooperative acts that, at the same time, impact the landscape in which they will make future decisions. This is the level on which a human cultural unit, a population or a human community, interacts with the other species that form with it a larger ecological community, or place. We might say that on this level communities articulate their long-term aspirations regarding what kind of society they will be and how that society will express itself on the landscape. I believe it is a shift to this expanded level—the level at which a community's aspirations inspire them to undertake preventive and restorative ecological practices—that Leopold refers to as "thinking like a mountain."

On this interpretation, then, Leopold advocates using hierarchical models, models that are sensitive to differences of temporal and physical scale to analyze both the impacts and the associated values that affect managerial decisions. This approach views natural systems from a human perspective, inside-out, as open systems embedded in larger systems that change much more slowly. Because humans are placed as actors in such an open system, it is not necessary to posit new entities such as "whole ecosystems" that must be

Temporal Horizon of Concern	Time Scales	Temporal Dynamics in Nature
Individual and economic	0 - 5 years	Human economies
Community intergenerational bequests	Up to 200 years	Ecological dynamics and interaction of species in communities
Species survival and our genetic successors	Indefinite time	Global physical systems

Figure 6.2 Correlation of human concerns and natural system dynamics on different temporal scales (from B. G. Norton, "Ecological Integrity and Social Values: At What Scale?" *Ecosystem Health* 1 [1995]: 239)

accorded moral standing. Open systems, with larger systems placing constraints on their constituents, exercise control from above. When human actions cause, or are expected to cause, these normally fast-changing variables to accelerate, they alter—often for the worse—the mix of options and constraints that will exist in the future. Such aggregated actions are maladaptive at the intergenerational scale.

What Leopold learned from his ill-fated experiment in wolf eradication was that scale counts; important human values will be missed—and destroyed—if we confine our concern to short-term considerations and impacts of our policies on economic individuals. Ecological systems, human value systems, and social, managerial systems are all complex. The best way to deal with these layers of complexity is by developing and refining hierarchical, space- and time-relative models of human actions affecting natural systems, on the one hand, and corresponding hierarchical systems that track human values and the impacts of landscape change on those values, on the other.

7

ENVIRONMENTAL VALUES AS COMMUNITY COMMITMENTS

7.1 Public Goods and Communal Goods

Leopold's insight that environmental problems have an inevitable scalar aspect can be seen, upon analysis, in another classic of environmental writing, Garrett Hardin's essay "The Tragedy of the Commons."[1] This essay, which is ostensibly about human population growth, has been used to explain and characterize many environmental problems. Hardin's argument is simple: imagine a pasture used as a commons, with each individual user of the commons free to decide how many animals he or she will graze there. Each individual, acting out of personal self-interest, will add animals, driven by the following reasoning. The value accruing from the additional animal will belong to the individual owner; if there are negative impacts of overgrazing, these negative impacts—or the costs of restoration—will be spread across the community. The individual receives 100 percent of the benefits of the added animal and only a portion of the costs, so it is always rational for any individual to add an animal, regardless of the impact of this action on the pasture. As rational individuals act in their self-interest, the value of the common asset, the pasture, is degraded by overgrazing.

Hardin's simple, inexorable, and "tragic" behavioral model has become pervasive in environmental literature and discourse because it is easily generalizable from the decision to have a child (Hardin's original instance) to other decisions affecting important public goods. In this sense, Hardin's analogy captures the driving "logic" by which individual interests inexorably threaten public goods in a commons. Hardin's analogy has been so influential because it exposes the logic—and the consequences—of human exploitation of common resources when left unchecked by a communally administered authority.

Hardin's analogy has been much discussed and much criticized; and important qualifications must be made in many directions. For example, some authors have shown that Hardin's tragic sequence is often avoided in practice; for example, communally organized cooperatives have successfully managed commonly owned pastures in the Swiss Alps for hundreds of years.[2] So the discussion requires at least a distinction between *common ownership* and *open access* to a resource, with the latter providing a better characterization of the problem than the former. Despite the possibility of mutual coercion, as noted by Hardin himself, Hardin's analogy remains a simple and powerful representation, a mental model, to explain the constant threat that environmental goods (often public goods) will be degraded in the pursuit of private interests.

I think it is most useful to think of the tragedy of the commons as a generalized model that applies to most, if not all, environmental and resource problems. The basic intuition behind the generalized model comes from ecology—it is, in its basic structure, a carrying capacity model. The size of the population of any species will be set by some limiting factor in the environment; in many cases it will be a limited resource like food or nesting sites. What is fascinating about Hardin's analogy is that one can treat the idea of a common pasture as a "variable" and plug in any limiting factor one pleases. Whether it is cattle running out of forage or suburbanites using crowded highways, the tragedy-of-the-commons analogy captures the basic structure of the problem of limited resources, public goods, and environmental quality in a more and more crowded world. In each case all individuals, similarly situated as users of a common asset, act to destroy the common asset. Assuming that the individual is predisposed to act in a self-regarding way, destruction of resources is unavoidable; for any limiting factor one chooses, the commons under individual control will lead to a tragedy.

The wary reader will at this point justifiably ask whether an endorsement of Hardin's model may commit us to the hedgehoggy, unidimensional thinking of apocalyptic ecologists and their IPAT equation, a position we found too rigid in section 2.2. My answer is, Not necessarily. One can agree with the apocalyptists that such a relationship provides a model that applies inexorably under the assumptions of common access and individualistic decision making, without succumbing to reductionism. But we can use the simplistic model as an argument for rejecting individualistic approaches to evaluation of environmental change. So used, Hardin's model highlights the weakness of self-interestedness as a basis for resource-use decisions. If one assumes that each herder will act selfishly and add more animals to the herd—if one assumes, that is, that the actors *must* act as self-regarding "economic man"—then individuals acting "rationally" by this standard will inevitably destroy a public resource. Fortunately, there is an alternative interpretation of the situation that avoids inevitable tragedy. We can, as Leopold did when he learned to think like a mountain, take responsibility for the larger impacts of our choices that unfold at a pace too slow to be understood as individual choices.

The core idea of Hardin's tragedy is so powerful because it captures, with elegance and economy, two separate but critically related ways in which environmental problems are problems of scale. In the first sense, Hardin illustrates how individual self-interest creates a force in the direction of growth and expansion and greater and greater demand on resources or "common" goods (the pasture). To use the literal base of the metaphor for concreteness, herd expansion represents an increase in the biomass of grazers feeding on the edible vegetation produced by photosynthesis. Since the capacity of any pasture to produce biomass and energy is limited by factors such as the amount of sunlight reaching the soil and the productive efficiency of the plants available, increases in herd size by individuals, acting alone but driven by the same logic, will eventually surpass the carrying capacity of the pasture. Here we have a simple, physical notion of scale as applied to a system of production and consumption of matter and energy; it refers to a measurable, physical relationship between production possibilities and rates of consumption by grazing animals.

There is another sense, however, in which Hardin's parable about decision making has a scalar aspect. Hardin concentrates, in generating the tragedy, on the reasoning of each individual as a self-interested individual and the inevitable negative impacts of following that reasoning on public goods, goods that cannot be provided by private property because they are nonexcludable. In the case of herders and cattle, the good can be conceived as a service provided to the community by the pasture. This good, though accessible by individuals at a given time, represents more than the sum of individual goods because the good of a healthy pasture can, if it survives, be of great service to the next generation of herders, even if each individual sells or otherwise transfers her or his individual access interest to another herder. This good of an ongoing opportunity for future herders I will call a "communal" good, distinguishing it both from individual goods, that is, the goods owned by individuals, and from public goods, goods that cannot be owned (in the sense that ownership involves the right to exclude use by others). Communal goods emerge and are counted on the scale of the community; they exist on a different temporal scale than do individual goods, and they can, in principle, survive many transfers from individual to individual. I am hypothesizing, then, in opposition to the economic model of value, whereby all values are aggregations of individual goods, that a proper understanding of the tragedy of the commons requires that we recognize *communal* values, values that cannot be represented as aggregations of individual values. If my hypothesis is correct, it will have serious consequences for economists' treatment of the tragedy of the commons because it hypothesizes a kind of value that does not exist on the scale of individual value and thereby challenges the central, individualistic concept of "economic man."

Economists usually respond to Hardin's tragedy by recommending that the commons be privatized—"enclosed"—creating exclusive ownership and an incentive for individual owners to protect their own fenced pasture. Economists believe that public goods are goods enjoyed by individuals but that these goods require special institutions to protect them. Therefore, the solution to Hardin's tragedy, for economists, is to protect public goods by transforming them into private goods. Because economists countenance only goods that are units of individual welfare, public goods can be replaced in any manner that provides an equivalent amount of individual welfare. Economists, then, deny the existence of communal goods as I have defined the term. Let us examine the implications of the economists' solution in some detail, because such an examination will clarify the nature of the goods at stake.

We can treat the economists' move to privatization as expressing the following two hypotheses: When a public good, such as a healthy common pasture, is threatened by overuse and degradation, (1) the creation of appropriate private property rights is a *necessary* remedy (no other solutions are available) and (2) creation of private property rights is a *sufficient* remedy for the threat (if accomplished, such a transformation can be expected to provide an effective remedy for the threat). Let us consider these two hypotheses in turn.

Is creation of private-property ownership of a resource the only way to protect it? Apparently not. As noted above, many commentators on Hardin's analogy have noted the importance of distinguishing between *common ownership* of a resource and *common access* to a resource. Many cases of cooperative arrangements show that creating private ownership is *not* the only way to resolve the problem of protecting commonly owned pasture. For example, there can be enforced agreements among users of commons to limit access, such as a limit on the number of animals any family can place on the pasture. Cooperative, enforceable agreements to limit access can, at least under some conditions, lead to protection of a commonly held resource. So recognition of the distinction between common ownership of a resource and open access to a resource apparently undermines hypothesis 1.

Some economists may not be convinced by such a simple argument against the necessity of privatization, however. An economist who is deeply committed to the value of privatization might argue as follows. Public goods, by definition, are goods from which consumers cannot be excluded. For example, if clean ambient air is provided for one person in the society, it will be enjoyed by all. Similarly, if the health and productivity of a common pasture are protected, presumably they are protected for all of the pasture's users. The problem, according to the economist, is that, given common access, the inexorable logic of self-interest will make such protection impossible, in fact, as each individual has an incentive to "free-ride." Limitations on the right to place more animals on the pasture, it could be argued, create private property rights by exclusion of some rights to use. If the ability to exclude is thus treated as creating private property for those who are not excluded, the economist might argue that, in effect, *any* limitation on access is a privatization. Each individual with a right of access for a given number of cattle "owns" a "license" to use the pasture; those lacking such a license are excluded. It is not hard to imagine a "market" in which such licenses are traded, as when commercial fishers, for example, are required to purchase a license to gain access to a fishery. Such tradable permits can lead to both limited access and a right that can be exchanged for value, including cash.

According to this line of reasoning, any enforceable limitation on access to a resource involves a privatization. There is perhaps no harm in such a conceptualization, provided one realizes exactly what is implied—that, for example, the families of Swiss herders who enjoy limited access to a commonly owned pasture have private property rights to a public good, the healthy pasture. The conventions, passed down over generations and centuries, by which such access is effectively limited are understood—by the definitional equation of ownership and excludability—to grant "private property rights."

We should be wary of this line of reasoning; it salvages the hypothesis that privatization is necessary to protect common resources by resort to definition and, in the process, reduces the hypothesis to a tautology. If all institutions that are capable of achieving limitation of access to a public good involve ownership by definition, then the apparently useful distinction between open access to resources and common ownership of resources is obliterated by semantic maneuver. Similarly, important political institutions, such as cooperatives and implied contracts, are treated as exclusively economic institutions, even though they clearly require important political and social institutions to thrive and continue. If open access is the problem, and any limitation on access, by definition, creates individual property rights corresponding to the excludability implied, then it is tautological that successful avoidance of the tragedy creates property rights. But this equation gains its plausibility by "theft through definition." It is a tautology based on specialized definitions introduced by economists and only plausible to someone who runs roughshod over ordinary distinctions. Worse, by defining privatization as excludability and then taking any effective exclusion to create property rights, the defender of hypothesis 1 co-opts any possibly effective solution to the open access problem as one of building effective *political* institutions, seeming to make privatization the only available antidote to tragedy. I conclude that hypothesis 1 is true if we grant the economists their specialized definitions, not true if we insist on the distinction—apparently important in public, political discourse—between access issues and ownership issues. Once we have considered the second hypothesis, that privatization is *sufficient* to protect resources from overuse, we will revisit this question of how we "should" define public goods and the relationship between private goods and public goods.

Turning to the second hypothesis, we can ask, When resources are threatened by overuse because of open access, is privatization of the resource *sufficient* to save the resource in all cases? In fact, it has been rigorously shown, by the mathematical bioeconomist Colin Clark, that there are many situations—involving not unusual economic and biological conditions—where ownership in a resource is *not* sufficient to protect the resource, and where the profit motive provides clear incentives to overexploit and destroy even privately owned resources.[3] To simplify a complex argument, Clark shows that if one distinguishes (1) maximizing "rent" derivable from sustainable use of a resource such as a fishery in which exploiters have ownership (in the form of equipment, licenses, and exclusive rights) over indefinite time from (2) maximizing profits from that resource by a particular owner over a limited time, the two goals lead to divergent policies. Although the total income *from the resource itself* might be maximized by taking only the sustainable yield, *total income on investment* may be higher if the agent maximizes income in the early years, exploits the resource to (economic) extinction, and then reinvests these large early profits in other high-return industries. The key variable, in other words, will be the "opportunity cost of money": if there are equal or greater opportunities to invest in other areas of the economy, then the best strategy may be to maximize profits in early years, exploit the resource to extinction, and reinvest in other regions or in other sectors of the economy.

This result from bioeconomics is relevant here because Clark's result shows that as long as there are adequate investment opportunities elsewhere in an economy (and today this refers to most of the world), property owners who act only on profit motives alone (the pursuit of which will maximize their *total* stock of capital over the short run), will often act to extinguish a resource by rapid early exploitation. These tendencies are greatly exacerbated in less developed countries, where much of the biodiversity of the world exists, because discount rates are typically high in less developed countries. Clark concludes his argument: "In view of the likelihood of private firms adopting high rates of discount, the conservation of renewable resources would appear to require continual public surveillance and control of the physical yield and the condition of the stocks."[4] In other words, pursuit of maximal profit for agents—even for owner-agents—diverges from policies that protect resources, and only governmental intervention will avoid destruction of productive resources. Clark's argument shows that privatization of resources is not, in general, sufficient to save a resource.

So where are we with respect to the economists' claim that the remedy for Hardin's tragedy is privatization? We have just seen that privatization—even within the economists' circle of definitions—does not, in many cases, protect a resource from overexploitation. Moreover, privatization can be called necessary for resource protection only if one adopts tendentious definitions of economists, definitions that obliterate key distinctions considered useful in ordinary discourse. Based on these arguments, we must turn our attention directly to the economists' definitions of private and public goods, and to whether these definitions will be useful and appropriate for characterizing real social goods or whether the reductionism inherent in economic models misses important aspects of the goods in question.

The easiest way to see the effects of this reduction is to contemplate the economists' model of *Homo economicus*. Economic man, of course, is a fabrication employed to simplify the description of human economic behavior. Economic man is considered to be rational and self-interested and to have full knowledge. It hardly needs mentioning again, since many philosophers and social scientists have said it already, that each of these assumed definitional characteristics is questionable when applied to real human individuals. Any attempt to predict the behavior of individuals based on these definitional, methodological assumptions is pure ideology and pure nonsense. Here I want to concentrate on another, less notorious, simplification implied by standard economic definitions, the assumption of "methodological individualism." This assumption, which is widely shared by economists, virtually all social scientists, and most philosophers, asserts that the legitimate object of social scientific study is ultimately the individual human being. Applied to normative analysis, the principle of methodological individualism implies that all benefits or goods must be understood as benefits or goods of individuals. This has been called the "person-regarding principle," and it is taken as a starting point for most human-centered ethical reasoning and analysis.[5] So economists have lots of company in supporting methodological individualism, which has come to be a cornerstone of economics and of most behavioral social science, as well as mainstream ethics.[6]

This simplification bears directly on a key claim of this part (part 2) of this book, however. I believe that methodological individualism, when combined with selfishness (an assumption that is constitutive of Hardin's dilemma and hence not at issue in this discussion) cuts economists off from an adequate analysis of public goods. I will show this by demonstrating that economists' analysis of individual goods and of aggregations of individual goods cannot capture what goes wrong in commons tragedies. Since their analysis is inadequate, it is not surprising that their remedy is off base. Let's have a deeper look at the seldom-challenged commitment of analysts and scholars, economists and others, to methodological individualism.

As an undergraduate major in political science in the 1960s, I learned that methodological individualism was not only in vogue in the social sciences; it would not be exaggerating to say it was a constitutive tenet of the paradigm of most empirical scientific study of human behavior, including political science, sociology, anthropology, and more. Indeed, from the fervor with which my professors advocated this methodological tenet, it was easy to surmise that more was at stake here than a simple methodological choice. And there was. As I learned in several classes on social science methodology, social scientists, justifiably horrified by the Nazi atrocities and Nazism's subversion of the sciences, identified for particular vilification one belief of the National Socialists: the primacy of the state over the individual. The whole generation of social scientists who taught me and my generation came to think of this "organicist" view of the state as the root of much political evil. The Nazis' commitment to, and obeisance to, the good of the culture and the state as something over and above the aggregate good of individuals was seen as integral with their willingness to commit atrocities in the name of the state. Therefore a whole generation of social science professors accepted methodological individualism both as a guide to doing good science and, at the same time, as an essential moral bulwark against evil itself.

Methodological individualism, in effect, functions as a constraint on the linguistic resources that can be developed and used for analyzing problems like tragedies of the commons. By prohibiting conceptualizations of goods other than individual goods and aggregations of individual goods, methodological individualists enforce a key reduction: any goods that count must ultimately be counted as individual goods. But consider what this means for our analysis of Hardin's tragedies: by mere definitional fiat, it follows that any tally of the total goods derived from a community enjoying any resource will include only individual goods. Anyone who argued that the transformation of the "goods" the community derives from the healthy pasture into individual, privately held goods destroys nonindividual goods *must,* by definitional exclusion, be talking nonsense. Given the multiscalar analysis of values developed in this book, such nonindividual goods are crucial; but they are ruled out by the economists' definition of all goods as individual goods. This is what I mean by calling the economists' definitions "tendentious." If I wish to say that a policy of privatization—the transformation of public goods into private goods by assigning private property rights (the act of "enclosure")—is unacceptable because nonindividual or communal goods will be lost, the economist, given strict adherence to methodological individualism, must reject my assertion as nonsensical. This is exactly the kind of ideological constraint on language that must be rejected if we are to employ an experimental method in which all claims are up to challenge by experience. It violates both the spirit and the practice of experimental pluralism. If we wish to develop a more comprehensive conception of environmental value—one based on the idea that human values emerge on multiple scales—we must challenge methodological individualism itself.

Methodological individualism, in effect, rules out by fiat the possibility of what I will call communal goods. Communal goods are understood as goods that emerge on a community level, a scale larger than that on which the goods of individuals are observed or calculated, and they are not commensurable with, or reducible to, individual goods or aggregations of individual goods. Such goods, I have argued and will argue, are exactly what is needed to properly understand—and avoid—tragedies of the commons. If Leopold was right that we must learn to think like a mountain, and if I am right to explicitly point out that thinking like a mountain must entail *valuing* like a mountain, then we must challenge methodological individualism.

The fact that Hitler and the National Socialists believed in the good of a state hardly impugns the usefulness of the communal-good concept in analyzing nonindividual goods. The methodological tenet of methodological individualism, which aggressively banished any experimentation with concepts such as communal goods, turns out to be one more form of ideology. It is a prohibition based not on experience but rather on wild generalization from one very special case in which a madman was trying to subjugate the world. For an experimental approach to evaluating changes in natural systems to work, we must reject definitional constraints and experiment with new conceptualizations and means of measurement. In section 4.1, we saw that environmental problems are best seen as problems of choosing a proper balance among multiple values and that environmental problems are therefore best seen as problems of cooperation, as problems of individual interests in competition with communal goods. In this section we have learned that possible reactions to, and analyses of, the famous tragedy of the commons have been constrained by economists' preexperimental prohibitions on nonindividual goods. Let us, then, throw off the straitjacket of economic jargon and see whether we can achieve a better understanding of common access problems and their solutions without the constraint of methodological individualism.

I therefore define, for further study, a category of values I will call communal values. These values are *human* values, but they are not, to use a technical term, defeasible into individual values. This means, to be less technical, that communal values, though they may be important to individuals, cannot be reduced to any set of aggregated preferences of individuals. After one has counted all of the values that accrue to individuals from, for example, maintaining a healthy and productive common pasture, there remain further values that are best interpreted as values accruing to a community or a society, values that unfold over multiple generations. These values emerge, as Leopold suggested, not at the pace of change of economic markets driven by individual preferences, but rather in the longer frame of time of the "mountain," which is the time frame in which a human community finds its proper niche in an ecological system.

Armed with this definition, we can pose the issue precisely: Are some non-individual goods deriving from environmental protection communal and emergent on a multigenerational scale in the sense just defined? If so, economists, as avid advocates of methodological individualism, would be forced either to deny the existence of such values or to admit that their framework of aggregative individualistic ethic is inadequate to comprehensively measure environmental values. The difference between the two systems of valuation, and its impact on Hardin's argument, can now be stated clearly. Being a monistically individualistic theory, economics cannot encompass values that are not individualistic; such values are invisible within the economists' conceptualization of value. To accept their framework of valuation is to assume, not to prove, that there are no communal values; and once we think of environmental evaluation as a pluralistic and experimental enterprise, we should put the ghost of Hitler to rest and at least *try* to articulate, and measure, impacts of our actions on communal values—values that emerge on a multigenerational scale.

The tragedy of the commons, I have argued, provides a simple model demonstrating the vulnerability of important resources; if decisions regarding those resources are made on a purely individual basis, with no checks placed upon the pursuit of individual interests, then the resources will be destroyed. We have taken this simple model to imply that there must be a "good" associated with the maintenance of the resource in question, and this good does not show up in individual, self-interested calculations. Thus, if one properly understands Hardin's analogy, it forces upon us the conclusion that "economic rationality" will not capture the kind of communal good that exists in a community that has "fertile pastures." The pastures, of course, represent productive processes that provide opportunities for humans to make a living and to pursue meaningful lives. If such processes are destroyed, important opportunities will be lost. A community can in this sense be impoverished, even if all individual members of its generations do well. For example, the herders might invest the gains they make in the destruction of the pasture wisely, perhaps sending their children to college, thus freeing those children from dependency on the pasture. But the community will be poorer as the pasture and the opportunities associated with it are destroyed. What can be lost to the next generation is a resource that supports an opportunity and the option to continue a way of life. Since any individual user of the pasture could decide not to be a herder, no individual good need be lost; but from the viewpoint of the community, the range of opportunities is narrowed. The community is poorer.

7.2 The Advantages of Democratic Experimentalism

What we need—and I admit that this is a mouthful and requires a great deal of explanation—is a *context-sensitive decision model* that is appropriate to open-ended, democratic processes such as an adaptive management partnership. Adaptive management requires *both* participation of stakeholders *and* buy-in with the procedural goal of finding a cooperative solution through the judicious use of management experiments. Although we have acknowledged that such buy-in is difficult to achieve in practice, we have also recognized that it has happened in many places that have developed long-standing management partnerships. We can now pose the central question of this chapter: What theory, conceptual approach, and model of decision making is appropriate (useful) in supporting and clarifying the kinds of decisions facing participants in adaptive management situations?

To answer this question, we can take advantage of the enormous flexibility built into the pluralistic viewpoint taken in this book, as the starting point for developing a useful evaluation procedure and decision model for adaptive management. Adaptive managers accept uncertainty and surprise as an unavoidable element of goal-setting and management decisions, so sustainability goals cannot be stated in advance. Rather, they must be approximately set, and adjusted as new information comes in from management experiments. The possibility of social learning is therefore the central driving force of adaptive management; and this driving force should sharply focus our attention on the deliberative and political processes associated with an adaptive management partnership. Further, because environmental problems—being "wicked" problems (see chapter 4)—almost always present unique situations, we must further focus on local processes, at least as our starting point. It follows that we must also be interested in the normal political contexts in which such local processes unfold. We know that every local context is different in important ways, but it is still useful to mention some general expectations that are likely to shape most such decision processes.

In almost all local contexts in modern democracies, one finds among affected parties considerable diversity of beliefs, worldviews, values, goals, and preferred policies. This diversity can be referred to as *the fact of pluralism*, and it is documented by numerous studies and by earlier discussion in this book.

We have acknowledged and incorporated this fact by adopting pluralism as a working hypothesis, and we accordingly stipulate that any decision process—and the associated model—that will be useful in an adaptive management setting must function effectively in situations where diversity of opinion pervades.

Accepting uncertainty and diversity of values and worldviews as unavoidable aspects of decision situations, we face a most important dilemma: How should we respond to this undeniable diversity? How we respond to diversity is crucial because this will determine the nature and limits of democracy in our process. Will our decision process favor open participation and group decisions; or will the voice of experts determine the "correct" policy, leaving the community with only a problem of "educating" those who differ? More specifically, we face an epistemological dilemma, a dilemma that raises crucial issues for any manager who seeks to initiate group-supported actions in the face of at least some opposition. The way one reacts to diversity of opinion has a considerable impact on the possibilities of truly democratic policy formation.

In adaptive management situations in which a group of interested parties seek cooperative solutions to local problems, the setting is perfect for the use of the methods of the pragmatics of language. We have assumed that our relevant population of participants share a commitment to achieve a cooperative solution to problems facing their community, and this minimal step toward cooperation signals a shared goal or purpose, at least a shared goal of working cooperatively. Even if the commitment deals only with process and involves little or no agreement on substantive policies, there is the shared goal of cooperation, which in turn places importance on communication among differing parties. Given these shared procedural goals, we have hypothesized that failures are often the result of failures of communication, and this hypothesis highlights our project of developing more adequate ways of describing and characterizing environmental values, especially sustainability values, and creating new policy objectives that will be favored by broader constituencies.

Our argument of the previous section led us to suggest that a comprehensive accounting of environmental values would include some communal values, values that unfold on the scale of generations and communities and are thus incommensurable with individual, economic values. They are incommensurable in the sense that they cannot be given a price in day-to-day markets, because these values are reflected in slower, ecological dynamics and emerge on a longer scale of time. By focusing attention on agreements to seek cooperation, we are signaling that we think problems of environmentalism are not *inherently* ideological; rather, we make them ideological by imposing a priori value assumptions upon our methods for reporting values. At this point we have passed a semantic Rubicon in the conceptualization of environmental values: we are waiving the a priori prohibition on terms referring to nonindividual goods and violating the strictures of methodological individualism in order to experiment with identifying and counting truly communal goods. Experimenting with communal goods is just one example of the "tolerant" attitude we are taking toward linguistic conventions for expressing human values.

By adopting the methods of the pragmatics of the language of science, and by applying these methods in an activist context of adaptive management, we can shift the debate about environmental values away from ideology and metaphysics and away from debates about what entities should be said to have intrinsic value, and toward the more constructive task of addressing failures of communication that block cooperative behavior. Further, we have a promising strategy for addressing this set of questions. Let us briefly review the advantages of an experimental approach to developing new linguistic resources within an activist context of providing a decision procedure for adaptive management.

We assume that our problem is encountered by individuals who, by dint of their participation as partners in an adaptive management plan, are already committed to seeking cooperative solutions to shared problems. In this cooperative context, our experimental attitude encourages freedom to develop linguistic tools that will comprehend the full range of environmental values. The pragmatic attitude encourages searching through the panoply of disciplines in search of methods and tools to use in understanding environmental values. The pragmatics of the language of adaptive management thus provides a tool to aid those activists who seek cooperative solutions.

The stage is now set for a fundamental reformulation of the problem of environmental evaluation. Here are the key elements of the new approach, which adopts as its units of evaluation various "development paths," or "scenarios," as introduced in section 5.5.

1. Pluralism requires multicriteria analysis. Since we have followed Rittel and Webber in understanding environmental problems as wicked problems—problems involving competing goods—I propose that we operationalize pluralism by evaluating development paths according to multiple criteria, with each criterion embodying an indicator that is believed by participants to track at least one important social value on some scale. If the values of our participants determine which indicators they favor, we can see our community's collective, political choice of indicators as an operationalization of a community process of value articulation, a statement of the type of community the participants want to be. Public discourse about values will manifest itself in an effort to choose what to measure and what goals to set for management. Likewise, we have learned from Rittel and Webber that the solutions sought will

not be an optimum of some single value but rather a temporary, acceptable balance among multiple criteria, and that this balance will be open to continual adjustment and refinement as conditions change.

2. Since we are addressing the problem of evaluation from within an activist, adaptive management process, we extend the idea of experimental management to the problem of choosing a useful method of environmental evaluation, a method that will be useful in making better decisions in real contexts. But since adaptive managers believe context really does matter in decision making, this means that evaluation must be "contextual"; that is, good decision making will reflect a deep understanding of the local context in which the action is taken. This context is also shaped by the evolution of a dialectical relationship between a people and the place those people inhabit. The values expressed by the participants, on this model, must be capable of reconsideration and correction, but the corrections will be more-or-less shifts in weightings of various criteria. Environmental values can now be seen as less ideological, more robust, and less ephemeral. Contests over environmental values, which take place in a context of value pluralism, with participants afforded the opportunity to express their values in multiple ways, now become a matter of assigning weights to various criteria associated with competing "goods" such as economic growth, clean air, a healthy landscape, and sustaining resources for the future. Notice that once the problem of evaluation is set up in this way, participants in an ongoing, iterative process of experimental management are expressing their values every time they argue for, or vote to use, a particular indicator. If participants in a community-based management process reject development driven by overemphasis on a single extractive industry such as forestry or mining, and favor economic growth that is more balanced and consistent with maintaining the landscape they love, then this expression of value can be reflected in a shift of emphasis away from simply economic measures of management success.

3. We can build on the hypothesized commitment of participants to act cooperatively despite differing values and interests, because this provides a basis for important "procedural norms." Procedural norms are quite useful because they exist on a different level than do substantive values like welfare satisfactions and intrinsic values. Persons with diverse values, believing that a negotiated, cooperative solution will be better than a solution imposed by raw power, can commit themselves to acting cooperatively even before they know the outcome of negotiations. Communities would often rather solve their own problems locally than have regulations imposed by federal agencies or federal legislation. A locally based, participatory process begins to look very attractive when, for example, the U.S. Fish and Wildlife Service threatens to invoke the Endangered Species Act to limit use of land in the area. The procedural

norms, justified by the goal of cooperation, carry with them apparent obligations to communicate clearly and to offer explanations and reasons for one's proposals for action. These obligations are explored in more depth in section 7.4. Here, however, we can already recognize how any kind of cooperative behavior among participants with varied commitments will require two "phases" or "moments." These correspond, of course, to the action phase and the reflective phase introduced in section 4.2.

4. The shared goals implied in the commitment to seek cooperation as preferable to force or chaos provide the impetus for a phase of discourse in which people who share a procedural goal of cooperation reflect on which goals they can support as a group. In the reflective phase, participants who are committed to cooperative action articulate competing and complementary objectives and propose and debate possible indicators and standards they could seek by consensus. In response to the admitted wickedness of many environmental problems, then, the adaptive management process offers an opportunity for cooperative participants to propose and discuss weightings of multiple goals and indicators. The weighting of multiple criteria is a central task. Since adaptive management is an activist, experimental enterprise, the action phase will be dominant; but if all beliefs are open to challenge under adaptive management, the discourse of adaptive management must also allow us to reflect on the weight we are assigning various criteria. Given the framework sketched so far, we are positing committed participants facing decision situations in which they and their cohorts express multiple values in multiple linguistic frameworks. Quite naturally, then, the reflective phase is entered when a functioning balance is called into question; in this phase participants are asking for a reevaluation of the weightings of the multiple criteria, or perhaps the addition of a new indicator and a criterion associated with it. Such reevaluations represent shifts in the emphasis the community places on various criteria. Notice that eliminating a criterion is equivalent to assigning it zero weight, so we can conveniently consider the question of which criteria count, and how much to count them, as problems of assigning weight (from 0 to 100 percent) to a set of $n > 1$ criteria. Value questions now appear as a matter of degree, not as all-or-nothing pronouncements, and we are on our way to achieving our goal of reducing the pull of all-or-nothing ideologies.

5. Finally, with this pluralistic, multiscalar framework in mind we are ready to consider a major experiment, true pluralism. Suppose we represent environmental problems as problems of finding a balance between individual goods and communal goods. Individual goods (units of individual welfare), let us concede, are fungible and fairly well captured by economic analyses; they unfold on a relatively short-term scale, the horizon of which is, say, less than five years. Then, following the reasoning of section 7.1, we can posit in

addition some communal goods, which cannot be reduced to economic goods but which participants find important, since such goods have to do with the kind of community they live in and what the community will become in the future. These goods will manifest themselves if appropriate options and opportunities that have shaped the community and its culture are preserved, protecting both the range of choice and the possibility of maintaining continuity of a community and its culture over generations.

Now we can ask a most interesting theoretical and practical question: How would a group of people proceed if (a) they were committed to acting cooperatively in conjunction with an identified group of other participants in an adaptive management process and (b) they were concerned to protect *both* individual, economic values (which are associated with the short-term, rapid dynamics of economic exchange) *and* communal values (which are associated with a slower dynamic that unfolds at the pace of an inhabited landscape)? These communal goods, and their association with multigenerational values, link human, plural values with the multiscaled dynamics of natural systems posited in the physical models of adaptive managers in our decision system. It is appropriate, then, that the multiscalar nature of the physical models used by adaptive managers should be matched with multiscalar systems of human values. Once the question is reframed in this way, a new set of theoretical and practical challenges faces us. We are almost ready to begin building a decision model that can encourage cooperative behavior even in the face of diversity of beliefs, worldviews, and interests in the community.

Before turning in the next section to the task of modeling environmental decisions using tools of decision science, we must decide whether to assume a commitment to democratic governance as a part of the management context. Should we strive to shape the decision process so as to achieve democratic outcomes as the only acceptable outcome of our decision process? I have throughout the book assumed that adaptive management would be embedded in a broadly democratic system of government, but it is still a valid question: What should happen when expert managers advocate one set of actions and the general public or most stakeholders or participants prefer another?

There is, of course, a well-known tradition in the literature on the politics of environmental protection that worries seriously about whether, as human populations grow and standards of living improve (so that resources must be spread ever more thinly), democratic governments will be strong enough to restrain overexploitation and avoid disaster.[7] According to this literature, increasing population and economic growth will push societies of the future to be more dictatorial and even totalitarian, in order to control access to ever-diminishing individual shares of natural resources. Most of the writers in this tradition are not properly referred to as antidemocratic; they do not argue that the destruction of democracy will be a good thing; rather, they question whether democracy will be *possible* in the future, given growing demands on resources. Most of these authors, indeed, bemoan the threat to democracy and write their jeremiads in order to warn the public and decision makers to restrain runaway growth and thereby give democracy a chance in the future. We will not discuss these positions in detail here because, given the project of this book, we know exactly how to answer these pessimists. We can best answer them by showing how communities can, operating in an open and democratic way, make decisions that will protect their environmental values. Of course, it would be more difficult to actually prove that our decision model will be equal to all the challenges to governance forced upon us by growing scarcity, but if we can show how communities can make some decisions cooperatively and democratically today, this demonstration should at least justify some optimism for the future. If we can build the institutions for orderly and democratic protection of nature today and encourage the social learning they embody, we may also learn how to sustain those institutions and practices indefinitely.

Since the authors just discussed prefer democracy but question its possibility in the future, their concerns only encourage us in the task of developing effective and sustainable democratic institutions for managing resources and the environment. A greater threat to our goal of making adaptive management democratic comes from two lines of reasoning that were initially distinct. One line of reasoning asserts that environmental problems are technical problems that should be left to experts to manage; if the majority of the public does not accept the solutions proposed by experts, then the public must be coerced or "educated" to accept the expert opinions. The other line of reasoning claims the right of experts and right-thinking people to override democratic governments when such governments fail to provide enough protection for the environment. Although these two lines of reasoning originate in very different intellectual viewpoints, their cogency as reasoning depends, surprisingly and illuminatingly, on a shared—and confused—key premise. I begin by explaining and critiquing a philosophical argument proposed to justify a right to override democratically chosen actions. What we learn about its epistemological commitments will help us answer the "technical expert" line of reasoning as well.

Although I believe she speaks for very few philosophers or environmentalists, the Canadian philosopher Laura Westra has taken an openly antidemocratic stance, defending the view that bad decisions made democratically should be overridden and made right regardless of public opinion or majority wishes. Westra explicitly claims that "democracy is insufficient to contain or eliminate the repeated threats to which we are exposed."[8] She argues that democracy is necessary, but not sufficient, to protect the integrity of nature, so we should "supplement" democracy with some "Second-Order Principles," which are based on a "law of peoples" and which override any democratic process.[9] According to Westra, her "second-order principles"—principles she explicitly refers to as "non-negotiable"—are to be enforced "even by coercive means."[10] Remarkably, there is no discussion of who, exactly, should exert the force in defense of this "law of peoples." "Supplementing" democracy with unnamed individuals who may coercively enforce their own view of right and justice upon the majority may seem too much like "supplementing" democracy with a führer and storm troopers.

Westra's cavalier dismissal of the hard-fought commitment to democracy is supported with an extremely skewed interpretation of the facts. She argues, for example, that "in some sense the plight [of residents of North American cities] is even worse" than the plight of peoples in developing nations under dictatorships. "Democracies are no different," she says, "from, say, military regimes or other nondemocratic states in terms of the severity of the environmental threats to which their citizens are exposed, although the threats themselves may be different in different communities."[11] These rash claims flout obvious observational evidence from the Soviet Union and eastern Europe with regard to dictatorships of the left. Similar evidence against the environmental benignity of right-wing dictatorships also abounds—in Nigeria, for example.

Westra's argument begins to lose cogency already. The key point is not the past performance of democracies and dictatorships in protecting natural systems and human populations. Given the evidence, both the democrats and the antidemocrats must admit that they have usually failed miserably in improving environmental policies and protecting environmental assets. The appropriate question is, Which system is more susceptible to reform and correction? On this point, I think there is no debate possible; for all its imperfections, democracy has sustained itself by its ability, often exercised only episodically, to "throw the bums out" and to effect a change in the public and political sphere of action.

The key weakness of Westra's argument, however, is more in what she does not say than in what she says. By ignoring the crucial question of how to decide when and under what circumstances an individual or group would be *justified* in overriding legitimately made democratic decisions, Westra—and anyone who chooses to follow her in demanding the privilege of overriding legitimate democratic actions—faces the epistemological dilemma mentioned at the beginning of this section. The question is, Can the decision to override democratic outcomes be justified by appeal to an objective principle? Or is such a decision to be justified on no more than the impressions and beliefs of a minority made up of individuals who, for whatever reasons, are convinced that their proposed policies are clearly superior to democratically chosen policies? To choose the latter alternative, of course, is to choose chaos. If every individual or minority reserves the right to override legitimately decided government policies, without justifying those policies intersubjectively, then there is in effect no government at all. If the group decision procedures that normally constitute legitimate government are then not binding, and no one has authority to enforce whatever decisions are made, are we not left with anarchy?

One thus suspects that Westra would grasp the other horn of the dilemma. She, but not others, is justified in overriding democratically legitimate policies of her government because she, unlike others, is appealing to objectively and interpersonally justifiable overriding principles that ensure that her alternative policies are superior to the democratically chosen ones. Unfortunately, as was explained in chapter 3, this objectivist epistemology is untenable. If one appeals to one's own experience, insisting that it is more valid than the experience of the other participants in a democratic process, then each and every individual should have the right to assert priority for his or her own experience as well, and we fall back into chaos and anarchy. Alternatively, Westra might support her position not on the basis of experience but on the basis of some a priori, reason-based principle. Since a priori appeals cannot, by definition, appeal to the particular experiences of those who believe in them, they must be supported as "self-evident" and as undeniable by any rational being. If the principle that separates Westra's position and policies from those of the majority is self-evident, but only to Westra and not to others, then the objectivist position collapses into subjectivity: we might all, on this view, have our self-evident principles, but unfortunately they are only evident to some selves and not to others.

This argument applies equally to "experts" of any kind, whether they are philosophers who divine some special type of rights that override legitimate decisions, or foresters, risk assessors, water quality experts, or some other kind of expert. If one appeals only to experience to support one's superior position, then the only basis for expecting experts to get it right more often than others is their track record. Those nonexperts who refuse to go along with expert recommendations might well be appealing to their own experience that, more often than not, experts *do not* get it right. Indeed, we have learned of many cases in which they get it wrong precisely because their "expert" training focuses their attention on the wrong aspects of the larger picture. So neither Westra nor technical experts can claim special privilege for their policies over legitimately chosen democratic policies, because all such claims are, ultimately, individual claims. They could be reasonably imposed on others only if those individual claims could appeal to higher authority than individual assertion.

Although pragmatists make no a priori appeals, they do not see experience as simply "individual" experience. Experience, because of the pervasive effects of language and social customs surrounding communication and cooperative action, is necessarily based in a community. The pragmatist does not, of course, concede that commonly accepted experience and community norms are necessarily correct; the pragmatist can claim, alternatively, that future experience will prove the minority opinion correct and the majority wrong. But the prediction that future experience will vindicate me in a current debate against a majority is hardly a basis for overriding the current majority unless, of course, I can produce the requisite experiences in the form of new experimental data and convince some nonsupporters to come over to my side. If I can convincingly present such experiential evidence, then the minority becomes a majority and I triumph *within the public, democratic process.* If I fail to convince, then I must bide my time and seek new experiential evidence to challenge the majority view at some point in the future. The pragmatist alternative thus avoids subjectivity and relativism by treating disagreement as based on differing predictions regarding what experience will show in the future. It avoids dogmatism and dictatorial tendencies by insisting that the beliefs of the minority, and the experiential basis they can claim for them, will be acted upon only when they are made convincing to a majority. On this view, diversity of opinion is not to be squelched, with one set of views overriding another, but is rather the beginning point—the raw materials—of ongoing debate and progress as new experiences give credence to one view over others.

In the end, I guess, we all face a choice. We must decide whether we are first and foremost environmentalists or first and foremost democrats. There is no point in fooling ourselves, as Westra does, by assuming that an environmentalist gestapo would enforce her particular, nonnegotiable view of the right in a way that is either more humane or more protective of nature's integrity than the way democratic governments do. Either we are willing to flout democracy, manipulate political institutions, and turn to coercion when all else fails to achieve nonnegotiable demands, or we embrace democracy, recognize its weaknesses, and accept the undeniable evidence that some democracies are environmentally destructive, at least in the short run. Having admitted this, we can set out to educate, to improve process, and to strive to improve policy through democratic means. For my part, given these alternatives, I choose democracy. I part company with those who urge us to go back to Nazism or to Stalinism, introducing coercion in favor of one's own goals over those goals that emerge from a legitimated political process. I accept it as a constraint, then, that policies proposed will be, and should be, possible to implement through democratic means. This is emphatically *not* to say that they will be embraced by all democracies, because most existent democracies are imperfect in various ways.

The commitment to democratic process, I conclude, should be a requirement of any exercise in adaptive ecosystem management. One cannot be a democrat and at the same time advocate use of coercive force to protect non-negotiable values. The antidemocrats need to recognize a very important implication of Neurath's analogy. If we are to keep the ship afloat *and* improve it as we sail, we have to accept some givens, such as a commitment of all parties to live with the outcome of a fair process of deliberation and decision, even when they disagree with the decision that emerges. But we must also be willing—and insist that others be similarly willing—to submit every claim to reconsideration and debate when it is challenged. Westra violates an essential precondition of open democracy: by holding her views to be "non-negotiable" and above challenge, she disconnects reason from the political process and is led to embrace coercion in favor of her principles over those of others who may disagree, even if they constitute a majority. I choose democracy, the pragmatists' imperfect servant of truth, over tyranny, the more perfect servant of egoism and self-flattery. We can memorialize this commitment to democracy by requiring, as a condition of considering a participative, deliberative process to be an example of adaptive management, that it must embody democratic ideals to some degree. Again, this is not to say that all participative public processes fulfill this condition, but rather to focus our attention on those processes that actually do fulfill these conditions. We simply stipulate that, by definition, a process will not be considered adaptive if it allows reversals of legitimate democratic policies by minorities who claim superior knowledge or expertise.

In the next two sections we will look at two very promising approaches to understanding interactive decision making in democratic contexts. In section 7.3 we explore decision theory as an analytic tool for helping communities to make better decisions. In section 7.4 we explore "discourse ethics" as an approach to dealing with values in a diverse community committed to cooperative behavior.

7.3 Environmental Problems as Problems of Cooperative Behavior

As noted at the beginning of the previous section, what we need is a *context-sensitive decision model* that is appropriate to open-ended, democratic processes such as adaptive management partnerships. Having discussed some general aspects of the context, we can now propose such a model by examining the broad literature on decision analysis and science. Actually, this area is composed of a number of subfields of study, using different methodologies and assumptions, and the subfields are organized and reorganized into changing amalgamations. The subfields include, at least, game theory, decision analysis, management science, decision science (a prescriptive approach to decision making), and the behavioral study of decision makers and decision making. We are searching for insight, then, in a vast ocean of writings. Fortunately for us, we can focus on a small cross-section of this work because our search for a decision model is constrained by three factors. These are, first, the nature of (wicked) environmental problems; second, our purpose, which is to develop a cooperative, democratically functioning decision model; and third, the constraints imposed by our use of the decision model as a general guide in an open, public, place-based process. We are therefore interested in context-sensitive decision making, which will require that we concentrate on some fairly narrow—and little studied—areas of the decision sciences, areas that are accessed by relaxing or denying some of the common, formative assumptions that shape most decision modeling.

We must also—if we hope to use the writings in this area to address environmental decision making contextually—sort through, or modify, much of this work because most of the work in decision analysis assumes that the purpose of a decision modeling effort is to get the "correct decision." This assumption of an objective, process-independent correct decision runs afoul of the same epistemological dilemma that led us to reject the claims of philosophers and experts when they seek to enforce their views on the majority of citizens. Indeed, emphasis on the correct-decision assumption can turn decision scientists into special cases of "expert managers," who arrogantly override politically legitimate decisions. To simplify things, I have separated this section on decision theory into two subsections. In subsection 7.3.1 I discuss the epistemological dilemma and propose (actually, encourage an existing trend in the right direction) a new way to interpret the conclusions and advice of decision scientists. After clarifying these interpretive points, I will focus in subsection 7.3.2 on a few particular insights from this vast literature, insights and techniques that will be very useful in developing a context-sensitive, process-oriented adaptive management model.

7.3.1 The Epistemological Limits of Decision Analysis

When one studies and evaluates decisions, it makes a lot of difference if one believes that there is a single best decision to be identified in any given situation. If one believes this, then even in a complex situation marked by pervasive uncertainty, one will expect that a correct decision exists. Decision science, ideally, finds that decision. When obstacles such as complexity and lack of important data obscure the correct answer, one can still think of the decision task as one of *approximating* an independently correct standard. Interestingly, this assumption, and the associated attitude that there is a preexisting right answer, colors most writings in the area of decision theory and decision science. In this subsection I will summarize the powerful conceptual arguments that can be brought to bear against the "one-correct-decision" dogma. These arguments have been around for decades, and decision analysts, seduced by the myth of an objectively correct answer, have failed to draw the almost-obvious conclusions from them for their own work. As noted in chapter 2 and the appendix, the Cartesian assumptions about reality and how we know it die hard; and it takes arduous work to replace them with anything better than abject relativism or a resort to force. One suspects that decision scientists, like most of us, find it hard to see an alternative to their dogmatic belief or a collapse into relativism. They might say, "if there is no 'correct' decision, everyone's choices will be as good as the next one, and we will fall into relativism," and decision scientists—devoted to how we find correct decisions—will have no subject to study.

Most writings on decision theory are thus predicated on there being a "best" decision, identifiable by appeal to a standard of "rationality." Practically speaking, this theoretical assumption of an objectively "right answer" has had a large impact on the way decision scientists understand their analyses and on the attitude taken toward the outcome of their analyses. When inserted into an ongoing process of negotiation, such dogmatic analyses are counterproductive because they break up ongoing deliberation in favor of "proofs" and formal modeling that derives from a specialized jargon and specialized assumptions. These analyses attempt to "close" an open process.

The decision theorists' assumption of a best answer is highly problematic; it has been devastatingly attacked, undermined, and refuted by several apparently unassailable arguments. I will survey several of the most devastating arguments against the right-answer assumption. One such argument, already introduced in section 4.1, shows that decision algorithms apply at most to "benign" problems, whereas environmental problems almost always involve "wicked" aspects. Although decision models are radically incomplete as decision tools because they cannot solve wicked problems with a single answer, one can nevertheless make assumptions and fix parameters until one can make any decision algorithmically soluble. But does such a manipulated problem remain the "same problem"? As Rittel and Webber point out, wicked problems can be reduced to benign ones only by assuming away most of the interesting issues and disagreements involved. If important issues are resolved by assumption in the problem formulation, a solution will not be a solution to the original, wicked problem.

Another theoretical literature, which is devoted to the application of computers to management decision making, expresses much the same point. Authors in this area, such as the highly innovative Herbert Simon, separate "programmed" from "unprogrammed" decisions, and Simon separates management organizations into three layers, defined by their tasks and roles: "an underlying system of physical production and distribution processes, a layer of programmed (and probably largely automated) decision processes for governing the routine day-to-day operation of the physical system, and a layer of nonprogrammed decision processes (carried out in a man-machine system) for monitoring first-level processes, redesigning them, and changing parameter values."[12] The lower layer, the system of productive processes, will function with automated decision making whenever possible; in the middle layer, decisions are progressively programmed as technologies improve, but this will depend on the state of the technology and the complexity of the tasks involved. Decisions on the highest level involve the most complex, judgment-laden decisions, and these decisions can have huge impacts, including revisions to operations in the first two categories. Although these third-layer decisions are clearly the most difficult to parameterize and program, Simon makes the optimistic claim that in principle, any decision in this category is susceptible of programming (it is "programmable").

Suppose we grant him his optimism. Does it follow that all decisions can be programmed, eliminating any role for managerial judgment? No; and the negative answer depends on a subtle point of logic. It is true that for any particular decision, if we care enough to make it programmable, we can do so. But Simon explicitly states that though *each* decision can be programmed *at some time,* this does not mean that *all* decisions can be programmed *simultaneously.* He is saying that we can isolate any particular decision, hedge it around with assumptions, parameterize it, and develop an algorithm for solving it. In so doing, however, we make judgments that necessarily shift some aspects of the original problem into the realm of unprogrammed "judgment." Just as Rittel and Webber say, we can force a wicked question into benignity, but we can do so only by abstracting our solution from the wicked aspects of the original problem. It is this wicked "residue" that inevitably reemerges when experts "solve" problems according to their special disciplinary algorithms. They solve the problem by simplifying it through assumptions that bury the most contentious aspects. To offer an everyday example: *If* the local garden club could agree on how to rank-order its objectives for the next year, its members could use a formula to allocate their income from dues to various projects. But since they cannot agree on objectives, the formula is worthless. Most environmental problems contain the same dilemma. What is needed is a reasonable method for agreeing upon objectives. Such a method, however, will always require judgment about what is most important, and that judgment cannot be programmed.

The consequences of this conclusion are both general and startling. As noted in section 4.1, it entails that the most important aspects of public policy decisions—questions about what goals and objectives we should seek—are not susceptible to optimization solutions. Optimization requires that all parameters of a problem must be measurable in a single system of accounting. In wicked problems, however, we have learned, the dispute is about how to formulate the problem and which competing interests to prioritize.

Since it is so hard to express all important outcomes of a decision in a single measure, why not propose multiple measures? This is a good idea, one I have endorsed above, and it is used extensively in what follows. I have proposed that in order to incorporate pluralism into our decision model, we think of environmental problems as multicriteria decision problems. An important clarification is in order, though. Many decision theorists, having chosen to posit multiple criteria, move beyond that to propose multi-attribute utility theory (MAUT), which promises to provide a formalization of how to use multiple criteria to maximize expected overall utility resulting from the decision. And here the wickedness of problems reasserts itself. MAUT might be helpful in separating several concerns raised in a decision situation and associating those concerns with measurable criteria, but it cannot tell us how, "objectively," to weight the multiple criteria. Once again, the idea of a correct decision or a programmable decision as a response to wicked problems proves elusive. The decision how to prioritize values and objectives necessarily involves a balancing judgment among multiple criteria. Choosing the best balance among competing goods cannot, ultimately, be "programmed" without changing the nature of the question addressed. Below, we will see that these weighting problems are best seen as "metadecisions," decisions that determine how the various criteria are applied in our decision model.

Perhaps the most powerful arguments against the view that the goal of decision science should always be how to identify or approximate the "correct" answer derives from a series of results originating in Arrow's impossibility theorem.[13] This proof has been followed by a series of similar proofs with slightly different assumptions and scope. The apparent outcome of this series of analyses and proofs is to considerably restrict the application of methods of problem-solution based on simple, aggregative methods, such as analyses that compute, from preferences, an "optimal" outcome. The restrictions, in turn, have been taken, more expansively, to question whether technocratic solutions to problems, based on formalistic and computational approaches to "rationality," are likely to yield solutions acceptable to democratic majorities.[14] Arrow, working within the tradition of "social choice," addresses problems of collective choice by aggregating the preferences and interests of individuals affected by the decision. Aggregation, however, can only be formalized within a system of analysis that posits certain minimal conditions of consistency, such as transitivity of preference (if *a* is preferred to *b*, and *b* is preferred to *c*, then *a* must be preferred to *c*). Once preferences are factored into a "rational" decision procedure and required to conform to these minimal conditions of consistency, Arrow showed, it is impossible to derive a democratically acceptable result. This conclusion has been taken to show a conflict between "enforcement" of a rational decision (as defined by an expert's method for aggregating fixed preferences) and free democratic choice. Accordingly, if an expert analyst derives a "rationally" correct outcome from a formal decision model, such an outcome will often be at odds with real decisions made by officials who are responsive to political pressures. Should a "rational" solution override a democratically obtained solution? If one sees no difference between the "rational" solution and the "best" solution, one will be tempted to answer this question affirmatively.

Another reason there is a systematic divergence between best solutions computed rationally from individual preferences and politically negotiated outcomes is that the analysts must make a number of assumptions that are necessary to allow the expression of disparate interests in commensurate terms, plus other assumptions for the purpose of simplifying calculations, but these methodological and disciplinary assumptions have no effect in the context of negotiated decision making. The methodological decisions, and there will be many of them in the construction of a computational decision model, may be influenced by the (perhaps unconscious) interests of the analyst, some of which derive from professional, not civic values. Whatever the reasons for the divergence, however, the divergence has been proved to be unavoidable. Worse, uncritical confidence in computational models leads to contempt for democratic and negotiated outcomes and reinforces the idea among analysts that they have "the right answer," against which negotiated outcomes can be critically compared.

Hopefully, the parallel between this conflict among experts and outcomes of democratic processes, on one hand, and our criticism, in section 7.2, of Westra's claim to a right to override democratic outcomes, on the other, is now clear. Westra can maintain her right to override democracy only if she has knowledge that is unavailable to, or more certain than, the information factored into the decision by other participants. Similarly, the expert decision modeler should override democratically chosen policies only if the expert is certain that the model proposed is free of taint and bias—if, that is, there exists a "correct" answer that can be known independently of any bias or viewpoint. To seek such an independently correct answer is to return to a Cartesian view of knowledge: that there is a truth that exists independent of any observer and that this observer-independent reality can be accessed by recognition of "self-evident truths."

In fact, however, the Cartesian envy that still afflicts many decision theorists is unnecessary to assuage their concerns regarding relativism in the absence of the correct-answer assumption. To fully explain this claim requires a bit of technical reasoning; fortunately, by once again hitching onto the research and storytelling talents of Louis Menand, we can introduce the technical argument painlessly. So once again, still shamelessly, I will abstract from Menand's account in *The Metaphysical Club*.[15] In this case our technical point can be illustrated by the tale of how Charles Sanders Peirce, despite a generally lackluster career, gained fifteen minutes of fame. Peirce was the son of a leading mathematician of his day, Benjamin Peirce, a disciple of LaPlace and a leader in the developing field of statistics. When Benjamin's second son, Charles, turned out to be a mathematical prodigy, he concentrated on teaching him probability theory and tutored him in mathematics. Charles wrote a history of chemistry at the age of twelve and introduced himself to logic by reading his older brother's textbooks when he was thirteen. Charles was also what they called in those days a "computer" (a description that is perhaps even more apt given the current application of that term to computing machines): he could compute extraordinarily complex mathematical problems in his head. Benjamin, the expert in probability theory, and Charles, the computer, had brief but spectacular roles as expert witnesses in one of the most celebrated probate cases of the nineteenth century, the Howland will case.

Upon the death of Sylvia Ann Howland, the Howland fortune, the largest derived from the lucrative whaling business in New Bedford, was by her will to be split between the last living relative of the Howlands, Hetty Robinson (Sylvia's niece), and other inheritors. Hetty, by all accounts a rather greedy lady, sued in an attempt to overturn the will and receive the whole fortune. I won't go into the details of the celebrated trial, but she produced a notorious "second page," an addendum to the will, apparently signed by her aunt, that revoked all wills made by the aunt, past or future. If the existing will was not probated, Hetty, as the last surviving relative of Sylvia Howland, stood to inherit the entire fortune. The case appeared to turn on whether the signature on the second page was a forgery. It was perfectly identical to the witnessed one on the first page, and the defense team for the estate claimed it was a traced forgery. The trial and associated depositions became a spectacle, as chemists analyzed ink, bank officers testified about signatures, and demonstrations that some people could exactly duplicate their signature part of the time were mounted: there were handwriting experts and penmanship teachers called, and there was nationwide news coverage. Benjamin Peirce was called to testify on the probability that the two identical signatures occurred by chance, and the elder Peirce brought his twenty-eight-year-old son Charles with him; both were deposed.

Charles, who went first, explained that he had obtained and used forty-two samples of the aunt's signature and had chosen thirty points in the signature to compare, and then he had calculated the likelihood that the similarity occurred by chance. From the samples, Charles found that about one-fifth of the positions coincided in the aunt's undisputed signatures, and then he tabulated the results of a total of 25,830 comparisons. Benjamin, subsequently, relied on a procedure that was later central to his book *Linear Algebra,* published a few years afterward. He surmised that there would be a statistical order hidden in the chance occurrences, an order that could be tested by Charles's calculations of actual coincidences. Benjamin then calculated what the coincident rate *should* be, provided the coincidences occurred by chance, and it turned out to be one-fifth. This calculation, based in probability theory, thus predicted what Charles found by rote comparison. Benjamin Peirce concluded therefore that since the hypothesis that the natural rate of coincidence was one-fifth was confirmed by both theoretical calculation and observation, one-fifth was the relative frequency of coincidences occurring without chicanery.

Benjamin Peirce then increased the fraction to the proportions necessary to take into account the fact that in the second-page signature, coincidences were found at every position. He dramatically concluded that the chance that the aunt had actually signed the second page was one chance in 2,666,000,000,000,000,000,000. "So vast an improbability," he argued, "is practically an impossibility. . . . The coincidence which is presented in this case cannot therefore be reasonably regarded as having occurred in the ordinary course of signing a name. . . . I declare that the coincidence which has here occurred must have had its origin in an intention to produce it. . . . It is utterly repugnant to sound reason to attribute this coincidence to any cause but design."[16]

So the Peirces were briefly in the national spotlight, though the general consensus was that they were engaged in mathematical sleight-of-hand, and the lawyers for Hetty Robinson ridiculed the performance as empty and baseless academic virtuosity. In the end Hetty's suit to break the will was thrown out on an unrelated technicality. Hetty, despite having to settle for part of the family fortune, became a shrewd financial operator and was eventually dubbed the Witch of Wall Street. She died in 1916 with a fortune of over $100 million. The lawyers and the public may have been disdainful, but the Peirces were indeed working from the cutting edge of statistics and probability theory; Benjamin's theoretical calculation was based on the "law of errors," which was discovered as the rapidly advancing fields of statistics and probability were applied (by Benjamin Peirce and others) to the field of astronomy. The law of errors was so important to both of the Peirces, who had accepted, from Darwin, the conclusion that the reality we encounter was created by chance in the form of spontaneous mutations, because the law of errors allowed them to claim that reality, however random in its origin, was nonetheless knowable on the basis of probability calculations.

Astronomers faced a problem in the exact placement of stars. Every observation taken of a star will vary slightly from others as a result of many unknown factors, including atmospheric changes, differences in equipment, and carelessness. Given that none of the deviations are intentional, any given measurement is as likely to be too high as too low. The law of errors in astronomy was based on the discovery that these measurements tend to form a bell-shaped curve. Menand cleverly uses the analogy of many archers shooting arrows at a target, trying in each case to hit the bull's-eye. One could, from the pattern of placement of many arrows shot, calculate (using a mathematical method of least squares) the "implied" position of the bull's-eye, even if no shooter actually hit it. The law of errors, in other words, uses the *process* of trying to measure the location of a star, or of shooting arrows at a bull's-eye, to progressively zero in on an unknown point. Provided the deviations can be considered random, their distribution can allow us to infer toward an exact location, even though we cannot—prior to the process of earnestly trying to locate the star or to hit the bull's-eye—calculate the "correct" location independently. Inaccurate observations can, provided they are combined with many other observations by many other individuals, point us toward "reality."

The law of errors, then, provided the foundations of Peirce's conception of truth. Truth emerges from many, many experiences; it is the inferential point of convergence of many separate shots at the mark in reality. Peirce emphatically denied, however, that such an approach suggested anything subjective or relative about our common reality. Reality, he said, "is independent of the individual accidental element of thought."[17] Reality is constructed out of experience, he believed, but not from individual experience. If there are many earnest observers who can communicate their observations, then the law of errors will allow us to infer the truth from the convergence of many observations made by many observers who all "aim" at the truth. In many ways the law of errors, which Peirce learned at his father's knee, was the unifying principle of the systematic philosophy that he could never bring himself to finish between two book covers. The law of errors allowed Peirce to maintain that reality is knowable, even after Hume and Darwin; to maintain this, he replaced individual perception with a communal process that provides the raw material for a constructivist account of reality. As time passes, observations converge around a hypothetical ideal—the point that would emerge from an indefinite process of observing. The real world, on this view, is the one the ongoing community gradually constructs from countless observations. Identification of the "correct answer" cannot be computed in advance of the process; but it can, in principle, be derived from the process as the inferred outcome of an indefinite number of observations of individuals whose characteristics vary randomly with respect to the aimed-at truth.

Peirce thus saw the law of errors as a confirmation that his forward-looking, community-based notion of truth provided an alternative to Cartesianism and the apparent requirement of Descartes's dualistic worldview that our language conform to real essences in nature. One need not believe that there is a truth "prior to experience" according to which our beliefs can be checked in order to believe in truth. Experience, as long as it emerges in a community of communicators, can approximate the truth by application of the probabilistic law of errors. The law of errors, for Peirce, replaced self-evident truths with a community-based, experientially guided process. Recognizing that our knowledge is always contingent and based on experience, he turned to probability and statistics to fill the void. Menand explains that Peirce adopted as his central philosophical task to show that the world, despite its randomness, could be known by humans because of the social tendency. Having accepted the contingency of the universe as we experience it, Peirce concluded, in Menand's words, that "in a universe in which events are uncertain and perception is fallible, knowing cannot be a matter of an individual mind 'mirroring' reality. Each mind reflects differently—even the same mind reflects differently at different moments—and in any case reality doesn't stand still long enough to be accurately mirrored. Peirce's conclusion was that knowledge must be social. It was his most important contribution to American thought."[18]

Here is a clear alternative to the correct-answer assumption of decision analysts; one need not claim that a problem has a computable solution that is prior to a community-based, participatory inquiry; one can seek acceptable outcomes for now and continue amassing experiences as part of a truth-seeking process, letting the law of errors guide us closer and closer to "reality." There is, then, a viable alternative to both relativism and the "prior best solution" assumption. That alternative is to consider the multiple techniques of decision analysts to be useful tools within a larger process that is based in experience and a commitment to experimentalism. Decision scientists do not, even if they accept Peirce's notion of truth-seeking, need to *do* their analyses differently; they need only to recognize that these analyses play a different role than they have assumed. Their analyses should not be understood as prior derivations of correct answers that the community should strive to match, but rather as one of the many arrow shots that, when factored together with the combined experience the community gradually accumulates through observation and communication, may help the community to "surround" the bull's-eye and thereby converge upon shared beliefs that are strong enough to act upon. Once acted upon, they will be constantly tested by more experience of the diverse participants in the community decision process. In subsection 7.3.2 I return to the importance of integrating decision theory and its insights into the ongoing, open-ended process called adaptive management.

The myth of a correct decision, despite the apparently devastating analysis of Rittel and Webber, despite formal arguments in social choice theory, and despite the persuasiveness of the experience-based, pragmatic approach that eschews self-evident truths, continues to bedevil decision analysts and decision scientists. Fortunately, this myth is gradually losing its hold on a new generation of decision theorists, who recognize that there will always be an arbitrary, assumed element in every decision modeling effort. This trend can be seen in a number of contexts, as analysts, for example, use several different analytic systems and present multiple "solutions" or ranges of solutions. Particularly interesting is a series of recent papers in which decision scientists have injected their methods—including many of the choices regarding which criteria to choose, how to weight them, and so forth—into an ongoing public decision process.[19] This trend, I think, is the key to developing a less dictatorial and more useful set of decision models. Implicitly, the decision analyst gives up the claim that the result of a single analysis will yield a uniquely correct truth; but that matters little, it turns out, once the decision model is placed in time and allowed to evolve with new information and the entry of new voices and new concerns into the public deliberation process. These new trends in the study of group decisions should not miss their anchor in correctness, because they no longer expect to apply a decision model at a given point in time and compute a best solution for all times. Correctness must evolve across time, one might say, even as the problem formulation and the knowledge base evolve.

By embedding important judgments in an ongoing and open-ended public deliberative process, the analyst need not justify a given decision model as yielding uniquely correct solutions; the analyst can, instead, understand the decision analysis as one way to analyze the problem, and participants can try out these models as one way of thinking through the problems they face. Freed from needing to show that these principles and rules will yield unique truth, and having embedded decision analysis in concrete, ongoing, deliberative processes, the analyst can thus suggest, for experimentation, a variety of analyses and conceptualizations of a problem. These can be put forward as useful, as worth experimenting with, and as illuminating possibilities. In a sentence, giving up the myth of a best solution in decision-analytic situations shifts decision analysis from application in isolated, "expert" contexts, into the open-ended, constantly adjusting process in which group decisions emerge from diverse interests and viewpoints articulated by the multiple voices expressed in a community. By giving up the myth of one correct solution, decision analysts become contributors to a process, and their skills can be applied, adjusted to the situation, and applied again. All that changes is the way we interpret and use their suggestions and recommendations. Once it is accepted that there is no decision that is correct, independent of community-based truth-seeking processes, decision analysts—who have studied many processes and have tried multiple techniques—can offer heuristics for good decision making. From these processes we can get an idea of what a good decision looks like, by experimentation and gradual evolution of techniques, perhaps even from generalization from multiple cases. This form of objectivity is enough for a pragmatist, for the most a pragmatist hopes for, epistemologically, is to gradually improve our ability to predict our future experiences.

In the next subsection, it will be shown that decision analysts can provide general guides—models—by which diverse communities, if committed to cooperation, can organize their decision processes to achieve cooperative outcomes that all can accept. I have already suggested one such heuristic: a procedural heuristic that encourages a more or less conscious separation of a community's discourse into an action phase and a reflective phase, as a means to focus group discussion on the right questions. The shift to expecting not "correct" solutions but heuristics means we no longer privilege the calculations of decision analysts over the outcomes of democratic process; we privilege, instead, the outcomes of processes that are fair, open, and cooperative and include attempts to model decisions as one of these processes.

7.3.2 In Search of a Decision Model

We are now ready to survey the literature of decision theory and science in search of insights that might guide us toward a decision model that will be useful in adaptive management partnerships. As noted, though the total literature on decision making is vast, interestingly, very little of this literature is devoted to the kind of decisions we expect adaptive management partnerships to face. Decision analysts have focused most of their attention on decisions that can be resolved by some kind of synoptic calculation. We just saw that there are irrefutable proofs that single-value aggregations cannot identify what most of us would consider good solutions to complex problems facing communities. Besides the deep theoretical problems emerging in collective choice aggregations as outlined in subsection 7.3.1, there is also the serious practical problem that even if we believe that a single measure of economic efficiency or expected utility from a decision is possible, we usually do not know how to translate values that citizens think are important into a common-denominator measure of individual welfare or of expected utility. Thus it is impossible for a single-measure analysis to provide a complete formalization of the day-to-day decisions facing adaptive management partnerships.

We noted above that decision science and decision analysis covers a vast, Balkanized area of intellectual exploration and that several subfields, usually organized around somewhat different methods, have emerged. Here, since we are mainly searching for useful ways to analyze a particular type of decision in a particular kind of situation, let us sort approaches to decision analysis according to the type of decision that is studied. Once we characterize a type of decision that we need a method to address, we can then ask what methods are appropriate and useful for better understanding that category of decision.

First, let us distinguish *interactive* from *noninteractive* decisions, with the former designation referring to decisions in which the chooser's decision and the outcomes of that decision are affected by the actions of other choosers. Alternatives in noninteractive decisions can be evaluated without reference to the actions of other actors. Noninteractive decision analyses are usually studied by abstracting and formalizing certain aspects of a real-world problem by stipulation. The analyses can be generalized to other decision problems with the same stipulations because the parameters of the problem are stated as the starting point of the analysis and are not open to correction or reconsideration; they formally *constitute* the problem under discussion. If one were to change them, one would change the problem itself. We can very quickly conclude that we should concentrate our search for appropriate decision processes and analyses by examining *interactive* decisions, because we are interested in decisions that are made within an ongoing and iterative process involving communication among participants.

This is not to say we cannot learn anything from the analysis of noninteractive decisions; it is rather to say that the types of decisions that interest us, as we study adaptive management partnerships, are interactive decisions. Formalized analyses of problems abstracted from their political context may be helpful as illustrations, and they can be invoked to make important points in some cases, but the basic structure of the decision situation remains the politically charged situation in which participants interact, disagree, and yet seek compromise in ordinary discourse. Formalizations in such situations function much like a momentary use of the machinery of symbolic logic within the flow of ordinary conversation, in order to make one specific point precisely. For example, in the course of a broader discussion about what to do, you might correct my reasoning as follows: "Your argument is invalid—it affirms the antecedent—you are saying 'if *p*, then *q*, *q*, therefore *p*.'" This can be a very useful ploy in argument, and its usefulness does not depend on a claim that the entire decision we are facing can be formalized in the symbolic, propositional calculus. Such formalizations are best thought of as useful but always partial methods of analysis. They can be inserted into ordinary conversation even though neither the speaker nor the hearer would know how to expand the formalization to capture all of the parameters of a real-life debate or decision. So we continue our search for a model of adaptive management decisions, focusing our attention on interactive decision models, and we do not expect that the models we use will be at the same time comprehensive and formalized.

Of the interactive decisions, perhaps the most discussed are decisions that can be considered "games" in the sense that outcomes of possible choices for deciders depend on the decisions made by others. Originated by mathematicians and economists, game theory initially emphasized "competitive" games in which it was assumed that by making the "right" decision, one player could come out better than the other in situations such as "prisoners' dilemmas." A simple prisoner's-dilemma scenario is as follows: two people are arrested for the same felony crime. Police question them individually but do not obtain enough evidence to convict them of the felony; the police do, however, have enough evidence to convict on a misdemeanor traffic violation. The police offer the following deal to each prisoner individually, informing her or him that the other prisoner has been offered the same deal: if one prisoner implicates the other, and is not in turn implicated, the accusing prisoner goes free and the accused prisoner serves a full sentence for the felony. If each one implicates the other, each gets a minor sentence for the misdemeanor. If neither implicates the other, both go free. Both persons, seeking to protect their own individual interests by avoiding the worst case, are led to choose the suboptimal outcome. This interactive decision situation obviously highlights the complexity of situations in which payoffs on one's individual decision vary according to the decisions of others.

Prisoners' dilemmas and other similar "games," in which one can count payoffs in dollars or years in prison, for example, and in which there are clear winners and losers in each outcome, have been of great interest because these properties allow the use of a combination of probability theory and a preference calculus (either welfare economics or expected utility theory) to calculate "best strategies" to assure the best possible outcomes for players. Thus John Nash, in his Nobel-winning work, showed that it would be rational for players not to cooperate, because this is the strategy that avoids the worst-case outcome; by not cooperating, players create an equilibrium at which the aggregated payoffs for "rational" players is worse than optimal.

Game theory has been very popular in mathematics, economics, and decision theory because of the precision with which one can compute given (unquestioned) preferences. Computability, however, requires that the game be structured in a particular way, with preferences of players assumed or given, and that it conform to a number of quite restrictive assumptions. Almost all technical discussions of game theory are based on the following five assumptions:

1. The set of possible strategies is given.
2. The outcome set is exogenously specified, and the outcomes are quantifiable in net present values.
3. Each player is rational (in the sense that each player acts as a self-interested maximizer) and acts on full knowledge of the situation.
4. Each player knows that every other player is rational and knowledgeable.
5. The rules of the game are nonnegotiable and exogenously given.

Would an analysis still be "game theoretic" if we were to relax or deny one or more of these assumptions? That is probably more a semantic than a real question: whatever one calls it, decision analyses that do not conform to the five axioms lose the feature of computability (because all of them are required by the assumption that each player has a complete and consistent set of preferences). Nonconforming games are of little interest to the usual theorists of games and decisions, who are mainly interested in games precisely because of the possible precision in analyzing them. They assume that a given decision situation has certain features so that the decision will be susceptible to precise solution. These artificial analogues to real decisions may yield some important insights, but the decision situations faced by adaptive management partnerships are nothing like the decisions posed to hypothetical "prisoners" facing a one-time, competitive situation.

Cooperative game theory is more likely to resemble the kinds of problems faced by partnerships in which most of the actors are committed to finding a cooperative solution to a problem. But cooperative games require some level of communication. A minimal form of communication is possible in *iterative*, multitrial games, in which individuals act alone in each trial but each player knows the other's plays in previous moves. In one of the most interesting outcomes in empirical game theory, it turns out that even in a game that is constituted as a competitive game, spontaneous cooperation can emerge. Indeed, Robert Axelrod showed that in an iterative game no strategy can consistently beat the "tit-for-tat" strategy, which says to cooperate in the first trial and then respond in kind to each of the other player's choices.[20] If one's opponent defects, then on the next trial the tit-for-tat player will defect; and if the opponent cooperates, a tit-for-tat player will cooperate. Assuming that the rewards are set so that consistent cooperation is better than consistent defection by both parties, spontaneous cooperation can arise. The success of the tit-for-tat strategy shows that even competitive decision situations can lead to cooperative behavior, provided that there is even minimal communication; this very general result supports our emphasis on both communication and cooperation, as well as providing a good reason to think about decisions iteratively (as the wickedness of environmental problems apparently demands, independently, that we do). Now we are moving in the right direction. We can begin to focus our attention on interactive decisions that are made iteratively, with at least minimal communication among parties, and are such as to allow—perhaps even to encourage—the development of cooperative strategies by the players.

Because we are positing that an irreducible plurality of environmental values exists in the community and that these multiple values will be represented by the various participants in our adaptive management process, we will focus our attention on decision models embodying multicriteria decision-making analysis (MCDM). I doubt that all environmental values can be expressed in a single vocabulary or measure, so I have suggested that we develop a multicriteria decision analysis in which each criterion is stated as requiring a certain level or standard with respect to a particular indicator. The indicator is a measurable variable that is hypothesized to vary with the production of a particular social good. We can thus operationalize value pluralism by conceiving decisions by the partnerships we study as attempts to find an acceptable balance among multiple criteria stated with respect to multiple indicators. I will return presently to the question of how such balancing questions should be modeled.

Given that the payoff functions are specified in the rules of the game by the analysts, we can distinguish between "positive-sum" and "zero-sum" games. In a zero-sum game, all combinations of plays result in the same total payoffs. In these situations, cooperation is difficult to establish. An academic illustration will clarify the point. Our dean is provided a fixed budget for the college for each upcoming fiscal year. For the next year's budget, the department chairs play a zero-sum game, directed by the dean, as they fight for the largest possible share of a fixed pie. But the dean also knows that the development of an impressive and exciting master plan can lead to larger budgets for the college, and perhaps even special funding from the legislature for a new building. In this situation the dean and the department chairs are involved in a positive-sum game: if they make the right "moves" individually, they will increase the total payoff, distributed among them. Assuming that these benefits are distributed reasonably fairly, everyone may be better off.

Are adaptive management partnerships, and the decision situations they face, most appropriately modeled as zero-sum or positive-sum games? It is obvious that, if at all possible, our decision model should allow for positive-sum outcomes; and I can offer two reasons to think the problems faced by adaptive management partnerships are often positive-sum games. First, since the adaptive management model assumes that we must manage on multiple scales and that different processes are likely to be associated with social values that emerge on different scales, it should be possible to identify policies that will protect or enhance both value-producing dynamics simultaneously. Such policies would create benefits on two levels; some benefits on each level would be missed if the policies were analyzed with a single-criterion evaluative system. The second reason it makes sense to treat our adaptive management decisions as positive-sum games is that our participants have already committed themselves to seek cooperative solutions. They must at least believe that by cooperating they will get a preferred outcome. But here we are not speaking strictly of the participants' individually preferred outcome—we have posited that our participants wish to protect communal, as well as their individual, values. They may prefer cooperation because it will protect their values more comprehensively, rather than maximizing their individual preferences.

If the proposal to consider only positive-sum games in which decisions made can increase the "total payoff" seems arbitrary, we can treat it as an empirical research question: will cooperative decisions made as part of adaptive management partnerships create win-win situations? We can test this hypothesis in at least two ways: we can ask participants whether the perceived outcomes of their decision process are more satisfactory than likely outcomes of noncooperative actions. Or, if we can agree on some standard, or several standards, by which we can judge the total outcome of decisions, we can compare actual outcomes of cooperative actions with model results assuming noncooperation. Here we will simply adopt the positive-sum hypothesis as our working hypothesis.

Now we come to the most important modification of standard game theory and decision theory that I will propose in order to develop a useful model for adaptive decision making. Most decision analysts, motivated by a desire for precision, generality, and computable results, exogenously fix the rules of the game to be analyzed. The rules written for the game *constitute* the game. A parallel assumption, however, does not appear to constrain the actual behavior of participants in an adaptive management project. Suppose, for example, that a watershed partnership has set reduction of sedimentation entering streams as a priority objective, and factions are arguing about what the biggest cause of sedimentation is: is it logging and logging roads in the privately owned forest lands in the headwaters of the watershed, or is it downstream livestock grazing near streams? In order to break the deadlock and respond adaptively, members of the partnership might propose two experiments. First, they negotiate a policy of riparian buffer zones with the timber companies, offering to support higher cutting ratios in nonriparian areas in return for a twenty-year moratorium on cutting and road-building within stream buffers. Simultaneously, they propose a list of best practices for grazers in riparian zones and support a government compensation program that offers tax incentives to farmers for adopting such best practices. These actions will appeal to adaptive managers because they involve actions that will set the stage for important scientific studies that can reduce uncertainty about the main causes of the problem, and the information derived from the experiments will also allow them to further refine their formulation of the problem they face. The results may lead to a community-based decision to change the rules and incentive structures by which individuals and corporations "play the game."

If we are to correctly model this kind of decision, which will often confront participants in adaptive management processes, we must separate two aspects of the decision situation, which can conveniently be referred to as "choosing the rules" and "applying the rules." Or, to shift back to the concepts I have been developing throughout the book, we can think of the decision problem faced by our participants as having a reflective phase and an action phase. In the reflective phase, the goal is to propose, discuss, and choose a set of indicators and an associated set of standards, or targets, associated with those indicators. Here advocates of plural social values compete and cooperate to get the whole group to endorse criteria favorable to their concerns and values. The process of choosing which criteria to apply can thus become a part of the management process. Once a tentative set of management criteria is developed, it is possible to seek actions that will improve the environment with respect to several indicators, to negotiate political agreements in support of those actions, and to design scientific controls to ensure that we can learn from actions taken.

Let us then concentrate briefly on the reflective phase, in which participants with multiple values, interests, and viewpoints deliberate and negotiate about what to count and how to set standards to guide their management activities. As noted above, we can understand these choices as articulating, refining, and choosing by negotiation a set of criteria that express the values the participants share as a community. But the proposal and refinement of criteria is only the beginning. After a slate of criteria is proposed, it is still necessary to determine what comparative weights—from 0 to 100 percent—to place on the various criteria. How, exactly, should such weightings be decided? If we knew the exactly correct weightings of the criteria, we could calculate a single best outcome. But we have rejected the possibility of such certain knowledge for wicked problems. In a multicriteria system, none of the first-order criteria (action criteria) are appropriate to determine the second-order question: how much comparative weight should we give each criterion?

If one approaches decision analysis with the correct-answer assumption, there must be some nonarbitrary way to set these comparative weights. In decision analysis, three alternatives are used. The first alternative is to arbitrarily set the weights, for example, as equal, and proceed to calculate. Too often, decision theorists choose this alternative, hiding their arbitrary rankings in methodologically driven constraints. But it is not plausible that an arbitrary specification of weights for multiple criteria will yield a uniquely correct answer, so decision analysts have sought to derive correct weightings in two further ways. They can ask experts in the field what their preferences are across outcomes and then try to specify weightings of the criteria to correspond to those preferences. Although there are many variants on how this kind of participation is solicited, we can classify this participation according to two parameters: Is the participation one-time-only, or iterative and ongoing? Second, is the participation by experts only, or is there open access to participative roles, including roles for lay persons and interested parties? Using this classification system results in four types of participation in criteria choice and weighting of decisions (see fig. 7.1). One-time participation by experts only (type 1 participation), is clearly inadequate for adaptive management partnerships. One-time participation in an open forum, such as a colloquium or a science jury proceeding (type 2), for example, may be helpful in bringing a variety of diverse voices to the table—a good starting point—but it is usually very difficult to get significant results in one-time meetings, because there is no possibility for improvement of communication or development of the trust that is necessary to believe that cooperative outcomes are possible. Type 3 participation consists of iterative expert judgment applied to the weighting of criteria; this type of participation is very useful in some situations, even within adaptive management processes, for example when a specialist committee might develop detailed indicators as proxies for ecological values. However, it is type 4 participation, iterative participation by a broad range of voices in ongoing deliberations, that is most important in adaptive management. Although type 3 participation may be very important—when expert committees are asked to do detailed work in order to propose measurable indicators and then report back, for example—such efforts should be considered supplementary to the ongoing, inclusive deliberations embodied in type 4, inclusive negotiations. Participation in an adaptive management process thus favors an inclusive process, in which experts report back with recommendations that are then discussed in nontechnical language to the extent possible.

	A **Expert**	B **Open Process**
I **One -Time**	Type 1	Type 2
II **Iterative**	Type 3	Type 4

Figure 7.1 Four types of participation

Type 4 participation, on the view of adaptive management, is absolutely essential if we are to develop, and assign weights to, multiple criteria associated with multiple indicators.

Here I apparently part company with many decision theorists of the "best-solution" school. They tend to make use of type 1 or type 3 participation, using expert judgment as an approximation of the correct ranking; they consider an iterative use of expert opinion to be preferable but fall back on type 1 participation if ongoing deliberations are impractical. Once weights are set, it becomes possible to calculate a best answer, algorithmically. However, these calculations count as a best answer to the original decision problem—when it still had its wicked aspects of trying to respond to multiple interests and viewpoints—only if we believe the expert participatory process got the weightings right or nearly right. But right against what standard?

Thomas Hanne, one of the few decision theorists who has worked explicitly on such "metadecisions," suggests that one might be able, despite the impossibility of formalizing metadecisions, to arrive at a synoptic decision. He says, "The basic idea in MCDM is that in many real-life decision problems several criteria are to be considered which are traditionally in most cases replaced by *one* hypothetical or constructed general criterion of choice." He offers economics as one example of a traditional approach to aggregating criteria and then adds, "Another simple and traditional approach is to aggregate the criteria using a weighted sum."[21] Hanne describes these problems as metadecision problems and notes that "method pluralism causes the problem of selecting a method for analyzing a given MCDM problem." He states that metadecision problems can be "formalized in different ways: as a problem of method choice or design, as a scalar or a multicriteria optimization problem," but that "in the literature, meta decision problems are in most cases studied as [multi-attribute decision-making] method selection problems."[22] Hanne acknowledges several problems with this approach, including the particularly damaging problem that if one is to achieve a uniquely correct result on the decision level, there is a "recursive" problem: before we can resolve the problem at the decision level, we must first resolve the metaproblem of comparative weighting of criteria. If, however, one sets out to assess that metadecision, prior to solving the decision problem, it would seem that we would need a solution on the meta-meta level, and a vicious regress threatens to leave the decision maker paralyzed, as action depends on prior decisions at ascending levels, with the uncertainties of each level being pushed upward into higher and higher orders of choices that cannot be solved algorithmically.

A good dose of pragmatism can cut through the paralysis. Rather than resorting to higher and higher levels of discourse in search of one that is comprehensive enough to allow full formalization, I suggest that we simply follow Peirce and search for the weightings in the reflective discourse of participants, who represent, in their diversity, a microcosm of the diverse evaluators in the community. Speaking epistemologically, the assumption of a uniquely correct solution to decision problems—understood as an independent standard that the decision analyst is aiming for—can be defended only on the outdated modernist and Cartesian/Newtonian assumption of a unitary, independent, and objective "reality" out there. The alternative is to embed these decisions in the broader adaptive management process as decisions to be made through multiple rounds of action and deliberation. The intellectual and practical cost of giving up the myth of the correct solution is thus nowhere nearly as high as most decision analysts fear. A pragmatist viewpoint suggests that as the community, united by its commitment to find a cooperative solution, amasses more and more experience, gathered and communicated by people with different values and different perspectives, and revisits its goals and its decisions again and again, its members will improve both their knowledge and the appropriateness of their values through social learning. A decision model that aims not at an experience-independent sense of "best solution," but rather at slow progress through social learning over time and many revisitings of a wicked problem, will best fit our needs as adaptive managers.

Environmental problems, then, are best seen as problems of cooperative behavior, behavior that creates value added (either by protecting some widely shared value or by reducing uncertainty about what to do or how to do it).

What, exactly, do I mean when I hypothesize that environmental problems are best modeled as problems of cooperative decision making? I mean, first of all, that adaptive management is about *acting*, and the process leading up to an action can thus be considered a decision-making process. Our task in this section has been to create a *model* that is appropriate for illustrating such a decision process as a feature of adaptive management. The characteristics of such a model—and these are all features that can be studied as important aspects of environmental decision making—are that it must be embedded in a democratic process of adaptive management, it must be iterative, it must be open to all voices in the community, and it must be receptive to multiple values and varied formulations of these values. The adaptive approach is experimental, both with respect to areas of uncertainty about natural systems and with respect to the goals and values espoused.

Because our decision model must function in a community with diverse values and viewpoints, and because of the failure of attempts to achieve single-value, computational solutions to wicked environmental problems, I have proposed modeling environmental decisions as open-ended MCDM problems. Members of the community, on this approach, can propose various indicators and associated criteria that they believe will track important values they cherish. Other members will propose alternative criteria, and some, no doubt, will be merged together; others will be rejected as too difficult to measure or as having no direct connection to a widely shared value. The actual decision faced can then be understood as a metadecision: which criteria—and which weightings of the proposed criteria—should be used as a guide to evaluate proposed actions and policies (and the development paths they will express)? Negotiations over these metadecisions, which are addressed in the reflective phase of management, thus reflect values felt by community members. But even if there remain differences in value commitments of community members and participants, tentative choices of criteria by which to evaluate policies allow action, however provisional and experimental. Action, in turn—given an experimental attitude—provides a context in which both actions and working criteria can be tested against reality in particular situations.

Finally, we have a method that is appropriate for addressing the wicked aspects of environmental problems. By embedding our distinction between active and reflective phases in an adaptive management process, we can now separate questions of what to do, given a set of rules, from questions of what rules to follow. Further, we have reconstituted the wicked aspects of environmental problems as problems of balancing or weighting multiple criteria, associated with indicators. This approach ties our approach nicely to scientific studies that associate unfolding "scenarios" (what I have called "development paths") with possible policy options. Policy options, in turn, are justified because they are expected to come out well on multiple criteria. If a proposed policy scores well on a number of criteria that the community cares about, we can say that it is "robust" over multiple values and multiple belief systems. As we switch from the reflective phase into the action phase, favored policies are tried—if possible in safe-fail situations where negative impacts of failed experiments can be controlled—and the current reflective balance of weightings of the multiple criteria is tried out in practice. Such probes of the system, based on even temporary agreements on actions worth trying, can be the source of social learning. Thus, by eliminating areas of decision analysis that make assumptions not appropriate for an adaptive, experimental approach, we have focused our attention on decision models that are iterative, cooperative, pluralistic, and driven by social learning.

Social learning, in turn, is driven by an open, deliberative process in which all interested parties bring their experience to bear on choices of experiments and pilot projects to undertake. The decision model gets its normative thrust and its validation by embodying type 4 participation especially in the reflective problem of choosing indicators, setting standards with respect to indicators, and choosing the proper weights to give the various indicators. This social learning decision model, I conclude, should be chosen for its usefulness in modeling decisions of communities facing wicked environmental problems. The process will not lead to an idealized "correct" answer; but it will, through a process of deliberation, negotiation, and experimentation, put the adaptive management partnership on a path forward, a path on which the community seeks problem solutions and policies that are robust over multiple criteria, even as community members discuss how changing conditions and new experience demand shifts in the weightings of the currently operative criteria.

Although this approach abandons the idea, so cherished by decision scientists, of a single correct solution, and with it the idea of an algorithmic solution to environmental problems, it fills the resulting void with methodological and process-oriented heuristics, heuristics that can guide participants to ask the right questions in their particular problem situation. These heuristics encourage experimental, deliberative methods, methods that can be expected to produce cooperative solutions in the face of diversity of beliefs, values, and viewpoints. The proposed decision model successfully operationalizes our guiding idea: to replace ideological commitments regarding how environmental values *must* be articulated with an open and deliberative process, based in experience and the experimental method and designed to encourage the creation of new ways to value nature and new ways to express environmental values. Decision analysis, on this model, will not be thought of as a parallel, "rational" computation of a best solution; it will rather be embedded in the adaptive process itself. There it will function as a tool in the social struggle for better solutions and more acceptable outcomes.

Before turning more explicitly to the normative aspects of the discourse that must accompany such a decision process as this, I will draw an important connection between our decision modeling and the importance attributed to Leopold's insight that we must learn to think like a mountain. Leopold's insight encourages us, in managing natural systems, to pay attention to multiple scales of ecological and geologic time as well as to immediate, economic considerations. In the first section of this chapter we posited that there are human values that are communal in nature and that communal values must be associated with longer, multigenerational scales of time and with large-scale, landscape-sized dynamics like the dynamics of wolf and deer populations. The multiscalar dynamics of nature, and indicators chosen to track their health and integrity, can thus be matched with decision criteria that are similarly "scaled" in time and space. I went on to argue that our decision model should be understood as embedded within the process of adaptive management; the re-modeling of nature that Leopold recommends occurs, then, in the process model, as new indicators and new standards are proposed, discussed, modified, and reformulated. If wicked problems are reflective of differing interests, and those interests are articulated and factored into the decision process through the development of appropriate indicators, this decision process can provide an appropriate decision model for adaptive management processes.

It is thus possible for a community to reshape and re-form its model of nature by paying attention to new physical dynamics that are hypothesized to be associated with values held dear by the participants. Deliberation about indicators and criteria, then, leads to social learning about both values and physical models. As noted in section 6.3, experiences that change one's perspective and reorganize one's perceptions often precede or coincide with shifts in moral ground. If we think of hierarchy theory as a more systematic way of realigning our perceptions of nature with what we, as human communities, really think is important, we can begin to see how community-based deliberation—once freed of ideology and unfortunate conceptual proscriptions based in a priori assumptions—can cut through the fact-value dichotomy. The community, by choosing indicators, is identifying natural dynamics that matter to it because they are associated with deeply held values. So we are most definitely not talking, here, about a disengaged and value-neutral science as the source of such revolutionary reorganizations. This science must be an engaged science, a mission-oriented science, and its committees of peers must include not only disciplinary scientists but also interdisciplinary scientists, representatives of interest groups, and concerned citizens.

Experts from all fields relevant to environmental management are thus being asked to exchange their special privileges for no more than a seat at what I will call the Table. The Table is simply a symbol referring to the sum total of public discussion and deliberation about what to, the ongoing discussion that goes into a community-based partnership program. Experts, in other words, may make major contributions, but only to the extent that they can translate their expertise into real tools that will encourage deliberation in the reflective phase and cooperative behavior in the action phase. The determination of what is useful in management must take place not in the special languages of the various disciplines, but rather in the comprehensive discourse of ordinary language.

The experts may, however, be surprised by the positive results if they do in fact sign on as participants and as contributors to a management process. If we then combine this role for experts with the argument, pressed above in subsection 7.3.1, that decision modeling must *generate* acceptable solutions to local problems out of a process of experimental efforts, rather than *calculate* a best answer independent of experience, then experts become important contributors to the open process. In that expanded context, with multiple experts and multiple interest groups present, social learning becomes more likely. As information flows in all directions across the Table, the chances for cooperative action increase. Furthermore, action leads to trust; and trust leads to more cooperative action. In the process, scientists will have many opportunities to study dynamics that are important physically, and they will also find that they can make an important contribution to managing the interface of humans and natural systems.

In general, everyone who claims special expertise or special, preexperiential knowledge, is asked to join the Table on an equal footing. Provided that there are adequate linguistic resources to encourage multidirectional communication across the Table, we expect this process to tend toward consensus as tentative, safe-fail pilot projects are tried, trust increases, experiments reduce uncertainty, and convergence on action increases. The Table is the best place to address wicked problems because, once the question of improving our linguistic resources and the question of which criteria to emphasize in choosing policies to affect the future are made endogenous to adaptive management, both questions of fact and questions of values and goal-setting will be addressed within a community-based process. At the Table representing this process, authority is gained by making good arguments in ordinary speech, and trust is built by contributing to successful cooperative action. Given the law of errors and Peirce's idea of truth as emergent from communication and cooperation, we need no authority, no self-evident truth, to invoke in order to identify best solutions. Better solutions emerge from a process, and we are now arguing that all participants should have equal status at the outset and that status in the process can be augmented, but not by appeal to higher authority; one's status depends on the trust one gains and the cooperation one contributes.

These conclusions, though consonant with many of the ideas developed so far, do raise a serious puzzle, one that is especially acute in this book on sustainability. If we abandon all claims to "higher principles" in the sense that we countenance no authority or expertise higher than our process, legitimacy of our decisions must derive from the openness of the Table. All who are affected by a decision must have a voice; and the experience of all earnest participants is to be respected as equal. The test of beliefs and values comes from counterclaims based in contrary experience. Each individual's experience, according to our favored decision model, is treated equally. I have argued, however, in section 7.1, that sustainability necessarily involves protecting communal as well as individual goods and that these communal goods are associated with long-term, multigenerational dynamics. The consequences of actions taken or not taken in an effort to live sustainably will fall most heavily on those who live in the future—and they will be conspicuously absent from the Table. They cannot participate in our open process because they are not here yet. Does the model collapse, then, into a "tyranny of the present" over the future? As each interest group seeks to uphold its interests, and a negotiated compromise results in action, is it action that protects the present and undermines the community's future?

This is an avoidable collapse, I will argue. The short version of my argument simply points out that members of the partnership—at least some of them—hold communal values along with individual interests. Individuals around the Table are not, as the economists' caricature of economic man suggests, deliberating only as individual consumers; they are also deliberating, to use Mark Sagoff's helpful distinction, as citizens.[23] How they so deliberate and how they determine what must be sustained represent an enormously important part of our theory of sustainability. Accordingly, the next chapter is devoted to this aspect of sustainability theory. Before taking up that aspect, however, I introduce useful tools for addressing explicitly normative issues in the next section and explore the role of the noneconomic social sciences in our process in section 7.5.

7.4 Discourse Ethics

We have adopted a pluralistic and democratic starting point for our examination of environmental values. In section 6.3 we saw how, on an individual (or group) basis, encounters with new information—or new ways of encountering old information—can reconfigure individual belief systems. As long as we keep this in mind, we can think of individual participants in our process as having values and maintaining these values or preferences over various periods of time but also being susceptible to important awakenings and reconfigurations of their belief systems, often in the face of new experience. Since, as pragmatists and adaptive managers, we do not claim to know the final outcome of the processes of belief-formation, information-gathering, deliberation, and truth-seeking among participants in an open public process, we are content to let these processes run their course within communities, recognizing that beliefs, interpretations, and values are all dynamic in the face of continually expanding experience. Changes in environmentally related beliefs and attitudes make a fascinating topic of study for psychologists; but we will not pursue that subject here, aside from offering heuristics that can help communities to focus on productive questions in their public discourse about environmental problems and goals. In this chapter I have shown that the choices to protect long-term, communally enjoyed goods can and must be understood as problems of cooperative action; they cannot be reasonably solved by examining or aggregating individual preferences.

Can we choose an approach to studying environmental values that is consistent with dynamic pluralism on the individual scale and still helps us to understand concerted and sustained cooperative action by communities, acting as communities? I believe we can. The secret to doing so is to focus on the development and preservation of fair and open procedures for deliberation at the community level—to shift value analysis from an aggregation of preferences to analysis of fair and open processes. Alternating reflective-phase deliberation with experimental actions makes it possible both to act and to reflect on the bases for action. It is also reasonable, in the reflective phase, to consciously consider new and improved rules of discourse and to suggest new linguistic expressions that will improve communication and reduce misunderstandings. We can therefore think of our two-phase decision model as constituting an attempt by sincere but diverse participants to approximate what the German philosopher Jürgen Habermas refers to as an "ideal speech community."

That said, we still need, of course, some kind of a normative method if we are to move from diversity of expressions of values toward consensus and legitimated public actions to support public values. Fortunately, there is a rich tradition of moral discussion, more prominent in Europe than in the United States, that offers a broadly pragmatist approach to analyzing social values; perhaps that tradition, sometimes referred to as communicative ethics, or discourse ethics, will prove a good fit with the goals and experimental procedures of adaptive management. This ethics tradition has absorbed the important ideas of American pragmatism into the influential European tradition of critical theory, which is centered in Frankfurt, Germany.

Building on Peirce's ideas of a science of semiotics, and incorporating breakthroughs by positivist and other philosophers of language, this group has offered a strong emphasis on public discourse, embedding this discourse in social praxis by focusing attention on the preconditions of intelligible language and discourse. By paying attention to the preconditions of intelligible discourse, communicative ethicists hope to emphasize the importance of situational conditions in the constitution of public discourse. This aspect of discourse ethics inaugurates and operationalizes Peirce's idea of a forward-looking, truth-seeking community, a community of members who express varied points of view but share a commitment to seek the truth.[24] And like Peirce, discourse ethicists argue that an adequate method for weeding out error, if applied by diverse voices in a community of truth-seekers, can claim enough objectivity to hold skeptics at bay. A right-seeking community, just like a truth-seeking community, gains through open communication; and as more voices are heard, the range of available experience brought to bear upon a question is expanded. Again, as Neurath's analogy suggests, we must remain afloat while shifting from one temporary consensus to another, always shoring up the weak links in our belief system and in our active coalitions, expanding consensus where possible, and undertaking relevant experiments and observations where consensus is not possible. This amounts to betting on social learning as an antidote to confusion and chaos in a pluralistic society.

Discourse ethics has created a rich reinterpretation of ethical reasoning and evaluative discourse by concentrating on the preconditions for meaningful speech and communication. Advocates of communicative ethics, or discourse ethics (we will rely mainly on the formulations of Habermas) examine ethics in terms of the various kinds of speech acts it involves. Although Habermas has been deeply influenced by Kant and Kant's idea of autonomy, it is useful to think of Habermas's viewpoint as a departure from Kant's reliance on individual autonomy as the basis of all ethics. According to Kant, the process of ethical reasoning was most realistically represented as individuals reflecting upon their motives and comparing their motives with the demands of practical reason, which enforce themselves—on pain of practical contradiction—upon the individual as unavoidable obligations based on principle. Habermas, taking his cue explicitly from Peirce, begins his analysis of normative discourse a giant step earlier with speech, speech that takes place in real situations, and thus ethics and morality gain their very meaning in a social, communicative setting.[25]

The unit of analysis, for Habermas, thus becomes a speech act. A true speech act is tied to a whole cultural pattern, however. A speech act, as opposed to merely mouthing meaningless sounds, presupposes a process of assertion and rejoinder, of claim and counterclaim, a process that binds all participants in a system of communication that is necessary for a community to participate in "communicative action." By distinguishing between communicative action and "discourse," Habermas recognizes the two somewhat separable phases of discussion we have referred to as the action phase and the reflective phase. Discourse is thus a specialized form of communicative action.[26] The action phase (what Habermas calls the realm of "communicative action") proceeds mainly on the basis of consensually accepted facts and shared norms. Communicative action represents the ability—so important in our story of Neurath's boat—for the community to stay afloat as a functioning unit. When the consensus in action is disrupted, however, a discourse, constituted by a process of claims and counterclaims, reason-giving, critique, and counterreasons, ensues. In this more reflective phase, members of the community air their differences and try to persuade one another to change crucial beliefs. Even when the members disagree in deliberations, they remain members of a deeper community that sets the stage for both their agreements (when these are sufficient to allow decisive action) and their disagreements (when arguments ensue). As members of this community they share presuppositions about the requirements of communication and of cooperative action. Habermas's shift from analyzing morality as an exercise of individual autonomy and individual moral reasoning to analysis of a community-level process ensures that normative questions will be addressed within a community, rather than as a matter of individual autonomy and reason. Now the constraints of a shared desire to communicate and get along together shape actions and also discourse, providing a richer social context in which to discuss environmental policy goals.

Moreover, Habermas supports our determination, in section 4.2, that the putative epistemological gulf between facts and values is artificial and should not be seen as restricting the forms of reasoning and deliberation by which participants cite experience and reasons for claims of all sort, some mainly descriptive, some prescriptive.[27] Habermas declares that ordinary language is the language of discourse because specialized languages exclude laypersons from deliberation, and he argues that deliberation must take place in the "open" language of ordinary speech. Ordinary language is crucial because Habermas understands political discourse in modern democratic societies as being constantly renewed and rejuvenated by communication back and forth with the "life-world," the underlying set of practices and conventions that constitute the community as a cohesive whole: "For institutionalized opinion-and-will-formation depends on supplies coming from the informal contexts of communication found in the public sphere, in civil society, and in spheres of private life. In other words, the political action system is embedded in life-world contexts."[28]

The practical benefits of embracing discourse ethics as a companion to adaptive management are several and important. First, the shift from individual reasoning to community deliberation allows Habermas to posit, counterfactually, an "ideal" speech community, a community in which the various requirements of disagreement, deliberation, reason-giving, and so forth are respected and practiced, and in which people respond to new information as expanded experience and—in the spirit of Peirce's community of truth-seekers—proceed to submit questionable beliefs to more and more experience through observation, deliberation, and experiment. We already noted, in the preceding section, that observing actual communities engaging in effective deliberation and decision making may provide a way to empirically test what actual communities, given a chance to deliberate, would conclude. Furthermore, this construct, however idealized, provides a framework for discussing and testing the general conditions under which community-based deliberations are effective in developing consensus.[29]

Another practical benefit of discourse ethics is that it provides an excellent basis for us to respond to moral skeptics, those who wonder whether they are obliged by the duties of morality to do anything. Habermas takes on the skeptic head-on, arguing that even in disputing a moral claim, the skeptic must accept the rules of discourse that are based in the deeper, communicative community. Even as a disputant, Habermas argues, the skeptic commits to the presuppositions of communication and is thus brought into the dialogue and assigned duties on a normative basis, just as the price of entry to the discussion.[30] Whether one finds this response to moral skepticism convincing or not, discourse ethics insulates us from the skeptic in another, more foolproof way. Since we are talking in this book about participants who are involved on an ongoing basis in an adaptive management process, it seems reasonable to assume that these participants have already committed themselves to the goal of finding cooperative solutions and doing so in a spirit of public deliberation. So even if someone remains unconvinced by the discourse ethicists' "ultimate justification" of ethics directed at the skeptic, we could drop those claims altogether and fall back on the everyday situation in which adaptive management takes place. No skeptics are likely to volunteer for thankless committees or attend meetings of citizens' advisory committees, so we can assume that all of our participants have accepted at least the minimal expectations of communicative discourse.

Another very useful outcome of the application of discourse ethics is that Habermas, who perceives layers of communicatively organized communities, also sees layers of normative obligations and commitments. I provide a brief summary of Habermas's view, because his layers will prove quite useful in characterizing layers of analysis of values in adaptive management.[31]

Habermas begins by interpreting *practical reason* as a basic and undifferentiated faculty that applies to pragmatic, moral, ethical, and legal reasoning—to all action-oriented choices.[32] When disputing a moral claim, the skeptic must accept the rules of discourse that are based in the deeper, communicative community. Habermas argues that, simply by entering the discussion, the skeptic is assigned duties on a normative basis. When one gives reasons in this mode, one is assessing "validity" of claims, where claims include, in ordinary language, claims of moral and legal prescriptions; "validity" then expresses "a nonspecific sense [of acceptability]." Here discussion and debate concern "action norms," which are very generalized behavioral expectations; and these discussions represent "rational discourse" at its most general level.[33] To put it simply, it includes open, inclusive processes in which a community of people practice reason-giving and operate on principle: "Just those action norms are valid to which all possibly affected persons could agree as participants in rational discourse."[34] This generalized conception of validity is the ultimate anchor of discourse ethics. It is important, however, according to Habermas, to realize that such generalized employments of the faculty of practical reason relate differently to three types of normative decisions, which he calls pragmatic, moral, and ethical. This tripartite classification is the key to a layered approach to normative reasoning—but it must not be forgotten that all three layers derive their rationality and support from the basic rules of discourse that unite the community, the presuppositions of open and inclusive discussions among affected parties trying to reach a decision in a situation demanding action. Here one might say Habermas is simply updating Dewey's situational logic and employing a notion very similar to Dewey's "warranted assertibility."

The first and most basic exercise of practical reason occurs when "our will is already fixed as a matter of fact by our wishes and values; it is open to further determination only in respect of alternative possible choices of means or specifications of ends." Habermas refers to this as "pragmatic" discourse. Here either a particular goal is already identified and pursued and action is directed toward choosing means to that goal; or, alternatively, several goals may be under discussion, but the goals in question are governed by well-established values and by options available at the time. Judgments reached in this mode can be referred to as "relative oughts": they involve judgments of what one "ought" to do if one wants to realize certain values or attain certain goals. The values, in other words, are accepted as unproblematic.[35] These aspects of practical discourse correspond to what I have called the action phase of a community-based adaptive management process.

Habermas also distinguishes between morality and ethics, a distinction that applies once value claims are problematized, challenged, and held to require justification. The distinction between morality and ethics has its origins in the concerns of Kant and of Aristotle. In Habermas's view, modern ethics has been unduly troubled by the conflict between the highly individualist, liberal tradition tracing to Kant's emphasis on the practical reason of individual, autonomous moral beings (which is often accused of descending into empty proceduralisms) and the richer exercise of reason that occurs in the search for a "good life" (which has often seemed to be based on specific, culture-based commitments and norms). Habermas attempts to avoid the conflict between these two general approaches to morality by identifying two separate realms for the appropriate exercise of these two distinct functions of practical reason. Morality, on Habermas's view, is best understood in a Kantian mode and is operative when the question is whether a given action is *fair;* it has to do mainly with impartiality in deciding among conflicting interests. This application of practical reason has to do with whether a "maxim"—a rule governing a "unit" of normatively charged behavior—could be willed as a general rule, applicable to all agents who might find themselves similarly situated and facing a moral quandary with the same relevant considerations. These considerations, Habermas believes, should be resoluble on the basis of a "universalizability principle," a technical version of the "golden rule." A *moral* judgment (as opposed to an *ethical* judgment) thus has universal validity traceable to the very core of practical reason—impartial reason-giving. Habermas says, "The question 'What should I do?' is answered morally with reference to what one ought to do."[36] Because many readers of Habermas find his use of "ethics" and "morals" in an opposed way to be confusing, I prefer to use the phrase *thin prescription* to apply to what Habermas calls morality (fairness) and *thick prescription* to refer to what he calls ethics.

Although all three forms of practical reason have to do with justifying choices of what to do among alternative courses of action, they each involve different kinds of answers and different kinds of actions. Ethical problems, Habermas says, do not have to do with universal judgments of fairness but rather with individual choices of what constitutes a "good life." Here the emphasis is not on pursuing an unquestioned end (as in acting upon "relative oughts"), nor is the agent appealing to universal principles of fairness; in the ethical application of practical reason, the agent is constructing an image of "the good life" and is faced with choices as to what actions are appropriate given this image. The issue is one of striving for self-realization and for resoluteness in one's commitment to that chosen ideal of the good life.

In all three applications—and this is an essential point—practical reason has its basis in communication, in the presuppositions of a truth-seeking community. Such a community thrives on claims, challenges, and reason-giving as a presupposition of communication and deliberation itself. When the problem is one of choosing an action, based on shared and unchallenged values and goals, communication is limited to discussions of means and of efficient techniques to achieve the unquestioned goal. Here we have examples in which, operating on Neurath's boat, the discussants accept the goals of staying afloat and of replacing weak planks. Discourse is limited to instrumental reasoning, and reason-giving is normally in the mode of marshaling empirical knowledge, asserting its applicability to the case at hand, and advocating for particular means to the shared goal.

Thinly prescriptive discourse requires an impartial standpoint that is necessary in order to set aside the subjectivity of the individual participant's individual standpoint. In the impartial standpoint, only those proposed norms that express the common interests of all can win assent of participants in the discourse. Habermas says that in this case the will is determined by moral grounds, and these grounds are completely internal to reason: they are norms that cannot be realized except in accord with impartial reason. As these norms are discursively explained and justified, the equal interests of all participants are revealed, and when unity of action is achieved without repression, the interest of all, as individuals, is legitimately expressed as the general will.

In areas involving thick prescription (the choice and pursuit of a good life) the individual is less constrained than in either pragmatic discourse or moral discourse. In ethics (thickly described)—choosing and living a good life—reasons and constraints stem more from the particular life history of the individual and the developmental process that has made, and is making, the individual the person he or she is. The roles of agent and participant cannot of course be so sharply separated: steps in one's argument about what constitutes a good life for oneself must be comprehensible to others. The individual must clarify and explain the choices that are made. Those who share an agent's life-world assume "the catalyzing role of impartial critics in processes of self-clarification." "Participants in processes of self-clarification," Habermas says, "cannot distance themselves from the life histories and forms of life in which they actually find themselves."[37] Decisions based in Habermas's moral reasoning are important because they prescribe fair and equal respect for the beliefs and experiences of fellow participants. Habermas's morality, then, addresses procedural questions of fairness, openness, and mutual respect in negotiations. "Ethical" reasoning is ideally suited to guide the members of a community toward a shared sense of who they are, what is important to them, and what must be saved to protect the integrity of their community and their "place."

Discourse ethics, as I hope this brief summary shows, derives its power not from "first principles" that can be known a priori, but from presuppositions that are embedded in the practices of language users who make claims and counterclaims. Discourse ethics thus accepts pluralism with respect to thick prescriptions, which are not based on unarguable first premises, but on the shared commitments of community members—commitments to deliberation rather than violence, for example—that are *independent of the particular beliefs and values of the participants.* There is no room in discourse ethics for "non-negotiable" "meta-principles," as Westra insisted. As on Neurath's boat, every plank will have to be inspected and improved eventually, but in the short run we rely on the multiplicity of participating voices to identify the weak links in our reasoning and justification. The goal of these procedural norms is to encourage an open process of claims and counterclaims, but to do so in a context that does not lead to paralysis of decision making. The value of discourse and deliberation, according to Habermas and the other discourse ethicists, exists on a deeper level than individuals' commitments to particular policies because, in the act of thinking in a communitarian way—the act of choosing a cooperative solution to a perceived common problem—the participants have accepted the preconditions of rational speech and of communicative action as the accepted mode of action.

The method is inseparable from the process in the sense that the process has to be stable enough to allow both action *and* reflection, and effective reflection provides feedback about the process, encouraging social learning that will continually improve the process over time. As Habermas says, these two aspects are "internally related" in a discourse that is alive and functioning within a community.[38] The process must develop enough trust to allow temporary coalitions for action and enough solidarity to follow through on experiments and reduce uncertainties over time. Existing norms function to provide order and predictability *only if* they have the legitimacy that comes with the promise of redress and correction in the face of new evidence and changing circumstances. This legitimacy, however, derives from no absolute or a priori source; it results from the implicit promise of reconsideration if new evidence is discovered or if deliberation calls goals and values into question. Within an open discourse in which yesterday's losers can again challenge the operating consensus today, the losers are willing to "buy in," to bide their time and continue in deliberation rather than turning to indifference or violence. It is this promise that all current policies and the scientific and moral supports they rest upon are open for criticism that creates a "deliberately acting community," that generates legitimate governmental action based on temporary agreements about values and scientific possibilities. Within discourse ethics, then, cooperative action can be legitimate even if some members of the polity oppose it, provided that those who disagree are confident that they have avenues of redress, including the "right" to challenge any assertion of the majority's justification, and including some means to gather evidence to refute doubtful claims.

I have outlined the basic ideas of discourse ethics, and I have noted its advantages as a supplement to the ideas and approaches of adaptive management. But now I want to make a bolder claim. I believe that discourse ethics shows the way toward a new, and more productive, approach to the field of environmental ethics. If I am right, we may begin to see the outlines of an approach to environmental valuation that provides a vital role for a much-transformed field of environmental philosophy, a role that will put it at the juncture of the many empirical disciplines that correctly claim to have policy-relevant information. Here we can refer back to the "epistemic community" (see section 3.5), which is constituted by representatives of various interest groups who engage in a broadened peer review of relevant science and who undertake experiments to resolve action-blocking issues on which they disagree.[39]

We have already noted in passing that Habermas's first category of discursive reason, the "pragmatic" application of practical reason, corresponds directly to what we referred to as the action phase in our discussion of phases of adaptive management in chapter 4. Cooperative action achieved through communication is indicative of broad consensuses—or at least large majorities accepting the goals and objectives of the group—and also of wide acceptance of the scientific information taken to support that action. Pluralism, in such situations, lies dormant. But when goals are challenged as unworthy or too costly to achieve, and when factual disagreements break out among factions in the community, with each citing "their" scientists in opposition to one another, pluralism becomes the force that drives the process forward to test contested hypotheses and to formulate new goals. And if our commitment is first to democracy, deliberation, and cooperative action, then the most fundamental task of environmental ethics is, as Habermas recognizes, to develop, explain, and implement procedures that will encourage deliberation within an "ideal-as-possible" speech community. Habermas's discourse ethics, and especially his theoretical construct of an "ideal speech community," thus defines a key role for philosophers at the center of the policy process.

One always worries—reasonably, I think—about unrealistic political constructs and the utopianism they encourage, so it is important to constantly remind ourselves that the construct of an ideal speech community is a *counterfactual construct*. It is not intended as a blueprint for a political system; it is expected to have no existence except as an idealization. It is an idealization, however, that reveals important, empirically testable propositions. We do not, that is, suspend our currently chaotic, messy discourse, waiting for someone to invent from whole cloth a *new* set of concepts and definitions for discussing environmental goals. Rather, we start where we are and embed our search for improved communication within an ongoing adaptive management process, using our imperfect language even as we try to improve it. The concept of an idealized speech community provides a basis on which to compare our current, inconsistent communication with an ideal, though imaginary, standard. That role corresponds to the task I have begun in this book: to consciously criticize, alter, and reconstitute public environmental discourse in ways that promote communication and cooperation.

I see Habermas's construct of speech communities that can be more or less successful in a shared goal—deliberation and cooperative decision making—as an invitation to study, criticize, and improve the actual deliberation that occurs in less ideal situations. Communities can be more or less successful, also, at the more reflective task of deliberating about present and future goals and disputing proposed policies based on new scientific evidence. It is the role of the social sciences, perhaps, to track the first success—success of communities in setting up institutions that encourage open and many-sided debates about policies and that lead to improved decision making through citizen participation. But the latter task—creating and sustaining a public dialogue about environmental goals and objectives—requires a dialogue that is open to all, proceeds through publicly made claims and counterclaims, and comes as close as practically possible to ideal deliberation. Considering necessary and sufficient conditions for such deliberations is a task for which philosopher-participants are well trained.

In fact, of course, the tasks of the philosopher and the social scientist cannot be separated; philosophers and social scientists will have to work together to better understand and better describe environmental values. The only arbiter is experience; the only ultimate arbiter is time indefinite. Having given up nonnegotiable claims and self-evident first principles, ethicists can settle into their role as students of the ways humans value nature, of the ways they express those values, and of the ways they might better express them in order to communicate with other participants in seeking cooperative solutions. Philosophers, in other words, should be especially helpful in encouraging and enriching the ongoing process of claims, counterclaims, challenges, and reason-giving that is the more intellectual side of the adaptive management process—the reflective phase. Social scientists, for their part, study the effects of public, participatory processes and can contribute to developing more effective institutions by learning what works in practice.

What is gained by posing these questions within a context of the assumptions of discourse ethics—in a context of seeking cooperative action—is that the obligation to "play the game" of public deliberation, once accepted, brings with it a large number of pragmatic obligations that can govern and guide the deliberative process. As noted above, these are "relative oughts," oughts that guide participants to behave in certain ways, *on the shared assumption that all participants accept certain rules of interaction that are justified by their common interest in solving problems cooperatively.* The commitment to use public discourse and public deliberation, both in the action phase and in the reflective phase, rests in the shared presuppositions of our communicative community. These obligations can be quite specific and would include, for example, a commitment to remain a discussant and not a provocateur if the democratic process, at some point in time, results in actions that one disagrees with. So the commitment to act as a community and to continue dialogue, rather than resort to force, is a procedural rule and does not depend upon any substantive claims about what has value, other than the value we performatively endorse each time we engage in reason-giving and challenging evidence. It is rooted, as Habermas would say, in the preconditions of our illocutionary, or communicative, acts. Skipping the jargon, it rests on our willingness to communicate at all, and this is no longer in question once one has entered the ongoing dialogue of an adaptive management process.

One can, as Habermas and his associates have shown, push this idea of "procedural" duties very far as a basis for obligation. For example, we can argue that it is an obligation of each participant to respect an opponent as a reason-giver—and therefore refrain from name-calling and ad hominem arguments; further, one can argue that the rule of universalization at the heart of what Habermas calls "moral" reasoning demands the respect of other participants as moral agents and therefore prohibits discrimination on the basis of irrelevant grounds such as skin color, ethnicity, or education level. Perhaps at some point the reader will become skeptical of loading too much freight onto Habermas's procedural level of moral reasoning. My point here is that following the basic framework of discourse ethics—at the cost only of declaring ourselves in favor of democratic discourse as the favored way to address common environmental problems—opens a whole realm of significant procedural obligations that bind participants. These obligations, if regularly respected in the *process* of environmental policy formation, would increase communication and trust among participants. Emphasis on process, in other words, allows us to build upon the procedural norms that can be derived, within discourse ethics, from the implied willingness of individuals to engage in deliberative discourse.

So Habermas's multilayered discourse ethics generates a large set of procedural obligations that participants have implicitly accepted by their willingness to participate in ongoing public deliberations over what to do. These obligations, once thought of as pragmatic commitments based on a shared purpose in communicating, are unshaken by changes in people's substantive beliefs or personal values. They are impersonal in this important sense, and thus they provide a very solid basis for the procedural rules we articulate as we—philosophers and social scientists—study the deliberative practices and decision-making procedures that evolve as we use them. More importantly, these rules, based as they are upon only the supposition of a functioning language community and a will to act cooperatively, can be affirmed as universal rules of conduct, rules *not* based upon the special cultural features and commitments of real individuals who live in a place and have a history.

One might also count it an advantage of discourse ethics that it separates from the universalist procedural claims of morality the more substantive choices that go into an individual's and a community's formation of an ideal of the "good life." In this realm, it is more comfortable to agree with Aristotle that values that count in such communities are highly contextual and dependent on accidents of birth and culture. This layer of substantive values is often precisely what is at stake when groups protest major changes in the landscape of an area, as when opponents of chip mills protest clear-cutting of hardwoods and their replacement with plantation pines, or when it is claimed that certain projects destroy the "character" of a place. These more substantive values are indeed important, and as we will learn in section 12.4, articulating these values—providing adequate language to communicate them within communities—will be a major task for philosophers of the future. It is also a task absolutely essential to the development of locally based ethics.

What we have learned from this little detour into discourse ethics is that, according to the quite reasonable conceptualization of normative problems developed by Habermas, there is one set of moral norms regarding the environment that are universal and govern all attempts to manage nature through an ongoing process of deliberation and open participation. These core procedural rules of fairness and unbiased treatment of all participants are based on a universal foundation and apply to all who engage in deliberative discourse; they exist independently of culture and they hold quite independently of the particular beliefs and values of particular participants. These rules provide the necessary support for normative obligations governing the interactions of participants who are committed to act cooperatively as an action-oriented discourse community. But discourse ethics leaves room also for a more particularistic layer of normativity, the choice—in conversation with others in one's community, of course—of a given ideal of the good life. Here, I would argue, choices of a good life are inseparable from choices of a good environment in a particular place.

Discourse ethics, then, offers environmental ethics a new role, or several roles. These roles begin with paying attention to the language available to participants to voice their agreements and disagreements. The task set for this book, in this sense, is a direct response to the problem that people in today's society are willing to discuss and deliberate about what to do, but they have as yet no language adequate to the task of articulating, much less fulfilling, the expanded responsibilities of technologized humans. But this is only the beginning. Once descriptive and prescriptive discourses are allowed to work together, the search for environmental values becomes a combination of (1) a creative act, of articulating an important value in a new and, hopefully, measurable way, linguistically; and (2) a social scientific task of measuring whether that expression of value is widespread in the community. In this way the problem of discovering and measuring environmental value provides a host of interesting and challenging questions for philosophers and social scientists. I hope that this brief introduction to discourse ethics has shown how all of these questions, all of Habermas's layers of value, could contribute to the ongoing processes of adaptive management.

7.5 Experimental Pluralism: Naturalism and Environmental Values

Remember, from chapter 1, that a major criticism of EPA is that its staff contains hardly any social scientists other than economists and that, whereas economists have more than one center of power—including especially the Office of Policy, Planning, and Evaluation—the few noneconomist social scientists are scattered and have almost no voice in directing the agency or the research it solicits and funds. Accordingly, a huge portion of the funding available for exploring environmental values goes to economists. Since scientists at EPA, including ecologists, risk assessors, and toxicologists (as noted in chapter 1) consider discussions of values to be specifically forbidden topics of conversation for them as scientists, it is unlikely that an open, science-based discussion of values will ensue within the agency. According to EPA's account, questions of value should be resolved by elected officials or political appointees, who act on behalf of the people. It may be a myth that political appointees somehow reflect voters' opinions, but in any case this belief encourages the separation of science from value-laden opinion. More importantly, however, the obsession to act as if science and valuation studies were entirely separate from value judgments means that decision makers receive scientific data and information with no assurance that these facts will be relevant to the social problems faced; the failure of social science at the agency ensures there will be no scientifically valid research on values collected—except, perhaps, some economic analyses—meaning that the decision maker can either appeal to economic values or make decisions with no research on public values at all. There is no forum, and no catalyst, for broad discussions of value questions at EPA or in the public discourse addressing policy formation. Again, this is partly a structural problem at the agency, which has never completed reorganization into functional units, but here we are concentrating on the associated problem of inadequate vocabulary for discussing environmental values. When social values do become the topic of conversation in Congress, elected officials—also lacking an intelligible, comprehensive language—fall into the polar extremes of economic value versus intrinsic values in nature. What is needed is a systematic approach to evaluation that is flexible enough to be used in many community processes and comprehensive enough to include the whole range of social values affected. In this section we will examine the possibilities of a constructive social science as a key contributor to adaptive management processes.

We have taken value pluralism as a starting point for our analysis; we live in a diverse society, with many environmental values being expressed in a variety of vernaculars, and I have therefore advocated an "experimental" approach to environmental values. We start by recognizing multiple values and inquire whether we can usefully group, rank, or reduce multiple values, and associated criteria, to simplify discourse about values. I think this policy encourages, in the long run, the development of just enough complexity. Terms and concepts will fall by the wayside if another vernacular proves more effective, and public management processes can be thought of as laboratories for studying how various values interact. In general, we here think of environmental problems as involving competing goods, as problems of setting priorities when public resources are limited, rather than as problems of choosing between good and evil. This is not, of course, to deny that there will be conflict, competition for limited resources, and sometimes bitter controversy about priorities and specific policies. The point is that a diversity of value positions and a diversity of environmental goals can, on a pluralistic approach, be treated as the beginning point, as rich soil for the growth of more cohesive environmental policies, provided that participants are committed to achieving cooperative solutions.

To say that diversity and pluralism of values represents a starting point is, of course, to suggest that it is possible to go somewhere from there. The purpose of this section is to explain how a community, inhabited by diverse groups with many different perspectives and value commitments, could engage in a process that results in improved cooperation among these diverse groups and also in improved environmental policies. In particular, I will argue that if good policy is to emerge from diversity, then we must develop a new kind of integrative social science, a social science that will find its role as mission-oriented science within an adaptive management process. Pragmatists and adaptive managers make much of praxis and of the importance of addressing real-world problems in a practical context. It is time to address head-on the problem of how to do evaluation, how to do objective social science research on values, and how to do it in a context in which the models chosen may be controversial. Agency employees and interest-group representatives have competing agendas, and what is called science—and what is called propaganda—often depends upon who advocates it.

Let us start by granting that economists, of all the relevant disciplines, have offered the most coherent, theoretically grounded, and technically sophisticated approach to measuring effects of policies and evaluating environmental change. The economists' system of evaluation was presented, and criticized, in chapter 6. My main criticism was directed against the claim made by many (but by no means all) economists that economic calculations of costs and benefits can in principle capture all legitimate environmental values. Once we get past that conceit, we can treat economists as offering one method among others of comparing the values of various policies and outcomes. On that basis, they have a lot to offer. For example, if one of the effects of a proposed policy can, in fact, be plausibly construed as affecting the availability of a commodity, then we can calculate these changes in economic terms. If policy A provides an increment in the availability of the commodity in question over that expected on policy B, then we have reason to prefer A to B, *other things being equal.* So the fact that an economic accounting of environmental values is not a complete accounting does not keep us from using it as a helpful tool, in many situations, to evaluate the impacts of policies.

Here, however, I want to look more broadly at the problem of evaluating policies and outcomes of environmental actions, considering what a comprehensive, integrative, transdisciplinary approach to such evaluations would look like. In this broader context, economists—despite the sophistication of their methods—must be seen as one set of contributors, among others, to the larger task. In order to solidify this difference, and to dispel widespread terminological confusion, I will stipulate that the terms *valuation, to value,* and *valuing* refer to the use of economists' methods, which measure changes to the environment in terms of impacts on aggregated individual welfare. These valuation studies are reported in terms of aggregated willingness of consumers to pay for improvements, or to avoid decrements, of the environmental good in question. By contrast, I will use the term *evaluate* and its cognates to refer to the broader, multidisciplinary process of comprehensively ranking policy options and possible actions on all dimensions and using tools from multiple relevant disciplines. Evaluation aims to summarize and integrate data, including economic data, to arrive at an overall judgment about the effect of an activity or a policy on social values. Given this stipulation, I wish to summarize what we have learned about such an evaluation process in this book so far, and then push forward toward a better understanding of the role of a yet-to-be-developed comprehensive social science in evaluating environmental policies.

Here are some of the conclusions we have reached, as we have looked at the problems of communication and evaluation, problems that confront Americans when they enter public discourse about environmental values. First, as just emphasized, we start with a pluralistic assumption. We expect that members of diverse modern societies will value nature in many ways and express these values differently. Second, to encourage more comprehensive evaluation, we consider the object of evaluation to be possible development paths, which are possible futures a community could move toward by choosing appropriate policies. Development paths can also be thought of as scenarios, and environmental discourse can be clarified and made more precise if proposed policies are related to various scenarios, which can then be discussed and evaluated. In section 11.4, we will return to the scientific art of constructing useful scenarios.

Notice that this choice of objects of evaluation does not rule out the use of economic valuation techniques for particular outcomes and effects. For example, if a given development path is expected to result in cleaner air and less risk of disease or more opportunities for outdoor recreation, it would certainly be relevant to know how much citizens are willing to pay for those changes and to compare the various increments of economic value associated with alternative development paths. An evaluation of a development path can, then, include valuation studies of various environmental commodities as an important part of an overall evaluation. We have also suggested that in evaluating development paths, more than one criterion will be used, implying that we adopt a multicriteria system of evaluation. For development paths that differ significantly in outcomes in the distant future, criteria other than purely economic criteria must be included. More will be said to support this result in chapter 8. Our general approach to evaluation is to develop a set of indicators and criteria that can be used to rank proposed development paths.

So far, I have characterized evaluations of possible development paths in rather abstract terms because our democratic commitments require that the criteria be in some way informed about what really matters to citizens in specific communities. We assume that a full-fledged evaluation of policy alternatives will be embedded in an organized process of adaptive ecosystem management and that this process will include some appropriate form of input from community members and groups, such as a stakeholders committee or a citizens' advisory committee. In that case, we can assume that various "voices" from the community will advocate policies based on values they hold dear. In the best cases, these concerns can be articulated as specific proposals of an indicator that could be used to track those environmental goals. An indicator is understood as a measurable index of change in valued states of a system. An indicator should thus be related to a state of the system—a state that may change given choices the community makes regarding development paths. Various participants in the public process can, then, advocate protection of their values by favoring a set of indicators associated with their favorite values and by advocating that goals be set to maintain or achieve specified standards for the indicators. The discussion of indicator choice provides a relatively concrete representation of values placed on the environment, because participants will favor indicators that they believe track values important to them; however, they will have to focus on measurable aspects of their environment. If coalitions form supporting particular indicators, it may be possible to enlist advocates of different values behind a single indicator. If so, this may be a sign of significant overdetermination of objectives by values and may suggest possibilities for win-win strategies directed at achieving goals stated in terms of robust indicators with broad-based support.

To move beyond abstraction, to begin to stipulate some actual indicators, criteria, and goals for evaluating development paths in particular communities, one must of course engage in a process. If environmental policy formation is going to reflect social values in a geographic place, it will be necessary to involve, in the development of indicators, citizens and stakeholders from that place who are affected by those policies. One issue that must be resolved in each case is how the process will be organized so as to engage public discourse and opinion on matters of considerable complexity that require technical knowledge. Again, it is difficult to generalize, but in general I follow Kai Lee and others who have argued that provided the process is open and ongoing, it is reasonable to hope for the emergence of the epistemic community we have discussed above—a group of interested citizens and scientists who serve on advisory committees, attend meetings, read reports, and offer opinions. As Lee points out, such communities have evolved in the past, but their emergence and nurturance is of the utmost importance; it is crucial that the people in the group develop respect for each other and a recognition that all parties prefer a cooperative solution, and they must be willing to engage in "experiments" to reduce uncertainty and ignorance. Again, we are talking, of course, not about sterilized, "pure" science, but rather postnormal, mission-oriented science that is undertaken in a place, within an activist managerial context, in response to uncertainty or disagreement among parties.[40]

How such a committee is constituted must depend on local conditions, but it is essential, from a democratic viewpoint, that these participants be able serve both as spokespersons for larger interest groups and as educators of such groups. Similarly, it is essential that the advisory committee include competent and interested scientists or have ready access to scientific advice. The model of democracy suggested here requires a form of two-way representation. Members of the citizens' advisory committee, who are interested enough to devote their time and effort, will represent the ideas of an interest group or a number of citizens. If some interest is not represented, it is hoped that the concern that decisions will be made without their input will encourage the previously unengaged parties to join the process. It is also the responsibility of committee members to go back and "represent" the evolving viewpoint of the citizens' advisory committee to uninvolved community members who share their interest. If new information, new studies, new considerations are brought to bear upon policy discussions in the citizens' advisory committee, the representatives can carry this information back to the rank and file, explaining changes in policy and the reasons for those changes. It will also be in their interest to do so; if they fail to "educate" other persons in their interest group, they will come to be associated with the "enemy" and ostracized from their own interest-based groups. By placing members of the epistemic community in the middle like this, one creates a ratchet by which social learning by individuals and by groups can occur. Of course all of this sounds quite optimistic—but the point is that we can, quite independently of specific, substantive values, pursue the ideal (however counterfactual) of an ideal deliberative process.

Let us assume, for the sake of continuing the exploration of evaluation processes, that all of the above preconditions are achieved in a community and that there is an active public discourse about indicators, criteria, and goals. Under these conditions, it would be reasonable to hope that such a lively public discussion—continued over months and years—of what indicators to use and what management goals to set would incorporate into the indicators public values as they were expressed by various stakeholders and interest groups and would also be scrutinized by scientists of various disciplines. One could also hope that out of such deliberations, measurable indicators would emerge, indicators that participants expect to track their values.

If all of these preconditions are in place—and I admit it is a lot to hope for—we are ready to proceed with a management process that includes (1) a multidisciplinary, public discussion of goals, supported by a policy evaluation effort that is comprehensive, integrative, and multidisciplinary; and (2) a lively tradition of "mission-oriented science." In this section I want to concentrate mainly on the possibilities of the evaluation process, but we must also respond to an important issue that cannot help but arise once we enlist the social sciences in the evaluation process. Is it possible for social scientists to contribute to a mission—which almost certainly involves some level of advocacy for one indicator over another, for example—and at the same time uphold the expected standards of scientific objectivity? After developing somewhat the possibilities of an evaluative social science, I will turn to the special problems of advocacy that seem likely to arise in mission-oriented social science.

1. I have argued that a comprehensive and integrative method of evaluation, a multidisciplinary method, would include methods of economic valuation together with methods developed by other social scientists. I have also argued that if such a method is to lead to improved environmental decision making, it will have to be applied iteratively, in an open, community-level public discourse that allows all voices and all affected parties to have their say, not just about goals but also about scientific evidence presented and about the indicators and criteria that are being used to judge success in management. What would be the role of social scientists who wished to study such a process or work as evaluators within this process, or both?

I turn again for contrast to the assumptions that shape economic valuation, which we have admitted is the best-developed methodology among those available to evaluators. Because of their interest in the aggregation of individual preferences, economists make a number of assumptions to avoid double-counting, including the avoidance of undue influence of powerful personalities, insisting that each individual be allowed to report her or his own individual and "sovereign" preference. In order to obtain aggregable data on individual preferences, economists take freeze-frame pictures of individuals' preference structures. The individual preference expressed by a purchaser or a respondent is taken to represent units of welfare equal to their willingness to pay to fulfill the preference. But the freeze-frame nature of this valuation prevents us from observing how participants' values change over time and in response to new information, deliberation, and so forth. And consumer sovereignty also leads economists to avoid letting respondents get together and discuss preferences, values, and goals before they are questioned in a contingent valuation study, for example; they are afraid participants will be swayed by strong personalities, whose answers will then gain weight through influence on the answers of others. For these reasons, economic valuers choose respondents who have not discussed their preferences with anyone and who are willing to estimate their willingness to pay for a good based on a set of facts explained by the inquirer. They seek individual preferences as felt at a given time, untainted by deliberation and accommodation to others and elicited in a one-time query designed to minimize bias.

In order to broaden the scope of inquiry beyond static preferences—which is essential if we are to use social science research to identify and illuminate emerging evaluations—I propose a distinction between economists' individual and static preferences, what we can call "felt preferences," and what can be called "considered preferences."[41] Felt preferences are the personal preferences that an individual feels and expresses at a given time, before careful examination. Considered preferences are the preferences one has after a significant process of examination, fact-finding, consultation, and deliberation. We can say that one object of an ongoing process of evaluation in the context of an adaptive management program is to move participants from unthinking commitment to their felt preferences toward thought-out, considered preferences. A considered preference may have exactly the same content as one's original felt preference, but the considered preference is based on careful attention to the matter at hand, including both the introduction of scientific information and ongoing, open discussions with peers about that information and about the goals of the community. The process of "considering," one hopes, holds open the possibility of social learning.

Evaluating policies, in contrast to valuing outcomes by aggregating felt preferences, requires an ongoing, dynamic, iterative process in which deliberation, social learning, and accommodation are expected to occur. Therefore, we relax the economists' attribution of sovereignty to consumers' preferences, and instead we expect preferences as well as beliefs to change as communities engage in social learning. Undertaking such a process will require that we place less weight on aggregation and more on trends, on changes, and especially on *hypothetical constructs*. These constructs are of two types. One type of hypothetical construct, which can be called an ideal-outcome construct, refers to the policy a community would eventually choose after engaging in an ideally democratic process in which key participants solicited the best science available, in which uncertainties and disagreements were turned into testable hypotheses, in which experiments and studies were undertaken—or at least planned—to resolve those differences where possible, and in which a long and open process of deliberation led the community to adopt a policy. In this case, we define the ideal outcome as the outcome that would result if the adaptive and deliberative process was ideal.

A goal of evaluation on this approach is to imagine a process that would allow us to confidently say that the preferences expressed are considered preferences of community members. We would know they were considered because they would be expressed after a process in which available scientific information was disseminated and explained, all interested parties had their opportunity to speak, and open discussion and deliberation occurred. Ideal-outcome constructs do not claim that the actual outcome of any process is in fact ideal—that the process resulted in the optimal environmental outcome. Ideal-outcome constructs say only that *if* the process was ideal, then the policies chosen in that process would be legitimate and worthy of support.

Ideal-outcome constructs, so defined and understood, thus have an ephemeral quality to them, much like a Peircean claim to truth: they assert that a policy outcome is ideal, *provided that* certain procedural conditions have been fulfilled; but it turns out that at any given point in the real process, we cannot know whether we have been stringent enough in applying our method to be sure our current policies are in fact ideal. Ideal-outcome constructs are thus useful only if we combine them with another kind of construct, an ideal-process construct. Ideal-process constructs have this form: if a public, adaptive process fulfills conditions 1, 2, 3, . . . , *n*, then the eventual outcome of that policy process will be considered the ideal policy for that community. Although these hypothetical constructs seem cumbersome at first, they have the advantage of pointing us toward at least three interesting interdisciplinary programs of research.

The first such program of research, which responds in social scientific terms to empirical questions associated with ideal processes, would probe the "ideal" conditions of discourse that encourage deliberation, social learning, accommodation, and cooperative solutions. This research would apply discourse ethics in the empirical study of public discourse, with special application to environmental discourse in adaptive management programs. Social scientists could thus help us to understand what conditions we must establish in order to expect that a public and democratic adaptive management process will encourage social learning, consideration and reconsideration of felt preferences, and so on. Here we can begin with the conditions already proposed by Habermas and his colleagues: openness, respect for others, and so forth. Our understanding can also be expanded by empirical social science of the type undertaken by Elinor Ostrom and others, who examine successful processes of cooperative action, such as processes by which interested parties manage common resources.[42]

Remember that we are here dealing with process norms, what Habermas refers to as moral rules, derived from the conditions of speech acts and of communication. These procedural rules depend not upon appeal to substantive values, but only on commitments of participants to achieving cooperative solutions and to the value of communication in reaching such solutions. Also, there are a number of empirical questions regarding what conditions promote communication. And beyond these general questions about the conditions of effective communication, there are more specific research questions about what conditions promote effective discourse in communicating complex scientific information.

A second program of research would propose and, with the cooperation of ecologists and other physical scientists, work with community groups to progressively articulate and refine indicators that might become the basis for a new criterion to be added to the multicriteria list that is used to evaluate possible development paths. It is important that indicators be relatively easily monitored and expressive of widely shared social values. Social scientists, working iteratively with participants and with physical scientists, can contribute to the articulation of values as indicators, by repeated rounds of polling about goals and by engaging in deliberation with the physical scientists to match measurable indicators with social values. This program of research, unlike the first, would need to be undertaken at the local level and would require ongoing interactions with real participants in locally based ecosystem management projects. This research would contribute positively to the critical task of developing indicators that are reflective of the most important values of the community, to the identification of criteria by which to judge development paths, and to the identification of various scenarios—paths forward that could become focal points for analysis, criticism, and revision. We will learn, in section 12.2, that the Dutch have successfully used comparative scenarios in a public, political process of reorienting national, provincial, and local policies toward a more integrative approach.

A third program of research would examine many local adaptive management processes and attempt to determine what works and what does not work in deliberative, consensus-finding processes. Since we do not know what an ideal policy outcome would be, we cannot of course compare the outcome of a local adaptive management program to an abstract ideal; but we can evaluate the process with respect to how well it succeeds in increasing communication and cooperative action. We can ask, for example, whether a particular local process fulfills various conditions of an ideal speech community; we can compare various local processes and learn which ones seem to tend toward consensus or at least cooperation; and we can try to determine what are the dominant variables in determining success in promoting communication, deliberation, and cooperation. Research of this type is already under way and is exemplified by several promising and effective research programs.[43]

2. I am suggesting that the social sciences, as well as the natural sciences, can assume an important role in the development of improved policies within an adaptive management process. Here the social sciences step across the line from disciplinary, curiosity-driven science to engage in mission-oriented science. Mission-oriented science, as we learned in chapter 3, is science that takes place within a public conversation about policies and difficult social choices. In particular, the relevant review panel for scientific work is not disciplinary but includes interested scientists from multiple disciplines and other interested parties who are willing to engage in critical reading and thinking about scientific presentations. When we think of mission-oriented science, we can imagine that disagreements about effective remedies might be explored using scientific models and pilot projects, reducing uncertainty and

encouraging cooperative behavior. But the role of the social sciences in mission-oriented science and in public management processes is sure to be more controversial. If social scientists, possessing expertise in understanding and measuring social values, also become a part of the public process, there will surely be questions about their "objectivity" and their ability to do unbiased research. In order to create a lively tradition of mission-oriented science, we must address the question whether scientists should also be advocates. I do not expect to be able to answer all of those questions and concerns here, but I think I can help to put them in a perspicuous light with a few comments.

First, I am not very worried—though perhaps I should be—that scientists who become involved in a policy process will become unconstrained advocates who cook data and consciously misuse and misconstrue research in order to support policies they favor. When critics—for example, Landy and colleagues, in their criticisms of activism at EPA—argue that being advocates undermines the credibility of scientists, I would reply that it does so if and only if advocacy leads to bad science.[44] Any scientist who values his or her reputation will not try to foist off bad science on other participants just to make a point in advocacy; advocacy, above all, requires trust and cooperative behavior. So to the extent that scientists can separate bad from good science, we can expect that scientists who are also advocates will have special incentives to avoid distortion and questionable scientific claims. This all depends, of course, upon the development of a multidisciplinary epistemic community that includes lay representatives and scientists who can review scientific methods and results.

Second, involvement with values is not a *choice* we make; it is simply unavoidable in public management discussions. In discourse about what to do as a society, problem identification cannot even begin without reference to what is important—what is valued. Given that encounters with, and appeals to, values are unavoidable, one might rather deal with a scientist who is also an advocate, because one knows where that advocate stands on values as well as on science, and as a result we have a better basis for evaluating the scientific contribution.

Third, and especially important, it must be remembered that our pragmatic epistemology has forbidden a priori pronouncements, requiring instead that all assertions be considered open to refutation by experience. What this means is that neither scientists nor philosophers can claim a special dispensation to override the public process. Each discipline, no doubt, can contribute to the process, but no discipline can "trump" the processes of public deliberation. As political philosopher Benjamin Barber says, there is no Archimedean point from which policy conclusions can be levered.[45] This means that, like other participants, scientists and philosophers are stakeholders, trying to influence the process by bringing information and value considerations to bear within public discourse. They will, and should, present their information and data from their studies, but they cannot expect to be granted a privileged role, because so many others also provide information. No single perspective can claim privilege in developing models and choosing goals for the management of resources and ecological systems. Because of this parity of disciplines, because members of all disciplines must plead their case in the ordinary language of public discourse, and because experience rules over a priori pronouncements, social scientists' role must be practiced within the management process.

Once all parties abandon the idea of preemptive principles, I would argue, most of the scariness is taken out of the warning that scientists may become advocates and thereby lose their credibility. No field can preempt all others in an open process, and each participant is seen as an advocate of a perspective and an interest. The key issue becomes one of trust. Scientists, in this context, become one important type of stakeholder among others. Can you trust the scientist sitting across the table not to fudge his data, even if he disagrees about the goals of management? If not, that is a problem of building trust in the cooperative process, not a problem of advocacy. Consider again the definitional characterization of mission-oriented science offered by Funtowicz and Ravetz: mission-oriented science is not reviewed by disciplinary peers only, but by the relevant review committee composed of all interested participants. Nobody can be denied input because of lack of degrees or certifications. The price of admission to the table of discourse is a willingness to dispute and to engage in social learning. If the dynamic of communication and social learning is functioning effectively, the fact that scientists—or real estate agents or park rangers, for that matter—advocate positions is no problem. The reasons they advance for their positions must be persuasive to people who are advocating in disparate directions. If they can express their special concerns from the perspective of their interest group and still retain the respect of other participants, they will have an impact on the decision process. For those who lack degrees and certifications, if they are willing to examine documents and question arguments, then by that commitment they are participants and "peer" reviewers of scientific input. Their peer-ship is a right earned by hard work and honest disputation. This openness to criticism from all fields and interests, characteristic of mission-oriented science, of course places tremendous pressure on the communication process. Broad participation by parties requires the reduction of jargon and begs for the development of models that are specially designed to improve communication and to assist in disseminating basic scientific information to participants.

This section may seem too optimistic in tone, even utopian. My purpose is not to spread unjustified optimism but rather to think through what an ideal process of public participation and social learning, including a systematic effort to develop a comprehensive evaluative process, would look like. So I helped myself to the assumption that the kind of process that would encourage open discussion, consideration of evidence, deliberation, and iteration of all of these across time can in fact be achieved. We all know that such perfect conditions will seldom if ever occur, but it is useful to imagine such an adaptive management situation in order to have an ideal to which our less successful efforts can be compared. We have put a lot of weight on the emergence of spokespersons for various interests, participants who are sufficiently interested to act as liaisons between their element of the community and the citizens' advisory committee. Again, I am not claiming that the emergence of perfect representatives or perfect processes is certain or even likely. Right now we are more interested in asking what would happen if they did emerge. We can also ask—and this can become an important topic for social science research—which shortcomings can be overcome and which are likely fatal to the process.

My point has been that all of these questions can be formulated as questions in the social sciences, broadly speaking. I believe that the social sciences, including economics but including also other disciplines such as anthropology, sociology, and political science, have important research and practical roles that fit nicely into the adaptive management tradition and evolving process. Since pragmatic philosophy claims no special access to a realm beyond experience, philosophers also find their role within this process as they concentrate on developing language capable of communicating and integrating knowledge from the various disciplines. I see the role of these social sciences as integrative, providing data that will help communities to better learn how to learn and how to integrate scientific knowledge with political and social considerations about values.

The social sciences can also improve deliberation and the decision process by studying and improving communication and processes of communication. This means encouraging participants to propose and experiment with various indicators and to test them as possible interpretations of the values they hold. To encourage effective social science, science that is effective in improving both processes and outcomes, I have proposed the articulation of an ideal process that encourages communication and cooperative behavior and which can serve as a touchstone for empirical social scientific studies of how to improve present processes. These would include studies that examine what is working—and not working—in particular situations and also comparative studies that can hypothesize general rules about what works and does not work. It is possible to use social science methods by developing more comprehensive evaluative tools, tools that will allow us to move toward an articulation of a particular community's sense of identity, including who the community, collectively, really wants to be. If the social sciences can be enlisted in the adaptive management process, and if they can address crucial questions like this, they will be contributing information that, when combined with economic measures, will contribute to a comprehensive method of evaluation for proposed policies. What is different about our pragmatist approach to specifying ideal policies, as compared to a priori approaches, is that we have conceptualized the ideal as the constructive outcome of an effective and self-correcting process in a particular community, rather than as an external standard by which the community's choices are judged.

8

SUSTAINABILITY AND OUR OBLIGATIONS TO FUTURE GENERATIONS

8.1 Intertemporal Ethics

We have emphasized the point that the term *sustainability* is given many meanings; but we should not go too far in this. It does, after all, have a clear, core meaning: sustainable living is forward-looking living; this is an idea we explored in chapter 2, by comparing sustainability to the pragmatist, intertemporal conception of truth. Sustainability is about the future, our concern toward it and our acceptance of responsibility for our actions that affect future people. Calls for sustainability evoke moral sensitivities; indeed, most commentators treat the problem as one of equity or fairness across time. Sustainability, whatever else it means, has to do with our intertemporal moral relations.

It therefore seems reasonable to say that sustainability concerns our obligations to future generations. But do we *have* obligations to the future? It has been asked, for example, Why should I care for posterity? What has posterity ever done for me? These questions acknowledge that our moral relations with the future are inevitably asymmetrical in an important sense. Obligations to the future therefore cannot be of the standard, contractual variety; they cannot depend upon assured reciprocity. And yet, despite the oddness of the obligations involved, most people do show a significant concern for the future and for the impact of their choices on the future. Opinion polls show that overwhelming majorities of people in modern democratic societies believe we have obligations with respect to the future. It is sometimes difficult to interpret this belief and the sort of commitment it implies because people have very different mental models of environmental change and of the human role in it. Nevertheless, the data is very clear: most people today, when asked to think about it, prefer to live in a society that cares for the future and limits actions that are likely to have negative impacts on the future. Surely this impulse is at least partially responsible for the widespread interest in, and acceptance of, sustainability as a public policy goal.

For convenience of reference, I will speak of the *bequest* that one generation leaves for the next. A bequest is the sum total of accumulated capital, technology, institutions, and resources that a given generation leaves for its posterity. The bequest of a generation, so defined, represents all the actual impacts that the generation has on all subsequent generations.

Some societies are quite explicit about what they hope to bequeath to their offspring; but it is unavoidable that there will certainly be a huge discrepancy between the actual bequest—the sum total of our impacts on subsequent generations, as they actually unfold through history—and the *intended* legacy that is spoken about and sometimes planned for. One thinks, of course, of monumental discrepancies, such as the difference between the promises of Communism and the realities of Stalinism and its aftermath. In all our actions there is the chance of unknown and unintended consequences; and we are notoriously bad at predicting outcomes of new technologies, for example. If, however, we admit that there exists a significant concern for the future and a willingness to act in ways to reduce the damage from our actions to the future, we must also have some means to project, with some reliability, the impacts of our actions across future generations. Accordingly, in chapter 3, we developed an epistemology for adaptive management. As noted there, both our ability to affect natural systems and our understanding of our impacts on them have greatly increased in the past few centuries. Today our ability to change nature outruns our ability to foresee the effects of the changes. It is clear, nonetheless, that accepting an obligation to protect the future from negative impacts of our own decisions requires some ability to foresee those impacts and at least some of their consequences for future people.

Historically, people relied upon tribal elders and religious wisdom to guide actions with widespread and long-term consequences; today, science and scientific modeling have become the favored means for projecting consequences into the future. Scientific models possess considerable potential for offering more precise predictions about actual impacts that will become evident in the future; they can also be used to generate scenarios that can be evaluated and compared as illustrations of possible development paths that a society might pursue. Despite future promise, our abilities to foresee the impacts of our decisions on it are quite limited. Every decision we make affecting the distant future is clouded in uncertainty and ignorance. Any successful account of our obligations to the future must somehow deal with the problems of ignorance and uncertainty. Any reasonable assessment of our bequest for the future requires us to recognize that however sophisticated our scientific projections and models, they can never be perfect in their foresight. So we cannot simply adopt an optimistic viewpoint that says science is advancing rapidly and eventually will allow perfect prognostication. No, realism is more useful here: we should seek an account of obligations to the future that makes use of the most effective available tools of foresight, but which can nevertheless be applied even in the face of considerable uncertainty about actual outcomes of our actions. It is certain *that* our current choices affect the future; what is difficult to predict is *how* they do so. Even if we act in uncertainty, however, we are responsible for our actions: if we believe in free choice, we cannot escape responsibility for the consequences of our actions.

Many societies, as noted, have worried to some degree about their legacy, but specific policies to improve the future have been as varied as the ideas of the good pursued by different communities. The most profound difference between attitudes toward the future today and in the past is a heightened awareness of ever-accelerating change. Because the acceleration of change is driven by human innovation and increasing technological power, rapid change is unavoidably our responsibility. Given that we daily unleash incredible ingenuity and power to alter and control the forces of nature, should we not show some concern for the long-term consequences of our present decisions? In this chapter we explore the puzzling idea, widespread in modern as well as ancient societies, that we are somehow accountable for the bequest we leave the future.

Assuming that we accept some level of accountability for our bequest, how should we conceptualize, itemize, and measure our bequest? I once had an interesting conversation on this topic with Brian Barry, an ethicist and social philosopher formerly of the London School of Economics and now at Columbia University, who has written on the topic as a part of his extensive work on issues of fairness. He said, "The problem of what we owe the future comes down to a simple choice: either you have to measure welfare or you measure some kind of stuff." Barry, if I understand his conversation and writings, leans—though not dogmatically—toward measuring welfare; whatever his position, his dichotomy is very useful in understanding the current interdisciplinary debate about sustainability.[1]

Barry's dichotomy cuts across disciplinary lines and divides contestants in the game of producing and defending one or another definition of sustainability. On one side are those who advocate a welfare definition. They see the problem of intergenerational fairness as one of maintaining constant or ascending levels of individual human welfare. On the other side are advocates of criteria that list various kinds of "stuff," which might include such items as intact ecosystems, adequate supplies of fresh water, unspoiled parks like Yellowstone and Yosemite, or simply "natural capital." It is remarkable that this hugely complex intellectual debate, with implications for multiple disciplines, can be stated so simply and clearly; but in this case Barry's dichotomy identifies the key fault line between "weak" and "strong" sustainability theorists (provided it is understood that weak-sustainability advocates count *only* welfare, whereas strong-sustainability advocates would include pluralists who count *both* welfare *and* "stuff"). Barry's dichotomy thus offers a rough-and-ready tool to sort sustainability theorists into two broad categories. This categorization, in turn, will help us to develop an appropriate "logic" for applying sustainability criteria.

Achieving weak sustainability requires maintaining a nondeclining stock of economic capital into the indefinite future. Weak-sustainability theorists and advocates reason that if the savings rate is nonnegative across time, each generation will enjoy at least as much opportunity to enjoy welfare as members of prior generations. The weak-sustainability criterion allows unlimited substitution among types of capital so that what is measured is undifferentiated capital—generic wealth. This approach, in particular, allows human-built capital to be substituted for wealth in the form of natural assets: there is, as economists say, fungibility across types of assets. Weak sustainability is weak in the sense that it requires the maintenance of no particular asset classes; action is governed by the comparatively weak constraint that total capital available must be nondeclining from generation to generation. In other words, achieving weak sustainability means that the savings rate never dips below zero; sustainability is achieved when total societal wealth is maintained across generations.

Strong sustainability, in contrast, specifies limits on substitution, requiring that there must be limits to the replacement of natural assets with human-built ones. It more strongly constrains action by delegitimating actions that destroy certain "stuff." Although Barry's welfare-or-stuff dichotomy helps us to keep our eye on the essential choice in defining sustainability, it does not, unfortunately, provide any guidance as to how to specify more clearly what stuff must be saved according to strong-sustainability theorists. Figuring out how we might reasonably specify—and justify—a bequest in terms of obligations to save certain kinds of stuff will occupy us for most of the rest of this chapter. First I must explain why the debate between weak- and strong-sustainability theorists is so important and say a few words about the status of weak sustainability, before passing on to the task required by this book's goals—to develop a suitably strong version of sustainability. By suitably strong, I mean that the definition must be sufficiently specific that we can use it to begin the adaptive process of managing resources, but it must also allow some substitution and enough flexibility that each generation can achieve its own aims to the extent possible, given technological means available.

The two strategies of counting stuff and counting welfare diverge with regard to the kinds of data and scientific modeling that are relevant to assessing bequests, so the disagreement has a strong turf-protection aspect: "The problem should be seen this-a-way, and so you need the services of [my research team, my graduate students, . . . fill in as appropriate]." Defining sustainability in terms of welfare puts its study firmly in the territory of neoclassical economists. Ecological and historical information becomes more important if sustainability is defined in terms of stuff, since intact and productive ecological systems are generally believed to be important in maintaining productivity that supports many services and amenities for humans.

These turf-protection aspects of the disagreement about how to define sustainability are undeniable and sometimes obvious, but it is possible to overestimate the importance of disciplinary boundaries in this case. The stuff-welfare distinction does not correspond precisely with disciplinary boundaries. Some ecologists prefer to measure welfare impacts of ecosystem services,[2] and many economists have agreed that some criteria beyond simple welfare comparisons must be applied. Welfare-counters come in many varieties, but one prominent form of the view is equivalent to the theory called Economism, which I discussed and rejected in chapter 5. Economism, when applied to problems of intergenerational economics and sustainability problems, expresses itself as what is called weak sustainability: policy decisions should be made on the basis of welfare effects only, with the understanding that neoclassical economic models can in principle provide complete and adequate measures of welfare both in the present and in the future.

The debate itself is far more technical than I have suggested, and I recommend diving into the literature only if one enjoys virtuoso mathematics (or, like me, has learned to get the idea by reading the sparse text passages scattered among the equations and proofs). Barry's dichotomy simplifies the intellectual landscape, however, organizing the battleground into two mostly separable concurrent disagreements. Accordingly, we can divide critiques of weak sustainability into *internal* and *external* critiques. Internal critiques of weak sustainability accept the formative assumptions of economic valuation and criticize the theory on economic premises. Accepting the mainstream assumptions and understandings of economic value, internal disagreements all concern how to measure and compare economic values. They do not question the use of economic techniques to measure welfare, nor do they question that improvement of aggregated welfare is the goal of economic and environmental policy. These internal critiques take place within the camp of those who count only welfare. In contrast, advocates of strong sustainability mount external criticisms of weak sustainability, attacking the fundamental assumptions and valuation techniques of neoclassical economic theory as applied to resource use and depletion; in doing so, they are questioning whether economic models based on intergenerational welfare comparisons are adequate to capture the full meaning of sustainability. These criticisms are directed at weak sustainability from the outside—by those who reject the methods and tools of mainstream economics in favor of physical measures and moral obligations to protect some types of stuff. Advocates of strong sustainability attack the paradigm that measures and compares utility, favoring a criterion of fairness that identifies some physical features and aspects of the human legacy that should be protected as a matter of obligation. This stronger view, then, challenges the premise of Economism, arguing that our bequest to the future cannot be measured using solely economic models: we must also have measures that track the persistence of key characteristics and aspects of the intergenerational trust. The internal debate, though complex and fascinating, need not detain us long; it is a debate among devotees of weak sustainability about the details of how to measure welfare across generations.

I will briefly summarize these internal issues, as background for the external argument of section 8.2. Internal critiques employ mathematical modeling to explore alternative formulations of the weak-sustainability approach. Critics in this camp accept the basic normative thrust of weak sustainability and share the goal of reducing the complex question of intergenerational fairness to one of developing intergenerational comparisons of aggregated welfare. At the heart of the appeal of weak sustainability, of course, is the promise of an operationalization of sustainability using a single, countable measure—a comparison across time of the total stock of economic capital.[3] Despite their general support for the weak-sustainability approach, internal critics have shown that the operationalizability of weak sustainability—its most attractive feature—is extremely limited and is based on unrealistic assumptions. Applying sophisticated mathematics to create many formal models, each of which provides a measurement of sustainability for a hypothetical and highly simplified economy, theorists and mathematical modelers have generated many models of sustainable economies. This work, besides showing that by varying assumptions and fine points, one can spin out multiple theoretical models of sustainability,[4] has mostly demonstrated how very restrictive must be the set of assumptions necessary to use neoclassical models to measure and compare welfare across multiple generations.[5] For example, operationalization requires an idealized world—very different from the real world—in which population and technology are held constant, where there is no international trade, where resources are given first-best use, where actual prices reflect full knowledge of future scarcities, and so forth. So operationalization of sustainability on the weak-sustainability approach does not offer a tool for measuring real-world sustainability; it offers instead an appealingly simple normative model that *could* be operationalized if we lived in a simple world. The worlds and models that allow operationalization of weak sustainability are so unlike the real world that only leaps of faith and wild use of counterfactuals could possibly make one think these models have any application to the real world.

The internal debates about how to measure weak sustainability within neoclassical economics have little direct relevance for the larger argument in this chapter. Since I reject the normative idea behind weak sustainability, I am little interested in debates about how to measure it. There is one important point to carry forward as an important contextual aspect of our argument, however: the greatest charm of weak sustainability has been from the outset its promise of practical application through operationalization. If economic data could be used to operationalize and measure sustainability of welfare, that would unquestionably be very useful information. Careful examination of the internal critique, however, shows that operationalization is a dream at best, and even "in principle" it could at best provide a simplified model of sustainability for a single economy with very few of the features of real-world economies. So we should take the conceit of Economists who are weak-sustainability theorists—that they are on the path of operationalization and formalization of models to measure sustainability—mainly with skepticism and bemusement, much as we would the efforts of a sincere inventor who believes he is making progress in creating a square wheel.

8.2 Strong versus Weak Sustainability

Historically speaking, the distinction between strong and weak sustainability was developed in the much-read and much-discussed 1989 treatment of sustainability concepts by Herman Daly and John Cobb, *For the Common Good.* Daly and Cobb contrasted strong and weak senses of sustainability, challenged the use of economic growth and savings rates as the sole indicator of sustainability, and rejected the assumption that human-built capital is substitutable for natural capital. They argued that in addition to maintaining economic capital, we owe it to the future to protect certain natural features and processes, which they refer to as "natural capital." Daly, in another treatment of sustainability, has argued that natural capital and human-built capital are complements and that they are imperfect substitutes for each other.[6] Is natural capital to be understood as specifying some stuff that must be saved? Does Daly qualify as a true strong-sustainability theorist using Barry's criterion? On my reading of this now-classic book by Daly and Cobb, there is considerable ambivalence on this crucial point. In the text of the book, the authors clearly seem to support constraints that cannot be explained or measured in economic terms, treating the protection of natural capital as a matter of specifying some stuff—natural capital, in particular—that is owed to the future. If this is their intention, they are, on Barry's categorization, advocates of measuring stuff; but in the appendix to the book, they seem to equate losses in natural capital with economic losses, requiring that future people be compensated for the destruction of natural capital by setting up a trust fund to compensate—substitute?—for natural capital. In other words, their Index of Sustainable Economic Welfare uses economic models to compare welfare across generations and seems to suggest the Economistic view that welfare counting is a complete accounting system for sustainability.

Without resolving that ambiguity in the work of Daly and Cobb, we can use their case to illustrate the effectiveness of Barry's criterion in exposing crucial differences.[7] Requiring the protection of natural capital does not, in and of itself, qualify a sustainability theorist as being "strong" on Barry's test. The term *natural capital* is at least as ambiguous, I would argue, as *sustainability* and *sustainable development*, in that this key term can serve as a counter for either a welfare-based measure or a stuff-based measure. If natural capital is defined as natural elements and processes that must be saved *because their loss will measurably impact social welfare in the future,* and if one uses standard economic measures of welfare to identify possible losses and to determine the size of the trust fund necessary to compensate the future, then use of the term *natural capital* does not trigger Barry's criterion. Specifying stuff that, if lost, will negatively affect welfare doesn't go far enough: if we save just the stuff that has a measurable impact on welfare, it is welfare that counts, not stuff.

According to Barry's criterion, then, some economists who claim to be strong-sustainability theorists don't make the cut. For example, David Pearce and Edward Barbier, authors of the influential series of Blueprint books, which advocate the use of economics to ensure sustainability, seem to me not to be strong-sustainability theorists on Barry's test. In the 2000 installment in the series, which examines the same theoretical and practical territory as the first version in 1989, Pearce and Barbier state, "We are firmly in the strong sustainability view of the environment and economic development,"[8] referring the reader to their chapter 2, where the distinction is set out as follows. They define weak sustainability as holding that "there is essentially no inherent difference between natural and other forms of capital, and hence the same optimal depletion rules ought to apply to both." Strong sustainability, by contrast, "suggests that environmental resources and ecological services that are essential for human welfare and cannot be easily substituted by human and physical capital should be protected and not depleted. Maintaining or increasing the value of the total capital stock over time in turn requires keeping the non-substitutable and essential components of natural capital constant over time."[9]

Notice that natural capital, on their conception, is that subset of the natural order that will "pay off" in terms of human welfare; it is what must be saved in order to maintain a constant level of capital, which in turn would signal, according to the Blueprint models, nondeclining welfare. In other words, they are counting welfare. Pearce and Barbier think of their position as strong because they doubt that risks to ecological systems and the "ecological services" they offer can be calculated using the marginal analysis of cost-benefit accounting. They believe that the standard methods of cost-benefit analysis ("still the best game in town")[10] ultimately will have to be supplemented with other policy instruments in order to take into account the importance of ecological thresholds and irreversibilities. The key point, however, is that these new approaches, which involve the use of threshold instruments, are all considered as means to estimate welfare effects, given uncertainty and variation in risk aversiveness. What they value, and what they count, is welfare. They do not count stuff as a part of their decision making. If they count stuff, it will be incidental to an estimate of likely impacts of a resource shortage on welfare—their determinative calculation. By Barry's criterion being used here, despite their concern for "critical natural capital," Pearce and Barbier are ultimately counters of welfare.

Am I just splitting hairs here? Is there an important difference between true stuff-counters and the Blueprint authors? There is a difference, and it is an important one for both theoretical and practical reasons. Theoretically, since all values that get counted in decisions are ones that have to do with individual human welfare, as measured by accepted methods of economics, this approach cannot accommodate the kinds of communal values I defined and defended as important in chapter 7. It remains an Economistic position—all environmental values are ultimately related to impacts on the welfare of individuals and aggregations thereof. Practically, the viewpoint of Pearce and Barbier creates a "filter" for the kind of information that is relevant to decisions. Any information that would pertain to more holistic social goals and indicators would be considered relevant only if there exists a plausible causal model that shows that changes in those indicators will affect aggregated individual welfare. Given the complexity of ecological systems and the irreducibility of communal values into individual welfare interests, we will often not have the necessary knowledge to connect a deteriorating indicator with changes in aggregated welfare. So I am taking a hard line, following Barry: all Economists (that's the capitalized version that insists that all environmental values are economic values) are weak-sustainability theorists. To pass the Barry test, one must specify some stuff that must be saved because it is associated with a social value—a value that may or may not be tightly coupled with changes in individual welfare.

According to strong sustainability as here defined, a full-fledged definition of sustainability will specify some stuff the loss of which will make people in the future worse off than they would have been had the stuff been saved, but worse off in some noneconomic way, in a way that affects an important social value such as the communal values discussed in chapter 7. The key theoretical divide here is over the question of substitutability of types of capital: weak-sustainability theorists assume there are no limits on substitutability among resources and pay attention only to welfare changes, whereas strong-sustainability theorists believe there are limits on such substitutions and specify stuff instead of welfare; they specify stuff to save in addition to welfare accounts. We can say, then, that whereas the weak economic sustainability theorists believe we owe the future an *unstructured bequest,* the strong economic sustainability theorists advocate *structuring our bequest,* differentiating special elements of capital-in-general that must be included in the capital base passed forward to coming generations.

Figure 8.1 offers a conceptual geography of sustainability definitions based on Barry's dichotomy that separates stuff-counters from welfare-counters. This dichotomy is shown as the widest vertical line—that between strong and weak sustainability—with Pearce and Barbier and similar thinkers placed on the weak-sustainability side of the line. Although it might be reasonable to describe theorists such as Pearce and Barbier as strong economic sustainability theorists, they do not ultimately reject the value theory underlying weak sustainability and the welfare-only approach to measuring sustainability. They only suggest what they see as more effective means to estimate aggregated welfare in uncertain situations. I will apply the term *strong sustainability* only to the theories that are strong enough to pass the Barry test, which requires the supplementation of the techniques of microeconomics, usually with obligations to protect some kinds of stuff as a basic requirement of sustainable practices.

Another useful division is to separate strong-sustainability definitions into "strong ecological sustainability" and "normative sustainability." This difference, to be explored below, is between those who favor an *ecological* strengthening—such as requiring the maintenance of "ecological resilience"—and those, including myself, who favor specifying *social and communal values* as guiding a community toward protection of ecologically measured features. This is an important distinction *within* strong sustainability, because some adaptive managers, perhaps because of their backgrounds in ecological science, feel uncomfortable referring to social values in the midst of management science. I, however, have long ago abandoned the search for a pure science and happily admit that a definition of social values is a necessary prerequisite for specifying which ecological features are to be protected. I hope that as

	Weak Sustainability	(Strong) Economic Sustainability	Strong Sustainability	Normative Sustainability
Home Discipline	mainstream economics	ecological economics	systems ecology	policy science / environmental ethics
Paradigm	welfare economics	welfare economics + natural capital designations	complex dynamic systems theory / adaptive management	complex dynamic systems theory / adaptive management
Definition	maintenance of undifferentiated capital	weak sustainability + maintenance of natural capital	weak sustainability + maintenance of resilience	weak sustainability + maintenance of options
Key Concepts	nondeclining wealth	maintaining natural capital	maintaining resilient ecosystems	integrity of place; community involvement
Key Advocates	Solow	Pearce and Barbier	Holling, Lee	Leopold, Norton
Weak Sustainability – Welfare Counters ←			→ Strong Sustainability – Stuff Counters	

Figure 8.1 A conceptual geography of sustainability definitions

ecologists and adaptive managers become more comfortable operating in the value-laden context of active management, the differences between ecological and normative sustainability will be rendered unimportant. This somewhat less sharp divide—this continuum from scientists who emphasize measurable ecological features and the normative theorists—becomes less important as all participants recognize that in the iterative and circular process we know as adaptive management, both values and facts must be visited and revisited.

It turns out, however, that there are two ways of going beyond economic analysis; one way is to simply shift into a more deontological mode, and attempt to explain intergenerational moral relations in terms of moral obligations, entitlements, and rights. On this view, sustainability would simply be understood and measured in noneconomic terms.[11] The second alternative is to develop some form of "hybrid," or "two-stage" system of analysis for evaluating bequest packages; on these views, economic analysis and reasoning are considered important but bounded, or subject to being overridden by commitments to protect stuff in some situations. For the remainder of this section, this difference is immaterial: both of these approaches require the rejection of utilitarian calculi as the measure of sustainability. In each case there is some stuff that cannot be traded off against other economic goods; sustainability is not simply a matter of welfare comparisons across time.

Strong sustainability—as I define it using Barry's dichotomy—directly conflicts with the central ideas of weak sustainability and with Economism more generally. So it is time to examine the arguments given to support the economic, weak-sustainability paradigm. In a series of lectures and papers, Robert Solow, winner of a Nobel Prize in economics for his work in growth theory, has championed the view that sustainability can be fully defined, characterized, and measured within the mainstream economic tradition of resource analysis. He clearly states the position that all we could possibly owe the future is that its people be as well off, economically, as we are. Solow's basic idea is that the obligation to sustainability "is an obligation to conduct ourselves so that we leave to the future the option or the capacity to be as well off as we are." He doubts that "one can be more precise than that." A central implication of Solow's view is that although to talk about sustainability is "not empty, . . . there is no specific object that the goal of sustainability, the obligation of sustainability, requires us to leave untouched."[12] Solow believes that monetary capital, technology, labor, and natural resources are interchangeable elements of general capital. Within this set of definitions, the future cannot fault us as long as we leave the next generation as much capital as we have had. These claims directly confront the defining tenet of strong sustainability, and Solow's conclusion at least indirectly challenges most environmentalists' programs, which apparently include many more specific items of obligation. Solow's argument is worth examining in detail.

First, Solow dismisses a "straw man" (sometimes referred to as "absurdly strong sustainability"): the theory that we should leave the world completely unchanged for the future. "But you can't be obligated to do something that is not feasible," he argues.[13] With the straw man out of the way, Solow asserts our total ignorance regarding the preferences of future people: "we realize that the tastes, the preferences, of future generations are something that we don't know about." So, he argues, the best that we can do is to maintain a nondeclining stock of capital in the form of wealth for investment and in the form of productive capacity and technological knowledge. "Resources are, to use a favorite word of economists, fungible in a certain sense. They can take the place of each other."[14] Because we do not know what people in the future will want, and because resources are intersubstitutable anyway, all we can be expected to do is to avoid impoverishing the future by overconsuming and undersaving. The ability of economies to find replacements for any scarce resource, if coupled with adequate economic capital for investment, will allow people of the future to fulfill whatever needs and wants they actually happen to have. Provided we maintain capital stocks across time, efficient production and real income will be maintained, and each generation will have an undiminished opportunity to achieve as high a standard of living as its predecessors.

This argument, if accepted at face value, promises to deliver a fairly simple, potentially measurable criterion of sustainable living. Theoretically, the result is surely attractive—it cuts through a lot of confusing issues and provides a clear and simple theory about intergenerational obligations. Even better, this sustainability requirement will no doubt be politically welcome—it implies that the costs of changing over to sustainable development as a social goal are virtually identical with the goal of having an efficient, constantly growing economy, with a savings rate of more than zero. Maybe it won't be painful to adopt sustainable policies after all. Pursuing the weak-sustainability path, it seems, greatly simplifies our thinking about the future; perhaps living sustainably is something that we'll accomplish as a by-product of our real work—creating economic wealth. Solow's line of reasoning, which ends in an endorsement of the weak-sustainability path—the path on which we simply keep track of economic growth and savings rates—is in fact a version of a more general argument that one encounters in the literatures of philosophy and economics. I will call it "the Grand Simplification" (GS for short). According to this view, since we do not know what people in the future will need—and since resources are substitutable for each other—the only thing we can do is to measure and compare welfare across time.

8.3 Philosophers and the Grand Simplification

Perhaps it is not surprising that Solow and other economists find the Grand Simplification attractive; it defines the problem of sustainability as one that can be measured by the very techniques he and his colleagues can offer, ensuring a strong and growing demand for graduate students. It is more difficult, however, to see why the Grand Simplification is so appealing to philosophers. Among the few philosophers who discuss the topic at all, most advocate something like the GS; and the list of adherents to this approach among philosophers is indeed impressive. John Rawls, in *A Theory of Justice,* formulated the problem of intergenerational fairness as a problem of choosing a "just savings rate" by assuming that provided this rate is maintained, each generation will be better off than the prior generation. Rawls saw history as a progression from poverty toward riches, so he did not even consider the possibility that an earlier generation could harm the future in some radical way such as destroying the conditions of human existence. Obligations to the future are thought of in terms of fair savings. As long as an earlier generation adds to, or at least does not detract from, the accumulated capital of the society, the generation will fulfill all obligations regarding resources and environment. This understanding of the problem apparently accepts the GS. Specifications of what each generation owes the future do not list specific resources or specific physical features of the environment. In Rawls's favor, he does emphasize that savings include not just wealth in a narrow, economic sense but also the development of just and fair institutions—a point we shall find important below.[15]

John Passmore, who authored the first book-length environmental ethics text, uses a variant on the ignorance argument to conclude that "our obligations are to *immediate* posterity[;] we ought to try to improve the world so that we shall be able to hand it over to our immediate successors in a better condition, and that is all."[16]

Brian Barry also subscribes to a viewpoint that is apparently equivalent to Solow's when, after noting that it is impossible not to reduce nonrenewable resources such as petroleum, he says: "The important thing is that we should compensate for the reduction of opportunities to produce that are brought about by our depleting the supply of natural resources, and that compensation should be in terms of productive potential. . . . In the absence of any powerful argument to the contrary, there would seem to be a strong presumption in favour of arranging things so that, as far as possible, each generation faces the same range of opportunities with respect to natural resources."[17] He also believes that "as far as natural resources are concerned, depletion should be compensated for in the sense that later generations should be left no worse off (in terms of productive capacity) than they would have been without the depletion." At first glance, it may seem as if Barry's statement of what is required for intergenerational justice is somewhat stronger than Solow's. To say that future generations should be left "no worse off (in terms of productive capacity) than they would have been without the depletion" seems to open the door to the specification of particular types of depletions that would be unacceptable. But the addition of the parenthetical "productive capacity" is key; as it turns out, Barry defines "productive capacity," as does Solow, in terms of accumulated capital; and although Barry's use of the substitutability assumption is slightly qualified, he seems to endorse a position almost identical to that of Solow: "Within limits, which over a long time period may be very wide, it is always possible to substitute capital [such as new technologies] for natural resources."[18] Barry places less weight on the ignorance argument than does Solow, but he nevertheless asserts that "uncertainty" is an unavoidable problem, "because, in deciding what technologies we ought to develop to compensate future generations for the depletion of resources, we must somehow deal with the fact that the risks and benefits are, to some degree, speculative."[19] Thus, with the substitution of Barry's "productive capacity" for Solow's idea of accumulated capital, we find the philosopher Barry expressing an idea that seems essentially equivalent to the GS.[20]

It appears that leading economists and philosophers are agreed that given our present ignorance of what future people will want or need, the best strategy available to determine our obligations to the future is to compare welfare opportunities across generations. Maintaining wealth and a "fair" saving rate—allowing liberal substitutions of technological and human-built capital for natural resources—fulfills all obligations to the future. By using accumulated capital as the measure of opportunity (because capital can be invested to increase productivity), a remarkable simplification is achieved. The goal of sustainability, taking this approach, can then be equated with a measurable feature of economic systems, the savings rate understood as the rate of capital accumulation. This strategy greatly simplifies the problem of intergenerational justice, avoiding controversies about the makeup of a fair bequest package, and it justifies concentrating our efforts on wealth production and economic efficiency.

8.4 Grandly Oversimplified?

Given the respect due Dr. Solow, the chorus of testimonials and supporting arguments from philosophers, and the neatness of the conceptual package offered by the economists, it may seem like futile troublemaking to dissent. Perhaps we should end the story here and turn the problem of sustainability over to the economists who study growth and savings rates, treating the problem of living sustainably as a matter of picking good investments. If sustainability is nothing more than maintaining a positive rate of savings and making enough good investments, we can pigeonhole sustainability as a subfield of economic growth theory.

I'm afraid, however, I cannot accept this account of intergenerational values. Although the theory exhibits the apparent advantage of simplicity, simplicity, by itself, is not the goal of analysis. The theory should be as simple as it can be, *given that it is adequate to the task at hand.* As Einstein once said, "Make everything as simple as possible, but not simpler." And the problem of intergenerational equity is more complex than Solow—or any of the ethicists who agree with him—have understood.

Defending his simplification, Solow could argue that there would be little or no difference between economists' and environmentalists' choices when it comes to choosing what to invest in, what to guard carefully against, and what to save. Weak sustainability implies a clear injunction against impoverishing the future, so it also implies that we should make good and efficient investments that are likely to pay off in the long run. If we can assume that people in the present will take future values into account in making investment decisions, then creating efficient markets—internalizing the costs of pollution and degradation of resources, and investing wisely—would be implied by the weak-sustainability commitment to fair savings and capital protection. If natural areas and other natural amenities are in danger of becoming scarce, economists would council that we invest in protection of those resources and that future generations might chide us if we fail to do so, since we will have missed out on some good investment opportunities. On these grounds, one might argue, economists would reach at least many of the same conclusions about what to avoid and what to save as environmentalists do, and economists would accomplish much the same goals as environmentalists by insisting on "good investments" in the resource area.

Nevertheless, I do not accept Solow's argument that the private pursuit of maximal profits will lead to policies that protect communal resources, for reasons summarized in section 7.1. There I cited Colin Clark's argument that maximizing profits on investments, in many situations, will not lead investors to protect a productive resource, because overexploiters who profit heavily in early years can invest wisely in other enterprises, and the exploiters will do better than more patient investors who attempt to limit use to maximal sustainable yields. The importance of this result by Clark cannot be overemphasized; it virtually ensures that recommendations made by advocates of the GS will diverge in many cases from the recommendations of environmentalists and other strong-sustainability theorists who identify some resources as requiring obligatory protection. The bequest packages that would be designed by most environmentalists would thus be sharply different from those that would apparently follow from Solow's recommendations, at least with respect to the degrees of protection offered specific productive systems and processes. Environmentalists would, following Clark, insist upon constraints on profit-seeking producers and perhaps also on consumers, in order to protect important productive resources. Free market behavior of investors and consumers cannot be expected, in general, to save the right things for the future.[21] Maintaining a fair savings *rate*—weak sustainability—may well be a *necessary* requirement of sustainable living. This criterion can be useful in reasoning about the necessary *size* of the investment devoted to the bequest.[22] Clark's result, however, provides good reason to be wary of leaving the choice of particular investments for the future to market forces and wise investments. Free actors, seeking to maximize present profits in an economy with many investment opportunities, may well choose to extinguish nonrenewable *sources* of resource flows, systematically reducing productive capacity in favor of maximizing profits for reinvestment elsewhere. Weak sustainability thus diverges from strong sustainability; only strong sustainability will protect stuff that is valued for noneconomic reasons.

Fortunately, there is another position available, a position that differs both from Solow's position and from that of Solow's straw man—the absurdly strong sustainability theorist, who hopes to save everything exactly as it is for the future. Solow dismisses an obligation to save *everything* and moves directly to the claim that there is *nothing* in particular that we owe to the future. He ignores a plausible alternative: what about "less-than-absurdly-strong-sustainability"—the apparently sensible view that there are *some things* about nature that we may change at our will and other things that we ought not to change? We can state it this way: There are some elements or processes in nature that are so important to the future that no generation is permitted to destroy them. And we can call this position strong ecological sustainability or, more simply, strong sustainability. If these essential resources are lost, future people will be worse off than they would have been had the items been protected—*even if they are more wealthy than their ancestors.* They will have suffered a noncompensable harm. This harm cannot, however, be assigned an economic value in terms of price. The lost resources are priceless. The nature of such noncompensable harm is further explored below in section 8.7.

We have already seen that even Solow's allies in favoring weak sustainability have been squeamish about some of the assumptions one must make to apply growth models to intergenerational comparisons; that was part of the internal critique. The external critique of Solow's normative position notes that it rests on three very questionable assumptions, in particular. First, as just noted, it assumes we know nothing about the preferences and values of future people. Second, it assumes that resources of all types are intersubstitutable. Finally, it assumes that present prices reflect future values—that future risks will be reflected in present markets.

The GS is an oversimplification of intergenerational moral obligations because the matter of what we owe the future is a cluster of interrelated problems, not a single, monolithic problem. For convenience, I have grouped these separable problems into four categories.

1. The problem of intergenerational trade-offs. How should an earlier generation balance concern for future generations against its own moral and prudential concerns?
2. The distance problem. How far into the future do our moral obligations extend?
3. The ignorance problem. Who will future people be and how can we identify them? How can we know what they will want or need, what rights they will insist upon, and what they will blame us for?
4. The typology-of-effects problem. How can we determine which of our actions truly have moral implications for the future?

I begin by briefly discussing these four aspects of intergenerational ethics.

1. The trade-off problem. In most discussions of intergenerational fairness—which, as noted, are carried out against a backdrop of utilitarian and often Economistic assumptions—the major focus is on the trade-off problem, the problem of how we should weigh the demands of the future against the undeniable and more palpable needs of present people, many of whom live in abject poverty. The trade-off problem appears to be the most important aspect of intergenerational ethics: how much sacrifice on the part of the present can be justified or required on the basis of obligations to the future? Eventually we must address this question in one way or another.

2. The distance problem. There is also an important philosophical puzzle about the "horizon" of ethical concern. One might, on the one hand, insist that our obligations are only to the next generation. According to this view, which we can call *presentist,* each generation should accept some considerable responsibility for the impacts of their actions on their children and their children's cohorts, because these persons are identifiable moral patients, known to us, in some cases loved by us, and worthy of our full moral concern and commitment.

A *one-generational presentist* might argue that all intertemporal obligations should be handled as simple bequests from one generation to the next: we accept responsibility for our children, both singly and as members of a cohort of persons we care about, and we leave the caring for the subsequent generation to our children and their cohorts, who will in turn pass responsibility to their children. In this view, although our children's children are identifiable persons whom we may care for and love, we can simplify the problem of intergenerational bequests by concentrating on obligations that reach only one generation into the future. A less presentist position—*two-generational presentism*—would argue that our grandchildren and their cohorts, as identifiable individuals with whom we have concrete moral relations, should also be moral patients for us, but that our obligations do not extend beyond those two generations, people whom we can know and love.

In contrast to presentism, one can imagine an opposed position, which we might call *futurist*. The futurist, impressed with the fact that there will probably be many people who exist in the future and that these persons will, like us, no doubt require inputs of natural resources if they are to achieve a reasonable level of welfare, would attempt to balance present entitlements to use resources against the demands of an ever-expanding class of future claimants. Especially with regard to nonrenewable resources such as stocks of fossilized energy, strong concern for the entitlements of the future—because there are so many potential claimants on this finite resource—might lead to paralysis or to severe limits on resource use today. At any rate, it seems essential to provide an answer to the distance problem if we are to know how to address the trade-off question. The future extent of our moral obligations delimits the set of individuals whose interests are to be considered in making any trade-off. So one cluster of interesting normative and conceptual problems surrounds the question of the appropriate temporal horizon for our concern.

3. The ignorance problem. The ignorance problem includes at least two variants, both depending on the difficulty of foreseeing the future. One variant—the ignorance-of-values variant, which will be very important in our discussions here—refers to the difficulty of knowing today what people in the future will want or need. This variant, which may be formulated somewhat differently depending on which moral theory one assumes, raises questions of whether and how we can know what future individuals will *want or desire* (if one is a preference-utilitarian), what future people will *truly need* (if one's moral theory emphasizes a distinction between simple preferences and basic needs), or what the future has a *right* or *entitlement* to (in the vernacular of rights theory). There is also the related problem that technological change can render many resources unnecessary, while creating a demand for others. Another variant of the ignorance problem, which will not be discussed in this book, has to do with the difficulty of knowing in the present which individuals, and how many of them, will be born in the future. This problem has to do

not just with ignorance about the future, but also with special problems of individual identity that occur in nonpresentist positions. If I accept obligations to people who will live several generations hence, I seem to face a sort of paradox: how can I determine my obligations regarding future people, and formulate actions and policies to fulfill these obligations in reference to their welfare, when my decisions may determine which people are born and even how many of them can survive?[23]

4. The typology-of-effects problem. If I cut down a mature tree and plant a seedling of the same species, it seems unlikely that I have significantly harmed people of the future, though there will be a period of "recovery." As long as there are many trees left for others, our seedling will replace our consumption, and no real harm is done. But if I clear-cut a whole watershed, thereby setting in motion severe and irreversible erosion, siltation of streams, and so forth, it is arguable that I have significantly harmed future people, having irreversibly limited the resources available to them, and that I have restricted their options for pursuing their own well-being. It is admittedly difficult to provide a general definition for characterizing cases of these two types and to offer a theoretically justifiable practical criterion for separating them.[24] A key impact of the GS and the ignorance argument that supports it is to obliterate this distinction and thereby to abandon the reasonable goal of separating culpable from nonculpable actions that affect the future.

We are now in a position to look in more detail at the results of Solow's blunt use of the ignorance argument to accomplish the Grand Simplification. What does it cause us to miss? The Grand Simplification (1) effectively foreshortens our obligations to only one generation, the next, imposing an apparently arbitrary cutoff in our obligations to the future;[25] (2) assumes the fungibility of resources across uses and across time, denying even the possibility of shortages or unfulfilled demands for natural resources; (3) apparently rules out ex cathedra the possibility that some courses of action we might choose will be economically efficient, and remain that way, but still impose uncompensated harm on future people; and (4) strongly encourages the hope that sustainability can be reduced to measuring economic activity and savings, thereby avoiding the need to specify stuff instead of welfare.

The Grand Simplification is so grand because it resolves the seemingly perplexing distance question, erasing any possible specific concerns for distant generations, even as it sidesteps the typology-of-effects problem by assuming the fungibility of resources. All we need do is to avoid impoverishing the future by overspending and undersaving, which can be achieved simply by maintaining a fair savings rate. The Grand Simplification therefore simplifies the question of intergenerational obligations to one of maintaining a nondeclining stock of general capital (which is taken to ensure nondeclining real income), understood as indicating nondeclining per capita individual welfare, avoiding the need to go beyond welfare measures. This simplified reasoning allows a direct comparison of welfare opportunities available to members of successive generations and allows advocates of this approach to prefer "unstructured bequest packages" for future generations.

What is remarkable is that this simplification is achieved on no better foundation than a simple declaration of an implausibly strong, even extreme, statement of the ignorance problem,[26] coupled with the unargued assertion that economists know that resources "can take the place of each other"—the fungibility assumption. This set of assumptions and beliefs, so presented, should not be considered an empirical theory, but rather a proposed conceptual—eventually, perhaps, an operational—model for judging the sustainability of proposed policies and activities. Solow's approach to intergenerational equity must therefore be examined not just for the verifiability of assertions made within the theory, but also with respect to the appropriateness of its assumptions and conceptual commitments to the task of understanding what we owe the future. Operationally, these assumptions in effect ensure the intersubstitutability of resources, committing one to the idea that for any resource that may become scarce, there is some substitute that will stand in as an acceptable replacement for the loss. What is equally remarkable is how few voices have been raised either against this particular weak-sustainability account or against the basic assumptions that are made in order to achieve such generality and simplicity of argumentation on this very complex topic.[27] It is time to confront the GS and its foundation, the ignorance argument, head-on.

For all its grandness and regardless of its acceptability to thinkers as diverse as Solow, Passmore, Rawls, and Barry, the Grand Simplification has serious problems both in theory and in practice. Let us see what happens when practical implications of the Grand Simplification are compared with ordinary intuitions as they emerge in concrete, possible situations. When we apply the simplified reasoning of the Grand Simplification to real cases of environmental decisions with long-term consequences, we find that it supports highly questionable moral outcomes and contradicts reasonable moral intuitions. I will sketch four classes of cases that apparently call the Grand Simplification into question and suggest a need for a more complex analysis of our obligations to the future.

1. The toxic time-bomb case. Imagine a proposition that the most dangerous of toxic wastes from technological processes of production be stored in a new type of container. Manufacturers guarantee that these containers will reliably isolate toxic substances for at least 150 years, eliminating any chance of exposure of people of the present generation and the next one. It is also known

that these containers are likely to explode unpredictably, spreading their contents throughout the environment at some unknown time after 150 years. Suppose that policymakers nevertheless pursue this policy, because it offers an affordable solution to the hitherto insoluble problem of disposing of toxic wastes. If anyone objects to the effects of this wanton proposal, apparently Solow and other advocates of the Grand Simplification will reassure them by pointing out that we know nothing about what future people will want. People of the future may prefer toxic wastes to clean water for bathing, so we may be doing them a favor by providing them a free chemical bath unexpectedly![28]

2. Greenhouse warming and gradual climate change. Many economists who have examined the costs and benefits of programs to limit the emission of greenhouse gases have argued that economic impacts on industrialized economies will not be severe and that the climate changes likely to occur in the first generation or two (after anthropogenic changes in climate are demonstrated unquestionably) will have neutral or perhaps positive impacts on economic growth, as lengthening growing seasons boost agricultural production, for example.[29] Since current policy recommendations by economic modelers factor in a discount rate, presentism is built into any approach that represents future values in terms of present dollars. Add the Grand Simplification to this line of reasoning, and it seems to follow that we would do no harm by setting in motion irreversible climate change, because anyone who might be harmed by this action is distant from us in time and beyond the purview of any specific ethical obligations. Persons within our ethical purview according to the presentist assumptions will be affected neutrally or positively, so we may even have a moral obligation to *ensure* global warming (even if it has disastrous consequences three generations hence).

The acceleration of greenhouse gases is similar to the toxic time-bomb case in providing an example of present activities that cause neutral or even beneficial changes for a short time, followed by a risk of irreversible and cumulatively disastrous outcomes beyond the horizon of one or two generations. Although there has been much uncertainty about what the impacts of anthropogenic increases in greenhouse gases will be, few discussants have taken the view that impacts on the third generation hence will be huge and disastrously negative but that such an outcome is morally irrelevant to our current choices. Again, it seems that many environmentalists and scientists believe there is a range of outcomes that future generations will surely wish to avoid if possible, and this commonsense position apparently implies that we need a typology of benign, neutral, and damaging effects if we are to address the complex policy cases faced today. Using the ignorance-of-preferences and the unlimited-substitutability assumptions to justify presentism runs roughshod over common sense and over important moral intuitions.

These two families of apparent counterexamples point to at least two serious weaknesses of the Grand Simplification. Our intuition that we would do something wrong by creating toxic time bombs suggests, first of all, that our intuitions strongly favor, at least in some cases, a broader viewpoint than narrow presentism. These cases also seem to call into question Solow's broad claim that we do not know what people of the future will want. It is simply ludicrous to suggest that we do not know whether people of the future will want to be doused unexpectedly in toxic chemicals. Indeed, to the extent that one has intuitions that these cases represent moral harm to the future, one's intuitions run counter to Solow's unqualified claims of ignorance. These intuitions seem, at least, to demand a more nuanced presentism. The following question is unavoidable: Given what we reasonably believe about future tastes, what effects of our activities can be predicted to be benign and which are likely to be harmful? Once we raise the question of what we do in fact know or justifiably believe, and which of our actions threaten reasonably expectable future values, the typology-of-effects question immediately returns to center stage. And now the Grand Simplification entirely unravels. We are back to trying to figure out what we owe and to how many generations, with a knowledge base containing some near certainties and a great deal of uncertainty.

Once we reject the extreme ignorance claim and recognize that we have a convincing basis for some expectations about what the people of the future will want, it is possible to reestablish a closer relationship between distance questions and typology-of-effects questions. It seems reasonable, once we have a typology of effects based on what we are sure will be harmful to future generations, to use shorter time scales for consideration of some risks and longer time scales for other issues (such as storage of nuclear wastes). And although the problem of knowledge and ignorance still drives any solution, a recognition of gradations in our knowledge and ignorance makes possible a more nuanced response to distance questions. In this book I have proposed multiscalar models that would facilitate the association of values with various time scales and encourage the identification of appropriate scales on which to address environmental problems. Hierarchy theory offers a tool for developing multiscalar models that help to associate social values with the physical processes necessary to protect those values. Once the question of what seems almost certain to harm people in the future is on the table, a hard-edged answer to the distance question is simply insensitive to key aspects of the choice problem. To answer it would require much more specific information than would be available in Solow's data on economic growth and savings rates. If we desert the Grand Simplification, however, specification of a typology of effects would be a major step toward defining a stronger sense of sustainability: it would open the way to specifying stuff that must be saved if we are to be fair to the future. Unfortunately, the prevalence of the Grand Simplification has virtually blocked—or totally confused—discussions of possible typologies of effects. So this will be an important task for chapter 9. But now we need to explore two more examples that illustrate another important divergence between environmentalists and strong-sustainability theorists, on the one hand, and Solow's GS on the other.

3. Old-growth forest and wilderness conversion. Suppose our generation systematically converts all old-growth forests and wilderness areas to productive uses such as farming and mining, producing wealth but making it impossible for future persons to experience unspoiled wilderness or other natural places. Since they are wealthy, they can perhaps provide for themselves Disney-type facsimiles of wilderness. As long as they have adequate income to be able to afford such substitutes, the economists tell us, they will have been adequately compensated for the unavailability of such places in reality.

4. Severe, but gradual, ecological declines. According to the Grand Simplification, there would apparently be no harm in degrading the Chesapeake Bay into an irretrievably polluted, anaerobic slime pond, provided we do so slowly enough that the most negative impacts will occur a couple of generations hence. In the meantime, people will be well compensated for the loss of the Chesapeake if an adequate portion of the profits from its degradation is invested so that the intervening generations retain an undiminished opportunity to achieve welfare comparable to ours.

All four cases suggest that we are not comfortable with Solow's unqualified claims of ignorance, nor with the inference he draws regarding our obligations to the future. But cases 3 and 4 also raise another set of puzzles having to do with what might be called intergenerational paternalism. Solow, at a different point in the essay quoted above, claims that not only do we not know what the future will want, but "to be honest, it is none of our business."[30] Solow's lack of interest in values of future people—in cases like 3 and 4—points up another divergence between his GS, on the one hand, and strong-sustainability theorists and environmentalists on the other.[31] If someone asserts, contrary to Solow, an obligation of members of the present generation to protect special places from severe degradation, the person is apparently making some assumptions about what will be valued in the future. She or he might, for example, be assuming that people in the future will greatly value the special places in question. But the environmental protectionists also believe that people in the future *should* value these special places. Imagine that our generation, through conscientious effort and some sacrifice, succeeds in protecting many of these special places; and further suppose that our children's generation continues the protection but that the next generation declares its preference for development everywhere and for degraded bays, and sets out to systematically destroy the natural legacy we have left them. If members of our generation committed to nature protection could somehow learn that their grandchildren or great-grandchildren would desecrate the heritage we so carefully preserved for them, nature protectionists would not accept this change in preferences as none of their business. On the contrary, their reaction would be to double and redouble efforts to educate today's population and to build lasting institutions that would perpetuate their deeply held values and ideals. Nature protectionists accept responsibility to protect places, and in doing so they also accept responsibility to avert, by any means at their disposal, the erosion of the social commitment to protection. The environmentalist accepts, as part of the obligation to save special places, an obligation to perpetuate the values and mind-set that find those special places worthy of respect and protection within our society and culture.[32] There is a paternalistic streak in protectionist thought. Protectionists hope to save the wonders of nature, but they also accept a responsibility to perpetuate, in their society and among their offspring, a love and respect for the natural places they have loved enough to protect.[33]

It was noted above that the natural protectionists would balk at Solow's unqualified claim of ignorance. But now it becomes clear that their disagreement with Solow may not be based mainly on a belief that they, as present protectors of future values, are infallible *predictors* of what people in the future will in fact prefer. It may be instead that present protectors accept responsibility for inculcating certain values and for ensuring that those values are perpetuated in future generations. This analysis of the protectionist mentality suggests—as might be argued on a number of other grounds as well—that wilderness areas and other natural wonders are not valued by preservationists simply as opportunities for preference-satisfaction and for welfare gain. To reduce the question of fairness across generations to comparisons of opportunities to consume across generations is to miss an essential part of the protectionists' program and commitment. Protectionists, in contrast to Solow's declaration, *do* care about the preferences of future generations, and their preservation efforts are a part of a larger project, that of shaping the values of the culture to include love and respect for natural things and to perpetuate these ideals for the future. Protectionists act to create a communal sense of caring, and they do so by creating and maintaining a community that expresses a deep and abiding value for nature, a community in a physical place, forming a kind of organic unity across generations.

Successful protection of wilderness and other special places such as the Chesapeake Bay requires not only protection of the physical aspects of the places but also the successful transmission of an attitude of love, respect, and caring for these places to persons of subsequent generations—including a sense of moral obligation to continue protectionist policies and ideas. Solow, who is committed to the individualistic, utilitarian view of mainstream welfare economics, sees the value of an object as identical to its ability to fulfill preferences that people, understood as individual consumers, actually have. Rejection of this purely individualistic value theory is at the very heart of the environmentalists' message. They not only work to protect natural areas and other resources in the present, but they also attempt to project their values into the future, as a reflection of their culture and its natural history.

If we are to understand the moral commitments of the nature protectionist, we must express them in a richer and more nuanced vocabulary than that of Solow's welfare economics or other comparisons of individual well-being. This vocabulary must be strong enough to express commitment of present people to value beyond economic and consumptive values; it must allow us to specify important stuff that must be passed on. It must also, however, project into the future the love of that stuff, because love and caring are, like saving the places themselves, viewed as essential aspects of an adequate bequest. Truly strong sustainability of the sort explored here asserts that if we destroy important aspects of natural systems, then we may harm the future, making its people worse off than they would have been, even if they are able to be as well off as we are in terms of comparative welfare.

Viewing value from within the economists' perspective of comparing welfare across generations, moral judgments about the preferences that future people express and act upon are simply irrelevant. But for the nature protectionist, it makes sense to say that people in the future who have lost all interest in nature are worse off in ways that have little to do with their ability to fulfill their actual preferences. Clearly, this additional claim by protectionists is controversial. Solow would no doubt argue that it is a meaningless question and that it is an advantage of his value calculus that questions such as this fall by the wayside. The fact remains, however, that nature protectionists accept a clear and at least initially intelligible commitment: to save some special stuff and to pass on to their successors the passion to continue the efforts. This commitment will seem odd to Solow, who—following the dogma of consumer sovereignty—thinks the values future people express is none of our business, because this commitment is not simply a matter of *fulfilling preferences,* but of *creating and shaping a community's values and preferences.*

8.5 Passmore and Shared Moral Communities

It was noted above that the philosopher John Passmore embraces a conclusion similar to Solow's Grand Simplification. Passmore states that "our obligations are to *immediate* posterity[;] we ought to try to improve the world so that we shall be able to hand it over to our immediate successors in a better condition, and that is all."[34] Passmore's argument, though sharing its conclusion with Solow's, differs in that he employs a lemma regarding membership in moral communities, so his argument provides a good point of departure as we search for a more communitarian conception of intergenerational obligations. Passmore, following the legal theorist Martin P. Golding, reasons that care for immediate posterity is based on "love" of a kind that justifies equal treatment with our contemporaries.[35] Passmore then asks, still following Golding, whether the present generation should sacrifice to protect resources for generations beyond its grandchildren. His answer is unyieldingly presentist: "When the posterity in question is remote, we can have no assurance whatsoever that they will form with us a single moral community, that what they take as good we should also take as good. . . . 'The man of the future may well be Programmed Man, fabricated to order, with his finger constantly on the Delgado button that stimulates the pleasure centres of the brain.' Towards such a being we should have no obligations."[36]

Although Passmore includes a reference to community membership, the ignorance premise again drives the argument toward presentism and very nonspecific obligations. For Passmore, obligations are owed to other members of our moral community. But future people may have very different values, values foreign to us, and we cannot know what their values will be. Therefore, he reasons, they cannot be members of our moral community and we cannot owe them anything. Starting from ignorance as a premise, he questions whether we share values and a moral community with people of the future, and he cancels all obligations to descendants in the "remote" future, settling for presentism.

On a second look at Passmore's argument, however, one cannot ignore an odd assumption or inference at work in it. By implication, he understands a "moral community" to be a group of people who share the same values and ideals. Suppose that the people who live two or three generations hence turn out in fact to be very traditional in their values and share most of their values and ideals with Passmore and the traditions he reveres. One would think, if such was to be the case, and if Passmore *could* somehow know this, that he would conclude that these people *would* form with him and his cohorts a moral community stretched across several generations. But in his argument Passmore seems uninterested in what values people of the future will actually hold or in puzzling about how to decide who, in the future, will be a member of our community. Instead, he cites one possible, worst-case scenario, reasons that we cannot be sure that future humans will prefer our ideals to self-stimulation, and banishes all remote generations from his moral community. This reasoning is remarkable: persons of the remote future are excluded from our moral community on mere suspicion of possible—and unknowable—infidelity to our values and ideals! What we have, in other words, is the ignorance argument woven right through an argument that seems to be about communities. The communities turn out to be mere collections of individuals, whose values are apparently entirely adventitious in the sense that no generation takes responsibility for shaping the values of its offspring or of the ongoing culture.

There is also an assumption, implicit in Passmore's unqualified ignorance claim, of impotence to affect the values and ideals of the future. It is not just that Passmore would drum future people out of the community because they *may* fail to embrace his values; it is also implied that the values they will in fact accept are totally contingent, unconnected to our activities or to the activities of institutions we create and nurture. So Passmore, no less than Solow, is missing a central, community-based aspect of the nature protectionists' moral commitment. Far from accepting our impotence to affect future values, the nature protectionist accepts responsibility both to protect special places and also to develop ideas, cultural ideals, and institutions—a community—that will, even in the more distant future, perpetuate the same ideas and ideals. The protectionist sets out to ensure, to the extent possible, that people of the future will share with us a love and caring respect for these same special places and for other places that become special for natural or cultural reasons, in their time. Nature protectionists, in short, see the protectionist effort as a process of community-building: they are not just saving stuff—special places, for example—they are also accepting responsibility for projecting the value they feel toward that stuff into the future. This added commitment includes both the development of institutions and the development of narratives and artistic traditions. At the heart of this value-articulation-and-transmission process is a particular dialectic between nature and culture, a dialectic that is unfolding in a given place.

Looked at from this perspective, Passmore's argument represents a self-fulfilling prophecy of disaster. If we believe we are impotent to affect the values of the future, and if we expect that people in the future will be Delgado-button freaks instead of nature lovers, then these beliefs would indeed argue against our devoting effort, especially sacrificial effort, to protecting special natural places for them. If we follow Passmore in assuming that the people of the future will be monsters or freaks, then it does indeed make sense to act on short-sighted, selfish motives, failing to protect special places that express our ideals. If we did follow Passmore, however, we would be sacrificing our identity as a culture. If we pay no heed to the legacy we leave, then we may in fact ensure that future persons are Delgado persons who share no moral ideals or sense of community with us. Acting on Passmore's pessimism is the most likely way to deprive members of future generations of their link, through us, to a rich, nature-supported and nature-loving tradition in our culture. If we, at Passmore's urging, break that link, we may by our very action doom them to a meaningless life at the Delgado button.

Nature protectionists should not, given their ideology, accept Passmore's passivity and pessimism. Environmentalists since Thoreau—including especially John Muir and Aldo Leopold, both of whom elaborated Thoreau's idea in detail—have believed that experience of nature affects moral commitments, builds moral character, and is increasingly important for the self-perception and moral development of modern, industrialized humans. Nature protectionists, in short, reject Passmore's pessimism, replacing it with a form of benign paternalism about values. They believe instead that if the present generation is diligent in protecting special places, in shaping the ideals and values of the future, and in building effective institutions for the protection of their place, then the people of the future *will* favor nature and special places over pointless self-stimulation and that these future people will also inculcate their values in *their* own offspring. Given this set of beliefs, Passmore's argument is simply irrelevant, because these future people would, as a result of their ancestors' efforts and their own receptivity, form a moral community with them. When nature protectionists build private institutions and land trusts, and when they lobby politicians to pass protectionist legislation, they are expressing and perpetuating their values; they are also doing their best to avert the intellectual and moral disaster that Passmore passively accepts. If our generation and successive generations act on these beliefs, it is reasonable to hope that humans of the future *will* share a community with us and that the special places that are preserved may remain for them shrines to cultural, intellectual, and moral ideals that unify and give meaning to our culture.

8.6 What We Owe the Future

We are now in a position to return to the trade-off question that, as noted in section 8.4, has been at the center of the public debate over how to define and measure sustainability. I have shown that the very formulation of this problem in terms of intergenerational welfare comparisons presupposes implausible answers to three less synoptic intergenerational moral problems and that, since these less synoptic problems set the boundary conditions for determining what would represent fair trade-offs across generations, the attempt to characterize sustainability in terms of utility comparisons necessarily begs important moral questions. We have not, however, directly answered Solow's key question, How can we decide what to save for the future if we do not know what people in the future will want or need? It was this question, of course, that drove his reductionistic argument that we need only maintain a constant accumulation of general capital, thus ensuring an opportunity to enjoy a nondeclining level of welfare. Addressing this question will point us toward a more positive view of intergenerational morality. Solow, who never questioned that our obligations can be understood in terms of welfare comparisons, found our inability to predict those preferences devastating to any project that would specify what a given generation should save for subsequent ones. Our ignorance of future people's preferences—and our indifference to what those preferences might be—forced him to fall back upon saving generic wealth as the best we can do to ensure that future people will be able to enjoy their preferences, whatever they turn out to be. Note, however, that the centrality of ignorance and our inability to predict future preferences is parasitic on the more general commitment to treating fairness to the future as a comparison of opportunities to consume or fulfill preferences.

However, if we clear the intellectual decks and reject Solow's formulation of the problem as well as his solution, the problem of ignorance loses its centrality and devastating impact on concern for the future. The collapse of sustainability into weak sustainability on the basis of ignorance is therefore preordained by the theoretical scaffolding chosen by economists to express the trade-off problem as a problem of intergenerational comparisons of welfare. By insisting that intergenerational moral obligations be measured in terms of comparisons of aggregated welfare, utilitarians have formulated the problem of intergenerational fairness so as to require information that cannot be available at the time crucial decisions must be made. The reduction of sustainability to weak sustainability—and the attendant relegation of future obligations to maintaining a fair savings rate—is simply a figment of the assumptions introduced in order to characterize the moral problem in utilitarian and economistic terms. With that negative conclusion established, it is time to address in more positive terms the problem of what we owe the future. We can now see that the whole problem requires a new formulation. Ignorance and uncertainty about future preferences are at the heart of Solow's arguments for weak sustainability: strong sustainability is impossible, he argues, because we would have to know the preferences of future people to know what stuff to save if we want to positively impact their welfare. If, however, we are counting something other than welfare, we bypass the need to know what people want. At its deepest level, the Grand Simplification rests not upon the *fact* of our ignorance about future values, but rather on a deep and unquestioned *commitment* to reduce all moral questions to descriptive questions, to questions that can be resolved on an empirical basis. This commitment puts extraordinary weight on our ability to *predict* future preferences. If we can be fair to the future only if we can predict future people's needs in detail, then there will always be an impossible task at the heart of all specific (strong) sustainability requirements. The impossibility of the task, however, is an artifact of the assumptions introduced in order to formulate the problem of intergenerational fairness as a matter of welfare comparison. It is Solow's problem—and the problem of other Economists and welfare-counters—but it need not be *our* problem, once we abandon the idea that sustainability is most basically a matter of protecting welfare opportunities.

There is a name for the mistake committed by Economists and many other utilitarians: it has been called the "descriptivist fallacy" by J. L. Austin, who was a leader of the Oxford school of "ordinary language philosophers." In his *How to Do Things with Words,* Austin argues that many of our sentences that look like statements have purposes other than to describe. As examples and illustrations, Austin mentions "I do" (when uttered in the context of a marriage ceremony), "I name this ship the *Queen Elizabeth*" (while striking the ship's bow with a bottle of champagne), and "I bequeath this watch to my brother" (in the context of a will). He says that in these examples, to utter the sentence in question (in the appropriate circumstances) is not to *describe* the doing of something, but rather, to *do* it. Austin proposes that we characterize such uses of language as "performatives," and he says they can come in many types, including "contractual" and "declaratory" and others.[37] According to Austin, "a great many of the acts which fall within the province of Ethics . . . have the general character, in whole or in part, of conventional or ritual acts."[38]

Reflecting on the views of environmental protectionists—those who try to save special places—we found that they embrace a *commitment,* not only to save these places but also to create and sustain the institutions and traditions necessary to carry on the commitment indefinitely. These acts include the creation of a place-based literature and narratives, as well as public and private trusts set up to secure, for example, habitat for indigenous species. All of these signal commitments to continuity between the past and the future; they are best understood in Austin's sense as "performatives." They are founded on the commitments that a community makes to continuity with its past, to its natural and cultural histories, and to a future in which its roots in nature are revered, protected, learned from, and cared for.

Applying Austin's idea, and based on the analysis of this chapter, a new way of thinking about intergenerational morality emerges. If we see the problem of intergenerational morality as one of a community making choices and articulating moral commitments, as expressing their values in the footprint they leave on their place, then the problem of ignorance about the future becomes less obtrusive. The question at issue is a question about the present; it is a question of whether the community will, or will not, take responsibility for the long-term impacts of its actions and whether community members have the collective moral will to create, through a *community performative act,* an identity reflective of the nature-culture dialectic as it emerges in "their" place. Will they rationally choose and implement a bequest package—a trust or legacy—that they will pass on to future generations? If they so choose, we ask, By what process might a community commit itself to ideals of sustainability?

If one wishes to study such questions empirically, there is important information available. One might, for example, study how communities engaged in participative ecosystem management processes achieve, or fail to achieve, consensus on environmental goals and policies. Although empirical studies such as these may contribute to the process of community-based environmental management, the foundations of a stronger sustainability commitment lie more in substantive articulations of moral commitments by the community to the past and to the future than in any *description* of welfare outcomes.

The shift from measuring welfare trade-offs to a search for responsibility and commitment makes a big difference in the way information is used in defining sustainability, and it alters our understanding of environmental values and valuation. If the argument of this chapter is correct, then the problem of how to measure sustainability, though important, is logically subsequent to the prior question of what commitments the relevant community is willing to make to protect a natural and cultural legacy. If that legacy includes some stuff the perpetuation of which cannot be related sensibly to economic measures, then so much the worse for economic measures. They will be inadequate as measures of a legacy so formed.

Here we face the challenging prior task of developing community processes by which communities can democratically, through the voices of their members, explore their common values and their differences and choose which stuff will be saved—achieving as much consensus as possible and continuing to debate about differences. These commitments, made by an earlier generation, represent the voluntary, morally motivated contribution of the earlier generation to the ongoing community. They should reflect the love and respect for stuff that is expressive of the communal values and aspirations that form a community in a place. Choosing measurable indicators is logically subsequent to commitment to moral goals, but the tasks of choosing measurable indicators can, and must, proceed simultaneously with the articulation of long-term environmental goals. It cannot be otherwise because the choices that are made by real communities regarding which indicators are relevant to their moral commitments represent, in effect, an operationalization of moral commitments. The task of choosing community values similarly cannot be sharply separated from the specification of certain indicators to track the extent to which actual choices and practices achieve those commitments. The specification of a legacy, or bequest, for the future must ultimately be a political question, to be determined in political arenas. The best way to achieve consensus in such arenas is to involve real communities in an articulation of values, in a search for common management goals, and to include in that process a publicly accountable search for accurate indicators—measures of the stuff the community cares about—to correspond to proposed management goals.

The advantages of the shift in perspective on the trade-off question are now evident: this approach suggests that the key terms *sustainable* and *sustainable development* are not themselves general *descriptors* of states of societies or cultures but rather refer to many specific sets of commitments on the part of specific societies, communities, and cultures to perpetuate place-based values and to project them into the future. This projection must include a strong sense of community and a respect for the natural history of the place in question. The problem of how to measure success and failure in attempts at living sustainably is now the problem, for each community, of choosing a fair natural legacy for the future and then operationalizing its commitments as concrete goals to be measured by democratically agreed-upon indicators. The problem of trade-offs is still a key issue, but it is more manageable because it is no longer dominated by the constraints imposed by our ignorance about future preferences. The trade-off problem no longer appears as a problem of comparing aggregated welfare at different times, but as a problem of allocating resources to various, sometimes competing, social goals.

It is undeniable—as the economists will be quick to point out—that ultimately people in the present must balance their concern and investments in the future against real needs today. Setting aside special places, or protecting traditional relationships between cultures and their natural settings, may compete with other values. The question is transformed, however. If we see sustainability as a commitment of people in the present to perpetuate certain of their values, the fact of our (partial) ignorance of future people's desires and needs—though a limitation in some ways—is not really relevant to the protectionists' case. The case the protectionist must make is that to the extent the community has committed itself to certain values and associated management goals, these goals are deserving of social resources and "investments" in the future. Certainly, the task for the protectionist is a daunting one, given the competing demands upon society's limited resources. To the extent that a community and its members see the creation of a legacy for the future as a contribution to their culture and its natural context, and to the extent that they accept responsibility for their legacy to the future, they have embraced a commitment that gives their lives meaning and continuity that cannot be found in an accounting of preferences that people happen to have. The meaning and continuity are tied, rather, to the communal values that constitute a place and create an identity for a people in their place. These meanings and commitments deeply affect community members' sense of self and community and are woven into the intergenerational fabric of the community. For more discussion of such commitments, see sections 9.4 and 12.4.

One aspect of the daunting task of protectionists, of course, remains the unavoidable present uncertainty about what will be important to the future. But that should not keep us from acting in cases where uncertainty is low (as in our prediction that people in the future will prefer that we not store toxic waste in exploding containers), nor does it keep protectionists from advocating more positive goals and values that they wish and intend to perpetuate. On the contrary, since we have shifted from seeing the trade-off as one of comparing welfare across generations to seeing it as a set of difficult choices about how to allocate social resources among competing social goals, short-term and long-term environmental management need not be paralyzed by lack of information about future preferences. Although this reformulation of the trade-off problem leaves many difficult conceptual issues unresolved, it opens up for discussion the task of choosing a reasonable distance, or scale, for the moral horizons applicable in various problems. Likewise, it raises the questions of how to deal with the inevitable uncertainty involved in saving things for a cohort of people who do not yet exist, and of how to decide what losses would be unacceptable. We may, for the present, be able only to muddle through with temporary solutions to these three boundary-setting problems of moral trade-offs across time, but it can only be for the better if these problems are discussed openly by philosophers, economists, policy analysts, and policymakers, as well as the public. Rejecting the Grand Simplification, in this sense, is only a prerequisite for a more rational formulation of the complicated and difficult nest of problems faced when a community attempts to conscientiously specify what obligations toward the future it accepts and to choose a fair trade-off between supporting the ideals its members would like to project into the future and the very real and present needs of cohorts in the present generation. We have arrived, then, at a better formulation of the question of trade-offs among present and future generations. It is *not* best formulated as a comparison of welfare across time—it is better formulated as a question of how to hold open options and opportunities for the future, as was illustrated in figure 3.1.

Environmentalists believe, we have seen, that it is possible to cause noncompensable harm to future people, even if they turn out to be richer than we are, economically. This harm, as was explained in section 7.1, is harm to communal values, values that cannot be broken down into bits of individual welfare. These communal values generate obligations not to destroy the natural and cultural history of a place where humans and nature have interacted to create an organic process that emerges over multiple generations. The communitarian, unlike the welfare economist, sees goods beyond individual ones. Once we have rejected the Economists' model of decision making as based solely on the aggregation of the individualistic values of *Homo economicus*, damage to the future can be understood as damage to communal values. If we destroy the historical connections between our culture and its basis in a physical place, if we fail to perpetuate a place-based and natural sense of ourselves, people in the future may be worse off than they would have been had we protected that heritage, *even if people of the future are, as individual consumers, more wealthy than we are.*

Nature protectionists and their program can thus best be understood within the broadly communitarian political ontology of the conservative philosopher Edmund Burke, who defined a society as "a partnership not only between those who are living, but between those who are living, those who are dead and those to be born."[39] What nature protectionists need to add to Burke's version of political community is a stronger sense of human territoriality and a more explicit recognition that both our past and our future are entwined with the broader community of living things, the living things and eco-physical systems that form the habitat and the context of multigenerational human communities. As citizens in Burkean communities, nature protectionists do not evaluate special places as possibilities for present or future consumption, but rather as shrines, as occasions for present and future people to recollect and stay in touch with their authentic natural and cultural history. This is a history of humans as evolved animals and also as beings who have evolved culturally within a particular natural setting. They are cultural beings who nevertheless cannot deny their wild origins.[40]

Here, I think, we have reached the nub of the matter: What holds all of the supporters of the GS together, despite their many differences in viewpoints and beliefs, is methodological individualism, the view that *the* good must be an aggregation of, or a function of, *individual* goods. Countering this, I have recommended that some noneconomic obligations to the future be considered communal goods.[41] We have these obligations because, as members of a community and a culture, we benefit from sacrifices and investments made by members of prior generations. These benefits include economic goods, but they are not reducible to such because they also include the political and cultural practices that give meaning and continuity to the culture. These practices and sensibilities form a kind of moral and cultural capital that Adam Smith, speaking of the "bonds of sympathy," considered an essential foundation of economic life.[42] They are bonds that cannot, themselves, be given an economic value. To try to do so is to commit an egregious category mistake.[43] Concerns for the future are not, then, best understood by mapping fluctuations in consumptive behavior, but rather by paying attention to the moral and cultural sentiments of persons and communities and by emphasizing the ways in which these sentiments form an essential part of a person's personal and community identity.

We can harm the future by failing to create and maintain a culture and a community respectful of its past, including both the human and the natural history of the common heritage. According to this line of reasoning, to reduce the question of fairness across generations to comparisons of opportunities to consume across generations is to miss an essential part of the protectionists' program and commitment. Protectionists, in contrast to Solow's declaration, *do* care what preferences people of the future generations will have and act upon. Today's protectionists' efforts are part of a larger, multigenerational and communal project, that of shaping the values of the culture to include love and respect for natural things and to perpetuate these ideals for the indefinite future. The preservationist acts to create a communal sense of caring and does so by creating or contributing to a community that expresses and perpetuates a deep and abiding value for nature.

A major target of the arguments of this entire book—against the position I have called Economism—is the view that all environmental values can be treated as economic, or "commodity" values. In chapter 7 I argued that a full accounting of environmental values must include communal values, values that accrue to the community as a whole rather than serving the interests of individuals. In this chapter that conclusion has been illustrated and expanded by showing that the difficult choices faced by any community that decides to live sustainably must include commitments based not on predictions about what individuals in the future will want, but rather on acts of self-definition by today's individuals who address the problem *as members of a community.* These choices will affect future individuals, to be sure, but they may not affect the levels of welfare or levels of preference-fulfillment of future persons. The choices may, I have argued, affect the range of options and opportunities available in the future and even the ways future people value these. Decisions as to what options need to be held open cannot be resolved by *predicting* what people in the future will want, because, to a considerable extent, what people of the future want will be shaped by the world—and the options—we leave open for them. Further, deciding *how* we should protect options that we decide are important—the ideas, values, and institutions we develop to support our value choices—will inevitably shape future decisions and choices. The present generation cannot, then, simply concentrate on leaving the means to fulfill the preferences people in the future will in fact have, because the decisions made today will help to determine those preferences. Living sustainably in the present requires choices about what to value; and these choices cannot be separated from the process of building a culture. Building a culture, though it will inevitably involve building an economy, cannot be reduced to economic terms; it will involve also creating a literature, arts, and ideas that instill current actions with meaning. To project those meanings into the future requires a commitment and also countless day-to-day acts that express and perpetuate those meanings. It is in this profound sense that sustainable living cannot be relegated to a matter of economic accounting; it is inevitably a process by which community values are articulated through the choice of what stuff to save.

8.7 The Logic of Intergenerational Obligation

Having rejected the Grand Simplification, and having argued that truly strong sustainability theories include obligations to save some kind(s) of stuff, we must now discuss the nature of the moral obligations that support this form of sustainability. Elaborating on Barry's stuff-versus-welfare dichotomy, I concluded that according to truly strong sustainability, there is some stuff the loss of which will make people of the future worse off than they would be if that stuff had been saved. And they can be worse off for our having lost those things *even if they are at least as well off economically as we are.* As noted in section 8.2, there are two alternative "logics" that fulfill these conditions; we can refer to these as deontological and hybrid approaches. We can briefly represent the deontological approach by citing Peter Brown, Edith Brown Weiss, Paul Wood, and one of Richard Howarth's papers.[44] Let us first examine the deontological approach, seeing what can be learned from it—but also noting its considerable problems—before concentrating on the hybrid approach and its advantages from the viewpoint of adaptive managers.

Peter Brown, in two important but underdiscussed books, has developed the idea of government policy as guided by the idea of a trust—not a trust fund to compensate for losses, as advocated by Pearce and Barbier, but an active trust, one that sets out to protect and preserve wildlife, biodiversity, and other broadly valued characteristics. Brown, following John Locke, argues that all peoples, including future people, have rights in three basic categories: "rights of bodily integrity, rights of moral, political and religious choice, and subsistence rights," and he treats protection of these rights as an obligation of present people and their governments.[45] Brown asserts that the "human institutions and natural processes necessary for satisfying the three basic rights . . . *should* be sustained . . . indefinitely into the future." Speaking of rights, Brown departs from Locke's theological grounding, arguing rather that rights represent "what one owes to another under role instability."[46] Brown discusses the duties humans have as individuals, and he posits a further obligation to create institutions to ensure that these primary obligations are fulfilled: "It is individuals who have the duties in the first place. It is the duty of government to see that these obligations are discharged by those within its borders toward others within its borders. It is the duty of other governments and global civil society to assist those governments that cannot, and to sanction those governments which do not, discharge these most fundamental obligations."[47]

I am very sympathetic to Brown's assertion of basic rights, including access to resources, but I do not find in his writing a convincing justification for these rights. And once he states that individuals and governments *must* fulfill these obligations, it seems as if either he must take refuge in a religious and authoritarian foundation for the rights and obligations or, as he suggests at some points, perhaps the rights approach can be justified as a useful expression of commitments that communities undertake because of the values they hold.[48] If this latter is what Brown intends, his methodological position is very similar to mine: he is advocating a theory of rights as providing a conceptual and linguistic basis for articulating community values. This form of the "trust" theory would, like my pragmatic theory, treat rights as a construction of communities as they form institutions and devise laws that express their values. Pragmatists cannot appeal to rights or obligations that are prior to the process of community government; rights and duties are ways of summarizing what actions legal institutions will sanction and which practices are permitted. In this sense, it seems Brown and I can agree that we need a new linguistic-conceptual framework for discussing and deliberating about what our obligations are. Brown's proposal is to include all environmental and resource values under a heading of individual rights. Rights extend to include access to what is needed to fulfill basic rights—and part of what is needed is for institutions to carry out their obligations to support those rights. I propose that we have a range of obligations based on a plurality of values, with sustainability being expressed in a variety of criteria of different types. If Brown's conceptualization proves useful in expressing values, we can incorporate it into our own pragmatic proposals, as long as it is recognized that rights and obligations are reflections of community practices, not ultimate reasons to justify actions.

Edith Brown Weiss, a legal scholar, argues for a trust conception of intergenerational obligations in a way that is parallel to Brown's, by referencing obligations that can be inferred from international law. She asserts that "each generation receives a natural and cultural legacy in trust from previous generations and holds it in trust for future generations. This relationship imposes on each generation certain 'planetary obligations' to conserve the natural and cultural resource base for future generations, and also gives each generation certain planetary rights as beneficiaries of the trust to benefit from the legacy of their ancestors." She cites international laws that recognize basic rights of all living persons to fair access to resources, and she extends these rights to future generations, specifying that these "planetary rights" are not owed to future individuals; rather, these are "generational rights, which can only be usefully conceived at a group level." Weiss explicitly rejects the idea of compensation funds, saying, "The compensatory approach is inadequate, both in deterring actions that harm the environment and in compensating fully for the injuries."[49] Instead, she roundly endorses the "stuff" approach, identifying a seven-category list of actions that "are likely to be viewed as infringing upon the basic right to live on a planet with as good environmental quality and natural and cultural resource diversity as previous generations had." Each of these actions causes damage to a resource or a process that earlier generations are obligated to protect, such harm including, for example, "destruction of tropical forests sufficient to affect significantly the overall diversity of species in the region" and "damage to soils so extensive as to render them incapable of supporting plant or animal life."[50]

Citing both Brown and Weiss, the economist Richard Howarth experimented with a deontological approach to defining sustainability, recognizing obligations functioning as constraints against destroying resources. Howarth suggests that equal access to resources of the earth can be understood as a property entitlement of future generations that results in strong deontological obligations toward the future. Again, there is the problem of grounding—Howarth cites Immanuel Kant, but many would question whether Kant's grounding of ethics in practical reason is convincing. Further, if one truly takes seriously the argument that fairness must be universalized across time, with each and every individual holding property rights, one might well end up in the position described above as futurist, a position that leaves the present paralyzed by the need to be fair to an indeterminate set of individuals in the future. Perhaps it is not surprising that, given these problems, Howarth has more recently, and more explicitly, advocated a hybrid position.[51]

These deontological approaches seem to me to clear the bar set by Barry's stuff criterion—here we have truly strong sustainability. These theories all recognize obligations of the present to protect various stuff—aspects and processes of natural systems—for the future. Intergenerational morality is thus understood in terms of obligations, and these obligations to the future cannot be traded off against monetary compensation. This interpretation gives clear meaning to the assertion that it is possible to harm the future, even if individuals in the future have the opportunity to achieve welfare equal to or greater than our own. We harm them if we destroy stuff that we should have saved; and this harm can be specified quite independently of any effects on the welfare of future individuals.

I have learned much from the deontological approach, and I share the normative sentiments of these authors, but I will now explain, by citing three considerations, why I think a hybrid approach is more appropriate as an element of an adaptive management process. First, assertions of rights immediately raise questions of legitimacy: on what are such rights-claims based? Although deontologists can offer a variety of answers to this question (rights are God-given, rights are implied by reason, rights are the common heritage of humankind), all such answers will add a burden of justification to our sustainability definition. If we adopt any of these explanations to establish the legitimacy and moral or legal efficacy of such rights and obligations, we ipso facto accept the need to provide justifications for some very controversial philosophical views such as theism, rationalistic ethics, and so forth.

Second, references to intergenerational trusts and property entitlements do not really provide any practical guidance regarding what needs to be saved. Assuming that we seek a form of sustainability that is weaker than absurdly strong sustainability and stronger than mere welfare comparisons, trust-fund and entitlement analogies get the logic of strong sustainability right, but they force the crucial question—what should we save?—and offer little guidance regarding how to answer it. Brown suggests that if we specify some basic human rights that *must* be granted, then we can specify what stuff must be saved by describing it as whatever is necessary for satisfying rights, but this puts tremendous pressure on the way rights are characterized.

Third, the assertion of rights that seem to be put forward as "principles" that are prior to experience is not in keeping with the pragmatic method I have placed at the heart of adaptive management. Remember from chapter 3 that pragmatists, though not opposed to speaking of rights and obligations, firmly reject any theological or metaphysical explanation of rights or associated duties. Rights are not principles that preexist social rules, supporting them; rights are simply the set of privileges that a society or legal system will support. This approach to rights will not yield strong obligations such as those the deontologist articulates, so advocates of deontological approaches face a dilemma—whether to appeal to theological or metaphysical principles or to leave their obligations without clear force and legitimacy.

A better approach for adaptive managers who favor truly strong sustainability is to follow the pragmatists and root their conservation and protectionist policies in community values—values that are expressed by a community as it lives within an ecological habitat. The key to an emergent expression of community values, I have argued throughout this book, is a two-phase system of action and reflection. Iterative phasing allows for social learning and adaptation through experimental action, and the possibility of such learning, in turn, supports the attention to democratic procedure that I am proposing. As this process goes forward, participants will propose commitments in the form of strong-sustainability criteria that are associated with ecological and physically measurable indicators. These indicators, as they emerge from an open and deliberative process, can be understood as specifying the stuff the community thinks is important enough to monitor and protect on the basis of a performative commitment. It may be helpful to note here that a performative commitment may not involve a single, conscious act of choosing to monitor a particular dynamic. Remember that the reflective phase need not, in reality, involve a separate "convention" or caucus in which indicators are chosen. They may emerge gradually in practice, and they may be measured and monitored before action is taken; it is possible, in other words, for the indicators and the criteria to evolve in practice, gradually, as an expression of what the community cares about.

Recognizing that the shape and contents of the bequest package we save for the future must be chosen on information that is available to us now, and given my argument that truly strong sustainability must protect some stuff, we can now see our way to a procedural specification of the stuff that should be saved. In this way the choice of community values becomes endogenous to the adaptive management process, and a hybrid system of valuation can emerge from such a process. A hybrid system—one that is pluralistic in the sense of counting multiple values, but also in the sense that the differing values may be formulated or measured by rules that have somewhat different logics—is one that makes use of both economic and moral criteria to judge sustainability. Endorsement of a hybrid system of course raises questions of what to do when multiple criteria give conflicting answers, and how to prioritize the various values included. Since adaptive management requires iterative problem solving, however, it provides participants and stakeholders an opportunity, during the reflective phase, to discuss and negotiate answers to exactly what is valued enough by the community to accept responsibility to manage it for the ages. In the reflective phase participants will propose and advocate various indicators that can serve as proxies for protecting important stuff. I have argued that deliberation and experimentation with such indicators, while the community engages in experimental action, can amount to an operationalization of what is valued and meaningful to the community.

So I am advocating a hybrid system of evaluation, which can include multiple indicators, even indicators of more than one logical strength. Indicators are to be chosen within an open, iterative, deliberative process; and, at least in ideal situations, it is not unreasonable to think that indicators that are chosen for monitoring and management should reflect a sort of cross section of the values of the community. Pluralism, if expressed within an ongoing, two-phase process, need not be chaotic. In the reflective phase, there can be discussion and negotiation about which indicators and criteria are appropriate and how they should be weighted. In different situations, different criteria may be emphasized. The evaluation of possible development paths, as a part of the two-phase iterative process, can cycle from trying out various indicators in the action phase to discussion, in the reflective phase, of which indicators to emphasize and give the most weight. In the reflective phase, a community contemplating its bequest chooses various indicators that reflect the community's social and natural values and develops, through deliberation and negotiation, a plan of action. The plan of action should aim at supporting the community's values by fulfilling criteria set with respect to indicators; an important aspect of reflection, then, is the community deliberation and debate about which criteria are to be emphasized and given most weight in the action plan.

Within this general framework, a hybrid system can mix differing criteria with somewhat different logics. For example, a simple hybrid system might include three classes of criteria. First, let us assume it has at least one criterion that tries to measure economic efficiency under various possible development paths. Criteria in this category operate according the logic of markets and measure the success of policies in delivering welfare to the population. Second, it may include some kind of equity criterion, based in an idea of fairness and designed to protect individuals from serious losses as a result of policies to make others better off. A third category of criteria, based on the several indicators that might emerge from a process of choosing the community's most cherished stuff, involves the acceptance of a responsibility for protecting that stuff and for transmitting to the future the love that supports acceptance of that responsibility.

These three types of criteria involve different considerations, rely on different types of information, and use that information according to different logical rules. Economic criteria—here understood not as a comprehensive measure of sustainability but as a partial measure of one aspect of sustainability concerns—can be thought of as expressing the community's concern about not impoverishing the future according to economic criteria. This requirement might incorporate the best thinking of the weak-sustainability theorists, but I am willing to leave the development and exact specification of such criteria, as expressions of various communities' commitments, to economists. I expect, of course, that to the extent possible they will use the usual analytic concepts and tools of their marginal analyses, such as cost-benefit analysis and discounting to reflect time preference, and that they will recommend actions based on aggregation of individual welfare.

I would also expect that most communities would include one or more criteria based on intragenerational equity and fairness in their evaluations of proposed policies and development paths. Many communities might wish to avoid having one individual or group experience huge welfare gains while others are assessed costs, or having the costs of environmental pollution fall unfairly on one racial or economic group. Application of fairness and equity criteria such as these, intragenerationally, is already enshrined in international law, as explained by Weiss. Given my understanding of rights as reflections of practices, I would expect most communities engaged in an iterative process of choosing management goals to develop and articulate rules protective of individual rights, including rights of access to resources. Since there are many treatments of this topic in the literature, I will not go into detail about how communities might specify rights to protect individuals from unreasonable policies;[52] our concern in this chapter is more with *inter*generational equity.

Finally, I would expect most communities—at least those whose process evolves toward truly *strong* sustainability and the specification of stuff that must be saved—to articulate, discuss, and refine a set of indicators that reflect important community values. After the indicators are determined, the discussion can continue as to how stringently the goals are to be set for the various indicators. Once set, these goals can guide the monitoring of how well the community is doing in maintaining its commitments. In the next chapter I will say some more about the content of these indicators and criteria and provide some examples that could be applied in the metropolitan Atlanta community, where I live. Here, however, my concern is with the logical status of these obligations, which are expressed as goals in maintaining a particular indicator or indicators at preferred levels, and which are based on performative commitments by which the community sets out to save some stuff—such as a bountiful bay or a beloved park—that community residents deem too important to their community and its place to lose. These three classes of criteria are summarized in figure 8.2.

I propose that these commitments and the action-guiding rules should reflect the logic of the decision rule referred to as the safe minimum standard of conservation (SMS). That decision rule has the general form "Save the resource, provided the costs of doing so are bearable." Applied to our case, the "resource" in question is either one item on the list of stuff or the sum total of the stuff that is listed in the bequest. A society acting on the SMS rule would save the designated stuff if the costs were bearable. Success in protecting stuff can then be understood as maintaining chosen indicators associated with the stuff in question at acceptable levels, provided the costs of doing so are bearable. The logic of these specific SMS rules is one of burden of proof: the despoiler of these resources must accept the heavy burden of proving that the costs of saving the stuff are unbearable by the society, which in effect negates the marginal analyses of cost and benefit counters. Saving stuff designated as important to a community's sense of self and history cannot be traded off

Type of Criteria	Type of Rule	Concern	Time Scale
Economic Criteria	weak sustainability	no decline in welfare (no limits on substitutability)	0 – 5 years
Fairness Criteria	intragenerational equity	fair access to resources	constant and timeless
	intergenerational equity	protect opportunities; develop fair institutions	rolling from generation to generation
Community Values	SMS save stuff if the costs are bearable	save stuff of great value to the community	indefinite

Figure 8.2 Three types of sustainability rules

against individual costs and benefits. Saving stuff is a community obligation and must be addressed by a community and its institutions; the discharge of this obligation will not be simply a matter of willingness to pay or of counting and comparing welfare. It is a commitment that trumps cost-and-benefit ratios unless the costs to the society become unbearable. Notice that this way of posing the question *assumes* the positive value of the resource and shifts the attention of economic analysis to costs. Economists' role should in this situation emphasize finding least-cost alternatives to fulfilling SMS obligations.

One advantage of the SMS rule is that it has at least a rough analogue in political discourse, especially in Australia and Europe, in the "precautionary principle" (PP). PP says, in situations of high risk and high uncertainty, always choose the lowest-risk option. I prefer SMS to PP because SMS gives the specific instruction to determine which risks should be avoided and when the risks can be acceptable: protect the resource "provided the costs are bearable." Although the specification of what costs are unbearable may be contentious, it is a debate about the right question.

It cannot be denied, however, that SMS is vague in both of its crucial terms, *resource* and *bearable costs*. We cannot say, "save everything!" and "restore everything!" because that policy would certainly entail unbearable cost, not to mention that taken and acted upon literally, this rule would freeze nature, interrupting its restless and creative processes through time. The first source of vagueness is addressed by the adaptive management process developed in this book: the resource that must be saved is the stuff that a community eventually decides must be saved, and that decision will be operationalized as a choice of a sustainability criterion stated in terms of maintaining set standards with respect to a measurable indicator. The second vagueness, the question of specifying unbearable costs, ultimately must have a political answer. In a democracy, the political will that is necessary to adopt these commitments must emerge as the legitimate expression of the community's will. Once it has done so, the community accepts the responsibility to save the stuff it loves, provided the costs are bearable—a determination that will be resolved politically.

I believe this logic is in keeping with the system of analysis developed here; such a burden of proof would be appropriate once the society has committed to saving some stuff for the future. Probably there are some costs—such as widespread famine—that might justify the reluctant destruction of some of the stuff in the bequest package, but the rule seems to support protection of stuff even if it requires significant sacrifice on the part of present people. Furthermore, this is what we would expect if it is judged that people in the future will, if this stuff is lost, be worse off than they would have been if the earlier generation had opted to save it.

I conclude that our hybrid system, and active participation by community members, provides a model for conceptualizing the choices facing a community that decides to create, thoughtfully and deliberatively, a differentiated bequest to future generations. I conclude, in other words, that—within the context of a well-functioning adaptive management process allowing social learning—a pluralistic and hybrid system of values can capture and express the core ideas of truly strong sustainability. We have available, then, a general process-model that a community committed to protecting its place might use to achieve strong-sustainability goals through a process of deliberation, discussion, and negotiation.

So far, so good. We now have characterized a process and a system of analysis that embeds the search for strong sustainability within adaptive management, and we have characterized the general class of sustainability options as embodying three types of rules, having somewhat different logics one from another. Concerns about impoverishing the future may be modeled using neoclassical economic models, as best one can, to estimate whether per capita welfare measures are flat or increasing across time; concerns about intragenerational equity effects of proposed policies will be measured using available or proposed measures of fairness across individuals; and the commitments deliberatively made by a community to value certain stuff as essential to the identity and future of the community and its individuals will be expressed as safe minimum standards, which require protection of the stuff so designated as long as the society's costs are bearable.

Critics, especially those who are enamored of various single-criteria and single-value systems of analysis such as Economism, may at this point object that the hybrid system is too complex and chaotic to work. My first answer to this criticism, originally addressed to Solow and his Grand Simplification, is that the system of analysis should be simple, but not too simple to address the question at hand. Having seen that the simplified system advocated by supporters of the GS is nowhere nearly as simple as it claims to be and that, despite this, it fails to express the real concerns of forward-looking environmentalists and conservationists, it seems reasonable to give another approach a chance, even if it seems dauntingly complex.

The instinctual abhorrence of monists for a hybrid system is based on the assumption that communities will express and advocate many values in a disorganized cacophony of multiple voices, that pluralism will mean conceptual and evaluative chaos. A pluralistic process may well begin in chaos. The chaos can be turned to advantage, however, if it encourages a creative process in search of indicators that will be widely supported in the community; it can be an idea generator for expressing values in more concrete and measurable goals.

We approach this chaotic expression of values with another significant advantage: hierarchical thinking, which is one of the key principles of adaptive management, provides a ready-at-hand tool for characterizing temporal and spatial scales, for beginning to address questions regarding the distance our obligations extend into the future, and for developing a typology of acceptable and unacceptable impacts we might have on the future. The adoption of hierarchical thinking allows us to go beyond simply describing a process heuristic by which a community can randomly consider any possible indicator associated with any possible stuff whatever as a candidate for protection. We can develop the heuristic to include, within the reflective phase, a hierarchical model that will help us to associate values—and environmental problems associated with the loss of those values—with the physical dynamics that create and control them. It becomes possible on the two-phase model with social learning to develop a scaled value system and to associate values with particular physical dynamics, so that we can sort values into categories according to their physical scale and their temporal horizons.

One reason pluralism sounds unacceptably chaotic to monists, whether Economists or IV theorists, is that they are used to facing every environmental problem as a zero-sum game. If either ideological faction wins, it is seen as a loss to the others, because they define their respective values in opposing terms. But pluralism, especially of the hybrid variety, encourages win-win situations and opens up the possibility of actions that may support several key values at once. Moreover, when environmental problems are modeled in a hierarchical system, and when important social values can be associated with a particular physical dynamic unfolding on a particular scale, it may be possible to tailor management plans to positively affect one dynamic while not harming another on a larger or smaller scale; and it may even be possible to identify policies that will positively affect more than one dynamic, making it possible to experiment toward win-win policies.

In effect, I am suggesting that defining stuff worth saving through a performative act will lead us to develop a typology of effects and the beginnings of a sense of responsibility for our impacts on all scales of the system, including changes to normally slow-changing systems that constitute the larger environment. Hierarchy theory provides a general guide to matching environmental problems with the scales and dynamics associated with them. The idea is that if we model natural systems as multiscalar—nested subsystems within larger, open systems, hierarchy theory provides a geography of spatiotemporal relationships in the system and gives general guidance regarding which criteria to emphasize in given situations. Larger systems change more slowly and constitute the ecological and evolutionary background within which organisms—including human organisms—act. This multiscaled aspect thus offers guidance as to how to organize pluralistic valuation into categories reflecting different scales of environmental concern.

By combining hierarchy theory with hybrid pluralism (multiple values expressed as criteria and multiple types of values with different logics), we can begin to see the advantages of pluralism in decision procedures. In the reflective phase of our two-phase process, we can discuss and deliberate about what we wish to do in the next action phase, but we can also deliberate about how to weight the various criteria and when to give priority to which rules in choosing and designing policies. These latter deliberations can be thought of as metadeliberations because they involve choosing indicators and criteria and making decisions regarding how to deploy the various rules of our pluralistic system. To the extent that we can bring order to the metadiscussion about what to count and which criteria to emphasize in various situations, hybrid pluralism can serve as a heuristic model by which a community can fill in the blanks in their sustainability definition. See figure 8.3.

In an effort to give some order to this metadiscussion, a few years ago I developed a little technique that I call risk decision squares. This technique can provide guidance in the metalevel task of choosing and weighting various of our multiple criteria by paying attention to the physical dynamics and the scale with which they are associated. First, represent these metalevel weighting decisions on a decision space defined by two parameters: the east-west dimension is defined by a continuum measured in years, and the north-south dimension represents spatial extent. Now imagine that we are considering a policy or an action that may carry some risk of harming people in the future, such as the decision to manufacture persistent poisons or to let some species become extinct. Imagine the worst outcome possible that could reasonably be feared from this policy or action. Now ask, If this worst outcome occurs, how many years will it take for the situation to be reversed? We can thus range various risky behaviors—including those that would harm people directly by exposure to poisons and those that would cause indirect harm by threatening the stuff the society has decided to protect—along a "reversibility" continuum, from impacts that can easily be reversed in a few years to impacts that could last for more than a human lifetime or even for periods exceeding the expected persistence of the human species (irreversible). Species extinctions are irreversible, that is, reversible only in infinite time. Next, consider the spatial impacts of the worst case. Will the impacts be limited to a small area? Or will the impacts affect huge areas, limiting the availability of good substitute sites. Will the impacts affect a whole ecosystem or region? Taking our cue from hierarchy theory, then, we can use this decision space, as represented in figure 8.4, to operationalize a partitioning of the space, guiding decision makers to emphasize differing criteria dependent upon the spatial and temporal

Reflective Phase:

Metadecision: A categorization of problems according to spatiotemporal scale of impact and social values threatened.

A. Classify the scale of potential impacts of a practice or policy
B. Judge the likelihood of impacts at levels that affect social values
C. Choose appropriate criterion from list below

Action Phase:

Several criteria are available to guide practice or policy in situations as classified in the reflective phase.

A. Benefit/Cost (BCA)	B. Safe Minimum Standard (SMS)	C. Precautionary Principle (PP)
A good policy is one that maximizes ratio of B/C.	Save the resource provided the social costs are bearable.	Take affordable steps to avoid catastrophe tomorrow.
Applies to decisions with small scale impacts and short time horizons.	Applies to decisions on ecosystem / landscape level with more than one lifetime reversal.	Applies to possibly catastrophic outcomes in the distant future.
Economic Sustainability	Strong Normative Sustainability	Sustainable Risks

Figure 8.3 The two-phase decision process in a pluralistic hybrid system

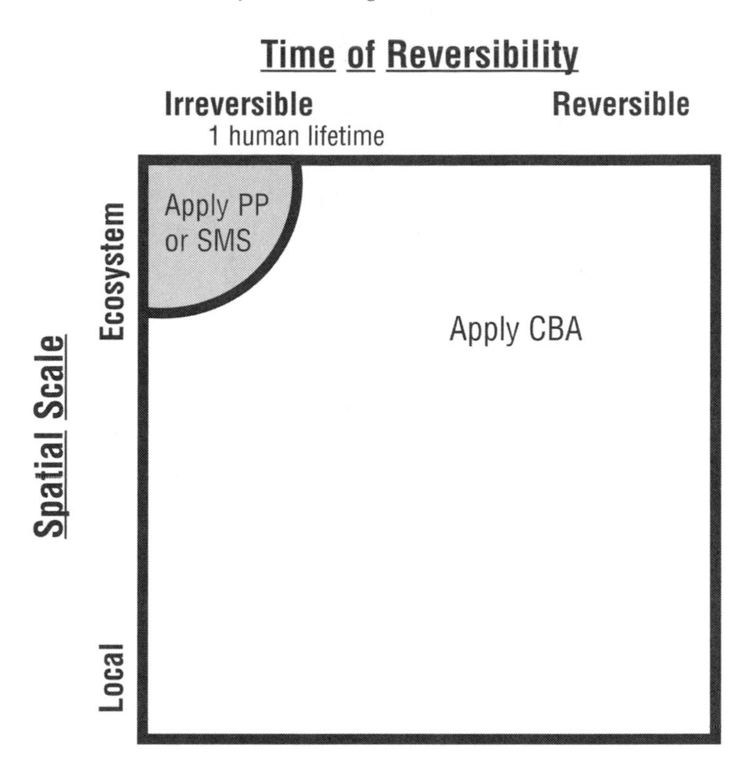

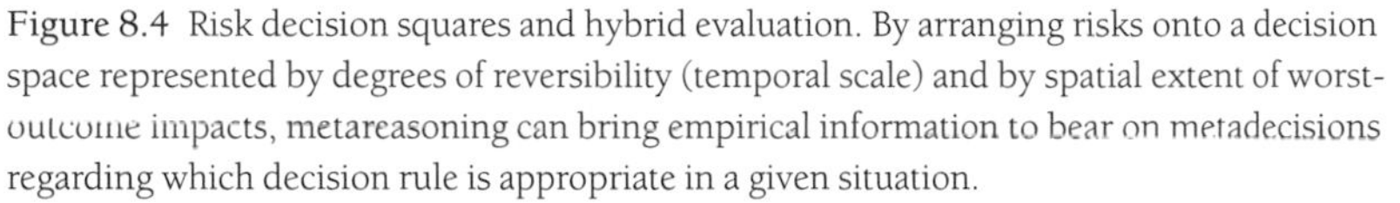
Figure 8.4 Risk decision squares and hybrid evaluation. By arranging risks onto a decision space represented by degrees of reversibility (temporal scale) and by spatial extent of worst-outcome impacts, metareasoning can bring empirical information to bear on metadecisions regarding which decision rule is appropriate in a given situation.

scale of the impacts risked. We can generalize that if outcomes risked by an action or a policy are easily reversed, we can decide about that action using a cost-benefit analysis; also, if the spatial impact of an action or policy is very small, this encourages us to use cost-benefit analysis because the small extent suggests that there will be substitute areas or resources. Decisions whether to take a particular action that carries risks in the northeastern quadrant, the southeastern quadrant, and the southwestern quadrant can generally be made using an analysis of costs and benefits. Decisions in the northwestern quadrant—what we can call the yellow zone—are decisions that may affect large areas such as whole ecosystems or may create losses to environmental values that cannot be easily replaced because of long reversibility times. In this area of the decision square, a prudent community will be careful, employing an SMS rule and protecting the stuff that is important to the community on a multigenerational time scale.

Risk decision squares thus help us to organize our thinking with respect to spatial scales and inform our choices about how to deploy our various rules and their various associated logics, in order to give considerable play to rules seeking to maintain economic efficiency while recognizing that empirical predictions about the size and likelihood of negative impacts can identify some decisions as more grave. These latter decisions might be based on the more restrictive decision rules of SMS or PP. The risk-decision-square technique can be useful to guide metadecisions regarding what criteria to invoke and how to weight them in various situations.[53]

I have tried to sketch a decision framework that can be thought of as a prototype of a two-phase, process-based system of management with a hybrid evaluation model. I am suggesting this as the starting point of a comprehensive evaluative and decision framework that might evolve as a community committed to cooperative action examines goals, engages in experiments, and learns through iteratively acting and reflecting. Since the hybrid evaluation model is built into the adaptive management process, decisions as to which of multiple indicators and associated criteria to apply, and how to weight them in various situations, are also part of the adaptive management decision process. These are metadecisions, decisions about which indicators to use to track protection of stuff deemed important in the society's bequest and also decisions about when to apply differing criteria in different situations.

A difficult aspect of this book is that although I have been able to say a lot about what a definition of *sustainability* will include and what it will lack, I have not been able to actually define the term. I have defended the view that environmental values are based in communities, and I have argued that the values that would guide a differentiated bequest package must emerge from an open process of involvement on a community-by-community basis. On these premises, there can be no single definition of sustainability because we have seen that defining sustainability is logically subsequent to deciding what is valuable enough to the community to save, even if it requires sacrifice. Communities, in choosing what they monitor and protect, express their environmental values; and the horizon of their concern for their place will be guided by the strength of their commitment to the future. Accordingly, I cannot offer complete and finished definitions. We have, however, learned a lot about the general shape of strong-sustainability definitions, and these insights will guide us to a schematic definition of multigenerational and multiscalar sustainability in section 9.1 and to a less abstract exploration of the processes by which communities identify indicators that express the values constitutive of their place.

I conclude this chapter on the nature of intergenerational obligations by noting that a definition of sustainability that is informed by multiple disciplines can be "strong" in a very precise sense and that any approach to sustainability that takes account of the full complexity of intergenerational morality will involve multiple criteria. We cannot understand what we owe the future by guessing what its people will want or by comparing their riches to ours; to understand what a community owes to the future, you have to know what that community would save—what stuff they would rally around and work to protect—if they engaged in a deliberative and iterative process that allowed experimentation with multiple indicators and multiple rules. This additional complexity can be mitigated by use of the heuristics developed throughout this book, especially that of the two-phase process and the hierarchical organization of space-time. There will predictably be conflict, once multiple indicators and criteria of sustainability have been proposed and advocated by different interests, but the conflict becomes more tractable when one sees the task as a second-order task of balancing competing interests in various situations. This shift puts ideology behind us, opens the way to dialogue about what is important to save, and can lead to negotiations—even to win-win situations that are made possible when citizens discover that some policies can protect multiple values. Coalitions can form as a result, creating trust and encouraging further cooperation. This little decision model, in my view, clears the way for a community to deliberate about what they really care about, to develop appropriate indicators, and to express their values in many vocabularies. The identification of important indicators and the development of goals associated with them, though not formalizable, can be made orderly by organizing plural and hybrid values by metalevel deliberation and decisions. In chapter 9, the last of this five-chapter part on values and valuation in relation to adaptive management, we will explore in more detail what a community might specify as the substance of a differentiated bequest package.

9

ENVIRONMENTAL VALUES AND COMMUNITY GOALS

9.1 A Schematic Definition of Sustainability

In the previous chapters I have tried to clearly articulate the communication problem that plagues environmental discourse and to survey some of the resources available—in economics, in ecology, in philosophy and ethics, in risk analysis and decision theory, and in philosophy of language and communication—for addressing these problems, especially as they regard environmental values and valuation. What is different about my attempt, as compared to most single-discipline attempts to define sustainability, is that I have begun and oriented the search from within many local, activist contexts in which communities have formed partnerships to manage their resources adaptively. Local orientation and emphasis on actual problems encourages practical and synthetic thinking. Problem orientation and practical attempts to be responsive to real problems causes us to look at the various disciplines, their models, and their theoretical terminology as tools to address the problems at hand. This led me to propose, in chapters 7 and 8, a bold new approach to evaluation in which social scientists, natural scientists, managers, and stakeholders participate in a process of articulating, clarifying, and eventually operationalizing the community's shared values as a system of multiple criteria. This formulation then encourages the asking of "metaquestions" about which criteria to stress in which situations, thus creating room for a reflective phase in which the community reconsiders and perhaps reformulates its goals and commitments. These questions, if embedded in an adaptive management process, can provide occasions for reflection, deliberation, and compromise regarding how well the community is living up to its commitments.

We have seen that scientists and philosophers, when isolated within their disciplinary walls, tend to assume their own core concepts—units of welfare, increments of risk, intrinsic values—and then attempt to expand the application of their core terms by reducing more and more phenomena to description in their terms. This is the disciplinary aspect of the broader problem we have called towering. What we have learned is that no value-analytic discipline—not economics, not environmental ethics, not decision science, and not risk assessment—is capable of fully capturing the social values that are involved in the most basic decisions affecting the environment. All of these disciplines provide some insights, suggest some models, and offer different ways to look at the environmental problems we face. But since environmental problems come to us in real-life, three-dimensional form, no one of these models provides the whole picture; it would usually be disastrous for environmental managers to adopt one of these models and make all decisions according to it. This would be equivalent to a pilot's deciding, for example, to steer by the altimeter only and turn off all the other instruments and gauges. The crucial difference between disciplinary sciences and the "science" of management is that in academic disciplines, the paradigm rules the actions of practitioners; judgments are made according to accepted procedures, and there are accepted norms for good experimentation and for supporting hypotheses. An activist science like environmental management—perhaps we should better call it an art? cannot let a single paradigm guide actions. The manager, who is reacting to problems as they arise, must rely on everyday observation, consultation, and common sense, although he or she may consult the more precise models of disciplinary scientists from time to time. For this reason, and because we want to develop an approach to defining and measuring sustainability that will be useful in real managerial situations, we seek a definition from the viewpoint of an environmental manager or activist. More specifically, we have examined both the concept of sustainability and the disciplinary tools available to describe and measure it, from the viewpoint of adaptive management.

Adaptive managers expect to act in uncertainty much of the time, so they adopt a scientific, experimental approach to management. When one looks at the environment from the holistic viewpoint of a manager who must digest the partial information, examine the models, decide on priorities, and act, one can see that all kinds of information will be welcome. What we have learned is that the many vocabularies and models of the special disciplines provide richness and variety of insight; what is lacking is a synthesizing discourse that would allow managers and the public to factor relevant scientific information into a larger, overarching discourse about social values, management goals, and priority actions. No single disciplinary discourse can fulfill this role. What is required is nothing less than a new, more holistic way of thinking about environmental problems and a new basis for evaluating environmental change, a basis that is useful as part of an ongoing process of public participation in environmental management. In this context—facing many problems and lacking an adequate synthetic vocabulary for relating these problems to each other, setting priorities, and so forth—we are exploring the possibility that the term *sustainability* might be clarified to function as a sort of keystone term in a new synthetic discourse about what to do to protect the environment.

In this chapter I try to summarize where our argument has carried us on our journey toward a constructive definition of sustainability and to assess how much further we have to go. We want to arrive at a definition that we can work with as an important assessment tool and which will also be useful as a tool of communication that can contribute to the further refinement of environmental goals and priorities. To start this process, I will provide what I call a schematic definition of sustainability. What I mean by the qualification *schematic* will be clear presently; the key point is that we are ready to develop a prototype definition of our target term. By producing and explaining such a proposed schematic definition, I hope to highlight the ways my arguments in earlier chapters shape such a definition and to provide an occasion for assessing remaining puzzles about what sustainability means.

The central focus of my argument, and the associated challenge of improved communication, has been to develop a new vocabulary for discussing environmental problems and goals, a way of talking that is useful to communities struggling to articulate a common heritage, a sense of who they are, and how they relate to the places that are important in their lives. We have learned to expect that in today's communities there will be great diversity of opinion; fortunately, adaptive management does not require resolution of all disagreements or consensus on all values to begin the process. The process, for any given community, must start where that community *is*, where it is located in space and where it stands in terms of the beliefs and attitudes that its citizens have. Movement toward sustainability from that place must begin with a commitment to work together to find cooperative solutions to shared problems. One lesson derived from our analysis thus far is that we must work from the bottom up to build new and more inclusive vocabularies that will further improve communication, especially communication around the practical nexus of what to do, given what we know and what we want as a community.

At the same time, this same central argument—that communities and community action must be built from the bottom up—makes it impossible to produce a single detailed definition of sustainability, applicable in all situations. This argument shows that many of the key details of a sustainable lifestyle will be idiosyncratic to a particular community located in a particular place with a distinctive ecology and will be intelligible only as a part of the history of *that* place, which we have interpreted as including both the landscape and the peoples who have lived there. So the details of a particular community's sustainability criterion will have to be filled in by the community itself, in the process of choosing indicators, goals, and priorities in an open, deliberative, and democratic process. No definition could include or dictate all of the locally driven choices.

What can be offered is a schematic definition, one that gives some general categories of values—variables to be specified in more detail by the community through its local processes—and thereby provides a sort of roadmap toward a definition of sustainable living for a community. These categories and their relations can also suggest to us the underlying logical structure of the sustainability idea and, more practically, might help communities to ask the right questions in their adaptive management processes. So it's time to have a shot at defining sustainability. Let us begin by summarizing what else—besides the local nature of the details of sustainability—we have learned about the term, its use in environmental discourse, and the context in which the term is used. These factors should shape our sustainability definition, and a review of them will provide an outline of a constructive, community-based definition of sustainability.

First, I have challenged at every frontier the insistence upon a sharp dichotomy between factual and evaluative discourse. We saw that slavish adherence to such a separation, whether in the form of an artificial distinction between risk assessment and risk management, as a "serial" view of science and policy, or as blind commitment to an ideal of value-neutral science, has wreaked havoc throughout public discourse about environmental policy, has paralyzed agencies, has discouraged social learning, and more. Accordingly, our definition of sustainability must include normative as well as descriptive elements in a complex mix. Understanding this mix is so important that I will devote a whole section (section 9.3) to the topic.

Second, although we have noted the flexibility and malleability of the term, we can state that sustainability stands for taking responsibility for the future impacts of today's activities, including impacts on natural, physical systems and also on the values and ideals of the community in the future. Since we have freed ourselves of monistic theories of environmental value, there is no need to represent present and future harms and benefits in commensurable and aggregable terms. In section 8.7 I proposed a hybrid system that combines the logic of the safe minimum standard of conservation rule (SMS) with hierarchy theory to create a linkage between important values and the physical dynamics upon which they depend. What we need is to provide guidance to communities that are trying to develop indicators and criteria into an operative definition of "sustainable for us." A schematic definition should serve as a guide for the creation of such definitions, and attendant evaluation methods, in various local situations.

Third, we hope that *sustainability* can express a broad commitment to future-oriented living, capable of including considerable diversity of goals and priorities but uniting a community by directing attention to longer frames of time. Leopold challenged caring communities to think like a mountain or like a river, rather than only at the pace of economic change; he wanted them to accept responsibility for their individual and collective impacts on all spatial and temporal scales. If *sustainability* is going to serve as such a general term, we will have to build many bridges between our definition and other vocabularies, including the vocabularies of important scientific disciplines that are essential to provide information for adaptive management processes; and connections will also have to be made to the values and aspirations of communities. Furthermore, these connections will have to be made transparently, in ordinary language, so that all stakeholders can participate in deliberation if they care to. Sustainability, if it is to be a unifying normative idea for environmentalism, must appeal to a diverse society in a language all members of the society can understand.

Fourth, we have embraced the pragmatist approach to epistemology, making experience the measure of epistemological acceptability and eliminating all claims to truths that are immune to revision by experience. Pragmatists understand truth claims—and sustainability claims—to be projections into the future; they are beliefs that will unlock future possibilities and can be validated by acting upon them and achieving results. Pragmatists are comfortable in situations where important questions are revisited repeatedly, where more evidence and experience is brought to light, and where all claims are open to refutation by future experience.

Fifth, although it can sometimes be very helpful to study present preferences, to determine how citizens value parts of their environment at a given time, elicitation of preferences for environmental goods viewed as commodities cannot capture all of the dimensions of value necessary to identify sustainable practices. In reality, people's preferences are highly dynamic, and we should think of the search for sustainable living as an educational process, a process of social learning that may lead to reductions in uncertainty but may also cause us to reconsider our values, our goals, and especially our preferences. Alternative environmental policies cannot be evaluated with only one-time elicitations of preferences; an open-ended, iterative process is required, in which new experience and public deliberation play a key role. Therefore I have recommended that we use multiple criteria to rank various "development paths" or "scenarios," rather than considering environmental elements as "commodities" that might be purchased, in order to encourage examination of more holistic aspects of management.

Sixth, by focusing on everyday communication—language used in everyday situations in which people are trying to act cooperatively—I have emphasized the multiple uses of language. Language is not used just to describe; it is also used to question, to deliberate, to persuade, to express emotions, to enlist allies, and to perform commitments. *Sustainability,* as a term for communication in public policy discourse, then, can be judged according to its usefulness in that broader context, as well as for the connections we forge between it and more specialized theoretical and scientific discourses. Our emphasis on communication, as we learned from Habermas and the discourse ethicists, allows us to build upon the presumed value of improved communication. Given that the community in question has agreed to attempt, through communication, to achieve negotiated settlements and cooperative actions, we have a clear justification for introducing new and more effective forms of communication.

Finally, I think our considerations in the first two parts of this book point to a new and much more important role for the social sciences other than economics. The social sciences can be strategically deployed to help communities choose indicators and criteria that are associated with broadly held social values. To this end I have introduced what I have called hypothetical constructs and have encouraged the formulation and exploration of ideal outcomes and ideal deliberative processes (see section 7.5). The point is not to set up unattainable utopian ideals, but rather to posit a basis against which to judge public processes of adaptive management as they emerge and develop in real situations.

Since we are trying to be inclusive of the whole range of social values in our definition of sustainability, we have not adopted the evaluative vocabulary of a particular discipline such as economics or nonanthropocentric environmental ethics. Even if these disciplines have identified important types of environmental values, we found quite implausible their claims to achieve both monism (reduction of all types of environmental value to a single ontological type) and comprehensiveness. We found that the vocabularies associated with these theories were not sufficiently rich to capture the full range of values by which people evaluate their environment, at least not without reducing many of these rich values to shadows of their original meanings. Recognizing the value of comprehensiveness, we chose to give up on monism, adopting pluralism as our starting point and allowing many evaluative flowers to bloom.

A plurality of values, if simply offered as a list, cannot of course tell us what to do; therefore, we must systematize and integrate these multiple types of value. This strategy led us to adopt experimental pluralism as our basic approach to environmental values. According to experimental pluralism, we expect members of any community to express a variety of values; the role of social scientists and philosophers is to help individuals and groups articulate their values and relate them to one another, to inform those values with the best available science, and to help in the development of indicators that are precise and reflective of the most important of these communal values. I have argued that a good way to seek such a systematization is to focus public attention on the problem of choosing indicators, setting goals associated with the indicators, and adopting a procedure by which various indicators of environmental well-being are proposed. The idea is that a lively public debate, with the careful attention of a committed citizens' advisory committee or collaborative stakeholders group who can encourage and engage in social learning, will encourage people to propose and justify various indicators and goals that should be set with respect to those indicators. In this way the indicators will come to be associated with the values that led people to suggest them as indicators. This strategy encourages public appeal to values, but the actual deliberation is over a concrete action—to choose a particular indicator as one that will be monitored and used in making management decisions. It is possible to choose such indicators without resolving all of the value differences and disagreements that occur in a diverse population. We will return to this topic in the last two sections of this chapter; here I am simply laying out the elements and the context in which to offer a schematic definition of sustainability.

In chapter 3 I argued that adaptive management, if supplemented with a more precise conception of options and opportunities, provides a very simple model for conceiving the difference between sustainable and unsustainable communities. I said that members of each generation make decisions on the basis of their individual, relatively short-term interests, as shown in figure 3.1. Their aggregated actions, however, can also be a causal force on a larger (ecological) scale on which the landscape changes through the interaction of populations and organisms. These larger-scaled systems normally change quite slowly; as Holling says, the ecological system is maintained within bounds by a relatively small number of key plant, animal, and biotic processes. Given expanded numbers and advanced technologies, it is possible that human populations can alter the structure of such systems. For example, the owners of forest patches might all clear-cut in the same year in response to an increased demand for timber. Their actions, taken as market values of individual actors, can of course be measured and aggregated in economic terms. But simple aggregation of the dollars in revenue does not capture the larger-scale dynamic that leaves a deforested countryside. Collectively, the choices of the forest owners change the landscape and thereby alter the mix of economic and lifestyle opportunities available to the next generation of actors. Not only will the actors have been replaced by their children; the stage upon which they played is altered as well. We summarized this kind of change by saying that the aggregated choices of one generation can alter (in the case of deforestation, reduce) the set of options open to choosers in the next generation. Further, in section 7.1 I showed how these reductions affect not just individual values but also *communal* values.

A schematic definition characterizes and relates the key components of a definition while leaving specification of the substance of those components open. Speaking schematically, we can say that sustainability is *a relationship between generations such that the earlier generations fulfill their individual wants and needs so as not to destroy, or close off, important and valued options for future generations.* If individuals in an earlier generation fulfill their needs in such a way as to destroy important options, leaving later individuals with a reduced range of choices and opportunities to fulfill their needs, then that earlier generation has lived unsustainably in that respect.

This definition is schematic in two senses. First, it refers to "important options" for communities. The general definition treats these options as a sort of variable, to be specified as particular communities articulate their values and decide what is important to save for their posterity. We can think of this as choosing a structured bequest by determining what options are important to the community.

Our definition is schematic in another sense, one that is not so easily dealt with, I'm afraid. Suppose, to continue with our simple example of deforestation, that the members of the first generation clear-cut the area as every owner responds to the same economic opportunity, as suggested above; but suppose, also, that these owners wisely invest their profits from the timber sold, and as a result factories are built and fine, productive farms replace the forests. The economist will quickly point out that the opportunity to make a living selling timber has been replaced by the opportunity to work in a factory or as a farmer. Opportunities have increased or shifted, not decreased. The economist would say that this community originally had an "immature" economy, still dependent upon exploiting its resources. Good investment has given the community a more mature, stable, and even more diverse economy.

In one sense the economist is correct: destruction of resources, even if it may eliminate some options, creates other options, and there is much to be said for great cities and productive industries. The story by which the loss of forest expands economic opportunities is not implausible, but treating it as a definitive refutation of my point amounts to begging the key question at hand. If the economist says the succeeding generation is better off because its members have more economic opportunities and can earn more money by working in a factory than their parents could earn through harvesting wood products, then we must point out that we think people of the future can be harmed according to noneconomic criteria and be worse off than we are in important ways, *even if they are more wealthy than we are.* This argumentative strategy works, provided we can go on to offer plausible noneconomic criteria by which we might say a later generation has been harmed by its predecessors, even if its people have equal or better economic opportunities than their predecessors did. Not surprisingly, we have circled back to the problem of specifying plausible criteria, based on useful, measurable indicators, that can be associated with important social values. Since we are looking at this task of specifying indicators and criteria from the community perspective, we must eventually describe the process by which communities specify these criteria and how they gain the kind of legitimacy that imposes a sense of obligation on advocates of sustainability. In order to give legitimacy to these obligations, communities must be able to associate a criterion with a legitimate social value, one that we can argue would emerge from open deliberation as a value the community feels obliged to protect. Our strategy is to concentrate public debate on concrete questions of what indicators to use and what goals to set, expecting that social values will emerge as part of the process as people use their values to support and defend the indicators they favor. Before we further explore the process of indicator choice as an articulation of noneconomic social values later in this chapter, it will be useful to survey the various types of values to which a community might appeal in order to achieve truly strong sustainability. See the next section for such a "catalog."

Where does our proposed definition stand now? Our definition of sustainability, we have concluded, cannot be fully specified in the abstract, because the definition we seek must be open to specification by local communities. So we have arrived at a schematic definition. We interpret sustainability within a hierarchical structure in which individual choice and landscape trends are seen as playing out on different scales of a complex system. Certain patterns in individual behaviors in the fast-changing system of human economics, augmented by technology, can affect normally slower dynamics that, if disrupted, create an ecological discontinuity and a loss of important options. Living sustainably is maintaining the mix of important options; living unsustainably is losing them, narrowing the range of options that subsequent generations can choose among in their attempt to adapt, survive, and prosper. To make this schematic definition a real definition, we must endow communities with the ability to choose what is important to monitor and what is important to protect. Choosing what to sustain is prior to choosing how to measure its sustenance. The persuasiveness, in public discussion, of any given definition that fits this schema will depend upon the plausibility of the arguments, provided by a community group or individual, that a given indicator should be monitored, and certain processes protected, because those processes are involved in generating an important social value.

9.2 A Catalog of Sustainability Values

To what types of values can we expect advocates of a particular indicator to appeal? In order to introduce a little conceptual order into our discussion of plural perspectives and multiple values, I now show how environmental values can be organized into four broad types that communities might try to sustain if matters are left up to them. Some of these types of value are associated with traditional disciplines that have paradigms quite different from those that other types of value are related to, and the types exhibit varied "logics," as explained in section 8.7. Although we have chosen pluralism and comprehensiveness over monism in our approach to environmental values, we can nevertheless use disciplinary approaches and methodologies to develop more precise representations of various types of value. The special sciences (such as economics), the other social sciences, and sciences such as ecology can contribute to these deliberations by offering ways to formulate indicators and criteria that are both precise and expressive of social values.

Having noted that there are a number of possible criteria and that communities must decide which ones to use in varied situations, we should inquire whether there may be some way to systematize these second-level decisions regarding which rules should be applied and how they should be weighted in a given situation. A goal of this chapter is to think through ways to rationalize—not formalize, which is impossible—this second level of decisions. Below, in section 10.5, we will return to a discussion of these second-level decisions. In this section I prepare the way for the systematization by offering a catalog of basic types of environmental values that we have encountered in our multidisciplinary wanderings. This classification of values can serve as a general guide to filling in the variables in our schematic definition.

The catalog simply provides a useful list and characterization of the range of values sometimes or often cited as justification for actions. It is an empirical question, one that can be answered only by observing particular communities, whether values of any or all of these types are in fact present and expressed in those communities. I propose, based on our survey of ongoing debates about sustainability and on the literature of several disciplines, four broad categories of values that seem likely to be important in discussions of how to live sustainably. These are (1) community-procedural values, (2) weak-sustainability (economic) values, (3) risk-avoidance values, and (4) community-identity values.

1. Community-procedural values. Environmental ethicists and environ-

mental economists have often overlooked values that emerge from the process of seeking cooperative action itself. Because these disciplines seek answers outside the process, by appeal to theory, they do not sufficiently emphasize process values. Our discussion of the pragmatists, Habermas, and the discourse ethicists, however, has focused our attention more on community processes of seeking agreement and less on the arguments of philosophers and economic theorists that certain environmental values *must* exist or *must* be measured in a particular way. It is an advantage of our approach to environmental values that, since we seek values that exist within communities, and since we have limited the scope of our policy analysis to communities that have undertaken an ongoing, community-based, adaptive management process, we can assume that participants implicitly accept obligations associated with communicative action. The community has placed a value on achieving cooperative solutions to shared problems. It may seem a bit odd to refer to such values as environmental in nature, but they are at least environmental in the sense that a commitment to seek cooperative solutions with one's cohorts, rather than to resort to violence, apathy, or raw political power, represents a value that can deeply affect the process of achieving sustainable living.

Beyond this, I will argue, the ideal of cooperative solutions to shared problems also has deeper resonance in the sense that cooperative activity to protect environmental assets—especially distinctive species, productive processes, and treasured places—*as a community* signals a commitment to achieve an appropriate intermingling of cultural and natural values. This deeper commitment of a community to an "authentic" relationship with its natural habitat, insofar as it is undertaken as a shared responsibility of citizens, represents an important aspect of how value is generated in the process of cooperation in protecting shared values. Given the implication that participants in adaptive management processes are committed to cooperation, discourse ethics provides a useful theoretical platform for posing the questions faced by a community seeking to live sustainably; we can accordingly use discourse ethics as one important aspect of our framework of evaluation. Incorporating discourse ethics—and especially its "ethic of discourse"—into the process of seeking sustainability provides, as just noted, a normative basis for justifying a number of procedural values.

One might ask whether these procedural values should be considered integral to the idea of sustainability. Developing a sustainability definition that is appropriate to a given community is an ongoing process; if that process is not fair and open, it will not embody a basic commitment to fairness in the decision-making process that determines land and resource. If someone says, "But these rules are all 'procedural' and imply nothing about the substance of what is valued," we reply that in a democratic process, substance arises from process and that sustainability commitments must, if they are to guide communities toward effective cooperative action, be perceived as results of a fair and open process.

2. Weak-sustainability (economic) values. I suppose it seems that I have been pretty hard on economists in this book (if so, I can only hope I have been just as hard on environmental ethicists), but it is important to remember that my criticisms do not point toward a rejection of economic valuation, but rather toward defining more carefully the area of values, and the situations, where the terms of economic valuation are appropriately applied. I have not attacked economic analysis as useless. Instead, favoring a pluralistic approach, I have rejected the claim of Economists that *all* environmental values are captured in their analyses. On this topic I have had much to say already. Here I collect the implications of these various arguments and comments and define a category of environmental values that are appropriately described using the terms and models of economics.

Economic uses of the environment and resources are almost too pervasive to notice; since we are interested in identifying benefits of environmental protection, we are mainly interested in what might be called sustainability values, those that can be associated with protection rather than destructive exploitation of natural processes and products. Although economists have their own classification of the values they measure—basically dividing them into "use" and "nonuse" or "passive-use" values—we saw in section 5.4 that this distinction is frightfully difficult to define behaviorally; moreover, such a division seems arbitrary and apparently does violence to ordinary conceptions by treating protection of resources for the use of future generations as nonuse values—because they are judged to be "altruistic." We can avoid these problems, which, after all, seem to stem more from the monistic Economism of some analysts than from economists' methods themselves, by listing economic values as (a) goods and services that are traded in (actual) markets and (b) goods and services that cannot be so marketed ("public goods" and "communal goods"). The first category will be unproblematic among economists. Consumers reveal their preferences (willingness to pay) as they make purchases. The second category includes some cases in which it clearly makes sense—and is probably useful—to infer a WTP for ecological features. For example, a park manager who must decide whether to support a reintroduction of bear in the area might want to know how much more (or less) the typical visitor would be willing to pay in entrance fees if there were bears in the park. There are also many environmental goods that are not so easily priced, and controversy abounds regarding the accuracy of WTP measures for many goods and the values they represent. Remember, for example, from section 5.4, that contingent valuation of "nonuse values of groundwater cleanup" turned out to be unmeasurable by any standard questionnaire means, because respondents would not accept the premise that water, if clean, would not be used. And there are many other values and experiences that most of us feel uncomfortable pricing.

In these areas of controversy, I intend to remain tolerant and encourage experimentation. I'm happy to see economists push their methods as far as possible and measure to the extent possible those ecological goods and environmental services that people would be willing to pay for. But I also reserve my own and others' rights to develop alternative means to capture and measure environmental values in terms of holding open options for the future, understood as generation-scaled commitments to a place, for example. Despite the attitude of tolerance, I proceed to examine in more detail the strengths and limits of nonmarket valuation techniques. We must ask under what conditions an economic analysis is useful in informing decisions that affect the environment. Fortunately, some thoughtful economists, Arild Vatn and Dan Bromley, have already asked this question and have given an excellent answer, one that I can accept with a few interpretive qualifications, which I will add after summarizing their answer.

Vatn and Bromley recognize that there are some goods for which it makes sense to estimate WTP by polling respondents.[1] These include goods for which people have experience with close analogues: for example, a survey question can ask how much one would pay in increased hunting-license fees for an improvement of habitat for wild animals. Vatn and Bromley go on, however, to provide examples of values that are not appropriately treated as commodities: in this category are values that are unfamiliar to respondents and values that are cherished for noneconomic reasons. Such values are beyond the reach of contingent valuation methodologies, according to these authors. Following their reasoning, I believe we must recognize a limited role for contingent valuation of nonmarket goods, which implies that contingent valuation has a useful range of application, but this range does not include important values that respondents are unable to comfortably understand as "market" values. These noneconomic values, I have noted, include communal values, which emerge on a scale larger than that of aggregable individual welfare.

I have suggested that the weak-sustainability criterion, which prohibits impoverishing the future economically, can be treated as a necessary but not sufficient requirement of sustainable living. We can also raise the question, though, whether this requirement *must* be included in every community's definition of sustainability. Can it be claimed that if a given community decides not to include weak sustainability among the indicators it monitors and the criteria it adopts, it has somehow missed out on an *indispensable* aspect of sustainability? Or should we simply say that because the weak-sustainability theory has been developed in conjunction with some plausible economic theory about economic growth, weak sustainability is one useful indicator that communities *might want to consider,* as they sort through and choose sustainability indicators appropriate to them and to their values. The former position is rejected here because it seems to appeal to an authority higher than that of community experience; it smacks of a retreat into a priori conditions on experience. I favor the second position, noting that it would not apparently be inconsistent for a wealthy society to intentionally reduce its standard of living in order to reduce its consumptive impact or in order to transfer wealth to desperately poor neighboring countries.

This may seem a terribly weak position to economists and weak-sustainability theorists, who clearly think they are doing more than specifying one condition that communities might choose, along with others, as a sustainability measure. What, then, could be said to a community that considers the question and decides that it is not so important to maintain the accumulated wealth of the society at its current level and, accordingly, refuses to embrace weak sustainability an essential aspect of its sustainability commitment? I respond that this is a question that each community must resolve. If economists believe this rule should be applied in all cases, they will have to become members of each particular community-based process and make the case for a commitment, by their community, to maintaining a nondeclining level of societal wealth. This will involve making an ethical case that a fair savings rate is a requirement and will also require a great many conceptual and technical explanations to convince other participants that the growth-and-savings-rate model is a useful one for measuring societal well-being.

So the category of economic values, as represented in the weak-sustainability criterion, provides one proposed criterion that all communities should consider and many will choose as a necessary requirement of sustainability. Whether they consider adopting such a criterion, and whether they do adopt it, given our assumptions about what drives public discourse on such matters, will depend on the iterative and deliberative public process of choosing sustainability criteria in a given community.

If such a criterion is chosen, if a community decides that avoiding impoverishing future members of the community is a priority, then we can expect that discourse about environmental problems in this community will be sprinkled with references to willingness to pay, consumer surplus, and nonmarket values. Economists, that is, will be active participants in the search for better policies; they will also be advocates for certain methods of measuring values, advocates for certain types of models, and advocates for funding of specific types of research. To the extent that they convince other participants that the values they tout and the methods they offer are important, to that extent they will affect the development of environmental policy. This, I submit, is all that most economists should expect; and I believe most economists will be content with this role.[2]

None of these qualifications, in my view, reduce the importance or the potential impact of economic analysis on the policy process. It will be a very unusual community that intentionally chooses to reduce the levels of individual welfare over time, so most communities will insist on monitoring economic trends and will favor policies that maintain the economic capital of the society. Economic growth and savings models will therefore be relevant to understanding this aspect of sustainability, and economic valuation studies will remain an important part of any comprehensive evaluation of impacts and policies. Even if weak sustainability is *not* required in some communities, if noneconomic goals and values are given greater weight, it still makes sense for economists to compare the costs of various means to accomplish the goals set, providing what is sometimes called a cost-efficiency analysis. In this pragmatic context, however, the skills of economists are being used to answer specific questions within a comprehensive, multidisciplinary evaluation process, rather than to provide a synoptic evaluation of alternatives. If we are right that there are important values at stake in sustainability besides economic ones, many communities may impose other criteria in addition to economic ones. There may be many different ways of achieving nondeclining welfare opportunities, and which of those development paths to pursue will be determined by the community achieving an acceptable balance between economic and noneconomic values, indicators, and criteria. Thus, the second broad class of values includes those that can be measured by the usual techniques of economics and can be reasonably and reliably understood as units of individual welfare and perhaps usefully represented as dollars that consumers are willing to pay.

3. Risk-avoidance values. Another set of values has to do with avoidance of risks of various kinds and spread over many different time scales. EPA, since the 1980s, has emphasized risk analysis—risk assessment and risk management—as the official policy of the agency, and there has been considerable pressure from within and outside the agency to use risk analysis as the unifying policy framework for analyzing policy and management concerns of the agency. As noted in chapter 1, however, ideological commitment to risk analysis could not make up for the huge gaps in the application of the models of risk assessors, models that usually required data that did not exist and would be prohibitively expensive to gather. Even worse, risk assessors had no

basis on which to compare immediate risks with ones that might have significant lag times. Much as with economics and with environmental ethics, we can ask two importantly different questions about values of risk avoidance. (a) Can some important environmental questions be analyzed using the techniques of risk analysis? And (b) Can all environmental problems be usefully construed as, or measured in terms of, increments and decrements of risk? The reader will by now not be surprised that I answer question a affirmatively and question b negatively. Risk assessment and risk analysis have their place, certainly, in setting standards and goals regarding chemical exposures and risks to humans and wildlife. But it is only in cases like this, I think, when we're trying to decide whether to ban or restrict new (or old) chemicals, for example, that we can appeal to these precise but narrow models to help us to decide what to do.

For the harder cases, which call for computing risks of nuclear accidents and their costs, determining what we should protect when we protect biodiversity, and measuring the problems of siting toxic storage facilities, I see little use for a formal risk assessment. In cases like these, one can expect public controversy, but it does not generally involve disagreements that can be resolved by formal decision models. At any rate, I believe most communities will adopt safeguards against inflicting dangerous and preventable risks on future people; and these commitments are an important aspect of developing a typology of effects. These commitments not to expose the future to undue risks operate according to the SMS logic, as noted in section 8.7. In this case the SMS logic dictates that we avoid or remediate risks, whenever the costs of doing so are bearable. In this application the SMS and the precautionary-principle (PP) rules seem to imply nearly identical patterns of behavior, based on similar intuitions. Most communities, if faced with evidence that their actual or proposed actions may impose terrible risks on the future, will, I predict, adopt commitments to reduce that risk.

4. Community-identity values. Finally, we come to the type of values that I have argued emerge on the community scale—the scale on which human populations interact with the landscape on which they are placed. These values express a sense of the good life as it is understood in a particular community. They spring up locally, as members of communities interact with their environment and develop skills and beliefs. Community-identity values, I have argued, are developed and passed from generation to generation, creating cohesiveness within human communities but also binding individuals and communities to their natural habitat. These values fall into the general category of values that philosophers have called "ideals of a good life" or "virtue ethics."

I have chosen to adopt the terminology of layered obligations, wherein a "thin" theory of justice includes obligations to treat other individuals with respect and "thick" obligations are obligations accepted as a part of defining and living a good life in a community with shared values. Thin obligations originate in universal aspects of the moral situation, whereas thick obligations are culture-bound and relative to particular situations. Aristotle, for example, wanted to define an ideal citizen of Athens, because on that basis he could characterize a good life in that context. Thick obligations are the special mores and commitments that give texture and distinctiveness to a culture; and whether of the moral or the economic type, these commitments can extend beyond obligations to particular individuals to include commitments to perpetuate certain ideals and aspirations of a culture and of a community. One kind of thick commitments would be commitments of communities to save what is distinctive about their physical surroundings, what they love about their place, as discussed in chapter 8. In my view, this category of environmental values is the most interesting category of all. These values are obligations that thoughtful community members, committed to cooperative action and loving their place, would articulate and impose upon themselves for the good of their community. In chapter 8 I called the process by which such values are articulated one of engaging in a "community performative," an act or series of acts that signal a commitment to perpetuate stuff deemed of great value to the community, provided the costs are bearable.

I cannot end this catalog without mentioning "intrinsic values" in nature: should we consider intrinsic values one type of values a community might set out to protect? This is a question that is sure to come from my friends among environmental ethicists: How will Intrinsic Value theories fare in a pluralistic adaptive management process? I have several comments in response to this question, and then I will make a recommendation—and a suggestion of reconciliation.

First, of course I do not dispute the fact that some environmentalists offer intrinsic values in nature as a reason for protecting natural areas and wildlife habitats; further, I have recommended tolerance of all expressions of value. So if the question is, Should we allow and encourage participants to express intrinsic values? my answer is, without question, yes. That is not the same as predicting that this terminology will survive indefinitely as a useful element of discourse about environmental values and policy, nor is it to adopt any particular theory about the nature of these intrinsic values.

Tolerance for multiple expressions need not lead to an overly inclusive and sloppy use of language. Indeed, I predict that as better linguistic tools are developed for expressing values that communities place on nature, the term *intrinsic value* will have its role replaced by commitments to more specific indicators and protectionist efforts. Intrinsic value will, I predict, give way to a rich pluralism, some values of which will be expressed as commitments to save objects and places and as acceptance of measurable indicators and precise criteria. Participants will eventually find references to intrinsic value otiose, as more specific goals are formulated. As noted in section 6.2, Intrinsic Value theory is based on an assumption that environmental values can be divided neatly into instrumental uses and noninstrumental valuings. In fact, of course, a pluralist will naturally recognize a range of values from consumptive to transformative to spiritual, and it is simply a philosophers' artifact to draw a sharp dividing line between these varied ways that humans value nature. We should be especially suspicious of this particular artifact, since it gains its importance from the Cartesian separation of spirit from body and its derivative, the sharp separation of humans from nature. To extend intrinsic-value status to nonhuman entities is not to break down the Cartesian dichotomy—it merely moves the divide over to include species, ecosystems, and elements of nature on the intrinsically valued side of the dichotomy.

Second, Intrinsic Value theory includes many differing and incompatible theories, ranging from strong ontological assertions (by Holmes Rolston III)[3] of the existence of value entirely independent of humans to much weaker assertions (by Callicott, for example) that intrinsic valuing is merely adverbial—a way in which humans value nature. In this latter view, intrinsic value has no objective existence independent of a human act of valuing. Given such vast differences, we must be careful regarding what kind of intrinsic values we are positing. Strong versions of IV theory have been dismissed, above, as incompatible with pragmatism and adaptive management, but one might think a pluralist could easily accommodate Callicott's view that humans sometimes value things noninstrumentally. And a pluralist might—if one purges Callicott's theory of monism—include noninstrumental values in the list of values often cited as good reasons to protect wildlife habitat, for example. The problem with this suggestion is that references to such values are not useful in telling us what to do, at least in the short run. Callicott's view of intrinsic values, which has been recently clarified, is that they are simply expressions of subjective value; they are based on individual feelings toward elements and processes of nature.[4] Callicott is concerned to separate his own conception of intrinsic value as far as possible from Rolston's strong Intrinsic Value theory. As I understand it from the article cited in note 4, and from other writings and discussions, Callicott fully embraces conventionalism of the type I have explained and defended in this book, and he intends to introduce and defend a type of intrinsic value in nature that is consistent with empiricism and conventionalism. Callicott says: "All value, in short, is of subjective provenance. And I hold that intrinsic value should be defined negatively, in contradistinction to instrumental value, as the value of something that is left over when all its instrumental value has been subtracted. In other words, 'intrinsic value' and 'noninstrumental value' are two names for one and the same thing."[5]

This solution to the problem of defining intrinsic value (introducing a trivial definitional equivalence) is actually a solution to a nonproblem: the pragmatic pluralist, who recognizes that people value nature in many ways, need not deny that humans value natural objects "noninstrumentally." For example, many people value nature spiritually; this value is surely noninstrumental, and it is clearly a human value. Once we accept pluralism and a continuum of types of values, spiritual and other noninstrumental values attributed to nature can be seen as differing types of *human* value. Intrinsic values, in Callicott's definition, are not independent values capable of overriding human values—they *are human values!* Humans must, furthermore, decide how to balance and prioritize these values. The only puzzle remaining is why Callicott thinks that embracing intrinsic values as the noninstrumental human values remaining after the instrumental values are drained out makes him a nonanthropocentrist.

There are many other problems with Callicott's valuations. Surprisingly, he claims that his valuations have no objective basis; he argues that they could as well be bestowed upon "old tennis shoes."[6] Notice that, on this interpretation of intrinsic values in nature, the values are merely reflections of subjective and uneducated feelings of individuals. They could not be useful in argumentation that we should protect nature—unless Callicott finds it convincing that he and other citizens should devote social resources to preserving the smelly shoes that have collected in my closet. So, if we are looking for value expressions that can be used as good reasons to protect species and ecosystems, Callicott's "old-tennis-shoe" variety of intrinsic value will not help us.

Perhaps, though, references to intrinsic value—if interjected into our discussions and deliberations—should be valued not for their argumentative force but for their clarity in identifying values that should be operationalized and measured as features of nature we would like to protect. Here we encounter another weakness of most versions of Intrinsic Value theory, including Callicott's. Intrinsic values link in no clear way to experience, so references to them do nothing to suggest possible indicators or other measurable characteristics of nature. References to intrinsic value sometimes sound as if such value is observable, identifiable, and verifiable, but at the same time, intrinsic value is associated with very general characteristics that aren't really measurable. In fact, of course, intrinsic values—whatever else they are—are highly theoretical entities. Furthermore, they are theoretical entities that have no clear correlation with any observable characteristics of natural systems. We are still left with the question, Should we expect or recommend that communities develop, among their other indicators and criteria, an obligation to protect all natural objects with intrinsic value?

Callicott has recently provided two arguments as to why references to intrinsic value are *necessary* in order to protect nature from developers.[7] He argues, first, that any anthropocentric theory will limit obligations to protect species and ecosystems to those that are useful to humans; many species are not useful to humans; therefore, an anthropocentric value system is inadequate because it would leave many species unprotected. Second, Callicott claims that references to intrinsic value are necessary in order to shift the burden of proof against developers who promise quick development in exchange for important environmental features like wetlands. I will respond to these two arguments in order.

I'm trying not to adopt a scolding voice in response to the first argument (scolding is not an attractive attitude in an adaptive management process), but Callicott should have known—given that he is one of the world experts on Aldo Leopold—that this argument was decisively refuted by Leopold in 1939:

> Ecology is a new fusion point for all the natural sciences. It has been built up partly by ecologists, but partly also by the collective efforts of the men charged with the economic evaluation of species. The emergence of ecology has placed the economic biologist in a peculiar dilemma: with one hand he points out the accumulated findings of his search for utility, or lack of utility, in this or that species: with the other he lifts the veil from a biota so complex, so conditioned by interwoven cooperations and competitions, that no man can say where utility begins or ends. No species can be "rated" without the tongue in the cheek: the old categories of "useful" and "harmful" have validity only as conditioned by time, place, and circumstance. The only sure conclusion is that the biota as a whole is useful.[8]

Following Leopold, we can thus answer Callicott's first argument by saying that, given the impossibility of picking and choosing what to save and the overwhelmingly powerful reasons to say that the whole system is of great value, any person who wishes to protect natural processes for the benefit of humans must save all species. It's too bad Callicott did not read his mentor a bit more carefully.

Callicott also argues, secondly, that intrinsic-value rhetoric will be important in the future because its widespread use would shift the burden of proof to developers and exploiters and away from those who would protect natural systems: "If something only has instrumental value, its disposition goes to the highest bidder. . . . If, say, . . . a wetland [has only instrumental value], conservationists must prove that an economic cost-benefit analysis unequivocally indicates that it has greater value as an amenity than it has, drained and filled, as a site for a proposed shopping mall." If intrinsic values of wetlands "were broadly recognized," he continues, "then developers would have to prove that the value of the shopping mall was so great as to trump the intrinsic value of the wetland."[9]

My development of community performative commitments to rules with an SMS logic in section 8.7 provides an alternative way to shift the burden of proof to despoilers. The SMS rules impose upon them the obligation to show that protecting the resource such as wetlands would entail costs that the community would find to be unbearable. So Callicott is simply mistaken that intrinsic-value attributions are *necessary* to shift the burden of proof from conservationists to developers. If a community is committed to sustainability values, spiritual values, including communal values such as those prescriptions that are based on the love of a place, then that community will already have shifted the burden of proof to anyone who would despoil those values for short-term economic gain. Callicott has failed to show—even if Intrinsic Value theory were *sufficient* to shift the burden of proof—that a multigenerational, sustainability ethic, fashioned to include a variety of amenity and spiritual values, *fails* to do so. Without that premise, his argument that nonanthropocentrism is necessary to environmentalism fails; it is one more piece of rhetoric and ideology. His second argument, in other words, reduces to the useless first one: it stands and falls on his false claim that even a broad and future-oriented anthropocentrism will not protect the full range of values humans find in nature. As for me, I will wait to see the evidence whether adaptive management programs that adopt Callicott's language go more smoothly and have more effective plans of action than ones that adopt a sustainability ethic, including a menu of multiple ways that communities value their home places and a process by which communities can set, and adjust, the priorities they place on them.

Callicott presses this second argument for embracing a nonanthropocentric vocabulary by offering rights language as an analogy: "Human rights—to liberty, even to life, may be over-ridden by considerations of public or aggregate utility. But in all such cases, the burden of proof for doing so rests not with the rights holder, but with those who would over-ride human rights."[10] I'm perfectly happy to accept this analogy, provided "rights" are understood in a way that is consistent with our conventional view of language and are not thought of as obligations that originate beyond the community discourse. I do not believe, any more than Justice Holmes did (see section 2.3), that "rights" are some prior existent object that is discovered and can be used to justify a given action.[11] Rights are internal to a system of politics and government; we create them by virtue of the many more specific "rules" we "follow." Rights, speaking morally, are the customs of respect and entitlement that are generally extended in a community; rights, speaking legally, refer to legislative language and to a pattern of interpretation of that language. Rules are not discovered "out there," exogenous to our moral discourse, but hammered out through precedent, argumentation, and standard-setting. This is the view of rights and of intrinsic values that would be consistent with the conventionalist view that Callicott apparently accepts. Applying this reasoning, analogically, to intrinsic values in nature, we should conclude that intrinsic value—if it were broadly accepted in a community—might shift the burden of proof in that community away from developers and toward environmental protectionists. But intrinsic value understood in this way is subservient to *practices*. Appeals to intrinsic values cannot provide discourse-independent reasons for acting to protect nature. Unless a strong majority of individuals express this affect, and their affect is, in turn, built through activism into legislation and policy, appeals to intrinsic value will remain mere ideology and rhetoric.

Notice, then, that the analogy with rights now seems much more tenuous and applies only hypothetically; we have a legal—and moral—system that has articulated, defended, and enforced human rights-claims (legally) for several centuries, and we have various forms of action that signal moral disapproval, including shunning, arrest, tort law, public criticism, and so forth, when rights normally accorded are infringed. The problem with extending predictions about what rights will be enforced or approved— or what will be protected as valuable in the future—is that there are no institutions, no particular guiding principles, and few precedents for linking intrinsic-value talk to specific behaviors affecting animals or ecosystems. There are no accepted practices to give meaning to the language in question. The only predictions that extensionists can make are that such institutions and ideas will develop someday and that people of the future will treat nature better because people of the future will value it intrinsically and nonanthropocentrically.

Given the adverbial and conventional view of intrinsic value that Callicott has accepted, his claim that attributions of intrinsic value can change policy by shifting the burden of proof is essentially a vacuous statement. It merely says that *if* the idea of intrinsic value in nature is widely expressed in a democratic process, and *if* people begin to act as if aspects of nature have intrinsic value, then decisions will no longer be dominated by economic calculations. *If* most of the population advocated policies based on intrinsic value, technocrats and economists would have less chance to institute their policies. Its hard to disagree. But does the improved decision making originate in nature—which seems a necessary condition for calling the theory truly "nonanthropocentric"? No, it originates as an expression of idiosyncratic, individual feelings of humans, according to Callicott. And, by hypothesis, this is the expression of a minority of participants in most situations today. What if the idea of intrinsic value does not gain broad acceptance? It apparently will do no good at all in the short run, given its minority status and its tendency to reinforce noncooperation between interest groups and people with diverse worldviews. At the current state of development of environmental values in the United States, then, it would appear—based on the rules of engagement implied by Callicott—that conservationists who attribute intrinsic value to nature must accept a burden of proof that will be shifted to developers only after a majority of the population embraces intrinsic values in nature and nonanthropocentrism.

Having said all of this to back up my skepticism that we should encourage communities to articulate and use criteria based on obligations to protect intrinsic value in nature, I would like to offer a possible reconciliation, a reconciliation that might find a role both for the development of an adaptive management approach to environmental goal-setting and for a parallel philosophical exploration of intrinsic values. Perhaps we could recognize that Callicott, who apparently believes that advocates who recognize intrinsic value—who are apparently in the minority today—could make a difference if they could become a majority and shift the burden of proof for degrading natural systems to the consumers and developers, has embarked on a long-term reexamination of social values. One could think of the problem he addresses as quite independent of the issues faced by problem-solving adaptive managers who begin in the community, with the values the community now has, and try to build a locally valid definition of sustainability. Callicott may be looking at the long run, hoping to encourage the slow process of replacing one worldview with another, hoping a shift in values will improve public discourse and communication. We could, that is, interpret Callicott as not discussing what values will be useful in today's practical decisions regarding what to do; he may rather be engaged in an attempt to shift the entire worldview of society from the top down, by introducing intrinsic value as a "great idea" that will change the world. On this interpretation, Callicott would be saying that over many years there will be a shift to nonanthropocentric value stances and that when this happens, environmental policy debates will be enlightened and more effective. If this was Callicott's position, there would perhaps be no incompatibility between his view and the pragmatist-pluralist view developed here, except that Callicott hypothesizes that in the long run, we will learn our way out of anthropocentrism and into nonanthropocentrism *because a nonanthropocentric framework of ethics will provide a clearer and more effective communication of environmental values.* I could be seen as simply expressing skepticism regarding the long-term impacts of such innovation on public discourse. On my side, I would cite the already-mentioned problem that intrinsic-value concepts do not seem well connected to any environmental measures or possibly measurable indicators. I would also point out that if Callicott is hypothesizing the usefulness of intrinsic-value language as a means to better communication among participants in deliberations about what to measure and what to protect, he must somehow come to grips with the point, noted from the outset of this book, that intrinsic-value formulations, being so isolated from experience, have contributed not to clarity but to ideological polarization of nonanthropocentric economists and IV theorists. In the development of the new, nonanthropocentric worldview, then, Callicott and other advocates of intrinsic value in nature have a lot of historical baggage to unload. IV theory, as speculation about the worldviews that will be popular in the distant future, may not conflict with adaptive management because the former addresses a long-term philosophical speculation—what one might call "thinking like a multigenerational culture"—whereas adaptive management describes a strategy that starts where we are and struggles toward better policies through social learning. On this understanding, both Callicott and I would be formulating hypotheses about what changes in our language will be useful, but my hypothesis deals with the here and now, while Callicott speculates about the future. Perhaps both of these are important philosophical tasks. My point, however, is that long-term speculation is not very useful if we have no way to get there from here; and I believe we cannot get to an effective nonanthropocentric viewpoint in the medium or long run because the concept of intrinsic value is not closely enough related to experience for this concept to guide social learning.

9.3 Beyond the Fact-Value Divide

I have insisted that attempts to separate factual information from value judgments have been the root of much miscommunication and dysfunctionality in environmental discourse; I have also insisted that sustainability cannot be defined in any purely descriptive vocabulary or using statements from a single descriptive discipline. It is time to bring these two lines of pragmatic reasoning together and train them on our keystone term for environmental policy: What do we mean by saying that the term *sustainability* is a *normative* term in public policy discourse? And how can a normative approach to defining sustainability incorporate the best science and still function normatively? These are fair questions, and I will try to answer them. Scientists and managers alike have attempted to keep these two discourses as separate as possible and have explicitly endorsed the separation of science and value analysis. But I follow the pragmatists like Dewey in believing that there is only one method of inquiry and one community of discourse, happily endorsing a normatively informed science and a search for scientifically informed norms. In the relevant discourse—public discourse—facts and values are inseparable.

Consider again the example/analogy offered in section 2.1, when it was argued that ecologists and environmentalists find themselves at a disadvantage in discussing environmental policy, especially as policy affects economic issues, because economists have succeeded in linking the Gross Domestic Product indicator with the ordinary discourse idea of economic growth. Thus, while economists can justifiably claim to be simply measuring economic activity according to accepted descriptive definitions—acting as scientists gathering data and factoring the data into a synoptic indicator—they can be sure that all discussants in public and political discourse share a general idea of economic growth *and that most of them also place a high value on continuation of growth and a low value on loss in economic activity.* Although there are many problems with, and objections to, using the GDP as a general measure of social well-being, the indicator helps economists by linking their scientific-descriptive models to an unquestioned—or at least widely shared—value of most citizens. Economists, then, can claim scientific objectivity, in that they define the GDP as a compilation of data that has simply been aggregated by their model, and at the same time expect their reports to be accepted within policy discourse as providing information about the performance of the economy and as implying a clear judgment regarding trends in the maintenance of social values. There is no question in most citizens' minds that indicators of the growth of the economy are relevant to public policy debate.

Economists can claim scientific objectivity and also have their research and models taken as directly relevant to the evaluation of policies because economists have created a measurable, operationalized indicator, and the public, including especially politicians and government agents, assumes an unquestioned connection between the economists' descriptive growth indicator (however imperfect it is) and the noncontroversial social value of economic growth. I feel no need to criticize economists for their role in creating and maintaining this advantage; all the economists did was to design a robust measure of economic activity, regularize their use of the data, and develop an indicator that could gauge increases and decreases in such activity. It was not the scientific accuracy or the precision of the GDP measure but the strength of the values that established the importance of the GDP indicator for policy discussions. Two lessons can be learned from this example: First, the *importance* of scientific data and measures is a function of the social context, of the value commitments of the population, of politicians, or of a particular management context, not of the descriptive precision of the data and the measures. And second, by the choice of a normative term, *growth,* in association with the GDP, economic jargon is united with apparent professional wisdom, providing an important linkage between economic models and social values.

Consider another example, from chapter 1, my frustrations in trying to help EPA to define *ecological significance* as a "scientific" concept. Remember, I was on a subcommittee that was assigned the task of defining and summarizing key aspects of "ecological significance" in a scientific background paper that was solicited by the Risk Assessment Forum. The paper was to serve as an aid in writing the protocols for ecological risk assessment. That particular policy context was governed by the myth that risk assessment could be a science functioning independently of risk managers and risk management. Our little subcommittee thus faced a quandary. The intention of the agency, in contracting with us, was to solicit *scientific* guidance on how to understand and measure ecological significance; but the agency had already assumed that scientific guidance would be limited to value-neutral *ecological* information. All of us on the subcommittee, three ecologists and I, realized that no such sharp separation could be made between ecological significance and social significance if the concept of ecological significance was going to function properly in policy contexts.

To see what I mean here, notice that the judgment that a process or event is "ecologically significant" is ambiguous. It could mean that the event or process is associated with very important ecological functions, with the clear implication that damage to these processes and events will have far-ranging ecological consequences; that is, it will drastically change larger-scale variables in the system. Or it could mean that the event or process, propagated by ecological change, is of great importance to humans who live in the area (because of its effects on ecologically provided services or cherished landscapes, for example). The Risk Assessment Forum wanted us to develop the concept of ecological significance along the line of the first meaning; they wanted a purely scientific definition, which would fit within their preconceived idea that risk assessors answer scientific questions to the extent possible before the problem is passed to the risk managers, who will act to maximize social values as they understand them, taking remaining uncertainties into account. To see how ascertaining the pure "ecological" significance of events or processes is insufficient to provide ecological risk assessors with what they need, consider the risk of major ecological changes resulting from shifts in the tectonic plates forming the earth's surface. Historically speaking, the ecological changes induced by the merging and disintegration of land masses are among the most significant changes that ecology can measure; but to rank such risks high on a risk assessor's list of important ecological risks would be ludicrous. The changes are orders of magnitude slower than the pace of our policy world and almost as disparate from any ecological changes that would affect our well-being; there are no policy levers we could pull, no technologies we could unleash, to affect the matter anyway. Risks on this temporal scale cannot be made relevant to a human concern, given the temporal scale of human lives and the normal span of civilizations. Because ecologists are humans and must judge significance from a human perspective inside a hierarchically organized system, the tectonic dynamic is irrelevant to the determination of ecological risk because it unfolds on a scale unimaginable to humans. Changes on that scale cannot be perceived by humans as a risk.

This is only one example of the role of scale in identifying significance. I can offer another counterexample, coming from the other extreme on the scale hierarchy. By comparison to what we know about the biology of mammals and their interactions with other animals, we know almost nothing about the "ecology" of the world of microbes. In many cases we treat very different microbial species as identical in ecological models, because they have the same function in providing some service to humans, such as decomposition of plant and animal biomass. However, in the special case of microbes that are "infectious" to humans—microbes that can live in the human body and cause disease or death—we generally study the microbes in the special "habitat" of the human organism or in other carrier species; we usually don't care what their "natural" ecology is—how they relate to other microbes on the microbial level (unless, of course, other smaller-level microbes might be capable of protecting humans from the infectious agent). Again, it is human concerns that direct scientific description toward dynamics that affect, or at least could affect, human interests and values.

Both of these examples—from very large to very small on the scale continuum—show that differences of scale can make some physical dynamics irrelevant to the human perspective. This outcome is in keeping with the assumptions of hierarchy theory: dynamics that are orders of magnitude faster or slower will be irrelevant to the system as viewed from the perspective of an agent acting on a much smaller or much larger environmental scale. Adoption of the human perspective, then, represents a narrowing of the searchlight of science that cannot be justified by observation; it is a matter of choice of a perspective. A "purely objective" description of ecological systems such as the Risk Assessment Forum wanted would give equal importance to every scale on nature's hierarchy, from the ecology of microbes to the mixing of species as landmasses are newly crushed together at the rate of centimeters per century. "Purely scientific" judgments of ecological significance cannot determine which scales and dynamics are valued by humans. The fact that ecological science is done by humans, with human body sizes and life spans, rather than by tiny microbes who die and replace themselves daily, necessarily skews any perception of the dynamics of the system. We notice dynamics that affect our lives.

These examples show that in order to maintain the illusion that ecology, or risk assessment, or any other science can "objectively" describe systems and make value-neutral judgments of the "significance" of events in purely scientific terms, one must conceive scientific description, as Newton and the modernists did, sub specie aeternitatis—from the perspective of an omniscient, all-seeing perceiver who describes the world from outside it, has no effect on it, and is unaffected by it. Such a view of the world, which might be purged of human perspective and human skewing of what scales and dynamics are "important," we now know to be impossible for a finite observer, a finite human ecologist who must observe and describe from inside a complex, dynamic system. Therefore, the Risk Assessment Forum's intended request, for a "purely scientific" summary of "ecological significance," cannot be fulfilled. Ecological significance must always be judged from a perspective and within a temporal scale.

So we could not and did not give the Risk Assessment Forum what it wanted. Instead, we produced a subcommittee report that argues for explicit recognition of the role of values in the formulation of risk assessment problems. But of course the bureaucrats can't be beaten at this game. What we produced did not fit their preconceived notion, so they published our report in an ugly purple cover and proceeded to ignore our advice. In the ecology of agency decision making, we were assigned to insignificance.

I believe my extreme examples at each end of the size scale can be generalized; we can say, more generally, that one cannot assign importance to a physical process without paying careful attention to the scales of the variables that drive that process. Humans, by being middle-to-large-bodied mammals, bipedal, and so forth, will care about some temporal and spatial scales and not others.[12] The perspective we must adopt, in talking about sustainability, is a profoundly human perspective, seen from human eyes and scaled by human concerns and capabilities.

Humans, as we have told the story, have expanded their ability to affect more and more scales of the ecological dynamic. If acting sustainably is accepting responsibility for the foreseeable consequences of our acts, then we are going to have to sort out the scales with which we are concerned by appeal to who we are and what is important to us. Expanding one's sense of responsibility in such a system is a very tricky thing, and identifying the scales and the dynamics important to human values is sure to be difficult.

Worse, it may seem as if this argument dooms the use of science in policy to a sort of catch-22: the ecological significance of a process, according to this argument, must be based on the human perspective and on human values.

But how can scientific judgments—that such-and-such process is ecologically important, for example—be claimed to be objectively true if the judgment depends not only on repeatable observation but also upon subjective matters like where we stand to take measurements or what we need and value? I suspect that some such worry as this is what lies behind the dogged commitment of scientists and policymakers to the impossible ideal of "objective" description and to artificial distinctions between risk assessment and risk management, as discussed throughout this book. We must, however, get over the fear that value-neutral science loses its scientific credibility.

What does the world of science look like, once we drop the artificial facade of value neutrality? Not all that different, except that a whole new range of interesting questions in the social study of science, questions about how scientists' values do in fact affect their science, are opened up for more explicit study. As for the forming and testing of hypotheses, it should be no secret that scientists are a diverse lot, with many different values and different perspectives, and it is precisely the diversity of values and perspectives of scientists—and the expanding opportunity to seek and find new information afforded by that diversity of perspectives—that guards science against bias. Bias is the inability to see things from some valid perspectives. In the greater community of scientists, the more perspectives and value sets incorporated, the more robust will be our science, and the more likely it will be that our accepted hypotheses will stand up to future experience. What roots out bias in science is not allegiance to the myth of an independent world, but the requirement that scientific results be replicable—under many circumstances, by many observers, from as many perspectives as possible. This limited form of "objectivity" is all that is required to ensure the progression of science and the gradual exposure and correction of error. In the real world of science, it is this diversity of perspective, this demand for replicability of results in more and more situations and constant exposure to new and varied experience, that filters bias from the system. Note that replicability is a temporally dynamic concept: it demands that something happen, that certain experiences be sought across time; it makes no reference to a cotemporal, external world as the standard of objectivity.

Replicability, of course, itself functions as a value in an important sense, and this outcome is what we should expect when we admit that facts and values are mixed together in our experience. Especially judgments of what is important, what is significant, and what is worth studying will affect what we learn first in science. We can expect that new hypotheses and new data will be challenged, all the more so if the new scientific results challenge beliefs that support values and policy positions. Thus, a healthy ferment, a lively public deliberation about values and environmental goals, should actually accelerate the process within science of questioning and replicating scientific results that affect policy controversies.

It is possible, then, even for mission-oriented science to filter out bias and to gradually make our theories more robust under varied experience. Surely there are opportunities for excess of subjectivity, when, for example, the models of scientists hired by automobile manufacturers directly contradict accepted global climate models, or when environmentalists sponsor opinion "surveys" that solicit opinions only from the biased sample of their membership. But bad science is bad science; if these models and claims are examined by others, including those with very different interests or ideological bias, they will not be replicated. Even in mission-oriented science, where decisions about what to study are frankly value-laden, there remains a difference between *observing* something and *wishing one observed it*. Not only do the advocates of these biased models or reports lose the scientific argument in the long run, but in addition their role in the policy debate is diminished as they lose their reputation for unbiased research and observation.

When one shifts from a view of science as exclusively an academic activity and begins to see science as a part of a larger social dialogue and deliberation—if one begins, that is, to see science as mission-oriented instead of exclusively curiosity-driven—relevance to real social values becomes one important determinant of what counts as good science. A decision of a scientist to do mission-oriented science does not, of course, free the scientist from satisfying disciplinary methodological requirements. There will be disciplinary scientists on the broader review committees that examine mission-oriented research, and they will insist that accepted methods be used. The addition of interested parties, stakeholders, scientists from other disciplines, and government managers to the committee that reviews scientific studies adds further bases for scrutiny; it does not release the scientist or scientific work from responsibility regarding disciplinary methodologies.

It is the diversity of experience and perspectives that helps us, as we sail on Neurath's boat of knowledge-seeking, to pick out which of the planks in our hull are weakest. Diversity of viewpoint, respect for multiple viewpoints, communication with people who adopt different perspectives—all of these are what expands experience and replaces biased viewpoints with solider planks of knowledge, planks that have been observed from more viewpoints (beliefs and values that are "unbiased" in the sense that they have withstood challenge by the broadest possible range of experience), guided by the broadest range of interests, and evaluated from every perspective possible.

The idea that science will somehow collapse into subjective rantings of solipsistic freaks if we admit the interpenetration of facts and values in our language and in our public discourse is just propaganda of the ideologically driven modernists. Get over it. The point here is that a healthy mix of evaluative and emotive concerns in our understanding of science is in itself a part of the solution for scientific bias, not a part of the problem.

So we place our bets on an open, admittedly chaotic process of public deliberation in which the traditional disciplines are thought of as nodes, perspectives from which various inquiries can be undertaken. But these disciplines, like meat and potatoes in a stew, retain a disciplinary life and pursue clarifications using their models. Because the various disciplines are inevitably in the stew, however, they cannot help but be affected by the broader intellectual and scientific milieu; funding for research, for example, flows to topics that are thought important by the government and funding agencies. In mission-oriented science, research must provide results that are both relevant and replicable, and study design, choice of scales, and so forth are explicitly affected by social values and by discourse and deliberation about social problems, including environmental problems.

We can return, then, to the questions with which we started this section: What do we mean by saying that the term *sustainability* is a *normative* term in public policy discourse? And how can a normative approach to sustainability incorporate the best science and still function normatively? The term's meaning, given the community process we have described to define it, is intimately tied to the values of the community that uses the term. This view is contrary, of course, to that of economists and others who seek a "purely descriptive" concept of sustainability. We can ask how they can know what is important for a society to save—the central question in defining a practical, actionable concept of sustainability—if they don't consider what is of most value to that community. Values, social context, social goals, and perceived problems can all affect what is important, what should be measured and studied scientifically. In mission-oriented science, it is essential that there be an ongoing, public deliberative discourse to determine what is important enough to study. How one studies it—whether one uses proper disciplinary methodology, describes replicable procedures, and so forth—determines whether it is good science. Its relationship to the ongoing discourse about social values and environmental goals determines whether it is useful, "significant," that is, in the broader context of environmental management and community development.

The articulation of a sustainability ethic by a community, then, requires the incorporation of both science and evaluation. I offered in section 9.1 a schematic definition of sustainability. This schema must be given substance and detail through the articulation of values, goals, and ideals by communities that hope to live sustainably. On this view of evaluation, the first step is to initiate an open and iterative process of articulating, defining, and refining a set of criteria for judging development paths. These criteria are then gradually further refined by explaining how adherence to them will preserve important social values and by proposing measurable criteria that can be expected to track the identified values. Once this process is begun, and as it continues through time, a number of scientific questions about natural dynamics and how they affect humans on various scales will arise. These questions will demand mission-oriented natural science to determine what dynamics and what processes are related to broad social values and the goals those values perpetuate.

At the same time, there will be opportunities and demand for social scientific studies about what goals and values truly are important: which values are negotiable and which are not. It is in this latter process of articulating, refining, and sorting values and goals that a new social science, of the type described and explained in section 7.5, will be important. Idealizations, such as positing ideal outcomes and ideal processes, can thus be useful because what we really want to know is not what people want at this moment, but rather what they would value if they had gone through as good as possible a process of public deliberation, including experiments in mission-oriented science and including social learning. To the extent that we create as-ideal-as-possible adaptive management processes, social scientific study of environmental values and how to achieve them becomes possible.

The question that Peirce taught us to visualize as a trajectory of belief in a community of truth-lovers—a community that is open to new and divergent experience—is, What will be valued after we gain more experience, after we engage in more deliberation, and after we have adjusted our goals in response to scientific probes of the systems under management? Sustainability is each particular community's understanding, based on the imperfect science of its day, of what should be saved for the future; it is a freeze-frame in an ongoing dialogue about who we are and why we care about our physical environment. Sustainability is thus most basically an acceptance of responsibility for impacts of our sometimes-violent actions on the future. When we state a set of ideals for what we want our community to be like in the future, we identify those options and opportunities that give meaning to life in a place. This amounts to understanding a place to involve a dialectical union of a people and their culture, within a physical context—their ecological habitat. In this chapter I have been placing more and more emphasis on the fourth type of environmental value, the set of values that make a community what it is and make a place what it is. These values emerge from the countless choices of community members, but these choices also, on a larger, community scale, add up to something: they add up to an expression of the defining values of a community that is adapted to a place.

9.4 Choosing Indicators as Community Self-Definition

From the beginning of our attempt to integrate evaluation of policies and discussion of social values into adaptive management, I have tried to focus attention on the choice of indicators that will be used in setting goals and in proposing comprehensive criteria for sustainable living. The indicators a community chooses to monitor says a lot about what its members care about and what is valuable to them. Once indicators are chosen, the deliberation of the participants will naturally turn to the question of setting goals and substantive criteria applicable to the indicators. The community's values will be operationalized as requirements that acceptable behaviors will not drive an indicator below a specified target level, for example. In this section I will show in detail how choices of indicators can fulfill the dual function of defining the self-identity of a community and incorporating community values into the environmental management process.

Explicitly choosing indicators is an excellent way to think about the community-based values that lead people to protect certain stuff, stuff that is constitutive of the place their culture will continue to evolve within. These values, I will argue, are the core of any community-based definition of sustainability. Such values involve a commitment—a community performative—to protect certain values that are so tied up with the community's identity as a place and with the shared sense of the good life in that community, that their destruction would render people who live in the community worse off than they might have been. This judgment, furthermore, could hold even if the later inhabitants are more wealthy than were the earlier generations who failed to protect and perpetuate those values. I have called such values communal values. They emerge as a community takes on character and are evident on the community scale. They are values that tie generations together into multigenerational communities. These values, in a sense to be returned to shortly, *constitute* the community by giving it unity across generations; they are all tied up with the community's sense of itself and its members' sense of themselves as individual members of that community; they give meaning to the life of individuals and of the community. Sense-of-place values refer to the special meaning that is given phenomena such as distinctive plants or distinctive landscapes because they are "ours." But I am most emphatically not using the term *ours* in the narrow sense of property ownership and deeds; I am referring to the sense of ownership a community has for its parks and its vistas. These values are "ours" collectively, as a community; they tie us to where the community came from—its natural history—and they also link present members, through the aspirations felt in the present, to a harmonious future relationship, one that embodies inevitable change but which retains its character and the intergenerational meanings stored in community tastes, attitudes, and rituals. Accepting responsibility to protect such values entails also an obligation to protect the context that gives them meaning.

In our decision model, I have treated these performative commitments as SMS rules: once a community commits to protect certain stuff, it is obliged, by virtue of those commitments, to protect the stuff if the costs are bearable. The commitments can then be embodied in criteria stated as maintaining a specified point on a measurable indicator. In this way an indicator tracks a deep, community-based value, and the deliberation about what level of the indicator is healthy and acceptable creates a forum for discussing the weightings and priority to be given *this* value, in competition with other criteria and other social values. In principle, this is exactly the discussion an adaptive manager hopes to create. Given the complexity of problem formulation in environmental policy, dealing with these issues can lead to both social learning and consensus building, if the issues are addressed in a context where all sides prefer cooperation, and if there is sufficient trust among the parties. That's the ideal. But how can I illustrate a process that has never been completed, as far as I know, in any community? Fortunately, there are, if not complete and ideal case studies, many well-documented examples of important, though partial, successes in collaborative decision making in support of cooperative action in an adaptive management spirit.

My discussion of sustainability has, throughout this book, been hampered by a self-imposed abstractness entailed by my theory that sustainability must be defined from the viewpoint of each community. As a result, I have been doomed to speak hypothetically and to hint at what might happen in some places, rather than providing a full-blown definition of sustainability for a specific place. Perhaps I can now take at least a baby-step toward more specificity by engaging in an imaginary exercise. I will assume that in the near future, the community that I live in, Atlanta, Georgia, comes together and decides to begin a dialogue regarding how the region can develop more sustainably and what goals of sustainable living would be appropriate in North Georgia. Then, to really stretch credulity, I will also posit that—by a miracle, I guess—there emerges in the city and the region something approaching what Habermas would call an "ideal speech community." That is, the many municipalities and county governments stop feuding and competing for more rapid development and, in a spirit of cooperation, set out to define sustainability for the region through an open, deliberative process designed to increase cooperation. If I may indulge in these wildly implausible assumptions, I can perhaps tell a hypothetical story that will illustrate how the process of developing new indicators can function to identify and articulate important community values. This little exercise can serve to explore some of my own aspirations for the region by projecting them onto the community as a prediction of what I think might come forth *if* the conditions for a constructive process were in place. My purpose in this fanciful undertaking is simply to illustrate the process by which new indicators can be identified and new and operationalizable goals can be set.

Since I am trying to be as concrete as possible, I should briefly mention what I take to be the important noneconomic values that are prominent in residents of Atlanta. Primary among these is the idea of many residents that Atlanta is a "city in a forest," or a "city of trees." Atlantans think of their city as one that has maintained a remarkable amount of tree cover (despite growing losses from rapid development). Values placed on the wooded aspect of the area are identifiable by noting that Trees Atlanta, a volunteer private group that plants trees for free in private and public places, is a very popular organization, and that most municipalities in the area have ordinances regulating and limiting the removal of trees. We can take this aesthetic value to be representative of the noneconomic values that are important to Atlantans. Even those who advocate rapid economic development are trying to shape that development so as not to destroy the value that is expressed when citizens refer to their city as the "city of trees."

This aspiration might also provide one possible approach to the sprawling growth that is now reaching its tentacles into the hills and valleys of the North Georgia piedmont area. Accordingly, I will use my little fantasy to propose two new indicators and associated criteria by which the quality of future development paths can be judged. More specifically, I propose one indicator that could unify activists and perhaps a majority of the residents of the metropolitan area and a second indicator that would track the quality of growth for the broader region as Atlanta develops and expands into rural areas and exurbs. The goal of these specific proposals is to show how deliberation and negotiation might result in what I will call a synoptic indicator. A synoptic indicator is some measurable feature of an area that reflects many of the deeply held, if inchoate, values and aspirations that residents feel for their area and region.

One of the misconceptions that monists may have about pluralistic systems is that once one opens the floodgates of pluralism, citizens will express many values and propose countless indicators and criteria. In fact, I would expect an initial spate of proposals and that the beginnings of such a process—even if undertaken in an ideal speech community—might indeed be chaotic. For example, most participants are likely to express the hope that economic prosperity will continue. Some will emphasize "smart growth," whereas others will argue for creating more open space, by government purchase of reserves. Still others will couch their aspirations in terms of developing improved public transportation. There might well be a cacophony of goals and proposals. Such a splintering of interests may prove difficult to overcome, but eventually—assuming as we are that there is a widespread attitude favoring cooperation—it will become clear that it is not possible to add one more indicator and one more criterion for each of the many values that people express. Once this realization takes hold, I predict that participants in the process will develop one or more synoptic indicators, indicators designed to track multiple values. For simplicity, I'll assume that there is, from the beginning of the process, sufficient support for developing measures of the long-term success of the local and regional economy and that some of the many economists who live in the region will become active in developing an index of sustainable economic growth. I will leave to the economists, working together with the participants in the process, the task of developing a measurable indicator of sustainable growth in the metropolitan area and in the piedmont region. How might my community supplement this criterion with additional criteria that track important noneconomic values?

Suppose, in the process, we get to the point where most participants, who are actively pushing their own values, agendas, and specialized criteria (such as miles of public transportation routes, or acres set aside in open space), recognize that there is not sufficient support for their own particular indicator to enshrine it as a stand-alone criterion. Then I'll make my move by proposing the following synoptic indicator: I think we in the Atlanta metropolitan area should rate proposed development paths on their likely impact on the total pervious surfaces in the area and in the region. This indicator would express the percentage of land cover that is not paved over, built upon, or in other ways made impervious to the penetration of surface waters into the ground. Why this particular indicator? First, this indicator is relatively easy to measure, using Landsat satellite images and analyses, perhaps supplemented by aerial photography. This indicator, then, can be advanced as a useful measurement that tracks one important aspect of the development process.

Second, and this is what is crucial to making the pervious-surface measure work as an indicator that reflects social values, the pervious-surface measure can attract a variety of participants who, though they may not initially think of themselves as committed to minimizing impervious surfaces, will through discussion and social learning come to see the pervious-surface measurement as indicative of how well their more specific concerns and values are being addressed in the development process. For example, individuals and groups who lobby for wildlife protection may see that an increase of impervious surfaces in the area would have negative effects on wildlife; and these groups will find that they can form alliances with other concerned citizens, including those who worry about conservation of water and protection of the aquifer, the volunteers and groups who work to keep Atlanta a city of trees, those who are concerned about water-quality problems caused by increased runoff, and those who advocate "smart growth" as an alternative to willy-nilly growth. Such a criterion might even appeal to advocates of public transportation, who will argue that centralized transportation systems would reduce the need for roads and positively impact the pervious-surface indicator. This indicator may also be favored by advocates of cleaner air, since smart growth that consumes less ground can also concentrate population to improve access to public transport and reduce the length of commutes. In other words, one proposed indicator can serve as a rallying point for citizens with very different interests to form coalitions and advocacy groups that will take a stand on what is a reasonable minimum—and ideal—ratio of pervious to impervious surfaces.

My vision for this process (recognizing that I have had to make some pretty strong counterfactual assumptions) would be that out of a pluralistic cacophony of voices, coalitions of citizens who may express their values differently and who may even value somewhat different aspects of their environment can support an indicator that can be expected to track several important values, including values and options that many members of the community would embrace as important to save as elements of a fair intergenerational bequest. Pluralism, in a context of deliberation and negotiation encouraged by the two-phase heuristic that recognizes both efforts at action and deliberation about goals of the society, thus opens the door to coalitions and to the development of synoptic indicators. What we look for in a synoptic indicator is a variable that is easily measured and has reasonable connections to a number of physical and ecological dynamics that support a variety of values. In this context—unlike the context created by ideological disagreements—citizens need not see values as competing with each other, nor as inviting confrontation. Rather, values will figure prominently in the discussions, as advocates cite their own values as reasons to choose the pervious-surface indicator as one important way to evaluate various development paths. Then, in the ongoing reflective phase, the participants can discuss and debate what percentage of pervious surfaces to set as the ideal and what level to consider a minimally acceptable ratio. People need not convince others to share their values, provided they can agree on measurable goals toward which to work, because those with differing values can all see how the measured indicator will reflect their own cherished values.

This process also has potential to create new alliances, or at least lead to compromises that are widely acceptable. For example, developers will presumably come to see the desirability of conserving water (so that development can continue), and they will also see the value of open spaces, wildlife habitat, and so forth, if these are widely shared values that will enhance property values. Developers who participate in the process, then, instead of attacking the goal of minimizing impervious surfaces, may start designing developments that have minimal impact on pervious surfaces, thereby integrating the broader values of the community with their own economic values. If a coalition of advocates of a variety of values should emerge, and if its participants were able to highlight a synoptic indicator that arguably tracks many or most of their values, then judging possible development paths could become less ideological and more cooperative. Despite recognizing how far Atlanta is from realizing a deliberative process such as this, I hope this example illustrates how accepting pluralism regarding values need not lead to an indefinite period of chaos and paralysis. Pluralism, because it avoids the either-or thinking of the ideologists, provides fertile ground for communication, negotiation, coalition-formation, and compromise. If Atlanta can get to that point—if the community can get over its tendency to see an irreconcilable battle of ideologies between developers and environmentalists—it may be possible for the community to choose a development path that will fulfill all or most of its citizens' cherished values. That development path, if it conformed to the approach to sustainability and intergenerational equity discussed in chapter 8, would be an implicit or explicit performative commitment to save pervious surfaces as one kind of stuff that the community feels is important enough to protect, provided the costs are bearable.

This fanciful story about my city of residence suggests, I hope, that a pluralistic value system better captures the actual values that people derive from their environment than would an ideological attempt to fit all values into one or another monistic system of valuation; and a pluralistic system opens up far more opportunities for cooperation and consensus than does a system of valuation that assumes gains in economic interests necessarily harm long-term community values. Pluralism and process provide communities with opportunities to agree as well as to disagree and to find actions and policies that serve fairly wide interests. The two-phase process, operating under these conditions, allows the emergence of new indicators and criteria that reflect broad social values within a community, enabling cooperative action even in the absence of full agreement about what is ultimately valuable.

I have also made much of the plurality of scales, pointing out that some social and environmental values are associated with larger dynamics that usually function independently of what occurs on smaller scales. This insight can be applied to the Atlanta story by adding another, larger scale of planning and management on top of the local perspective just discussed. After the Atlanta metropolitan area becomes (in my fantasy) a community in search of sustainability at the level of the city and its suburbs, we might also begin to think about how Atlanta fits into the larger landscape, and also about what kind of indicator would be appropriate to measure and rate various possible development paths for the hills and valleys of North Georgia. Can we choose another indicator, one that would track the quality of development at the regional level, thereby opening up a discussion of which criteria should be applied to rank various development plans on this larger scale?

I think so; to make my point, I will build upon ideas that were suggested to me by a local environmental activist from southern Tennessee. Although he was talking from the viewpoint of a resident of the Tennessee River system (which drains to the Mississippi, rather than to the Chattahoochee and Flint River systems of North Georgia), I think we can modify his ideas to propose a regional-scale indicator that could be transferred into the upper Chattahoochee Basin. I met the gentleman at a regional forestry conference. I never even got his name. He was there as an interested party, listening to papers and trying to get professional foresters to take a stand against chip mills on the Tennessee River. The activist was earnestly expressing his frustration at a series of governmental and private decisions that seemed to make it more and more inevitable that large multinational corporations would be allowed, even encouraged, to construct megamills along the Tennessee River. These huge mills for grinding hardwood forests into chips for the international trade in paper are of such extraordinary size that it would take only a few of them to ensure that virtually all of the remaining hardwood forests in the southern Appalachians would be "chipped out." The activist said to me, "If they let the big chip mills in, they'll scour the Southeast, and replant fast-growing pines in straight rows. I grew up in a hardwood forest. We like our hardwoods. I'll fight to stop them, but it seems pretty hopeless, with the government talking 'jobs,' and the big Japanese money behind the mills." The chip mills no doubt offer the fastest economic growth of any options currently available for the region, so it seems doubtful that the value the activist was expressing was an economic value. He feels that he and his family will be harmed if his community goes down the development path that big international corporations are offering through investments in chip mills.

Perhaps the activist was expressing an aesthetic preference for mixed hardwoods over a pine plantation landscape, but the values to which he appealed against the chip mills certainly went beyond any simple aesthetic taste in the "consumption" of landscapes. Speaking generally about these very specific experiences, we can say that there is a range of experiences or options that are especially important for the activist, experiences that are somehow essential to his sense of self and his sense of family and community. Perhaps he is a hunter or a birdwatcher or a hiker or . . . The landscape and habitat in which he learned to love those activities is the context in which they take place. If these options are removed as a result of the destruction of the hardwood ecosystem he grew up with, he will become poorer by the loss of options that give meaning to his life, that connect him to his past, and that give him hope for the future. Following Per Ariansen, a philosopher friend of mine from Norway, I will call the values associated with these options "constitutive values," because if they are lost, the integrity of a place—its identity as a place that humans of a particular community call home—is diminished, as is community members' sense of self.[13]

Obviously it is impossible to capture scientifically all of this detail about experiences and particular bonds that form between individuals and communities. What is needed is a simple indicator that tracks, in a general way, the values expressed by our activist. If we can articulate such an indicator based on his arguments and concerns, we may be on our way toward the development of a regionwide indicator that would allow us to articulate a criterion that could be applied, probably alongside economic indicators, at a regional level. Assuming that our activist is likely to favor at least some economic growth and increase in standard of living for the region, and given that we have identified the values that motivate his activism as noneconomic, our activist illustrates our claim that some values are too important to be determined by economic calculations, and his case therefore provides a useful instantiation of our pluralistic model. The situation faced by our activist can be characterized as follows. Development interests have proposed to pursue a particular path toward economic development, a path that would positively affect economic activity and likely increase income levels in the area over the coming decades. According to economic criteria, then, the chip-mill path scores high, perhaps higher than any other development opportunities, if projected over a few years. But our activist also knows that there will be predictable ecological and landscape effects if that development path is pursued. The chip-mill path will eventually eliminate certain options that he values highly in a noneconomic sense having to do with his personal, family, and community identity. If these options are gradually obliterated as his community pursues the chip-mill path, the related longer-term and more personal values will be obliterated as well, reducing the continuity he feels with his children, his parents, and the community that will evolve in his place in the future. Logically speaking, then, the loss of these valued options, which support important values constitutive of the activist's sense of self and community, can be understood as losses that are noncompensable economically.

Our activist's objection to the chip-mill path—if I can speculatively put words in my activist's mouth—can now be given expression as follows: "Allowing the chip mills is a development path that scores high on economic criteria, but it has unacceptable consequences for other social values. We should seek a development path that scores reasonably well on economic growth measures and is also able to hold open important options that give meaning to my life and to my social interactions; it is important to me, and to my community, that these options be held open for the future—participation in them represents our identity as a family and as a community." This point might be expressed as "If my community changed in *that* way, I'd no longer want to live here," or "If that happened, it wouldn't feel like home to me any more." The activist and his family, he believes, will be worse off with the chip-mill development than they will be if a different path toward development is followed. This can be described as a noncompensable loss because it is felt as a loss even though the chip-mill path to development is likely to make the community richer. This set of concerns, if expressed by an activist, is not perspicuously discussed using a single-criterion system. It seems more like a problem of finding a development path that comes closer to fulfilling two criteria, based on independent variables. Neither criterion need be given absolute priority; but it might make sense, if other residents of the area agree with the activist, to set de minimis standards for each criterion and restrict serious consideration to development paths that achieve minimal levels for each. The proposed noneconomic criterion would function together with the economic-development criterion on the regional scale, but it would also have to be balanced against other, more local goals and criteria in a multiscalar array of rules. Decision making would take on the form of finding a fair and reasonable balance in the pursuit of multiple goals. Under the circumstances, discussion might be contentious as different groups at different levels agitate to give priority to their own concerns; but given a pluralistic value system and commitment of the participants to cooperative action, the disagreements would be re-solved politically. In this situation, one could say that an approach had evolved that would allow the community to gradually tame wicked problems by finding more-or-less-stable balances among opposing and competing interest groups.[14]

If our activist was a member of an advisory committee in an adaptive management process, we can imagine him proposing that his community should choose "percentage of area in mixed hardwoods" as a useful indicator, and he could explain that the hardwoods are a useful proxy for many of his personal values and that he thinks many others of his community feel the same way. Again, this indicator can be measured using satellite data, allowing a relatively precise measure of types of vegetative cover. Assuming that there was also a participant representing local business interests, we can expect that she would make a case for setting a goal of consistent and robust economic growth. Discussion and negotiation would now become a matter of trade-offs between at least these two goals, and degrees of achieving them, within a democratic process. Although the values advocated by the varied stakeholders would not be commensurable, both would be at least roughly quantifiable and representable as matters of degree. The value of our evaluative heuristic is now clear: if we maintain multiple criteria throughout the participatory process, it will be possible for the participants to discuss the usefulness and importance of multiple indicators and decide which indicators should be emphasized in various, or particular, situations.

Now I admit this is a favorable case, in several respects. First, I happen to have found an activist who expressed his concerns for his community by reference to a single, measurable (by satellite) feature of the landscape. In the case of most values, identification of a measurable landscape feature would require a long process of discussion, scientific observation, and deliberation. Further, it simplifies my case that I spoke to one individual with well-formed concerns; if we place our activist within a complex public process, the purpose of which is to choose several indicators that express or track values shared by a highly diverse community, our task will be much more difficult. Our activist, in other words, would be only one voice in a cacophony of voices and interests, and it would be an open question whether he would succeed in enlisting sufficient support for his proposed measure. Nevertheless, I think this case provides some insight into the process by which sense-of-place values might be expressed by an indicator—in this case, percentage of forest in hardwoods. Such an indicator could be fashioned into one criterion that would function, together with other criteria, as a more significant way to evaluate changes in a community's landscape.

If we can encourage our activist into a public participation process, then he and his neighbors—some appearing as plain citizens and others as representatives of various interests in the community—can begin to articulate which outcomes and risks are unacceptable and to play off economic criteria against other criteria in search of acceptable compromises. The identification, articulation, and measurement of these important values must be undertaken, I have argued, within a broad-based, participatory, iterative process, a process that must be begun, and pursued continually, within a larger adaptive management process in each particular place where community members resolve to live sustainably, according to a definition they have actively chosen.

North Georgia, as noted, faces somewhat different issues than do residents of Tennessee, especially in that large-scale chipping of hardwoods is more economically feasible in Tennessee because the Tennessee–Mississippi river system supports large-scale barge traffic to the open ocean. In North Georgia, however, two economic forces are likewise threatening native vegetative patterns. Large tracts of land owned by timber companies are being transformed from hardwood forests to pine plantations by clear-cutting. Plantation pines have a much shorter growth-to-market cycle, but in the process the landscape and the biota are being radically altered. At the same time, some of the timber companies, impatient even with fast-growth timber rotations, are selling tracts of land to developers who will clear the land and build residences and second homes. Both of these development trends threaten the native vegetation of the area and greatly reduce habitat for wild species. So perhaps we can transfer—and modify as necessary—our Tennessee activist's proposed criterion, percentage of land cover that remains in native vegetative patterns, as a regional-scale indicator that could be overlaid onto the local criteria developed for the Atlanta area. We would then have a system of spatially nested criteria: at the local area level, an economic criterion will be balanced against the pervious-surface criterion in the search for a balanced form of development in the metropolitan area. Other synoptic or specific criteria might also be adopted as the community progressively articulates its aspirations for and commitments to the future. Development planning according to these criteria would also be judged according to the spatially broader (and temporally slower) dynamic that determines the vegetative mix in the wider area. Notice that although the pervious-surface criterion at the local level measures something different from what the native-vegetation criterion measures at the regional level, one could reasonably expect that doing well on the local criterion—disturbing as little area as possible in the process of development—will contribute to the broader, regional criterion. This confluence opens up the possibility of win-win policy initiatives through the creation of greenbelts, riparian and other reserves, and wildlife habitats. If we can devise policies that encourage smart growth at the local level, those policies will also have a payoff on the regional scale. We can hope that in the future, development of multiple criteria, and better understanding of the dynamics associated with them, can lead policy discussions away from bitter contention over ideologically opposed values and toward cooperative actions that serve multiple values.

Thus, although in actuality growth in the Atlanta area today seems out of control and it is feared that rapid growth will destroy many of the amenities that make the area and the region so attractive to residents new and old, we can—by stretching our imaginations—envision a process whereby development would be shaped and controlled by insisting that proposed development have no negative impacts on criteria stated in terms of a few key indicators that are associated with widely held values in the area. If I could see even the beginnings of such a process, I would feel a lot better about the future of the Atlanta region; not seeing even the seeds of such a process yet, I still nourish hope that the region will come together and act not as a community that values only maximal short-term development, but as a true community that will adopt an array of indicators and associated criteria that could track the many values that residents associate with this place.

Speaking more generally, then, sense-of-place values to which a community commits itself by choosing measurable criteria of success in protecting them are best understood as constitutive values, values that express the character, distinctiveness, and continuity of a community that occupies a place. We have acknowledged that these values are not universally binding; they represent commitments, in the form of community performatives, by cohort members of a community to protect certain values—and associated physical, biological, and ecological processes—for the future. Because our analysis of the nature of intergenerational obligation showed that these obligations cannot be based on anticipating what people in the future will want—this is something we cannot know in detail—we took it that these commitments must be based on something that can be known in the present. One thing that can be known about a community in the present is its history. We can learn, by studying its natural and social history, what it is about a place that has been cherished by former members of the community and what current members cherish about it today. It can also be determined why and how such things are cherished, what their meaning is to a culture, by studying its poetry, its art, and its literature. Finally, we can learn what values people in the present want to save for the future, provided those people participate in an open, iterative, deliberative process of self-examination and value articulation, and provided the community can commit to the goal of protecting those values that constitute their community as a community through time. We thus hypothesize that communities with a sense of who they are as a community will, if given an opportunity to articulate the distinctive values that unify them with their place, choose to accept responsibility to protect those values as an expression of their sense of community identity.

Acceptance of responsibility for the bequest package one leaves the future includes, besides commitments to protect the economic well-being of future people, a commitment to protect them from avoidable risks of bodily harm. These are acts of community self-definition. By deciding what to protect, and what indicators to monitor in order to protect these things, a community operationalizes its community-defining values and, if its members follow through and act, they will rewrite the values they have inherited from earlier generations back upon the landscape where they were formed by the interaction of nature and culture in that place. Each generation inherits a sense of place from its forebears. The current people amend this relationship to apply to the ever-changing context and then leave their own mark on the place. As this process continues, if each generation in a place interacts authentically and lovingly with its place, the community's growth will mark a trajectory of authentic interactions between the culture and its context, a trajectory that both expresses and reinforces itself across time. If that happens, the community will have learned how to think like a mountain.

PART III

INTEGRATED ENVIRONMENTAL ACTION

10

IMPROVING THE DECISION PROCESS

10.1 Decision Analysis and Community-Based Decision Making

This book is about finding a middle ground between the ideologies of Intrinsic Value theorists and of free marketeers, and between locking up resources and leaving them open to overexploitation. In part 1 I show that good environmental science is important but also that an overemphasis on science in environmental decision making leaves us unable to identify what is important to sustain. And in part 2 I reject attempts to reduce all values to a single measure, arguing instead for an approach to environmental values that is pluralistic and aimed at finding an acceptable balance among competing legitimate values. More specifically, in chapter 2 and in the appendix, I undermine efforts to establish a prelinguistic reality against which specific empirical beliefs can be held up and tested. I assert that the varying conventions of language that we adopt for the task of description inevitably shape our world; there is no single reality corresponding to our descriptive utterances. In chapters 4 and 7, in particular, I criticize decision theorists for approaching their subject with a mythical belief that the decision problems they study can have uniquely correct answers and, at the same time, capture the essence of complex, real-life environmental problems. In this, the last part of the book, I must, while accepting these apparently unavoidable limitations, show how careful attention to evaluative procedures and consistent application of the methods of adaptive management can guide us toward better environmental decisions, even in communities that embrace many values, including competing and conflicting ones.

The practical problem to be addressed here is whether it is possible to design and implement a deliberative decision process that can operate effectively in complex and pluralistic situations. The decision process we design must, first, encourage enough solidarity in the articulation of constitutive values to allow members of the community to accept responsibility for perpetuating the values in spite of the diversity of attitudes and preferences expressed by community members. Second, the process must lead to a management plan and procedures that will succeed in perpetuating these distinctive and constitutive values across generations. I will briefly survey some important trends in the fast-changing fields of decision science and risk assessment, trends that are resulting in changing views about public participation in decision processes. Paying attention to these trends will allow us to assess the likelihood that the values I have called constitutive values will emerge from an active, participatory, adaptive management process.

What is today called decision science collects at least three mostly distinct approaches to analyzing decisions. The first one is game theory, invented in the 1940s and developed by John von Neumann (a mathematician) and Oskar Morgenstern (an economist). Game theory exploits the idea of a "bet" (a combined or "expected" utility based on both the expected benefit of winning and the subjective probability introduced by the possibility of losing). As noted in section 7.3, this tradition is of immense importance, theoretically, because it established the possibility of measuring subjective probability calculations and of assigning cardinal utilities to outcomes. Game theory, however, has not been extensively applied to particular decisions, largely because economists continue to measure preferences (willingness to pay) rather than employing cardinal measures of expected utilities. Preferences are considered by economists to be behaviors—tendencies to act in certain ways—and the study of them need not refer to subjective probabilities. Welfare economists, in other words, have essentially assumed a behavioristic conception of the human organism as a consumer, and this conception does not encourage an explicit examination of subjective probabilities. Although mathematical game theory, which computes cardinal utilities, can be considered a subfield of economics, it does not connect directly to much of the rest of contemporary economics. Detlef von Winterfeldt and Ward Edwards state that

> economists have in a sense turned their backs on the intellectual consequences of cardinally measurable utility. Welfare economists of the present day, for example, debate with decision analysts about how to assess the value of some socially important development, like a public park or clean air. Welfare economists want to measure that value in dollars, using observed willingness to pay as a basis for the measurements. Decision analysts see numerous problems in this, primarily in that it ignores nonmarket effects. They would much prefer to measure value as utility, in just the contemporary version of the von

> Neumann—Morgenstern sense. This is the intellectual partition that divides cost-benefit analysis from decision analysis.[1]

The central ideas of game theory, underutilized in economics, have instead become the basis of another decision science, which is called decision analysis and which represents, in the cardinal utility tradition of game theory, applications of methods for analyzing utilities resulting from actual or hypothetical decisions. Decision analysis emerged in the 1960s and has developed into an exciting and rapidly evolving field that has had some success in making applications of utility-based analysis to complex problems. Decision analysts include risk analysis as one of their tools; this tool can be made more flexible as it is combined with multi-attribute utility theory, to which I will return shortly.

But the decision sciences also include a third, relatively independent, discipline: the psychological study of how individuals actually make decisions. Here, as in economics, the methods are empirical and the topic is human decision-making behavior. Today, however, most psychologists have gone beyond behaviorism, recognizing the value of including (among other factors) at least subjective probability judgments as an element of decision making, and so their models of individuals making decisions diverge sharply from that of economists. This is surely one of the most exciting aspects of modern decision sciences, as psychologists with diverse theoretical frameworks—from economics, physiology, the cognitive psychology of inference, logic, the study of artificial intelligences, and so forth—direct their attention to human choice. Obviously, because of the complexity of these three approaches, we cannot study this area in detail. I will simply offer a few remarks about the importance and role of such studies, theoretically, and then make some opportunistic use of these tools to develop a new and more eclectic approach to community-based decisions.

Given pluralism, and considering all models to be useful tools for varied purposes, I propose to use all of these tools simultaneously. For example, I referred to the formalisms of game theory to show how environmental problems—complex in structure and including community-level values—are a type of game that can be "won" only through cooperative behavior. At this point there is little reason for us to pay much attention to disciplinary boundaries. We are ultimately addressing a practical problem: How can diverse, democratic communities develop procedures that encourage cooperative action to protect their environment? In order to answer this question, we will survey the decision-making tools that are available to be employed in various adaptive management situations.

Here is the complex situation as I see it. Von Neumann's mathematical work on game theory in the 1940s spawned several research traditions. Two research traditions that are directly relevant to our task are game theory itself and decision science. Game theory, which can be thought of as the study of the "logic" of choice games, thrives as an important theoretical tradition with the goal of understanding decisions in rational terms. Though a minority among economists, game theorists also include applied decision analysts and philosophers, cognitive psychologists, and so on. We found this tradition useful in demonstrating the impossibility of solving problems as complex as that of protecting the commons through analyses of individual sovereign preferences. That proof helped us to locate environmental problems in the "logical space" of cooperative decisions involving communal values. Environmental problems are complex problems that must be solved from the perspective of a community's place in a multiscalar physical system, using iteration as a tool to enable community members to act cooperatively and adaptively.

Cognitive psychologists and other cognitive scientists are interested in many of the same questions about how people make decisions, but they approach the problem empirically, attempting to develop hypotheses and explanations that will allow us to predict what people will choose in well-defined situations. What most of this empirical work shows is that formal models do not have a very good record in predicting actual human behavior. Actual people, once a problem gets complicated, or when dread risks become involved, tend to fall back on heuristics (like "better safe than sorry") rather than computing expected utilities. So this field of psychology and related decision sciences—the study of how people actually do make decisions—can help us a lot in the search for better public processes. It also tells us, substantively, that by far the most important aspect of problem solving and decision making is problem definition—which corroborates our findings based on an analysis of wicked problems in section 4.1. Again, this is an invaluable insight, one that I will try my best to build into our general approach.

Finally, there is the somewhat independent tradition of risk analysis, which has grown up in the shadow of EPA and other regulatory agencies and which defines itself as a "pure" science of exposure and risk, a science that can analyze a risk, once the "relevant" scientific information has been gathered. We have portrayed this field as being in a period of transition. As EPA's objectives broaden to include regulation of ecological and other risks, the field of risk analysis must respond with useful models for exploring ecological risks and other systemwide phenomena. In essence, this field has fallen victim to overly optimistic assessments of its potential to act as a unifying framework for all environmental policy. Its failure has long been masked by a confusing, and confused, attempt—by EPA and risk assessors themselves—to perpetuate the myth that risk assessment can and must be studied separately from the evaluative and political processes. A complete analysis of a risk situation must include some way of determining what is important enough to be worried about; and such an analysis cannot be completed within a discipline that does not encompass the discussion of values.

Again, as with our criticism of economics, it is important to avoid misinterpretation: the conclusion is not that risk assessment tools are useless, but only that they are misused when they are treated as monistic and comprehensive. We must consider risk analysis to be one tool among others, useful for some tasks, not for others. If risk assessment does survive as a set of tools, its practitioners will have to join the dialogue with value and political disciplines so that their tools can be absorbed into the practical process of policy choice. Risk assessors must be integrated into mission-oriented science and contribute to the larger adaptive management processes. Time will tell whether risk assessment survives as an independent science. My prediction is that it will gradually be swallowed up as one fairly precise model for decision making that can be applied only in rather narrow situations, where important value questions can be agreed upon and certain kinds of relatively unusual data is available.

The whole area of decision science is changing so rapidly because it must do so. Much of the prior work in this field has been shaped by the myth of value-neutrality. Once the full implications of the impossibility of separating facts and values in actual real-world, problem-filled situations are appreciated, I predict that there will be a major shift in the field, with disciplinary boundaries being relocated. These changes, precipitated by a changing understanding of the role of values and value-neutrality in risk analysis and decision science, are already under way; but in most areas they have not gone far enough in the direction they are headed.

To show these trends—and also some limitations in them—we will look at three stages in the understanding of risk and decision making over the past two decades. Let us break into the ongoing story in 1983, the year the National Research Council published its "Red Book," formally named *Risk Assessment in the Federal Government.*[2] The Red Book was so important to the development of environmental policy in the 1980s and early 1990s because it articulated and solidified the standard view of risk analysis accepted at that time. This conceptual model was adopted by Reilly and many of his deputies at EPA as they tried to reshape the practice of the agency to use risk analysis as a unifying method of agency decision making. The authors of the Red Book insisted upon a sharp distinction between risk assessment and risk management, based on the assumption that values can be confined to the purview of the risk manager.

A second stage can be represented by reference to recent developments in decision science, developments that continue even as I write and which are still not fully accepted by all practitioners. Somewhat arbitrarily, I will freeze-frame these developments in 1986, by examining the widely respected text *Decision Analysis and Behavioral Research,* by Detlef von Winterfeldt and Ward Edwards.[3] This text, despite being almost two decades old, remains a standard general text in the field. The book also (as noted by the authors in the preface) undertakes some unorthodox treatments of values and utility in its analyses, and thereby it allows us to illustrate changes and trends in the ongoing—and clearly incomplete—transformation of the decision sciences.

The third stage in the process of rethinking the role of decision sciences and the risk assessment/risk management (RA/RM) process is represented by another NRC report, *Understanding Risk: Informing Decisions in a Democratic Society.*[4] This 1996 report explicitly distances itself from the 1983 report, explicitly criticizes the RA/RM distinction, and urges the importance of process over narrow modeling efforts. I hope this surfing over disciplinary boundaries is not confusing—the point of my narrative is to show that as we learn more about how people make decisions, including the importance of meaningful public input to decision processes, the roles of various tools are changing. In the case of risk analysis, tools that were developed for narrow tasks associated with regulating chemicals that affect human health have proved unable to address problems of ecological risk. As the shortcomings of risk assessment in the broader arena became obvious, the rationales for many of EPA's regulatory responses and priorities were increasingly called into question. Little progress has been made in implementing risk assessment for ecological risks, and even less in reordering priorities according to comparative risk studies. The resulting situation is exactly what one would expect, given the linguistic and communicative problems plaguing environmental discourse.

In this context of failure and confusion, the 1996 NRC study seems to me to get it mostly right and pulls no punches in acknowledging the failures and problems associated with sharply separating risk assessment from risk management. Although this study does not go as far as I would go in wiping out the descriptive-prescriptive line, it goes a long way in that direction and thereby has set the stage for an important step forward in both theory and adaptive management practice. Let us examine the trajectory of the fields of decision science and the intertwined question of the role of risk assessment in the process of policy evaluation, by briefly summarizing and analyzing these three documents.

10.2 What Does Not Work: The Red Book

The Red Book was the report of an expert panel convened in the early 1980s to examine and codify the theory and practice of risk assessment in federal government agencies. The field of study called risk assessment was defined narrowly, as "the assessment of the risk of cancer and other adverse health effects associated with exposure of humans to toxic substances." The report states at the outset, without argument: "Regulatory actions are based on two distinct elements, *risk assessment,* the subject of this study, and *risk management.* Risk assessment is the use of the factual base to define the health effects of exposure of individuals or populations to hazardous materials and situations. Risk management is the process of weighing policy alternatives and selecting the most appropriate regulatory action, integrating the results of risk assessment with engineering data and with social, economic, and political concerns to reach a decision."[5]

The Red Book describes risk assessment as culminating in a final stage, called "risk characterization," defined as "the process of estimating the incidence of a health effect under the various conditions of human exposure described in exposure assessment. It is performed by combining the exposure and dose-response assessments. The summary effects of the uncertainties in the preceding steps are described in this step." The passage is illustrated in figure 4.1, which shows risk characterization as the culmination of the risk assessment process, as forming the "final package" of factual information that is passed to the risk manager.[6] The sequence of tasks, as shown in the figure, begins with the research stage. In this stage scientists prepare "laboratory and field observations of adverse health effects"; these are passed on to risk assessors, who identify hazards, estimate dose-response assessments and exposure assessments, and summarize the data in a "risk characterization." This information is then forwarded to the risk manager, who develops regulatory options and evaluates public health and other effects of regulatory options, leading to a decision. This, of course, is nothing but the serial view of science and policy diagrammed and applied to decisions regarding risk! The Red Book was the document that solidified the seriously misleading idea that good decisions would result if decision makers were to have "objective science" on which to base their decisions. Pure science, however, must be translated into useful applied models, including dose-response models and exposure models, so that a decision maker can analyze the consequences of regulatory options. According to the Red Book, risk characterization includes "no additional scientific knowledge or concepts,"[7] and judgment was to be held to a minimum. This approach minimizes the role of determining what the problem really is, and it seems to preclude a meaningful discussion of the values at stake, what type of science is important, given these values, and so forth. These questions, hardly even discussed, would fall under "risk management." This model informed the Risk Assessment Forum—which I failed to please with my reports advocating more front-end discussion of values and goals—and it has dominated federal agencies since.

In particular, the Red Book model still maintains a stranglehold on EPA. Throughout Browner's tenure at EPA and well into the George W. Bush administration, the agency has held on to a definition of risk characterization as a summarization or translation process, built upon scientific analysis and then passed to "decision makers" or "risk managers." This explains, at least partly, why there has been so little attention to the noneconomic social sciences at EPA. Since the agency, even under Clinton, was committed mainly to minimizing economic impacts of regulation, this meant the only social science applicable to risk managers was economic science. Consequences, mentioned in the second box of the final column of figure 4.1, have been counted as economic costs only. As a result, the agency has developed no expertise—and has funded precious little research—on noneconomic ways of valuing natural systems or changes in the human environment. Even the original advocates of this model, as well as long-time critics of the "serial view" such as myself, have found the 1983 conception of risk assessment severely deficient, as will be shown in section 10.4.

10.3 Heading in the Right Direction: The Changing Field of Decision Science

Before we continue our critique of—and improvement upon—the old RA/RM decision model, it will be helpful to consider briefly some important trends and developments in decision science more generally. Rather than try to summarize this fascinating field in a few pages (which would of course be impossible even if I knew enough), I will simply discuss one important text that for our purposes can be considered representative. The book by von Winterfeldt and Edwards is the latest reasonably successful attempt to put our knowledge and the possibilities of decision science between two book covers.[8] The authors offered a thoughtful analysis of the assumptions of their own field and suggested major revisions in goals and methods of decision science, as of 1986. Many of these revisions are truly important, and their proposals have come close to orthodoxy today. We are using this book as our representative of the middle period in the trajectory of change in decision science and risk analysis because the authors' reassessment of the goals of their field sets the stage for the major reassessment of regulators' use of RA/RM models that was eventually embodied in the 1996 NRC report, *Understanding Risk*. The viewpoints summarized in von Winterfeldt and Edwards, in other words, pushed orthodoxy regarding the nature and role of risk assessment in the broader field of decision science in a critical new direction, a direction in which further evolution continues to occur.

Let me begin by mentioning several specific innovations proposed by von Winterfeldt and Edwards, to provide a better understanding of the trends that I hope will eventually lead to a more integrated way to study environmental values in the context of adaptive decision making. First, although they were certainly not the first to notice that there was surprisingly little cross-fertilization between behavioral and physiological psychology and decision science, the authors undertake the huge task of integrating empirical research on decision making and decision processes with the more formal and theoretical work of decision scientists. One of the main aims of their text, as implied by its title, was to assess the relationship, in the mid-1980s, between decision sciences and behavioral sciences and to improve communication across this disciplinary divide.

A major methodological change was also introduced in the book as the authors found themselves "uncomfortable with the usual distinction between value and utility," on which "value is typically a transformation (which may be linear) on some physical scale, and utility is a further transformation on value, intended to take into account the decision maker's attitude toward risk." They therefore rewrote their field without one of its key distinctions: "Though we recognize that attitudes toward risk are observable and important, we prefer to think about them in ways other than as visible evidence of the value-utility distinction."[9] Moreover, the authors increased the extent to which they considered the context of decision making in addition to formalizations of aspects of the process, and they accordingly considered whether different attitudes toward risk would be more appropriate in some situations than others. This attention to context implies more up-front emphasis on the role of values in problem formulation and risk characterization, a departure from the approach taken in the Red Book. In all of these ways, Winterfeldt and Edwards's book seems to me to be a huge stride forward, not just in how decision analysts should proceed to do their work, but also in identifying realistic goals for the discipline.

Another important development in decision analysis is the introduction of multicriteria analysis, or multiple-criteria decision making (MCDM). Following Harold Glasser, I will use *MCDM* as shorthand for a variety of related techniques as exemplified in several disciplines and approaches, such as multiple-attribute decision theory and multiple-objective decision making.[10] MCDM, so understood, was not an innovation introduced in *Decision Analysis and*

Behavioral Research, since the authors and others had been experimenting with this innovation for more than a decade by 1986, but MCDM is presented as an important technique for analyzing decisions that have multiple value dimensions. For example, in considering an offer of a new job, one will surely consider the salary offered, but one must also compare the job to other job opportunities with respect to benefits, convenience of access to one's home, and the likelihood that this job will lead to further opportunities in the future. We have argued at several places in this book that the economic model for decision making, which can be applied only if one can assign a dollar figure to these various characteristics, imposes upon the decision maker an awkward and sometimes impossible task of expressing all of these dimensions in economic terms. MCDM is an important and sometimes useful technique for modeling decisions in which multiple, apparently competing, values are at issue. The technique, to simplify greatly, begins by listing as complete as possible a set of value-relevant objectives and then attempting to assign weights to attributes associated with those objectives. Choices available, such as the various job opportunities available at a given time, can then be evaluated according to these various weighted criteria. MCDM computational techniques can then be used, based on these weightings, to rank the job opportunities. In group and collective decision making, one can sometimes involve governing bodies in the process of assigning weights. MCDM thus differs from economic analysis by avoiding the difficult task of reducing multidimensional values to a single dimension; but it shares with economic analysis the broader goal of developing a method of decision making that, given an initial elicitation of preferences, can mechanically compute the best outcome based on those preferences. MCDM therefore constituted an important new tool for examining decision processes.

Despite these innovations and new directions, however, von Winterfeldt and Edwards also seem to me to have held on to questionable assumptions, which limit what they can see as future possibilities. Most importantly, they retain a rather positivistic insistence on a sharp separation between descriptive and normative science. For example, they imply that work in the empirical, behavioral sciences is "descriptive," whereas the field of decision analysis is "normative" in the sense that its goal is to improve decision making rather than to describe and understand it as a human behavior. The authors acknowledge that descriptive models of decision making, nevertheless, have led to changes in the content of normative models over the years:

> The earliest normative model(s) of decision making in effect attempted to prescribe not only how one should go about implementing one's value system, but also to some extent what that value system should be. Such prescriptions

> turned out to be so different from the actual behavior of reasonable people that the content of the normative model was reexamined and much of the prescription of values was removed. Contemporary normative models for decision making are little more than sets of rules designed to ensure that acts will be coherent or internally consistent with one another in the pursuit of whatever goals the decision maker may have. Nevertheless, this requirement of coherence and internal consistency is a very strong one, so strong that no one seems able to satisfy it fully.[11]

I take this comment to show that its authors—even while explicitly stating that the distinction "between normative and descriptive models is much less clear-cut than it sounds"—are adhering as closely as possible to the myth of a sharp separation of normative and descriptive tasks. This is shown in their pullback from specifying substantive values that must be attained as a part of the evaluation of decisions. Because they continue to maintain the artificial distinction between factual and evaluative discourse, specification of values is treated as exogenous to their decision analysis. If a given specification is challenged, they have no techniques to examine the values and the challenges, so the only strategy they can countenance is to retreat, eliminating substantive values from their analysis.

In retreat, they restrict their rules and decision analyses to the development of formal analyses of decisions, implying that they can criticize and improve decision processes simply by appeal to formal norms of consistency or coherence. Empirical psychologists, however, have found that most people, including very successful decision makers, do not act as predicted by the formal models in many cases. Worse, as noted in section 4.1, most important environmental policy problems are wicked problems, and no formal model can compute a single best outcome for such decisions. This outcome leaves the decision scientist, whether of the descriptive, the normative, or the formal ilk, in a quandary about what, exactly, he or she is doing. To see this, consider three possible answers to the question, What do decision scientists do? (1) decision science is a science that describes how people make decisions in actual situations; (2) decision scientists provide decision makers with an account of what a good decision is; (3) decision scientists tell decision makers what processes are more likely to lead to better decisions.

Once they have purged substantive values from the normative study of decisions, Von Winterfeldt and Edwards have apparently rejected the second role for their science, because it seems highly unlikely they can provide a general account of which decisions are good decisions without reference to substantive values involved. So the quandary is between description of actual decisions and the prescription of processes—between answers 1 and 3. One would like, of course, to say that decision science can answer both questions by descriptively studying the practices of good decision makers. But unless one can provide an independent characterization of which are the "best" outcomes, one cannot nonarbitrarily separate good decision makers from bad decision makers. Decision scientists, in other words, cannot proceed toward specifying good procedures as those resulting in good decisions without having a process-independent means to determine when a decision is a good one. They ultimately cannot bring values analysis into their science, because they are committed to value-neutral science; but without values, they cannot proceed normatively to evaluate the substance of people's actual decisions.

Interestingly, this same argument applies to MCDM because, despite its ability to evaluate decisions on multiple parameters, it is still dependent upon an initial elicitation of unquestioned preferences. The decision, then, is evaluated by a mechanical computational technique. In MCDM one obtains weightings of specific criteria in addition to preferences over outcomes, but the basic structure of the analysis remains unchanged: inputs representing values are elicited and taken as givens for the purposes of analysis—no deliberation or reassessment of those initial values is built into the process—and the decision is analyzed by computation and aggregation over those given values and preferences.

The decision sciences, in other words, by adhering to the outmoded belief of positivists that facts and values should be addressed in different discourses, leave value analysis exogenous to their science, frustrating their hope to provide a full-bodied normative service to potential decision makers. They fall back on weak criteria like internal consistency, which are clearly inadequate to guide decision makers who must make substantive choices among conflicting values.

I believe von Winterfeldt and Edwards recognize this problem—if they do, it would explain why they insist on the importance of decision context and why they begin to advocate more emphasis on process. They recognize, that is, the impossibility of a decision analysis that fully formalizes reasoning processes, including value conflicts, within an abstract system of inference. This impossibility apparently implies that decision analysis cannot be accomplished by comparing actual decisions to a formalized "argument" that relies only on reports of individual preferences and "logical" aggregative inferences from those reports. Descriptive reports of individual preferences, supplemented by the rules of logic, cannot achieve anything approaching completeness over the range of human values and achieve internal consistency at the same time.

Von Winterfeldt and Edwards respond to this quandary with a crucial distinction between "instrumental values" and "ends." They note that "few people are willing to commit themselves firmly to the goal of rationality, and even fewer are willing to claim that the goal is attainable." They say they do commit to this goal, and they set out to explain in what sense the goal is attainable: their book is intended to help people be rational in making inferences and decisions. They say that "One of the several reasons for the unfashionableness of the notion of rationality is the failure to distinguish between rationality in the selection of ends and rationality in the selection of means," and they state that their book "is almost entirely about the latter." That is, they set out to analyze "instrumental" rationality, the rational pursuit of an end by the most efficient means, but they have almost nothing to say about how to choose ends rationally, and they indeed even suggest that no rational analysis of substantive values is possible. But the qualifying phrase "almost entirely" gets introduced because of their experience in actual consulting, when processes for improving instrumental rationality are encouraged: "In almost all applications I have felt that decision makers benefited from clarification of all of their values . . . insofar as they were related to evaluating the decision to be made." Although decision analysis focuses on instrumental values, "in the process it also helps decision makers think about and clarify the values here called 'ends' as well."[12]

This practical "admission" should have led von Winterfeldt and Edwards to look again—and more critically—at the artificial separation of descriptive and prescriptive discourse and the associated distinction between instrumental values and "ends." Trying to create a science that merely makes inferences from values to actions as a way of sealing oneself off from discussions of true ends is a valiant try at a hopeless task. The only way to separate values and facts is to create an artificial, formal language that arbitrarily enforces the distinction. But since such languages are so different from ordinary discourse, in which facts and values are all mixed together, they are of questionable usefulness in practice. If, however, one turns this problem on its head, if one notes that in actual decision-making practice, values (ends) will always be present explicitly or implicitly in the context, then decision analysis can be broadened to make the articulation, justification, and disputation regarding "end" values a part of the analysis. In such a broader analysis, carried out in the more realistic conditions of public discourse, end values will be all mixed together with the language of "instruments." In short, if we follow pragmatists in emphasizing problem context and decision analysis in real situations, it will be impossible to separate proximate from ultimate ends in any sharp fashion.

We are concerned here with the role to be played by the abstract formalisms of decision analysts. Interestingly, the issue perfectly parallels the problems struggled with by Bertrand Russell and by Ludwig Wittgenstein and Rudolf Carnap in their early careers.[13] Their emphasis on the tools of symbolic logic encouraged them to create formal languages, but they paid too little attention to the context in which languages operate. They tried to "model" ideal thought and inference—which is of great theoretical interest—but they had no way to get at the meaning and the impact of language in its communicative, social context, which requires semantics and pragmatics in addition to formalization ("syntax"). Repeating this mistake, early decision scientists have overemphasized getting formal rules right, without paying enough attention to decision contexts and problem formulation. Any language that is rich enough to consider, within a diverse community, all aspects of a cooperative decision is too rich to be fully formalized. Only a language as rich as ordinary discourse can give voice to all of the considerations necessary to reach a "rational" decision. The formalized languages can be advanced as "models" of a person's decision process by validating appropriate inferences that would justify certain behaviors on the basis of "bets." But modeling a decision from within one person's decision process is of no use if the language in question cannot be a tool of communication as well. For mere symbols to capture the complexity of human action, they must be embedded in rich, natural languages that are suffused with values, perspectives, and meanings. But once one moves into actual decision arenas—adaptive management programs with public involvement—context becomes much more important than formalized inferences.

As noted, von Winterfeldt and Edwards clearly recognized the importance of context in 1986, and they explicitly considered whether attitudes toward risk vary situationally. Once the full implications of these considerations are known, a revolution in the field of decision science will be forced. In particular, the attitude toward models must change as one moves into actual decision-making contexts. We have seen time and time again in this book that only conceptual confusion results from attempts to take narrow, disciplinary tools and force all available data and information into the form necessary to fit the tools. To recast decision science as a practical discipline that addresses real problems in their decision context, it will be necessary to reconsider the role of formal models. Formal models, in a decision-making process, are tools that can be consulted, but the decision context must be part of a larger discourse that can include deliberation in a real, full-blooded language. A multivoiced, deliberative discourse cannot be reduced to a system for formalizing inferences because in a real decision situation, important questions turn on interpretation, meaning, and applications to particular situations, and these questions are, in turn, inseparable from problem formulation. This process of interpreting information and weighing priorities cannot be represented as an inference based only in syntax; it must always occur in the richer world of purposes and communication.

Again, we see that the separation of value analysis from the techniques for evaluating decisions can be accomplished only within an artificial, formal system that necessarily misses the rich context in which people actually deliberate and interact in the face of real problems. In real problem situations, values are present, either implicitly or explicitly, in the problem formulation; and one cannot get the problem formulation right without an iterative process of comparing, restating, and recombining values. It is this broader, richer discourse that is essential to getting the problem formulation right, and what we have seen is that no formalized, axiomatic process of reasoning can capture the full richness of this broader, open-ended discourse.

Success in community-based adaptive management can be defined and pursued only in this broader discourse. The decision sciences therefore face a dilemma. Either they can continue, as they have generally in the past, to insist on a sharp separation of values from factual information, in which case the tools they develop can at best be useful tools, perhaps useful in specific situations as specified in the broader conversation; or they can expand their discourse to include discussion of values and full-blown, contextual and problem-based deliberation about what to do as a part of their decision analysis. In the first case, they must apparently leave important questions of context and problem formulation out of their "science"; such questions must be left to others (or to themselves, after they have removed their scientific hats and become plain citizens). In the latter case, they can make value analysis endogenous to their discussion and plunge into the richer discourse of public deliberation and debate. In that case they must give up any suggestion that they can evaluate public decisions by generating inferences from preferences to actions within a logically consistent axiomatic system. In the latter case, they can expand their discourse and their "science" to include problems of context and problem formulation, as von Winterfeldt and Edwards clearly wish to do. To fully accomplish this goal, however, they will have to desert their positivist commitment to value-neutral science, and they will have to give up their misguided search for "correct" outcomes that can be formally derived. Good decision making must be based on an understanding of good decision-making *processes*, not on deriving measurably superior *outcomes* within a formal system of analysis.

10.4 Getting It Mostly Right: *Understanding Risk*

I have portrayed decision science as a discipline undergoing rapid change and rethinking of basic principles and goals. There is no better illustration of this shift than to return to the narrower question of how the specific techniques of risk analysis can be used in public decision processes, and to examine the results of the more recent National Research Council study on risk. As mentioned above, from the early 1980s until the middle 1990s, the official position of the federal government vis-à-vis risk analysis was the one stated in the Red Book, the 1983 NRC report briefly described above. The NRC has conducted several studies over the past two decades, examining broad issues at the intersection of risk science and policy; in the early 1990s, a consortium of federal agencies agreed that "the way the nation handles risk often breaks down at the stage of 'risk characterization,' when the information in a risk assessment is translated into a form usable by a risk manager, individual decision maker, or the public."[14] This formulation of the problem is a natural one, given the thinking at the time. As in figure 4.1, EPA turned problem formulation—"risk characterization," as the agency called it—into a virtual "black box" by referring all of these issues to "risk managers" and carefully purging its analysis of any references to values.

But now the risk managers themselves were asking by what method they should evaluate the risk, once characterized. The implicit answer, of course, is that the risk manager, furnished with the best science available, can "evaluate" the risk by reference to "public values." This means, however, given the serial view of risk assessment and risk management, that all of the science will have been gathered *before* problem formulation is undertaken. This unrealistic model, forced upon EPA by its commitment to value-neutral science, virtually ensured that there would be no systematic way to gather and organize information about social values. Worse, it means that most of the scientific work relevant to an environmental problem is completed before we have a clear understanding of the problem or of the conflicting values involved. Scientific studies cannot be designed to resolve uncertainties that really make a difference. In a dynamic management situation, however, there will be quite specific disagreements among different interest groups who have quite different mental models of their common problems. What is needed is targeted science to turn previously ideological and perspectival disagreements into testable hypotheses. The best way to test such hypotheses is to incorporate research right into the management process, encouraging design of studies relevant to determining just what are the driving forces behind the problem—risk assessment, in other words, can be working on the right problems only if it is embedded in a process of adaptive management.

But adaptive management is mission-oriented management, and mission-oriented management, obviously, needs a mission. Given that EPA was spending most of its research budget to create a new highly specialized, value-neutral science, risk assessment, it is not too surprising that the question of how to articulate, evaluate, and prioritize social values had been given short shrift. And these are precisely the issues—sure to be controversial—that are crucial in the problem-formulation stage.

In response to requests for help from those charged with "managing risk," the NRC convened a committee of seventeen representatives of various specialties, including risk assessment, epidemiology, toxicology, ecology, public policy, economics, decision science, social science, medicine, public health, and law. This committee, upon meeting with its sponsors, which included eight federal agencies and commissions and the Electric Power Research Institute, requested a broadening of the narrow charge that they examine risk characterization. In the end the committee was given a very broad charge, to address "technical issues such as the representation of uncertainty; issues relating to translating the outputs of conventional risk analysis into nontechnical language; and social, behavioral, economic, and ethical aspects of risk that are relevant to the content or process of risk characterization."[15] This charge opens up the broadest questions of value, and these questions, of course, quickly spill out beyond the boundaries of risk assessment and even beyond risk management more generally. Any discourse rich enough to encompass these issues would necessarily take place in ordinary language, augmented from time to time with the technical models of particular disciplines, including risk analysis. Simply by insisting on such a broad charge, the committee preordained that they would have to examine the "context" of risk assessment as well as the content. But the more context is included, the less the topic can be confined to any technical models of any profession—what we have called an EPA tower. The broad charge entailed that the subject matter of the study would be the whole process of decision making, not just a technical issue about how to "characterize risk." And that is exactly what was recommended by the committee, as will be illustrated below, when I briefly summarize its five recommendations. First, it will be useful to look at what this committee said after having looked again at the controlling idea of risk assessment—the sharp separation of risk assessment and risk management—in the federal government.

On the first page of the first chapter of the study, the committee states that it has uncovered a "basic misconception of risk characterization and its relation to the overall process of comprehending and dealing with risk." It goes on to say, "Our approach involves a substantial change from the formulation of risk analysis that many federal agencies and other organizations have been using for more than a decade." Once the committee had interpreted its charge broadly, it could not avoid a central dilemma of risk decision making in a democracy: "Detailed scientific and technical information is essential for understanding risks and making wise decisions about them, yet the people responsible for making the decisions and the people affected by the decision and who may therefore also take part in them are not themselves expert in the relevant science and technology."[16] Although the study noted that in such areas as airline and automotive safety, this dilemma had proved manageable and that the public had trusted the agencies to regulate in these areas, it indicated that in other areas, such as siting decisions in the management of toxic and radioactive wastes and in many areas of environmental regulation, the regulative agencies involved had been stymied by lack of public trust. In these latter areas, special interests had been successful in blocking preferred policies of relevant agencies.[17] This outcome, of course, is what would be expected if the viewpoint of this book is correct: in automobile and airline safety there is no question that the goal is to reduce casualties and that to do so one must reduce crashes and impacts of unavoidable crashes. This provides enough unity on goals and on intermediate objectives to turn most controversial issues into technical questions of how best to achieve the goals and objectives. In areas where there are sharp conflicts of interest, however—siting of a toxic processing or storage plant, for example, since wherever it is placed it will lower property values—the choice of goals and objectives expresses itself in competing understandings of the science; participants identify different models of the dynamics as driving the problem. In such situations, political and regulatory stalemates are to be expected, because various parties construct mental models of the "problem," models that often minimize the impact of their own activities and point the finger at others.

The committee, in order to address the dilemma in those areas where science is contested, found that "one must consider other parts of the risk decision process, particularly the various analytic activities that provide the information used in characterizing risks. The purpose of risk characterization is to improve understanding of risk, and everything that goes into such understanding is necessary for effective risk characterization."[18] With this broad understanding of the problem, reliance on "technical" solutions and experts' models will never be adequate. If the goal is to improve decisions in a situation in which value conflict creates scientific uncertainty and lack of trust, attention must be directed to the decision process, rather than to abstract models.

The summary recommendations of the committee speak eloquently about how far risk analysts and associated disciplines had progressed in their thinking since 1983. The committee begins its report with a ten-page summary explanation of seven recommendations. I summarize these here, using the committee's own words whenever this is consistent with an efficient expression of the recommendations.

1. Risk characterization should be a *decision-driven activity,* directed toward informing choices and solving problems. As such, risk characterization can

be achieved only through an analytic-deliberative process and can be successful only if, through deliberation and analysis in asking the right questions, it is able to correctly characterize the problem faced.

2. Coping with a risk situation requires a *broad understanding* of the relevant losses, harms, or consequences to the interested and affected parties.
3. Risk characterization is the outcome of an *analytic-deliberative process*. Its success depends critically on systematic analysis that is appropriate to the problem and responds, through an open public process, to the needs of the interested parties, who communicate their concerns and who also gain greater capability to participate in the ongoing process.
4. The analytic-deliberative process leading to a risk characterization should include early and explicit attention to *problem formulation* in an open process in which the full spectrum of interests is represented.
5. The analytic-deliberative process should be *mutual and recursive*. Analysis and deliberation must be ongoing and complementary.
6. Those responsible for a risk characterization should begin by developing a provisional *diagnosis of the decision situation* so that they can better match the analytic-deliberative process leading to the characterization with the needs of the decision.
7. Each organization responsible for making risk decisions should work to *build organizational capability* to conform to the principles of sound risk characterization. At a minimum, it should pay attention to organizational changes and staff training efforts that might be required, to ways of improving practice by learning from experience, and to both costs and benefits in terms of the organization's mission and budget.

In keeping with this summary, which embodies in different language many of the recommendations developed in this book, the committee offers a new definition of risk characterization: "Risk characterization is a synthesis and summary of information about a potentially hazardous situation that addresses the needs and interests of decision makers and of interested and affected parties. Risk characterization is a prelude to decision making and depends on an iterative, analytic-deliberative process."[19] Further, as the committee follows this new and expanded approach, it explicitly acknowledges that though the sharp separation of risk management and risk assessment is sometimes useful for important purposes, "such as insulating scientific activity from political pressure, . . . a rigid distinction of this sort does not provide the most helpful conceptual framework" if the goal is to better understand and communicate risk in an ongoing, deliberative public process.[20] These important changes are represented in figure 10.1, which emphasizes the recursive and iterative approach to problem formulation, value inputs, and public deliberation.

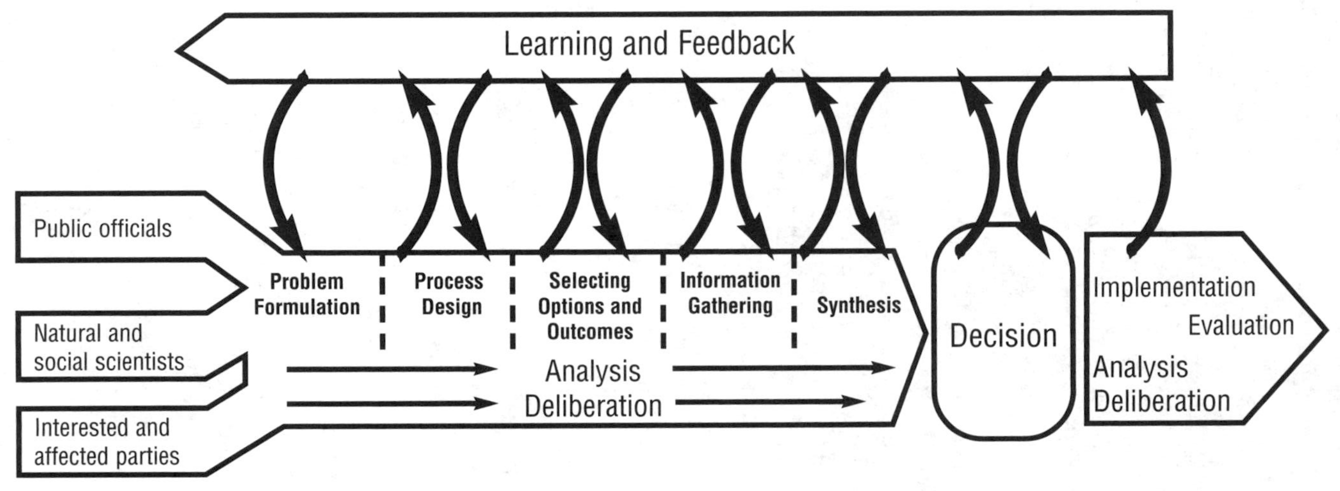

Figure 10.1 A schematic representation of the risk decision process (redrawn from National Research Council, *Understanding Risk* [Washington, DC: National Academy Press, 1996], 28)

The key log that has been released to set in motion these huge changes in accepted views of risk assessment has, in my view, been a change in the way we think about public involvement in decision making. Public participation was required in the National Environmental Protection Act, which called for environmental assessments of many federal projects, since public hearings were mandated at the point at which a draft assessment was completed. What we have learned from thirty years of experience is that this kind of participation is too little and too late to ensure that the public has a meaningful input into the project. This approach led to what is today disparagingly referred to as the "DAD" approach to public participation: "Decide, Announce, Defend," implying correctly that in processes so designed there was a tendency to present to the public the results of an internal technical process completed within an agency or firm, based on a problem formulation into which most participants had no input. The result was often projects that did not respond to the needs of the interested and affected parties. Project plans, delivered ex cathedra, often provoked severe criticism of projects after the decision process had moved so far that changes and improvements proposed by the public faced bureaucratic inertia. Therefore, the trend in advocacy of public involvement is justifiably directed at ensuring that public input is solicited earlier in the process, with greater representation of interest groups and stakeholders. Furthermore, it is recognized that this early input must be used to shape the problem formulation. This recognition undermines the idea of a value-neutral risk characterization; and it signals an end to the slavish obeisance of risk analysts to a sharp separation between the analysis of scientific fact and the analysis of social values.

What is ironic is that the theoretical outcome we are seeing, so decisively stated by the 1996 NRC committee, can be read either as a story of liberation of risk assessors from their imprisonment in an inadequate analytic framework, or as the death knell for their discipline. In a sense, it is both. If risk assessors continue to insist, as they have done in the past, that they are value-neutral scientists who do predictive models based on transport, exposure, and dosage levels, they will become irrelevant in the new, process-oriented approach to environmental management; funding will dry up, and the discipline will gradually die out. But if risk assessors desert their artificial, value-neutral models and develop new methods that are more responsive to public values and more likely to contribute to cooperative solutions in societies with differing values, they will find an important mission-oriented role in adaptive management studies. Either way, it seems doubtful that risk assessment will continue as a distinct discipline. In the first case, it will fade away; in the second case, it will merge into a more comprehensive methodology for talking about environmental values and decision making. I doubt, as I have expressed before, that the lingo and concepts of risk assessment, and even those of the broader field of risk analysis, are sufficient to articulate many of the values about which communities dispute and negotiate.

So I do not expect risk analysis—now that the artificial distinction between risk assessment and risk management is abandoned—ever to achieve the high hopes that some environmentalists and many bureaucrats had for it. It will never fulfill administrator William Reilly's dream for it as a systematic and unifying framework for analyzing policy at EPA. If we interpret risk technically, as it can be modeled by scientists, most of the things people care about (that is, the things that should be deliberated about in an open, democratic management process) cannot be included in the analysis. If we interpret risk more broadly, we could consider the "risk" that our sitings of facilities will be unjust and the "risk" that constitutive values of a community may not be protected. In this case, given the importance of communication among many affected parties and interest groups, it will be necessary to drop the jargon and adopt ordinary language as the comprehensive language of environmental policy discourse. Once ordinary language is installed as the comprehensive language of mediation among interests, it can then be enhanced with the introduction of more technical tools, including conceptual and quantitative scientific models as aids to understanding and communication. Ordinary language has, after all, the advantage of allowing us to say pretty much anything we have on our mind, however inelegantly and inefficiently. And for those spoilsports among the technicians who point out that it is easy to generate inconsistencies when using the loose rules of ordinary language to reason, we can (I hope it won't sound too cheeky) point out that it is likewise possible to generate inconsistencies in the formal, axiomatic systems of rational choice, unless one somehow impoverishes them so much that they cannot articulate all that must be involved in a real-world decision.

If the argument of this section is convincing, however, risk assessment must be understood as only a tool of adaptive managers, who are much better positioned to provide a comprehensive, integrative outlook and decide strategically, given the public values at issue, what science is most important and relevant. Given that adaptive managers have a long-standing tradition of activism and of involving public groups in adaptive monitoring, modeling, and management efforts, I prefer their framework of open-ended, iterative, and multiscalar experimentalism over the stilted language of risk analysis as the basic framework for understanding management problems.

My prognosis for decision analysis and decision science is far rosier. I believe that these sciences, if they ever break through this debilitating commitment to the fact-value myth and the associated belief in a single correct outcome, have a great future. As was noticed in our discussion of von Winterfeldt and Edwards, good practice seems to ignore ideologies about methodology, anyway, and decision scientists began in the 1980s to emphasize the importance of communication, context and problem formulation, and process. Indeed, even though decision scientists have been slow to see the full implications of the impossibility of modeling good decisions purely in terms of outcomes and the theoretical necessity of characterizing good decisions partly in terms of process, they have in their actual work—when responding to real problems—expanded their framework of analysis and provided useful analyses of decision processes and important insights about problem formulation. For example, Robin Gregory and Ralph Keeney of Decision Research Inc. have used a process-oriented approach that embeds their formal, MCDM techniques in a stakeholder workshop format.[21] These researchers worked in Sabah, a Malaysian state on the island of Borneo, as part of a preliminary environmental assessment of a proposed coal-mining development. An Australian firm had discovered substantial deposits of high-grade coal in the Maliau Basin, a pristine and hardly explored wilderness area that is the habitat for an unusual diversity of plant and animal species, including the rare Sumatran rhino. The company sought permission from the government to explore these reserves and requested a drilling permit to allow them to assay the deposits. Recognizing that granting this permit was a first step toward allowing a coal mine to be opened, and realizing there were no well-defined or accepted practices for making such decisions, the Malay government hired a Canadian firm to prepare a preliminary environmental impact statement. This firm hired consultants from Decision Research to help structure the problem in order to define a focus for the impact statement.

The danger, they quickly learned, was that the two sides, the proponents of protection of the reserve and the proponents of development, would address the question only in all-or-nothing terms, whereas there were other possibilities including developing coal mining in some areas while giving increased protection to other areas by using much-needed revenue from coal extraction to address the already-difficult problem of poaching in the remote area. So, concentrating on problem formulation in a complex decision situation, Gregory and Keeney clearly did the right thing; they turned their attention to designing a process that included stakeholders from the outset, using a workshop composed of representatives of stakeholder groups to develop a broader set of development alternatives. They formed an advisory workshop, which they called a public value forum, to provide information about stakeholder objectives and to search for alternative possibilities. They described their role as follows: "As analysts, we find that our key role in the process is insuring that the decision context is cast broadly enough so that all stakeholders can agree on the context. Disagreements tend to occur when the initial statement of the decision context explicitly or implicitly rules out either objectives or alternatives that certain stakeholders consider important."[22]

They used the technique of eliciting objectives, separating fundamental objectives from means objectives, and settling on five major headings of fundamental objectives; workshop participants were then asked to set priorities within each of the five major categories of objectives, and much of the workshop was devoted to articulating alternatives. What emerged from these deliberations was that the original formulation of the "develop or protect" choice seriously misstated the problem because excessive poaching and illegal logging was already taking a huge toll on wildlife in the area; the flora and fauna, especially key animals like the rhino, were in danger. What emerged was an alternative involving some development of the coal reserve and use of revenues from the reserve to protect the largest share of the basin; this solution would reduce poaching and protect the option of developing a small, high-end eco-tourism project and other forms of nature-based development. Although no conclusions were reached, a list of alternatives and a set of questions for research and deliberation were derived (prior to any scientific analysis or any use of the specialized techniques of MCDM) from the list of alternative development paths that emerged from the analysis of objectives and values. These approaches seem to me to offer great promise, and work of this sort by decision scientists and decision analysts will be crucial as adaptive management proceeds.

Notice, however, that these contributions are made by starting in a real situation, by eliciting multiple objectives from a diverse set of stakeholders, and by working progressively toward better formulations of the problems. There is hardly any role here for nifty technical models or flashy mathematics. The greatest advances in decision science, I am suggesting, are occurring in the non-formalizable aspects of the field, especially in our understanding of the process of properly formulating a problem in a given situation and the emphasis on designing processes that are responsive to particular, local concerns. In this broader sense of decision science, which includes examination of values and objectives in real situations and emphasizes effective process, I expect that the field will make an important contribution to adaptive management, as we have already seen that it can do.

More particularly, I see a role, but a much-altered role, for the systems of analysis here referred to under the general heading of MCDM. This ingenious technique is designed to improve decisions in cases where there are multiple competing objectives by providing a formal procedure for ranking alternatives based on elicited preferences. Like contingent valuation researchers, MCDM advocates generally speak as if they are working with fairly stable preferences that are usually elicited only once prior to the use of their analysis. To their credit, then, MCDM advocates are likely, because they elicit a variety of objectives from participants, to capture a much broader range of values than would, for example, risk assessors or economists, but they still treat the choice of "end" values and the formation of preferences as exogenous to their process of analysis. Creating a ranking based on a formal method of analysis of participants' objectives still leaves values as unanalyzed inputs into a computational decision process.

The difference between our use of MCDM in this book and usual uses is that our criteria are given a life of their own in an ongoing, dynamic management situation. We do not factor in information regarding many objectives in order to "model" ideal reasoning in the form of aggregating fixed preferences. Rather, we emphasize decision processes and introduce several indicators and multiple criteria as part of a dynamic process. We do not expect that we can model the decisions a community eventually makes in an open-ended, dynamic process by representing these decisions as examples of computational rationality. Rather, we set out to learn more about processes that actually lead—even in difficult and divisive situations—to better decisions. Whereas MCDM gives its multiple considerations a baptism in formal logic, I throw our multiple criteria into the many different rivers of process by which environmental policies are formed, implemented, and re-formed in many particular communities. Admittedly, the waters are muddy, and we don't know how to swim well; and our analyses are unlikely to be as elegant as MCDM. We set out to create an ongoing, open, public process of deliberation, experimentation, and further deliberation, and we include not just discussions about what to do (what we have called the action phase of adaptive management). We also initiate, in the reflective phase, a dialogue about what is important to measure—by choosing indicators—and about what goals to set with respect to those indicators. Perhaps, at this point, we can return from our excursion into the literatures and the lingoes of the special decision sciences to our own framework of analysis, and restate what we have learned in this and preceding chapters about how to develop value and policy analysis that will actually serve the kind of open-ended, situation-specific problems that environmental managers face today.

10.5 The Two Phases Revisited: Putting Multicriteria Analysis to Work

I have tried, in my struggle to defeat ideology and towering at EPA, and more generally in environmental discourse, to survey most of the special disciplines that have a legitimate claim to provide input into the complex, overarching process of articulating and implementing a publicly supportable and scientifically informed environmental policy. In many cases we found these disciplines to embody insights about particular environmental problems and about aspects of other, more complex environmental problems. It therefore makes sense to think of these disciplines as important contributors to a comprehensive attempt at adaptive management in a community. What we have also found is that in most cases these specialized disciplines have tried to extend their analyses to apply to all types of environmental values. This strategy, of trying to "reduce" multiple phenomena and many human values to fit into a single disciplinary framework—of striving simultaneously for both theoretical monism and practical comprehensiveness—has failed in every case. The idea behind the strategy is seductive; but it makes no sense to start our actual study of environmental values by assuming we already know how to execute this reductionistic strategy and therefore limiting discourse to its present confines.

We have avoided this unsuccessful strategy by accepting pluralism as our starting point. I do not, however, wish to advocate indefinite continuation of chaotic, unintegrated pluralism; I advocate instead an experimental spirit and careful attention during the reflective phase to ways to simplify and increase the generality of our value measures. I have thus proposed an iterative, two-phase deliberative process in which communities sometimes focus on what to do now, given what they know and value, and at other times reflect on their goals and how they might measure progress toward those goals. I have thus formulated the overall problem—the problem to which we hope a unified definition of sustainability will provide a solution—as one of choosing more than one criterion to form a multicriteria system of evaluation. This multicriteria system of evaluation can be applied to proposed development paths, considered as possible paths from where a community now is to where it might be in the future if particular choices are made and particular policies are chosen.

Given this broad framework, an abstract schema for an adaptive management process in a particular community, I have also offered a taxonomy of four types of sustainability values, values a community might wish to include in its own articulation of sustainability goals: (1) community-procedural values, (2) weak-sustainability (economic) values, (3) risk-avoidance values, and (4) community-identity, or sense-of-place, values (see section 9.2).

Finally, we have explored a further refinement of this last kind of value, by suggesting that one special case of place-based values, "constitutive values," can be associated with commitments to hold open certain options as essential to maintaining continuity of culture and nature in a given place. This concept of options is very useful because it connects holding open specific options, like hunting or hiking in a hardwood forest, with possible actions to protect them, and it gives concrete meaning to the idea of these substantive, normative commitments. These commitments can be freely made by members of a generation to their community and its future inhabitants. Also, when we believe that if these options are lost, people of the future will be worse off than they would have been had the options been protected, we clearly commit ourselves to the idea of noncompensable losses.

Despite the work already done to provide a comprehensive evaluative framework in which such noncompensable, constitutive values can have a role, I have not yet said enough to clarify how such options and associated constitutive values can be identified and justified. In this section I must show how the two-tier process, embedded in an adaptive management procedure and recognizing multiple values unfolding on multiple scales, can provide a comprehensive evaluative process that could be embedded in a broader management process that is driven by iterative deliberation. The two-phase system, developed in earlier chapters, is a system in which public discourse is alternately focused on what to do, in the action phase, and on choices of indicators and goals, in the reflective phase. Speaking historically, most attempts to provide analytic frameworks have addressed decision making in the action phase only; indeed economists and risk assessors seem implicitly committed, by their drive to achieve a single, complete decision-analytic framework, to ignore all of the messy prior questions that arise in the context of problem formulation. These questions are usually isolated and hidden away in a "black box" labeled "problem formulation" or "risk management." In chapter 8 I showed how a community that is concerned about the future as well as the present might, through a performative act of commitment, identify "stuff" that is so important to the culture and the social identity of a community that its loss would be considered noncompensable; recognizing this, the community might consider itself bound by a safe-minimum-standard-of-conservation (SMS) decision rule: save the stuff, provided the costs are bearable.

Rather than black-boxing problem formulation and values analysis, I have proposed a two-phase process that formally recognizes the difference between action-oriented discourse and the context-setting decisions that shape the problem and introduce the concepts and measures that will be used in evaluation. Once this separation is made, it becomes possible to submit the context-setting decisions to more careful analysis. Our goal, in sharp contrast to that of risk managers, decision scientists, and economists, is to focus precisely on these second-level decisions—what to measure and what goals to set—decisions that are usually hidden within the methodological assumptions of technicians and their tools. From our problem-oriented perspective, these hidden questions are crucial if our process is to be able to generate cooperative action in the short run *and* ongoing reflection, criticism, and re-formation of goals and objectives in the longer run—what we call social learning. The goal of our two-tier analysis, then, is to increase our understanding of problems of what to measure and what goals to set and, hopefully, to make these decisions a transparent part of an open and democratic adaptive management process. We have seen that a part of this process is making explicit the commitments that are implicit in the willingness of participants to struggle together toward cooperative solutions despite their differences.

I don't want to be (merely) a cockeyed optimist. I'm sure my justifiably skeptical readers are keeping track of my assumptions, and the favorable conditions I have attributed to processes, to point out how unlikely it is that all the stars will align properly, that an epistemic community of scientists and serious stakeholders who trust each other will emerge, and that there will be an open and contentious dialogue about what is important to preserve. The bad news is that it *is* a slow and arduous process for a community to form an epistemic consensus and a consensus on management goals. The good news is that most communities have everything they need to start on a more cooperative and adaptive course. Adaptive management requires no more than that a community respect science and experience as the best arbiter of differing opinions. The place to start is to begin to ask, openly, what is most important to this community and what kind of measures might track the production of that good.

Operationally, this formulation of the problem focuses attention on public involvement in the choice of indicators that can unite a community behind policies that will protect, on various scales, all or most of the dynamics that generate human values. Again, my presentation suffers from lack of specificity. See section 9.4, however, for a fanciful exploration of how some very effective synoptic criteria might be introduced as a way of identifying and summarizing feared impacts in the Atlanta metropolitan area and in the northern Georgia region. Such synoptic criteria, if carefully crafted through ongoing discussion and debate, can unite multiple constituencies behind important long-term commitments and provide crucial elements in a community's struggle to define what sustainability means in its place.

In cataloging sustainability values that a community might want to consider, I also suggested that most communities will adopt some sort of policies and commitments regarding dangerous risks to future people on grounds of fairness, and that most communities will adopt some safeguards against inefficiency—such as maintaining a fair savings rate—as an expression of their sense of fairness in economic matters. I have treated the deliberations and choices that any community makes in these regards as important questions that must be addressed in a reflective mode of thought, because good decisions here involve looking at the bigger picture and balancing varied considerations. My goal has been to hold these deliberations up for special examination, for the introduction of heuristics to encourage the asking of the right questions, and for iterative revisiting of problems and possibilities. I have introduced discourse ethics and other tools as helpful in making these decisions, but I have tried to embed these philosophical and analytic tools in a public discourse in which these subjects are discussed, iteratively and openly, using philosophical and other arguments. In the end, however, I have left these reflective decisions to communities operating in concrete contexts and facing real problems.

This commitment to analyzing and reformulating problems and criteria by which to judge them is especially important when we turn to the fourth type of environmental value, sense-of-place values. Whereas communities may be affected by the very general arguments, based in discourse ethics, that affect the fairness and openness of the process, it is in the determination of sense-of-place values that the true substance of a place-based environmental policy must emerge. Now I must say more, here and in the following chapters, about how public process can lead to commitments to certain indicators and to certain goals of sustainability for a place-based community.

The problem, put most simply, is how to count, and how to value, options. I have argued that it is useful to understand sustainability as measured by a multicriteria system, with economic concerns (usually) represented in at least one criterion. Additional criteria might include fairness requirements and limits on the types of risks we can knowingly impose on future people. I have made a lot, however, of this type of values called sense-of-place or constitutive values, which we have hypothesized to be values that are associated with community inhabitants' sense of who they are and what they feel is distinctive about them and their place. I have insisted that logically speaking, these values can be defined only in terms independent of economic calculation, because (I have hypothesized) losses to these values are noncompensable in economic terms.

Further, we can tell from listening to ordinary discourse when people fear that such values are threatened, because these values are all tied up with people's sense of who they are and a community's sense of itself and its place. So expressions like "If that happened, I suppose I'd move away" or "If that happened, this wouldn't seem like home to me any more" can be interpreted as invoking noncompensable, sense-of-place values. Finally, and most importantly, I have identified these sense-of-place values more concretely with the value communities might place on holding open, across generations, certain sets of options, options that are essential for people to partake in the history of their place and the meanings they and their family have created in that place. These options are necessary if future people are to have a chance to be members of a real community with an authentic relationship with its past, including its natural history. Lacking these options, they will never have the chance to be full-fledged members of the multigenerational community that has evolved in their place. Their set of choices will be impoverished in a way that makes them worse off; and they will remain worse off, no matter how much they are compensated economically. So the specification and explanation of sense-of-place values comes to this: for each community, provided its members are committed to preserving their community's sense of place, there are certain options—opportunities to engage in certain behaviors and authentic practices—that must not be lost if the community is to co-evolve with its place, adapting its practices to the cycles, as well as the capabilities and the limitations, of local environmental systems.

One aid I have at hand, as I try to be more precise in describing these sense-of-place values, is hierarchy theory. Hierarchy theory, remember, is a way of thinking about temporal and spatial relationships. It organizes these relationships from a point within a complex dynamic system, a place; and it structures and sorts spatial scales by treating smaller, faster-changing elements as subsystems of a larger, slower-changing system that provides the environment for the smaller elements. I choose the group of conceptual models defined by hierarchical assumptions because, if human decision making is interpreted in those multiscalar models, a definition of sustainability falls out as a bonus. Sustainable activities are ones that can be carried on in the present without negatively impacting the range of important choices that should be left open to the next generation. Short-term choices of individuals, on this interpretation, are modeled on the scale of individual choice. These choices have "environmental" implications only if—because of the violence, rapidity, and scope of human means to enact these choices—they threaten to adversely shift the large-scale, usually slower-changing, system that creates the context in which the next generation of individual choosers adapt. Certain kinds of actions by individuals and by collectivities create a situation that we call unsustainable, meaning these actions so change the environmental context that a viable choice today will no longer be available tomorrow. So we have a nice little conceptual model of sustainability, interlaced with a theory of how to make sense of space-time relations when making decisions from within a complex, dynamic system. To operationalize this definition of sustainability, a community need only specify those options which it associates with important sense-of-place values, a specification that should be carried out within an open, democratic, adaptive management process.

The question is whether sincere and determined participants in a public process—despite differences in both values and beliefs—can agree upon which options must be held open if they are to pass forward the opportunity for future generations in their place to continue an authentic relationship with the place they call home. This is indeed a difficult problem, and there are no easy ways out. Some options are positive (for example, the option to live a healthy life) and others are negative (such as the option to die of pollution-related illnesses). And we cannot simply count options or choices available—either positive or negative—because the desirability of options is surely weighted. The opportunity to have an ice cream cone today is hardly comparable to having an opportunity to go to college, for example. Actual communities have the difficult task of identifying the options that are to be associated with their sense of place and self-identity. I am certain of one thing: this task cannot be successfully undertaken without a deep examination of the community's values, aspirations, and sense of meaningfulness, along with a great deal of deliberation and social learning. What I am proposing is a process—a method—for bringing citizens' real sense-of-place values to bear upon problem formulation and goal setting. Being a pragmatist who rejects the fact-value dichotomy and doubts all appeals to preexperiential principles, I propose that we not think of environmental decision making as the deductive "application" of general, ontologically justified, monistic principles, but rather as an iterative and constructive process.

The process will not start by trying to achieve widespread agreement with a single value or a single way of measuring value; it will rather proceed as participants propose, discuss, and deliberate about what trends and features of their environment should be monitored and which of these can be treated as indicators that correspond to various management goals and objectives. In this way the values of community members will be extremely important in public discourse and deliberation because people will appeal to their values as they argue for the importance of particular trends, features, and indicators. They will be saying that, given the values they hold dear, given their aspirations for their place, they think certain goals should be set, as tentative starting points for management actions. They will also be recommending that these goals be stated explicitly in terms of a physical, measurable indicator that allows assessment of the management process over time.

Fortunately, a good example of such a process, including the choice of a widely affirmed indicator, came very early in the attempts to save the Chesapeake Bay from degradation in the face of rapidly escalating impacts of regional urbanization and agricultural intensification. Following a large EPA study in the early 1970s, formation of a multistate compact to address the bay's problems, and considerable discussion and deliberation, the goal was set to reduce the flow of nutrients into the bay by 40 percent, a goal that was recently reaffirmed. Here we have an example of a process—suffused with values and love for the bay and for the many distinct communities that exist there—generated by a public discourse concerning turbidity in the bay. The urgency of this discourse drove scientists to do basic research that led to the conclusion that the main continuing threat to the bay was widespread sources of nutrients: from sewage outflows, and from runoff from farmers' fields, suburban lawns, highways, and parking lots. Through public discourse over a period of a decade or so that included ongoing involvement of scientists and managers, bay residents evolved a broad "mental model" of bay pollution based on the hypothesis that the decline of submerged aquatic vegetation was a result of explosions in planktonic populations that were living on excess nutrients and threatening to turn the waters anoxic when they died and decomposed.

This public discourse led to an important reconceptualization of the problem of pollution in the Chesapeake. Tom Horton, a regional journalist, said in the late 1980s, when the process of reexamination and scientific studies was in full swing: "We are throwing out our old maps of the bay. They are outdated, not because of shoaling, erosion or political boundary shifts, but because the public needs a radically new perception of North America's greatest estuary." The new maps included much of Pennsylvania, New York, and other northeastern states whose waters flowed downhill toward the Chesapeake Bay, the main source being the mighty Susquehanna, which winds through rich farmland and past major industries. "The new kind of Chesapeake Bay maps show how we must regard it if we are to redress the pollution that has closed two major Bay fisheries . . . this decade."[23] The public discussion eventually worked; through the contributions of scientists, politicians, the Chesapeake Bay Foundation, and many others, to research, interpretation, public meetings, and so forth, the public developed a new spatial model that related values placed on the bay to a new, watershed-scale dynamic. People gradually learned that to think like a bay, one has to first learn to think like a watershed. This process exemplifies social learning at its best, because communities that once related to the bay locally were able to add another scale to their understanding and to their sense of responsibility.

The emergence of this mental model of bay degradation resulted in the identification of an important indicator—bay water clarity—which was, in turn, related to a landscape-scaled dynamic, the rate of nutrient-loading from various sources. Water clarity, as an indicator of success of efforts to save the bay, was then related in many different ways by many different people to their own values and feelings about what was important to them. The choice of water clarity as a key indicator not only solidified action and resolve on the part of the public, the states, and the agencies; it also expressed concretely the many ways the communities around the bay valued it. Taking aggressive action to reduce nutrient-loading, hypothesized to be driving the increase in turbidity, was a positive expression of values placed on a variety of bay-dependent options, including fishing, boating, and maintaining tourism-related businesses. The indicator of water clarity, as was pointed out by scientists during the public deliberations, can be expected to track reductions in nutrients entering the estuary. This variable, in turn, is important in many ways. For example, submerged underwater grasses, which depend on the penetration of sunlight, are the foundation of the complex bay food web, which supports populations of fish and shellfish. Water clarity is essential to the widespread practice of "crab dipping," and of course it affects the quality of boating and swimming experiences. People's evaluations, in other words, were summarized and expressed in the choice of a key indicator that could be scientifically or otherwise related to important social values. It is, like percentage of pervious surfaces, a pretty good measure of broad processes that affect many important social values in this region. Rather than measuring the economic value of all of these activities and then aggregating all of the ecosystem "services" derived from the process, trying to achieve a uniquely efficient or welfare-maximizing outcome, the process involved choosing a measurable physical-ecological indicator and setting a specific goal regarding reduction of nutrients by a specific date. In this setting ecological, toxicological, biological, economic, anthropological, and sociological evidence was relevant and could be brought to bear upon the public discourse in which indicators and goals were proposed, advocated, criticized, and reformulated.

No formal, deductive discourse can capture this complex process. Fortunately, if we have an open process, which alternates between the action phase, in which the community seeks cooperative, collective actions that reduce uncertainty and result in social learning over time, and the reflection phase, in which we discuss, debate, and deliberate regarding what indicators are most closely associated with what we value, social values will be embodied indirectly in choices of what to monitor and what goals to set. We have shifted the approach, then, from trying to model an ideal decision process by which to represent a "rational" decision based on the best science and aggregations from individual valuations, to actually immersing the choice of goals and multiple criteria and requirements for sustainability into an ongoing public process, relying on democratic discourse and people's values and love of their place to encourage the use of scientific studies to reduce uncertainty and serve consensus-based community goals. Processes such as this, though messy, have the potential to encourage social learning, reduce conflict, and result in the choice of synoptic indicators that can be supported by coalitions of individuals and groups who may be expressing very different values. Values function in this process as important elements in the public deliberation, but often indirectly. They function, first, as motivations for participants to propose and support specific criteria and goals and, second, as reasons to criticize proposals of criteria by other individuals and groups through arguments that important values will not be protected by monitoring a proposed indicator or pursuing a particular goal.

Identification of options that are important for a community to protect is thus a public process—an endogenous part of our adaptive management process. This is an explicitly and openly value-laden process, and all participants are encouraged to weigh in and let their voices be heard. In this way, proposed indicators and goals are evaluated from many different perspectives and based on many value systems. If we can find an indicator that is consensually accepted as expressive of many of the shared values of the community, as in the case of water quality in the Chesapeake, and if we can identify physical dynamics that drive the problem, then we can begin to set goals associated with the indicator and begin to act collectively and experimentally. As we so act, it becomes possible to test both scientific hypotheses about how things work and hypotheses about what goals and objectives truly support our values. Starting from wherever they are, communities can begin a process of cooperative action and experimental management, focusing their attention on reducing uncertainty and on the ongoing articulation and development of goals and objectives. This process of articulation, ideally, should embody its citizens' evolving sense of themselves and their place as an ongoing community. A proposed set of criteria, all of them necessary conditions of sustainability for that community, can be then be proposed as a set of hypotheses—specific to that time and that community—about what is important enough to the community to save. Similarly, specific management objectives and goals can be considered to be hypotheses that if those goals are achieved, then sustainability of the core, constitutive values of a community will be on track. It will remain an open question whether any given set of multiple criteria is sufficient to capture all such values; therefore, achieving a complete characterization of the community's values in terms of measurable indicators will no doubt be the subject of ongoing discussion and debate. Some environmentalists, concerned about risks and negative impacts of technological change, will continue to try to add more rules to the regulatory scheme by insisting on adding more limitations and requirements to the core concept of sustainability. So debate will surely continue.

We never promised that adaptive management would by itself produce answers. It merely provides, by using citizens' advisory committees and scientists as teams, a forum in which the various interests can question, criticize, and propose amendments to proposed actions and to objectives pursued. Open debates between environmentalists about specific indicators will give scientists an opportunity to introduce studies that will resolve important disagreements about management. It is the commitment to act together to resolve outstanding scientific uncertainties that animates the early experiments and starts the process of action and deliberation. If public discourse in the community is successful, intractable disagreements between interest groups may be replaced with cooperative attempts to learn. If these various elements fall into place, it is possible for a process of adaptive management and cooperative action to emerge. If a communal process of value articulation occurs as a part of the process, it may lead to a successful operationalization of a community's environmental values. The values people express give meaning and support to the commitments the community makes, and the ongoing process of articulating, criticizing, and re-forming our beliefs and our goals in the light of new evidence provides a basis for ongoing criticism and improvement of policies. Once this process starts, and there is an opportunity for communities to act, even as they recognize that their current actions are based on uncertainty, their action becomes gradually self-correcting. The process, essentially, is a process of seeking wider experience. We want to create, then, a schematic criterion that includes some core sustainability requirements that we expect most or all communities to incorporate in some form or other in the multicriteria list they employ as indicators of sustainable living. These core criteria all derive from a general concern for justice and fairness, both substantively, with respect to intragenerational fairness and intergenerational bequests, and also procedurally, with respect to fair and open process.

I view it, then, as a sign of a healthy adaptive management process when multiple criteria are being proposed and indicators are debated in a spirited manner. This deliberative process is what drives social learning; as Habermas recognizes, it also allows communities to engage in cooperative action even when they have diverse interests and values. The possibility is always there to reopen a question of policy if new evidence is unearthed or new values are invoked. Those who lose in any given round of objective-setting can accept objectives chosen by the group so that legitimate cooperative action can take place, even when there is disagreement both about scientific models and about what is really most valuable. They accept the experimentation because they are assured that when new evidence comes in, the choice of objectives can be reopened. Adaptive management, in each community, sets out to develop a process in which these questions can become the focus of public deliberation, with the participation of scientists, interested parties, policymakers, and the public. The process should be designed to allow decisions and actions provisional on new evidence and also to allow public deliberation and criticism that can lead to social learning. By embedding the study and the discussion of indicators in a public, democratic process as representative of values, we have paid off on our promise to break down the barrier between fact and value. In this public discourse, participants will invoke values when they think large numbers of others share those values, and they will use these as justifications to undertake studies to reduce scientific uncertainty about forces critical to ecological processes associated with the community's values. At times apparently irrefutable facts will be invoked to show that certain objectives are misguided. The process is chaotic; but it has the virtue of ensuring, over time, that all of our policy proposals will be evaluated from many viewpoints and that much diverse experience will be invoked regarding those proposals. Ultimately, then, the two-phase system of analysis we have proposed works to winnow out disagreements by finding actions that are favored by diverse groups, creating a robust policy that can in the future be further modified in an open-ended process of criticism, analysis, and experiment.

Ultimately, it is this open-endedness that we have insisted must characterize evaluative as well as scientific discourse. In public deliberation, value proposals must survive in much the same sense that scientific hypotheses must survive. Communities of forward-looking inquirers, acting on those values, are likely to ask productive questions, to learn more, and to prosper as they gradually improve their policies and their performance regarding their sustainability criteria and goals. It is this same open-endedness that distinguishes our approach, which is to use multiple criteria as tools of understanding and communication within an ongoing, open-ended process, from the use of MCDM by decision analysts. Our criteria are treated as hypotheses about what the community values, hypotheses that may well be revised as new evidence is fed into the process. These criteria are understood as "variables" that express management goals and objectives. They are sometimes used to justify decisions and actions; they are sometimes themselves the object of criticism and revision, in the reflective phase, because some groups believe the current criteria do not adequately capture certain values or because proposed objectives are judged to be impossible.

Unlike MCDM, which is an exercise in deriving better outcomes by computation, based on a one-time expression of preferences and on a one-time choice of weighting principles, the proposed approach is iterative and proceeds by considering and using multiple criteria, which are both used and continually refined in the face of new experience. The evidence from logic and decision theory shows that such derivations cannot be completed, without contradiction, within a formal system strong enough, semantically and pragmatically, to allow communication about environmental problems. MCDM, as it is usually used, computationally, artificially closes a process in order to apply as many of the mathematical techniques of decision science as possible. This use of multiple criteria, however, cannot provide a deductive solution to real decisions because no formal language can do what people, in communication and in search of cooperative action, can do. They can, for example, creatively seek solutions that are not answers to the original problem formulation, as Gregory and Keeney found in Sabah. The important point in the big picture is that such broader searches cannot be modeled as a deduction carried out within a system. Context is as important as content in the development of place-sensitive environmental management. Problem formulation can be improved only iteratively, through deliberation, and with the addition of careful, mission-oriented science. Our intent is to put the technique of multiple criteria to work by fostering a public process of value articulation, attempts to operationalize criteria by proposing and discussing various suites of measurable indicators, and integration of these choices into a public process of ongoing public deliberation and social learning. By opening this process to articulation of new values, as well as new facts, these indicators become useful means of communication across "towers" in the policy process. Formal decision procedures do not solve complex problems; people solve complex problems; formal procedures can at best be aids to people in solving those problems.

When we try to take language out of its dynamic, everyday context, the context in which it serves as a tool of communication, and artificialize it—formalize it—we lose the absolutely essential aspect of creative reformulation of problems, goals, and objectives across many iterations as a result of deliberation. Only a conversation animated by values can make the creative leap from obstinate opposition to negotiated compromises. Our two-phase process, and the linguistic innovations and heuristics associated with it, is designed to encourage just such creative leaps, resulting in concerted action.

11

DISCIPLINARY STEW

11.1 Beyond Towering

This book has been all about integration in environmental policy—integration of analyses and integration in the service of successful environmental policies. I have shamelessly used the old EPA building as an emblem of our failed policies and inadequate analyses, because the inappropriately towered structure of the building both symbolized and reinforced the structural and conceptual problems at EPA. But I have mainly exempted the employees at EPA from my barbs; I think the problems are at root conceptual and linguistic, affecting all of environmental discourse, and cannot be blamed on individual bureaucrats at EPA or anywhere else. In this chapter I address the positive task of designing a new discourse, one that has potential to broaden and deepen public deliberation about environmental problems without resorting to preexperiential commitments and to ideology and empty rhetoric. The task is daunting; we cannot accept Callicott's view that change can occur simply as the result of a trivial redefinition of a term in use. Nor can we make strong, metaphysical claims about the "nature" of natural value, as Holmes Rolston III does, because these claims would commit us to principles that cannot be justified by experience.[1] Neither subjective, unverifiable feelings nor a priori metaphysics will do. If we are to have a science-based environmental policy, then the solution is for ethics to become more scientific, not more metaphysical. It is crucial that public discussions of values be linked to observable environmental qualities in order to integrate evaluative and descriptive discourse in the discussion of environmental values.

Disciplines, too often thinking of themselves as the main course, have insulated themselves by creating in-group jargons and techniques that make them inaccessible either by the public or by scientists in related fields. This insulation reinforces towering, and towering reinforces insulation and failure of communication. As the title of this chapter (recalling a metaphor briefly introduced in section 9.3) would suggest, I view environmental discourse, including scientific disciplines, as integrated but lumpy. By thinking of the particular disciplines as the meat and potatoes in a rich stew, rather than main courses available à la carte, we can conceptualize ordinary public and political discourse as the broth that both flavors and is enriched by the special models of the distinct disciplines. I have represented this relationship in figure 11.1.

This analogy is important theoretically, because it embodies a crucial insight of pragmatism: when it comes to articulating agreements and differences among the many interest groups that wish to affect policy, pluralism and communication are better than monism and isolation. The key to this insight is that the language of policy deliberation in a democracy must be ordinary language: otherwise, discussions of what to do will be dominated by experts. Disciplinary science, of course, has its place, and most scientific studies should be peer-reviewed according to the criteria of an appropriate discipline. But management science, the science of adaptive management, ought not to strive to develop its own disciplinary language and criteria. The language of management science, because it is a language of public deliberation, must be the inclusive language of common discourse. Public debate among individuals and groups with different experience and no shared disciplinary language must ultimately be engaged and resolved in the common, ambient discourse of the shared, public language.

Speaking more practically, this insight embodies the central idea of mission-oriented, or postnormal, science. Management science, being public and mission-oriented, must not only maintain connections to the special disciplines; it must also provide a nondisciplinary, action-oriented forum for the discussion—and the nondisciplinary peer review—of the special models and detailed studies being completed within the disciplines. As Funtowicz and Ravetz argue, mission-oriented science can function in the public interest only if its findings are presented in (or translatable into) information that is, in principle at least, understandable to lay persons. Disciplinary peer review, the touchstone of disciplinary science, is inadequate to multidisciplinary, action-oriented deliberations. For example, though a specialized disciplinary language may be useful to communicate a particular causal connection, one must step outside that disciplinary language if one wishes to question the relevance of that causal connection to a particular management problem. Since adaptive management is action-oriented, and since it requires open and inclusive deliberation about goals and the means to achieve them, forays into the specialized sciences must be regarded as useful interludes, temporary side trips into detailed, disciplinary analysis of specific questions. If these forays are to

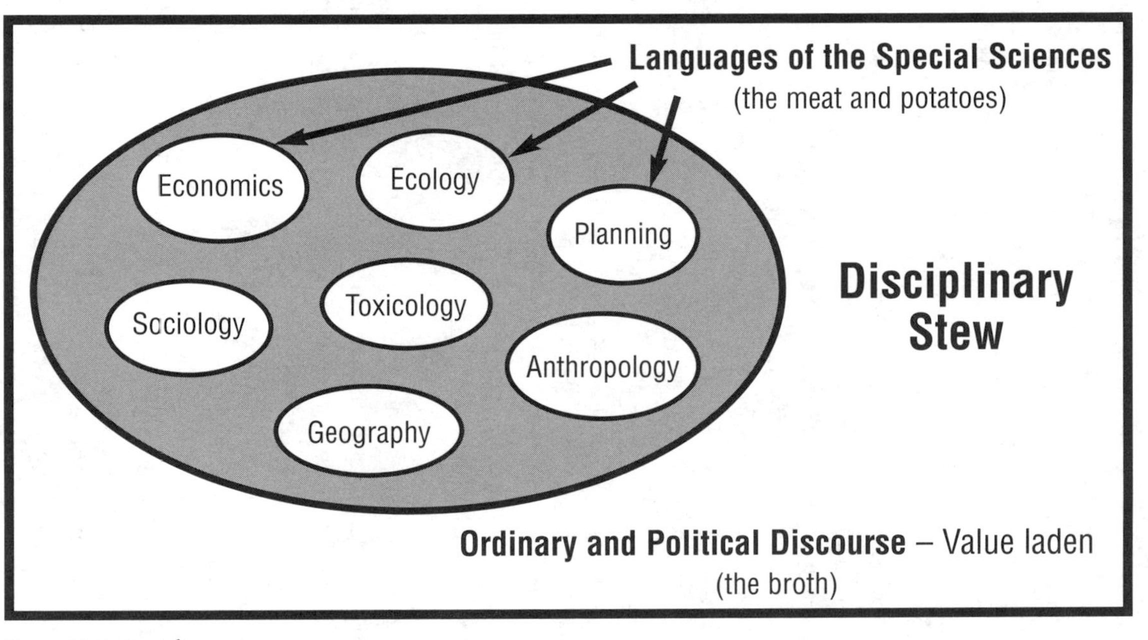

Figure 11.1 Disciplinary stew

inform a public, democratic decision process, they must be translated, and evaluated, within the ordinary language (perhaps temporarily augmented with specialized symbols) spoken by the community's citizens.

So this chapter provides a recipe for disciplinary stew: create a rich linguistic broth, using ordinary language as the stock, and add, for substance, specialized disciplinary models and careful fieldwork to clarify particular points of confusion and disagreement. Here we examine the intellectual and linguistic resources available for creating such a rich stew.

My analysis of the problem at EPA is that the agency lacks a shared conceptual framework for thinking through complex environmental problems or for relating one set of environmental problems and policies to another. Public discourse is currently disrupted by ideological language and problem formulations. Consequently, the agency has failed to present to the public an intelligible and positive framework for addressing the complex problems of environmental impacts unfolding on different scales. This analysis has two applications. First, there is an internal problem, some of it resulting from the irrational bureaucratic organization foisted upon EPA as a result of its politically charged creation. The internal problem is exacerbated by the inability of different offices and arms of the agency to communicate, there being no common "language" for developing and integrating environmental policies because different offices use different disciplinary jargons. Policy discourse is, accordingly, fragmented and incoherent. This problem in communication encourages the towering that has been so devastating to coherent discussion within EPA and other policy-oriented agencies.

Second, there is an external problem, in that professionals at EPA and other government agencies have failed to articulate for the public an integrated program that makes sense as a plan of action to protect the environment and the benefits citizens derive from it. I have suggested that both the internal and the external problems are most essentially problems of communication, caused by the lack of an adequate language to initiate and support intelligent public discourse about what we *should* do. It is perhaps a chicken-and-egg question whether the structural and organizational problems at EPA "came first," or whether the conceptual problems have isolated offices and programs of the agency. Similarly, one could argue either that EPA's internal confusions have left the public confused or, alternatively, that the failure of public discourse to develop an integrative language has kept EPA from reflecting a coherent mandate for action. Whichever way one runs this analysis, the unavoidable conclusion is that conceptual isolation has prohibited EPA from communicating an intelligible program to the public, a program that is based on an integrated approach to public values, just as it has failed to achieve most of its chosen objectives. Since I can do little to solve the organizational problems at EPA, I have addressed the conceptual and linguistic problems. The case of wetlands policy may serve as a vehicle for final integration, because it turns out to share many features with other failed EPA policies.

Our analysis in chapter 1 of the wetlands banking policy, administered by EPA and the Army Corps of Engineers and favored by both Republican and Democratic administrations, helped us to identify several specific failures in the development of wetlands policy. First, many contracted substitute wetlands have not even been built or are so inadequate as substitutes that they offer hardly any of the services of the natural wetlands they replace.[2]

The second point is that the wetlands banking policy was developed without any procedures for evaluating the success or failure of the program. It thereby violates the most basic principle of adaptive management. In the absence of scientific knowledge as to what is sustainable, we must understand new policies as experiments, as opportunities to learn what works and what does not work. At last, a few studies are being done; the evidence is that the policy is failing, but we have learned little in nearly twenty years about how to do better.

The third characteristic of the failed wetlands policy is a failure to develop a scientifically based and comprehensive means to evaluate various wetlands according to function, especially since their functions affect social values. Indeed, we found that anxiety about mixing value questions with scientific study caused—or at least provided an excuse for—trained academic ecologists to avoid getting involved in wetlands evaluation at all. This reluctance has left the choice of which values and which trade-offs to pursue up to politicians and, worse, to politicians who are unarmed with any solid scientific guidance to cite in opposition to the political pressures of interest groups.

Fourth, and even worse, these politicians have never had the benefit of public deliberation carried out in a truly integrated environmental discourse in which all the disciplines provide the best they have to offer in the analysis and solution of environmental problems. In fact, social values involved in wetlands protection were hardly discussed in the entire process of developing a policy. The argument about what to do and how to set goals has bounced back and forth between "science" and "politics," with the discussion of social values limited to assessment of commodities that might respond to a social "demand" for various products or services. What might be called ecological values—the very values that emerge in an intact and healthy ecological system and depend inevitably on interactions with a larger landscape—are least likely to be captured in an economic analysis of commodity values. For example, many wetland animal species do not complete their full life cycle in a wetland, so although the context of a wetland may be crucial to its ability to function as wildlife habitat, a purely atomistic and commodity-oriented concept of valuation is certain to miss these broader, contextual—but all the more important for all of that—values.

We also saw that towering emerges as a sort of implicit conspiracy between bureaucrats, who would like to have what they can call value-neutral science to cite in supporting their actions and choices, and scientists, who find it more comfortable to pretend that their science is value-neutral, despite unassailable arguments that such purity is impossible. So professional ecologists and bureaucrats—for very different reasons—fall back upon the myth of value-neutral science as the justification for not looking beyond economists' accounts of individual preferences and aggregations thereof.

Finally, besides making the general point that wetlands policy has been discussed without an adequate vocabulary for reasonable deliberation on what is important about wetlands for public values, I recorded a very specific observation and criticism that the vocabulary available for introducing ecological considerations into policy deliberation is lacking a particular kind of term, the function of which would be analogous to *Gross Domestic Product* in discussions of economists and economic policy analysts. For all its weaknesses, the term *GDP* has the important characteristic that it summarizes a great deal of economic data. *It also suggests, because of its historical role as a measurement of economic growth, a clear connection to positive social values of importance to all citizens.* I hypothesized, based on my understanding of the wetlands case study in chapter 1, that communication of ecological information in policy contexts would be enhanced if we could introduce and explain terms that have scientific, descriptive content measurable in ecological terms and also express important social values.

In 1995 when I delivered a talk at meetings of the Ecological Society of America in Snowbird, Utah, I learned in conversations that this specific contention was controversial. In my presentation I responded critically to a subcommittee report that had criticized the use of "new" and "unscientific" terms; the report had mentioned "ecosystem health" and "ecological integrity" as offenders. I first noted that these terms were perhaps not best thought of as "pure-science" terms—they may be, I suggested, terms more closely connected to policy than to ecology. Afterward I had an interesting exchange with David Policansky, a respected ecologist who works at the National Research Council. Policansky said he thought it was important to do our best to segregate factual reports from evaluative prescriptions and that he would avoid these terms because they inevitably mix the two. "I'd rather get the facts independently of evaluations," he said, "because then I can make up my own mind about what I value." I quote Policansky here because I suspect that many ecologists, other scientists, and perhaps also some nonscientists would agree with him. One might think of this book as a detailed response to the attitude and set of beliefs reflected in Policansky's comment.

Let me summarize why I reject the segregationist and serial views of science and policy. First, the approach is based on an impossibility; real science is not done in value-neutral contexts, and especially practitioners of mission-oriented sciences such as conservation biology and conservation ecology cannot seal themselves off from political issues and social values. Second, even if a value-neutral science were possible, it would not be desirable, because value-neutral science has no way of telling what is important information, what dynamics to monitor, or what indicators indicate something important. Addressing policy issues often requires negotiating past "wicked problems": issues that require formulation of disagreements and problems as to what to treat as causes, and what as symptoms. Wicked problems, because they arise at the intersection of competing values, have no single, correct answer and must be negotiated on an ongoing basis. What I ask is that scientists recognize this prior, important subject of discourse and open it up for public discussion about values and proper problem formulation. Further, we can note the implication that little could be gained by a public discussion of environmental values and goals, that individuals are ready to make such decisions without deliberation and careful consideration, and that there is little "method" involved in making decisions once the available facts are presented. Nothing could be further from the truth, of course, and the reconstruction of a more robust, publicly informed process of evaluation, held together by the keystone concept of sustainability, is the subject of section 11.5. But first, I must expand on some philosophical themes.

11.2 Philosophical Analysis and Policy Choice

Because there are already a lot of books on sustainability, I have felt a need to justify the addition of another one by noting that I am offering a new kind of book, one that is distinctive in its view of *language*. Most books on sustainability address, in one way or another or from the viewpoint of one discipline or another, the question of what sustainability "really is." The authors of those books imply, implicitly or explicitly, that there is a possible state of affairs called "sustainable living" that can be discovered and given a linguistic label. The work is in finding this state of affairs; labeling it as sustainability is the relatively passive act of applying a label to a preexisting concept. This book is based on a different view of the relationship of language to the world we live and talk in. Unlike dualistic theories that assume there is a reality, preexisting, to which we, as mentally active beings, attach labels, my approach to language assumes that linguistic acts are as important in constituting entities and relationships as are the objects of our experience. Indeed, it is not wrong—though possibly misleading—to say that language *constitutes* our reality in an important sense. Defining a contested concept like sustainability, then, must involve a construction. My book fulfills its promise of a new approach, it turns out, just in case my constructivist approach to language results in an integrative model for evaluating environmental impacts of human activities.

I show in chapter 2 and in the appendix that, given the evolution of linguistic analysis, it is indeed possible, if one adopts a pragmatic attitude toward language and communication, to make strong recommendations about linguistic forms without making frightful metaphysical commitments or appealing to transcendent categories or principles. The traditional, essentialist view of language held that linguistic elements correspond either directly (by reference) or indirectly (through "ideas") to a prelinguistic reality. This view has always been justified by arguing that only essentialism can rescue us from a relativist and subjectivist world in which we would be trapped in our own "constructed" realities. This fear is based on a crucial alternation, the view that *either* "reality" must be constituted independently of human language *or* reality will be merely socially constructed. This is a *false dichotomy*. The middle ground, which I stake out, says that the reality we encounter is a *joint product* of convention and of experiences that come to us independently. According to this view, we are "free" to propose or use new forms of speech to report and explain our experiences; but the requirements imposed by *communication*—the reality that linguistic forms accomplish their intended purpose of communication within a community of language users—also impose constraints on our formation of linguistic artifacts.

This intermediate position may seem, at first glance, to lean heavily toward the "social constructivist" alternative, since language and communication are decidedly social acts. This is where Charles Sanders Peirce and the pragmatic turn come into the picture and push our form of conventionalism back toward the side of objectivity and realism. As noted above, Peirce's greatest contribution was his idea of the semiotic—the view that all acts of cognition necessarily involve an indivisible triad, including an actor (complete with a language, goals, and purposes), a signified object of experience, and a community of speakers of the local language. If any of these elements are missing, *social meaning* is missing. And language, above all, lends meaning and the possibility of communication to human social life. Language also affects the questions we ask and the alternatives we consider.

Since Peirce also realized that communication is a social action with consequences, he saw that these consequences opened up a new understanding of objectivity. He reasoned that since humans are evolved as social beings, and since language functions within social communities—communities *use* language to accomplish shared tasks—there will be a connection between people's reports of their experience and an "outside force," as explained in sections 2.4 and 3.4.

Peirce, having recognized the bankruptcy of Descartes's dualistic epistemology of correspondence between speech and a prelinguistic and "external" reality, located the "outside force" in human communities, especially communities engaged in cooperative action in a shared, problematic situation. If an actor in a situation calling for cooperative action offers reports of experience that mislead cohorts and result in the failure of joint projects, that actor's reports will consistently be challenged. Also, when a group has unexpected and confusing experiences, challenging some of its members' beliefs and thwarting an action they have undertaken to achieve a common goal, they feel uncertainty and doubt. According to pragmatism, the group then faces open choices: they may alter their beliefs, but they may also innovate linguistically in an attempt to reconcile recalcitrant experiences. This activity makes sense only within a community where cooperative action is sometimes sought. It is this relation—this struggle of individuals and communities to act together—that serves to isolate and eliminate error, however slowly and indirectly. It is the same relation that leads, sometimes, to linguistic innovation, in which new theories are introduced and new logical relationships are explored. Either the incremental movement of belief systems through addition of new facts or the more radical movement toward linguistic innovation is driven by shared community goals. The search for cooperative action, then, provides a (somewhat indirect) guide forward, epistemologically, and allows us to embrace a constructivist account of defining sustainability as part of the broad social purpose of "adapting" to our habitat. In this broad social purpose, truth-telling will serve cooperative action; wrong accounts of experience are in this sense maladaptive for a society of people who are attempting to live together cooperatively.

Here I have tried to steer as clear of metaphysics as possible, concentrating on the policy implications of the pragmatist view of the world and emphasizing linguistic conventions that connect to observable experience. In this action-based context, linguistic acts are judged not by their correspondence to a prior reality, but by their effectiveness in particular situations demanding action. Peirce's triangular relationship is the key to avoiding both relativism and rank social constructionism; it is also the key to the pragmatist view that language functions as a tool that can, with care, be sharpened and made more functional. What animates the irreducibly triangular relation is that language is used to communicate and pursue goals in the real world. What unifies this triangular relationship—what binds it irreducibly—is the act of communication in service of a shared social goal. Although our belief structures are, on this view, socially constructed, the constructions themselves are not the property of one individual or any limited segment of the community; they are constructs out of the experience of communities of truth-loving inquirers. Language, as it functions in the service of communication within real communities, is constrained by the common experiences of the other members. Experiences in these communicative situations, though individual in origin, become shared building blocks of experience as they are rendered in language. One cannot separate the linguistic component from the experiential component of a sentence; nor can one purge values out of individual experience or the expression of it. These two philosophical realities doom the correspondence theory of truth and the representational approach to epistemological justification. Good riddance.

But we find, also, a kernel of truth at the heart of the failed strategy of construing truth as a matter of correspondence. Objectivity is, indeed, imposed from outside of us in the sense that we do not *control* our experience, and this independence of our control is the basis for what I have called limited realism. The resistance of our experience to our desires, our inability to shape our experience to be what we *want* it to be, is reinforced by the linguistic acts of others, who see things from a different perspective with different values and interests. Successful linguistic acts of communication create a shared, communal experience. When there are disagreements between two speakers of the same language, they use their shared vocabulary to describe their individual experience and try to agree on how to describe their shared experiences.

Thus, as Habermas has noticed, challenges by others to our beliefs will lead us to cite our experience and compare it with that of others. The process of claiming is ultimately anchored in the process of experience-becoming-communal. Subjectivism fails because we cannot choose our experiences and because our claims based on our experiences are open to counterclaims based on the experience of others. Our linguistic acts, in this indirect way, are continually being submitted to experiential tests. Although our belief systems, so understood, are inevitably relative to a given linguistic community, this relativity is attenuated significantly by our ability to learn new languages and to create translations of descriptions of experiences from one language to another. Translation itself allows considerable slippage of meaning, but the pull toward objectivity is nevertheless driven forward whenever speakers of different native languages try to work together cooperatively.

This form of limited realism, relying on the involuntariness of our experience coupled with the objective pull implicit in community discourse—especially discourse in the service of cooperative action—is all we need to begin an adaptive management process. Adaptive management does not depend on first principles or on a priori truths. It starts from uncertainty and builds toward action by trying to articulate shared experience and to agree upon shared goals. One casualty of the rejection of correspondence is the notion that sentences are the unit of truth or falsity; and a single sentence cannot convey a moral evaluation. When we challenge each other's experience, citing our own, we must challenge underlying complexes of beliefs and interests as well. Speaking logically, as Quine has shown, data always underdetermines interpretation and theory. If we encounter an unexpected experience, the recalcitrant experience challenges our beliefs, taken as a whole; but it cannot dictate which of those specific beliefs *must* be changed, or what it must be changed to. The question whether to hold on to empirical beliefs, even in the face of countervailing evidence, is always a matter of negotiation. One can alter one's assumptions, shift emphasis, or flat-out deny prior beliefs. Since evidence from experience underdetermines any theoretical commitments, if we eschew a priori principles to give necessary shape to our belief structures, this decision becomes a matter for the community to resolve, at least temporarily. The point where agreement to act cooperatively is reached is the point where our beliefs matter publicly; beliefs support actions. But all such resolutions, in an open situation, are temporary and open to continual challenge. New vocabulary and new meanings for old terms may be introduced or may evolve in response to experience. Theories and descriptions, likewise, will be fine-tuned in the process of communication in search of cooperative action.

This conception of language has the advantage of including a means to challenge falsehood: the counterclaims of other truth-seekers who have contrary experience. It also recognizes the importance of language and linguistic innovation in the service of common social goals. The triangular relationship is irreducible and inseparable because we cannot be aware of an object that exists independent of a communicative act, and the communicative act presupposes the shared conventions necessary for language and discourse. So much for metaphysics; we can proceed without these deep claims by simply restricting our discussion to communities that have already committed themselves to cooperative action. Nevertheless, it is nice to know that Peirce's theory of semiotics can support a limited realism, which is all that is necessary to support the adaptive management approach as a method of truth-seeking and goal-testing.

Speaking more practically, and assuming we ply our trade of pragmatics in a community with a shared language and some level of commitment to common goals (at least the goal of acting together cooperatively), it is now possible to see the transcendence of representational epistemology and linguistic essentialism—both bad seed from the common tree of dualism—as a liberating outcome. If language is a tool of communication in search of cooperative action, and if both misinterpretation of experiences and linguistic confusions can gradually be repaired by communicating about our experiences, then it is perfectly legitimate to propose, experiment with, discard, or revise the terms we use to frame common problems and to seek cooperative solutions.

Relying on a conventionalist view of language and the science of the pragmatics of language, which allows development and choice of linguistic forms that improve communication, I have chosen the term *sustainable* and the related phrase *sustainable development* as unifying and integrative concepts that can anchor a new, more useful linguistic framework for discussing and evaluating environmental policy. I choose these terms not because of the strength or clarity of their current definitions-in-use, nor because they have an essential connection to some particular disciplinary theory. My goal, rather, is to propose—to construct—a definition that is useful in a democratic process of deliberation about what to do. The advantages of the term *sustainability* are of two kinds, political and practical. Politically, this term has been accepted in global forums discussing global problems and policy. Some governments have committed themselves to sustainable policies and to achieve development that is sustainable. So, I reason, there is no doubt that *sustainability* will be a useful term in public discourse, provided it can be assigned a clear and communicable meaning.

Practically, the term *sustainable* has the advantage that it is clearly forward-looking, and it expresses humankind's acceptance of responsibility for actions that have impacts on long- as well as short-term scales. The term is also very useful because, like "economic growth," it inevitably has normative content: *sustainability* is just the kind of term that might serve the dual role of communicating important descriptive information and doing so in a way that points toward understandable social values and goals. It thus fills the gap in current discourses, serving as a bridge term that can summarize a lot of data *and* express an important value commitment. I do not, of course, expect that *sustainability* can do all the communicative work alone. It is a sort of "capstone" term that gets specified more fully from the bottom up, as a system of less inclusive and more specific evaluative terms are defined and used in real communities. These less inclusive evaluative terms may employ criteria derived from one of the special sciences such as economics or ecology, including especially the choice of important indicators as expressive of the values of particular communities. We may also need midlevel bridge terms, such as *health* and *integrity,* because these encourage us to attend to systemic goods.

Further, I start with the assumption that members of modern democratic societies will describe multiple values that they derive from nature, using multiple vernaculars and appealing to a variety of basic values and to various worldviews. This is a descriptive statement; but it also represents a strategy, described above as one of "experimental pluralism." The strategy of environmental pluralism—in which various styles and methods of evaluation are studied and used in particular situations—begins by characterizing environmental values eclectically and then attempts to integrate the study of values into specific contexts with particular ecological, economic, and cultural givens. In this way, our categories of value will be tested against the cumulative experience of members of the society. In keeping with this strategy, I try to develop a definition of sustainability that incorporates and integrates data from the special sciences to the extent possible. In choosing a definition of sustainability, a community articulates what really matters and incorporates its values into this normative definition. It would then be possible to use disciplinary conceptions as more formal expressions of some of these felt values.

Although I have been very skeptical of claims that narrow, specialized disciplinary conceptions capture the whole of what we mean by sustainability, a more comprehensive definition of sustainability should, of course, build upon and incorporate these narrower measures. Disciplinary variety and innovative potential is important, but true integration will have to be achieved in the rich, value-studded context of ordinary discourse, where communities, through communication among members, decide what to do.

I have, strictly speaking, provided only a definitional schema, with variables to be filled in mostly based on local, situational factors. Accordingly, the real work that can be done in a philosophical book is to develop a process whereby particular communities are likely to ask the right questions and progress toward consensus or at least toward cooperative action in search of better understanding of their place. A philosopher, on my pragmatic view, can offer no self-evident principles, but a philosopher *can* offer heuristics. Accordingly, this book has unfolded as a multidisciplinary exploration of a process of constructing a sustainability definition in many particular places in many particular communities, a process that can be guided by important heuristics. As noted above, recognition of the importance of language in all meaningful discourse implies that the language we use determines which questions we ask. As we learned from Rittel and Webber and their idea of "wicked problems," problem formulation is in many ways the most important step in getting on the path to cooperative behavior (see section 4.1).

One problem with current public discourse about policy is that there is little agreement about what criteria should count in evaluating policies; every question of how to evaluate an environmental change requires a contentious decision regarding what to measure and how to evaluate changes. Worse, these practical questions get all tangled up with ideological disagreements about which values should count, take precedence, and so forth. And these latter debates are linked, ultimately, to very basic ontological commitments that are built into the paradigms and constructs of various disciplines. I have therefore eschewed "applied" philosophy—philosophy that contributes by first establishing an ontological principle in theory and then proceeds to derive policy commands that apply to specific and local situations—in favor of a more pluralistic view that takes the goal to be achieving cooperative action in response to shared problems. We can therefore think of philosophy as a tool to improve communication within democratic processes.

The social problem, as defined in the system of analysis and action proposed here, is to achieve cooperative action in response to a shared problem. The focus is on action, not on the truth of ontological claims. The pragmatist, then, can concentrate on the practical question of choosing which indicators will be used to flesh out the local definition of sustainability. Indicators thus *express* values but do not *represent* values. One might, for example, argue for the importance of economic criteria by appeal to an ontology of the world as "resources" and potential commodities. Or one might argue for the preservation of a wilderness reserve on the grounds that it has intrinsic value. What matters in the search for cooperative action, however, is that participants be able to work together to choose indicators, which identify certain processes and changes in the system as important enough to monitor. Once these processes are chosen for monitoring, it will be possible to identify goals that can be stated as desired levels to be achieved and maintained with respect to the chosen indicators.

The shift from abstract arguments regarding the ontological priority of ultimate values to a discussion of which indicators to monitor and regulate may seem to minimize the importance of values. It does not—it merely minimizes the importance of *unanimity* regarding values. This shift from debate over values to debate over indicators allows individuals in a pluralistic society to choose indicators that, to their way of thinking, will protect the values they hold dear. The shift to a dialogue regarding what to *do* can deflect attention away from *disagreements about ultimate values* and focus it on concrete, measurable *objectives that serve multiple values.* This shift is from ideology to pragmatism, and the strategy works *because it assumes pluralism of values,* not by insisting upon *consensus regarding ultimate values* among discussants.

Philosophers may see this as a sad demotion for them; but it is not. The proposed shift of focus does not prohibit abstract discussions of value in broader contexts, or even in activist contexts. What is proposed is not a rejection of abstract philosophical reasoning, but a shift in the role attributed to such reasoning. Philosophical and moral principles now take the role of reasons that can be adduced for choosing a given indicator; and moving forward with action requires only a majority or a strong coalition in favor of a policy, so experimentation is possible. As Habermas argued, such action can be legitimate, provided the public discourse is open-ended, and today's losers may, if they can introduce new supporting experience and experiments, be winners tomorrow. Particular policies can be supported by a number of alternative value frameworks.

The shift toward a more practical approach to cooperative choice actually opens up a new role for philosophers: as practical philosophers, who improve language as a means to improve communication and encourage cooperative action to solve environmental problems. This study, which has been called the pragmatics of language, involves a normative search for more perspicuous and transparent ways to deliberate about what to do. Here the philosopher seeks to improve language in order to serve a shared human purpose; there is no need to dispute about ultimate values. Communication is justified not by ultimate values but rather by the proximate values individuals come to share when they join a discourse community, and especially when they engage their cohorts in a search for cooperative action in response to common problems.

The task is to create new and more transparent ways to represent both problems and solutions, including the development and integration of new and more precise models to sharpen the search for better solutions. Given a conventionalist view of reality, philosophy becomes a creative discipline, a discipline capable of literally rebuilding the world. For it is language—especially language exercised in pursuit of a social goal—that ultimately expresses new and creative ways to articulate problems and to conceptualize solutions. And it is deliberation and the struggle toward cooperative action that provide the test of new philosophical articulations of problem formulation and solution. In this context I, a philosopher, have offered my services in creating an improved vocabulary for characterizing environmental problems and for deliberation about environmental actions. And in this context I propose to help, through various heuristics and logical clarifications, guide communities through a process by which they "fill in the blanks" in a schematic definition of sustainability in the context of their particular community. The philosophical aspect of this community-based process, one might say, is the act of focusing attention, reflectively, on linguistic choices, including choices of indicators as an important phase in an ongoing, open public process of deciding how to act in a particular situation in which ca community faces real problems. The reflective, more philosophical, phase of policy analysis and formation, in other words, is embedded essentially in the ongoing process, which has practical action as its ultimate goal.

11.3 Scale and Value: The Key to It All

Okay, the stage is set. I've explained what a philosopher can offer, however modestly, in the large and necessarily multidisciplinary struggle to integrate scientific and evaluative knowledge in a framework that can reasonably guide action. It's time to do some philosophy, to offer some integrating concepts, to innovate in the development of new ways to talk and to think about environmental values.

When innovating, it is best—as Newton noticed—to stand on the shoulders of giants. I refer back to section 6.3, where I showed how environmental enlightenment came to the great conservationists such as Muir and Leopold through shifts in perspectives on the world, rather than as a result of conversion to a specific moral principle. I am not belittling the attendant moral impulses, of course; my point, however, is that both Muir and Leopold described their conversions not in moral terms but in terms of a changed architectonic of their universe, a new perspective.

The single most dramatic, creative, and productive idea in the history of environmentalism has been Leopold's simile, thinking like a mountain. I say this because this idea has integrative power by virtue of its combination of evocative and emotive connections, on the one hand, and its potential for formalization by association with hierarchy theory and general systems theory on the other. Given its stated emphasis on temporal horizons of human impacts, Leopold's simile links human actions and their impacts to more than one natural dynamic unfolding on different scales in time and space. Today's decisions, often evaluated in the short-term calculus of economics, can lead to long-term impacts that change the system subsequent generations will encounter. Leopold's simile thus evokes dimension and scale in connection with human responsibility, a recognition that there are rhythms and dynamics in nature that we do not experience as immediately relevant to us, but which affect our world by changing dynamics we have hitherto taken for granted. Leopold reconstituted the world we experience as a complex world, where impacts of our actions unfold on different scales and dimensions and where humans are capable of more and more "violence" in the management of natural systems. It is a world in which human responsibility and concerns are expanded.

This is, as Leopold quickly realized, a chaotic world, a world that is not at all Newtonian. For example, Leopold spent several years studying ecology to learn how to distinguish naturally occurring variation and cyclical changes in systems from true "damage" caused by humans.[3] He never got the answers he was looking for; he never solved the problem of separating natural from human-caused changes affecting landscapes. The best he could do was to say that the difference is a matter of degree of violence; human actions, enhanced by technological prowess, can change nature irreversibly, quickly, and pervasively. Coming to think like a mountain is accepting responsibility for both the short-term and local scales and also for long-term and larger-scale, ecological impacts. But this reasoning yields only a more-or-less judgment. It provides no single, yes-or-no answer applicable in all situations. For this and other reasons, Leopold never got past an outline of how to solve the integration problem.

Nevertheless, I see no better place to start than with Leopold's sketch of a multiscalar, adaptive model in which activities of individual actors can affect their personal well-being on a short scale and at the same time, when aggregated, can set in motion slower-scale changes that will change the decision context in subsequent generations. To think like a mountain is to think hierarchically, organizing space-time relations from the inside out by choosing a perspective (for example, here and now) and then understanding space-time relations as smaller-scale, faster-paced dynamics embedded in larger-scale, slower-paced systems. Leopold thus had all the elements of a formalization embodied in his simile of thinking like a mountain, but more than thirty years passed before hierarchy theory, a form of general systems theory, was introduced as a formal and reasonably complete method for relating multiple scales to each other. When it was articulated, first by Allen and Starr, it was proposed and developed on a conceptualization of space and time very like Leopold's. Further, its advocates, like Leopold, see the problem as one of *constructing* a multiscaled system for organizing space-time relationships in a complex system.[4]

More specifically, however, one must recognize an important divergence between Leopold and Allen and Starr. The latter authors state quite specifically that the systems and models chosen to represent systems hierarchically are chosen for their usefulness in *understanding* and *describing* nature. Here again we see the impact of disciplinary science; Allen and Starr, as ecological scientists, implicitly assume that cognitive or descriptive aspects of language can be separated from its real use in communities already immersed in problems, projects, and goals and that hierarchies can be constructed by reference to the categories of a purely descriptive language of science. And here Leopold-the-manager's thinking departs from the pure-science line. Thinking like a mountain is a simile that emerges from Leopold's managerial perspective; he experienced the failure of his deer policies in an active, not a purely descriptive, context of inquiry. Thinking like a mountain, in other words, is ecological thinking, but it is value-charged thinking nonetheless, because it occurs within a management context, where competing goals and values are already implicit in the scalar complexity of the management system.

Leopold's hierarchical thinking is far more than an abstraction from a class of descriptive models. Thinking like a mountain involves a shift in perspective, a shift that is evocative of a new set of concerns and values. So Leopold's mountain thinking, formally identical to that of hierarchy theorists in its approach, is much richer and emotionally evocative. The ecological content, the recognition that plans have backfired and that important values have been left unprotected, is charged with a recognition of failure; Leopold's invitation to think like a mountain involves a much broader reevaluation of our main concerns and goals. Thus, although Leopold's approach to space-time relations and their organization anticipates the formalisms of hierarchy theory, it incorporates them in the richer, value-laced language of management; Leopold's simile of thinking like a mountain is rich enough to include reconsideration of goals, even deep values. Adaptive management, by incorporating Leopold's simile, embodies the formalizations of hierarchy theory within an action context in which social values direct our attention to relevant experience and experiments. Social valuations and science are both endogenous to Leopold's management system.

Let's stay with Leopold's evocative idea for a bit longer before committing to particular formalizations, because it is through this evocation that the meaningful connection to values is established. In a wonderful essay, "Marshland Elegy," Leopold said he saw the world as usefully organized into three scales. The first is the usual scale of human concern and human perception. This scale is naturally very short, its duration symbolized by a long wait on a cool March day for the arrival of cranes—audible, but not visible—in the morning fog. We are capable, however, through the development of an aesthetic sense, of sensing the majesty of limitless time and the slow march of evolutionary change, which takes place in geologic time. Leopold mainly emphasized the middle scale—the ecological scale, on which species compete for space in an ecological system and develop dependencies and interrelationships with other species. Leopold was most interested in the scale on which cranes adapt and reproduce themselves within an ecological niche that is constantly changing but is regular enough to allow genetic evolution. It is also the scale on which humans, from their humble beginnings, struggled to maintain a niche. For millennia, human impacts were small and local. But now these age-old relationships have changed: humans have developed stronger technologies, and with these have come the potential for more pervasive violence and more rapid transformations of ecological systems. The ditch-diggers and marsh-drainers destroyed the evolved relationships of the "Arcadian Age"; Leopold mourned the loss of the day when "man and beast, plant and soil lived on and with each other in mutual toleration, to the mutual benefit of all."[5] The central lesson of this change, Leopold learned and expressed in his simile, was that humans in the modern era must accept responsibility for all of the impacts of their actions, especially if those impacts are irreversible and pervasive; and if they do not know those outcomes in advance, it is their responsibility to learn, to engage in pilot projects and limited experiments with time frames long enough to provide information on the intergenerational implications of their actions.

I think Leopold achieved a remarkable integration; all the essentials of a pragmatic, activist inquiry were present in Leopold's multiscalar approach to thinking through proposed policies according to economic impacts (short-term dynamics) and also according to the long-term dynamics affecting a mountain, a river, or an ecosystem. Some have read Leopold and thought that when he went over to mountain thinking, he stepped beyond analysis and experience to become a mystic, that he gave up on a rational and scientific approach to management. I see the breakthrough differently. I think that learning to think like a mountain is additive to, not discontinuous from, his understanding of economic management. One doesn't stop thinking like a hunter or a consumer when one learns to think like a mountain. Thinking like a mountain incorporates our local and day-to-day concerns; but it adds also the awesome responsibility that comes with the recognition that our decisions today, in a technologically powerful society, can have impacts on longer and larger scales.

What Leopold saw was that if you want to manage for human goals from within a dynamic, open-ended ecosystem, you cannot look at only one level of impacts, at only one dynamic to which your managerial actions will be directed. Leopold continued to assume that individuals would express their preferences and that economists would aggregate them according to "demand" and "supply" models. But Leopold recognized that no single-level rule could adequately characterize a decision in a complex system. One must also think about impacts of one's actions, and the collective actions of one's cohorts, on the ecological systems that form communities, places, and habitats. Leopold realized, in other words, that good management means more than correctly following "a" rule (maximization-optimization); good decision making in complex systems necessarily involves both applying rules and being reflective regarding which rules are appropriate and necessary to apply in given situations. The latter process requires altering and rebalancing our rules when experience tells us our actions are too violent, too pervasive, or too difficult to reverse. Making such determinations requires both good judgment and good science. Scientific models can help, for example, to identify indicators that are associated with important dynamics, but it requires good judgment to decide what is important. Leopold recognized that if environmental management is a game, it cannot be a game with a single payoff; and it cannot be a game played according to a single set of rules. If, however, there are multiple rules associated with different scales of space and time, then we must—to continue to claim to manage rationally—extend rationality to include not just the application of rules but also the choice of which rules are appropriately applied in given situations.

Thinking like a mountain, on my interpretation that it *adds* rather than *shifts* levels of analysis, implicitly includes a requirement that we develop a second-order rationality that provides guidance in decisions as to which rule should be given precedence in various situations. This result should not be surprising, given our interpretation of results from decision theory and planning theory, explored in chapters 4 and 7. To use Rittel and Webber's apt explanation, in decisions—"wicked" decisions—that require the balancing of competing goods, there is no optimal solution. We are stuck with balancing multiple indicators associated with multiple, sometimes competing, values. Following in Leopold's footsteps, then, I am proposing nothing less than a new approach to management discourse, a metadiscourse in which we reflect upon and balance and rebalance our rules of thumb in response to feedback information from the action phase. In this second-level discourse, we have no single currency by which we calculate winners or losers, whether dollars, years in prison, or expected utilities. Instead, we institute a two-level game, with iteration. On level one, deciders face a problem armed with limited information (high levels of uncertainty) and several "decision rules." We can think of these rules as methods for evaluating decisions. What methods are in fact available?

We assume that individuals will often do an accounting of decisions and policies in a self-regarding way Exercising this level of analysis would involve submitting the policy to a cost-benefit test, which encompasses individual levels and short temporal horizons. This is a useful criterion to apply to many decisions, but in keeping with the pluralism developed in this book, the cost-benefit test must be balanced against other considerations in many situations. This first scale of individual valuation will be strongly skewed toward individual consumption and impacts on the shortest temporal horizons. To follow Leopold, we would add at least a second scale of responsibility and concern, the scale on which our collective actions have impacts on communal goods, especially those associated with the ecological dynamic affecting the mix and distribution of species on the landscape.

Hierarchy theory (HT) can be united, as briefly explored in section 8.7, with the technique of risk decision squares to provide a quantification that can be helpful in considering these metaquestions of what rule to apply and which rules to emphasize in particular situations. For example, HT provides, at least in principle, applicable quantitative tools that can help to determine the proper *scale* on which a given environmental problem should be addressed. It also allows us to use empirical estimates of the extent and reversibility of possible impacts to provide significant input into whether a decision can be resolved by cost-benefit analysis, or whether the risks to social values are of such extent and difficulty of reversal that a more stringent burden of proof, such as the safe minimum standard of conservation (SMS) rule or the precautionary principle (PP), should govern some decisions. In this way, HT points toward a possibility of empirically testing hypotheses about the areas of potential impact of various decisions and the time of reversibility of those impacts. We can look forward to the day when a manager could argue, for example: "Evidence gathered from isolated pilot studies suggests that if this particular plant pathogen (which is proposed for use in weed control) were to be used, it could attack native vegetation and negatively affect a whole ecosystem irreversibly. Therefore, whatever the cost-benefit analysis of the use of this pathogen says, the risks are too great and we must—based on the evidence of broad extent and irreversibility of impacts—apply the SMS or the PP and ban the use of the pathogen." To back up this reasoning, the adaptive manager can say: "You are proposing to put at risk—for longer than a human lifetime—important features of our ecosystem. Decisions with that kind of risk may negatively affect future generations, irreversibly narrowing their options to draw services from the ecosystem; so this decision requires a demonstration that the costs of avoiding this risk are not bearable. Since there are other means to manage weeds, this burden of proof is not fulfilled, *based on scientific evidence* about the spatial extent and irreversibility of impacts." Thus, although metadecisions about what rule to use cannot be formalized and solved according to an algorithm, we can bring scientific evidence and testable hypotheses to bear upon such decisions by paying special attention to empirical information about extent of impact and irreversibility. This feature of metadecisions can protect them from becoming merely ideological.

Leopold's multiscalar thinking—and his acute observations about multiple scales and dynamics affecting, and affected by, human activities—are thus the most central aspect of environmental decision making. Leopold gave us the evocative stimulus; it is up to us, in each local community that values the future and accepts responsibility for effects of current activities on the future, to build a locally viable concept of sustainability, one that involves thinking like a mountain, a forest, or a watershed. What we have learned from the conventionalist view of language is that thinking like a mountain will require talking like a mountain, developing, that is, new conceptualizations and a new vocabulary for describing long-term impacts of our actions and a new vocabulary for expressing intergenerational concern. Leopold's breakthrough, and our hierarchical interpretation of it, provides us with the guidance needed to start thinking and talking in new ways. For pragmatic philosophers, acting on this guidance is the highest philosophical calling.

11.4 Disciplinary Stew: The Prospects for an Integrated Environmental Science

Adaptive management is scientific management under the assumption of uncertainty; uncertainty plagues us especially when decisions are forced, when not acting may be the most destructive act. Real situations are value-laden, with alternative, sometimes competing or even contradictory, values and commitments expressed, and we can expect private interests and public agencies to jockey for position and control in order to further specific agendas. Can we hope for unbiased and useful science in such situations? Throughout this book I have emphasized the importance of integrating descriptive and normative discourse in public discussions and deliberation about environmental policies. Adaptive management can work only if good science emerges within public discourse; only if science in the service of policy can reduce disagreement and contribute to cooperative behavior. Our next task is to look ahead at the future of mission-oriented science, as it could function within adaptive management processes (see sections 3.3–3.5). I have suggested that an illuminating analogy for management science is a disciplinary stew, with ordinary language and commonsense policy discourse making up the broth, while the meat and potatoes, the really nutritional part, are the empirical scientists who are capable of moving management science forward by engaging in careful and publishable experiments in management.

We can, along the way, assess the prospects for a postdisciplinary science and whether such a science could encourage cooperative behavior by contributing to social learning. In effect, this assessment requires no less than an exploration of the future of environmental science. The question is whether environmental scientists can move beyond the limitations imposed by the putative fact-value divide and develop a unified, pragmatic method of experience as applied to both the reduction of uncertainty and the re-formation of goals and values through ongoing reflective deliberation.

A manageable way of discussing the very large topic of whether and how science can better serve adaptive management in the future is to focus specifically on the use of models designed to clarify issues in specific management situations.[6] Three key, related points will be made about these models, which are treated as one promising approach to postnormal, mission-oriented science for adaptive management. First, I discuss the purposes and motivations for models and suggest a functional analysis of available models for the aid of environmental policy deliberation. Given that analysis, I urge the development of models specifically for the purpose of improving communication among participants in adaptive management and other participatory processes. Second, I discuss a multiscaled participatory exercise, the European

VISIONS project, as an illustration of the current state of participation in modeling exercises, and make the point that models can be and have been used as aids in public participation and deliberation about policy goals. Third, I briefly examine a Canadian exercise in developing scenarios for sustainable development, because this group addressed the fact-value issue effectively, illustrating quite clearly the role of values discourse in the future of mission-oriented science. From these three points, we will be able to conclude that a self-conscious use of models to improve communication and deliberation among participants, and to improve adaptive management processes, can represent an important contribution to public discourse. The important point to see, however, is that although such models make a *contribution* to the broader management dialogue, they cannot *replace* ordinary discourse. This approach to model-building as a scientific element in public deliberation provides one example of how a postpositivist, postdisciplinary science could flourish as an essential part of effective adaptive management processes.

11.4.1 Some Ways to Think about Models

Model is a very elastic term. Sometimes it is used narrowly and precisely, to refer to a set of equations that describe the behavior of a system. But we can also call any simplified representation of a complex relationship a model. According to this broad definition, complex mathematical equations that describe relationships are models, and so are many of our day-to-day proverbs, such as "a stitch in time saves nine." What models have in common, on this broad view, is that they somehow describe "how things work" in simplified ways: they represent a set of causal relationships that allow users of them to make sense of new information and project likely outcomes of actions, however loosely. Models, of course, are always *simplified* representations: they allow us to ignore some information as irrelevant and to consider certain possible outcomes as not worth worrying about because they are unlikely. Models, because they are simplified representations, can function to avoid information overload and can provide a structure into which to integrate new information; they thus help us to interpret and evaluate our experience.

As simplified representations, models can be thought of as a special category of symbolic, even linguistic entity; a model, once in use, conveys information by virtue of conventions, as do languages. We thus apply the reasoning of pragmatists since Peirce, assimilating all representation, including scientific representation, to the role of symbolic behavior, behavior that gains meaning only within a community whose members are capable of using that language. Assimilating models to linguistic behavior also brings them under Carnap's "principle of tolerance," which ensures that they will be judged for their usefulness in particular situations, and encourages us to consider a variety of forms that might be useful in different situations and contexts.

Notice that the term *language* has meanings that parallel the narrow and the broader senses of *model*. Whereas *languages* usually refers to what are called "natural" languages (for example, English, Spanish, Japanese), logicians and philosophers also speak of "formal" or "constructed" languages. Languages of the former kind resemble the everyday proverbs I just called models—they have evolved to fulfill a function in a culture; the latter, more formal models are constructed by scholars, built up of basic elements and explicit transformation rules, like logicians' languages, and these can be developed for various purposes. The key point is that both models and languages are best understood as conventionally formed symbolic systems. These systems can be formed either "naturally," in the sense that they arise and evolve as an integral part of a real-life, cultural practice, or they can be consciously "constructed" in order to accomplish a purpose. Natural languages, and the expressions they contain, survive if they are useful for communicating in everyday situations; these linguistic practices become embedded in ordinary life processes and survive, evolve, and adapt in place because they make cooperation within social groups possible.

The pragmatic method, and the choice to employ ordinary language as the language of management science, allows us to combine the best features of both natural and constructed languages—and both natural and constructed models—in a single discourse. If ordinary language is the language of policy deliberation, this fact does not rule out taking advantage of the special sciences and their specialized models. As participants encounter more complex scientific issues that require more formal modeling, scientists can be commissioned to study particular dynamics and impacts of choices, and these findings can be fed back into the process. It should be a goal of mission-oriented science—once such specialized models are patched into public discourse, creating information in a form usually consumed by experts—to provide reasonable translations of specialized jargon into ordinary language, whenever possible. A commitment to transparency is an essential precondition for the practice of participatory model-building (see subsection 11.4.2).

We can use these ideas to suggest some useful vocabulary for analyzing environmental discourse. Let us consider the term *mental models* for naturally occurring models that are used by individuals to understand a natural phenomenon or an impact on a natural process; we can use this term in the same way it is used by cognitive psychologists; mental models are simple representations of complex systems that people use to understand and predict the future and to interpret new events and information as they encounter them. In studying such models, we may seek to represent, to the extent possible, the way individuals understand "pollution" of a lake or a watershed. We can also, following cultural anthropologists, define *cultural models,* which refers to the common-denominator beliefs and understandings that allow individuals to function as a social group. We can then modify the idea of cultural models a bit and apply it to stakeholder groups, hypothesizing, for example, that members of a stakeholder group tend to share mental models with other members of the group, whereas they have little in common with the mental models of members of other groups. This hypothesis, if true, would explain why some environmental problems are unyielding, even in the face of careful attempts at participatory solutions: members of different groups not only speak differently; they also have different mental models associated with their speech patterns. Although a mental model is an internalized, operative model and is defined by the ways it is in fact used by a person in a real situation, we can imagine the development by adaptive managers, over time, of a model that is designed specifically to help the various stakeholder groups acquire some facility and comfort in using that model as a tool of integration and as a means to improve communication in the discussion and evaluation of proposed policies. Such a model might well derive its value for adaptive management from its contribution to communication and cooperative process, rather than from an ability to predict the future.

Once we begin thinking about models as linguistic tools, it also seems natural to expect models to be as diverse as the uses to which they are put in both everyday and scientific situations. A tolerant and experimental attitude toward the usefulness of models for quite different purposes can be further refined by following Jan Rotmans in distinguishing between supply models and demand models.[7] Supply modeling attempts to push forward the frontiers of knowledge in a particular disciplinary direction; supply models are almost always formal structures, constituted by their rules of construction. In supply modeling, a model is proposed, based on the best available techniques, and designed to increase our understanding in important ways, where importance is understood according to accepted disciplinary paradigms. Supply modeling, to once again apply the distinction of Funtowicz and Ravetz, is "curiosity-motivated"—motivated and shaped by a disciplinary paradigm. Demand modeling, by contrast, is motivated by a social problem or issue. Professional modelers in this case bring their techniques and frameworks into a practical situation, and success or failure will depend upon whether proposed models improve the decision process.

It is also useful to distinguish between the activities of integrated assessment (IA) and integrated assessment modeling (IAM). IA is an activity that seeks integration across disciplinary and political boundaries; it must therefore include qualitative factors and, at best, can aspire to quantify key relationships as much as possible. IAM seeks quantitative models and therefore tends to ignore important factors that are difficult to quantify. According to Dale Rothman and John Robinson, "the 'science-driven' view of IA . . . tends to result in IAMs and forms of analysis that de-emphasize, or even exclude, qualitative issues such as trust, power, credibility, status, governance, social organization, morality, etc. that are typically the subject of the social sciences and the humanities. However, since it is precisely these qualitative factors that often dominate human behavior, both quantitative and qualitative information are needed for an IA to be truly useful for decision making." They go on to mention the obvious advantages of fully integrated quantitative models for incorporating feedbacks and dynamic effects, but they warn about pitfalls and biases that can be introduced by overemphasis on quantification.[8] This attitude supports the thesis of this section (11.4) that the more basic activity of IA—integrating information in a current situation choosing policy interventions—should be carried out in ordinary language; attempts at quantitative modeling are thus seen as important additions but not aimed at achieving the comprehensiveness necessary to guide decision making.

One way to combine the search for integration (IA) with integrated modeling (IAM) is to recognize two somewhat different roles for two types of scenario-building. Rotmans and colleagues, who are interested combining the strengths of an integrative process with the development of more powerful and useful modeling techniques, propose moving back and forth between *participatory* and *expert* scenarios in order to access the advantages of both demand and supply modeling approaches. They distinguish between the two types: "Participatory scenarios refer to approaches in which stakeholders (non-scientists), such as decision-makers, business people and lay people play an active role: the participants are co-designers of the scenarios. Expert scenarios are developed by a small group of technical experts, responsible for the design and development of the scenarios. Participatory methods are advocated because a diverse group of stakeholders with different knowledge, expertise and perspectives provides a greater richness to scenarios."[9] Rotmans and colleagues did not, however, succeed in replacing participatory scenarios with more precise, quantified, expert scenarios. They end by calling for new and improved methods to accomplish this task. One can applaud these researchers' intent to merge public participation and science-based modeling and their optimism in thinking that improved methods may allow the completion of such tasks, but a realistic assessment of the current status of demand modeling suggests that such models will not achieve full quantification in the foreseeable future.

11.4.2 Participatory Modeling

I am especially interested in demand models because these models—despite the extreme difficulty of developing a good one—seem to be just what is necessary if science is to usefully support adaptive management processes. Demand modeling can be very useful in creating a common model to enhance participatory dialogue among managers and stakeholders and to create a context in which citizens can deliberate about their values, goals, and objectives. A useful demand model would be helpful, for example, in discussing and introducing new information and assessing its importance in decision making. To illustrate this use of demand models, we can briefly examine the IAM effort called VISIONS, undertaken in Europe from 1998 to 2001. This project had as its goal to develop a shared, science-based model to improve communication between experts and lay participants in the search for an acceptable approach to sustainable development in Europe.

The "ambition" of the VISIONS project was "to raise awareness of sustainable development by increasing the understanding of the many links between socio-economic and environmental processes and by improving the assessment of the consequences for Europe from an integrated viewpoint."[10] The VISIONS project was based on two methodological "pillars," the first being the adoption of a scenario approach that places events in a "*factors, actors and sectors*" framework, and the second involving "*integration across scales.*"

VISIONS included two scales, investigating sustainable development of three representative "regions" (Venice, Italy; the "Green Heart" area of the Netherlands; and the northwestern United Kingdom, including the city of Manchester) as one scale, and all of Europe as the other scale. A "storylines" approach, which uses a combination of participatory methods, including presentations by experts, discussion groups, and various techniques to encourage creative thinking, developed narratives, or storylines, that described a sequence of events, offering "rather unconventional future pathways, which go far beyond the business-as-usual perception." These storylines, refined by analytic techniques that allowed managers to link them in a logical and consistent manner, were incorporated into multiple scenarios that were constructed for each of the three regions and for Europe as a whole.[11]

Using participation and the storylines approach, group leaders were thus able to piece together multiple narrative scenarios on each of the two scales. The scenarios were not, of course, treated as tools for deterministic predictions, but rather as possible ways the regions and the continent could develop in the future. Researchers working on VISIONS recognized two broad approaches to the process of scenario building. One approach, which relies heavily on supply models, is to try to interpret the storylines or scenarios within an existing model. In this way, the informally developed scenarios can be made more consistent, both internally and with known causal models that define the range of real possibilities. The second approach, which has the advantage of encouraging creativity and imagination, is to leave the scenarios as possible narratives, unconstrained by judgments about their consistency or likelihood.

The managers of the VISIONS project chose a combination of the two approaches, a mixed strategy that is dictated by the current state of demand modeling. Since many of the managers were professional modelers, attempts were made to reflect elements of the narrative models in existing (supply) models. They used, for example, a general equilibrium model for the world economy developed by the Dutch Central Planning Bureau and other modelers. Several other off-the-shelf models were used to formalize and improve the narrative scenarios at the regional level. The results of these attempts were, according to their reports, mixed: "It is clear that models should not be expected to mimic the different narratives with a high degree of precision. We did not want to limit the richness of the narrative scenarios by forcing them into a model. At the same time we did not want to unjustifiably stretch the scope of the model. This became even clearer during the actual analysis. Thus, in the end we focused on using the models to analyze certain aspects of the VISIONS scenarios by undertaking targeted experiments."[12]

Here we see both the promise and the great difficulty involved in doing true demand modeling. By involving participants in an ongoing process of developing narrative scenarios, managers of VISIONS were able to create a useful dialogue in which information flowed both directions, from experts to stakeholders and from stakeholders to experts. These efforts were clearly undertaken with the intention to develop something like a demand model, one that functions as a touchstone for public discourse and deliberation. Demand modeling, however, which takes as its starting point the messy dialogues of ordinary speech, is not well enough developed to build *from* the information and value base represented in ordinary discourse *toward* a more precise and useful model of how the world actually works in management situations. No demand models yet developed can achieve plausibility as a comprehensive model of environmental forces on the regional or continental scale. So managers of VISIONS tried to call upon existing supply models in an attempt to plug them into the evolving narrative scenarios to give the latter greater precision. Forty modeling studies were reviewed, with ten of them used intensively to evaluate "the state-of-the-art in European scenario-building." In the end it was admitted that this strategy was not satisfactory, and the project managers concluded that there was a need "for further methodological advancement in this area, . . . for new types of scenarios that integrate spatial scales, economic sectors and societal institutions."[13]

To summarize, then, a consortium of research institutes undertook to involve the public and stakeholders in a comprehensive attempt to integrate knowledge from different disciplines into the shared understanding of environmental problems and sustainable futures. They produced "narrative scenarios" for three regions and for Europe as a whole. These scenarios were useful for improving communication among experts and stakeholders, but the modelers were unable to build the imaginative scenarios created by participants into precise, "constructed" models that would allow the precise evaluation of proposed policies. So they fell back upon using existing (supply) models to improve the narratives but never succeeded in building a demand model that started with the problematic situation of European management for sustainable development and developed a working, integrative model.

This effort is nevertheless very useful, because it points the way toward more work in developing demand models; the outcome, however, shows that the development of true demand models is still more a matter of hope than of reality. Nonetheless, exercises similar to VISIONS may become more common, and scenario creation and model-building are likely to become staples of the adaptive management processes.[14] For some time, we will not have the techniques to create a fully functioning demand model for any particular management situation; it will be necessary, in the meantime, to do as the managers of VISIONS did—they left their scenarios in imprecise narrative form, used available supply models to check the narrative for consistency and plausibility, as possible, and lived with the fact that they could not "connect" the integrative "models" they were building from the demand side to a fully functioning, precise model built from disciplinary pieces. This situation apparently cries out for some timely research in methods to improve demand modeling, but as noted above, there are many aspects of complex, real-world problems that cannot be formalized.

Despite the limitations of demand models, I still believe it is important to foster the spirit of this activity. Demand modeling, I predict, will become increasingly important in adaptive management, because it encourages the development of place-based, problem-oriented, and integrative models in response to the needs of specific, problematic situations in particular places. Ordinary discourse, together with the models that might be built into its fabric to improve it, on the one hand, and the specialized discourses that include formal or constructed models taken from disciplinary supplies, on the other, have their unique strengths and weaknesses. The most reasonable strategy therefore is try to achieve the best of both linguistic worlds. We make sense of everyday problems and respond to them in ordinary language. When specialized disciplines and formal models are useful, we call upon them, trying to improve our scientific understanding of natural processes and human impacts on them. But those models must also be interpreted, and fed into the public dialogue, by "translation" into an integrative model that can contribute to public understanding and social learning. The integrative model should be understood as a demand model: the model should respond to the needs of a community devoted to social learning about how to care for its place. One of the key roles of such models, provided experts commit themselves to striving for transparency in their appeals to science, can be to improve communication by developing a shared cultural model of ecological processes and human impacts affecting a watershed or an ecosystem. As noted above, members of disciplines and interest groups—stakeholders, that is—may in many situations have their own, in-group models of what is happening and why. If there is a shared, integrative model, failures of communication due to clashes of mental models can perhaps be bypassed, improving opportunities for cooperative behavior. I represent this situation and the value of demand models in figure 11.2 as a table that signifies the coming together of stakeholders and interest groups in an attempted collaboration. Building a demand model, in this context, can contribute to cooperative action by offering participants, who come to the table with different mental models, a shared model that can aid in communication. The demand model is placed on the table as a means by which the group can gradually work toward a shared characterization of "the problem" by reference to the model, or to subsequent versions of it.

11.4.3 "Backcasting" from Goals: The Role of Values in Demand Modeling

Whatever the future of demand modeling, we must address one more issue about the future of scientific modeling—how, exactly, do social values get built into a demand model? I said above that the adaptive management approach must provide a key role for both facts and values, descriptions and prescriptions, normative analysis as well as scientific hypotheses, in its action-oriented discourse. By examining another example of building a model to increase social understanding, we can get a better understanding of precisely the role of values in demand modeling and in adaptive management processes. With support from the Canadian government, a group at the Sustainable Development Research Institute at the University of Vancouver, beginning in 1988, carried out the six-year Sustainable Society Project, which resulted in a number of publications including a book.[15] Many other such projects have occurred since (President Clinton's Commission on Sustainable Development in the United States is another example), but I choose the Canadian project and writings about it to illustrate the role that values can play in a truly postnormal, mission-oriented science.

More than most active modelers, Rothman and Robinson share my view that environmental problems and environmental management goals must be understood within a situation that is already value-laden. They say that "attempts to portray IA as being science-driven are misleading. They are also potentially dangerous. To do so perpetuates the presumption of an objective sci-

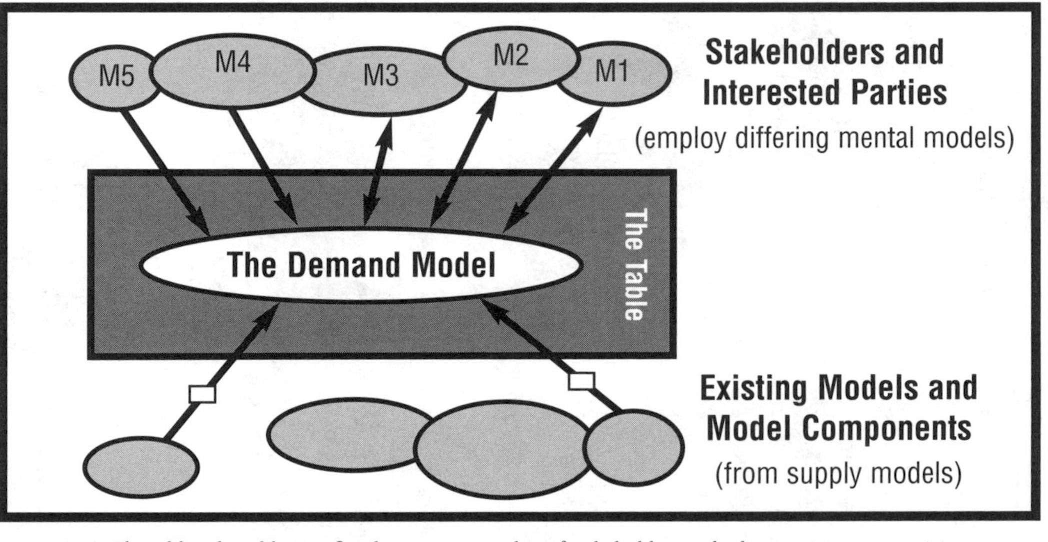

Figure 11.2 The table. The table signifies the coming together of stakeholders and other participants; participants have differing mental models of the problem and what is driving it. The demand model can be placed "on the table" to aid in the communication among diverse groups and interests.

ence that provides legitimacy to political decisions, often reached for quite different reasons. . . . Instead, there is a need for a more explicitly self-conscious and reflexive approach to analysis that is deliberately defined in terms of the real world problems that it is intended to address."[16] Or, as other authors put it, such assessments should "begin and end with the policy problems, not the scientific problems."[17]

The key method, used with great effectiveness by Robinson and his colleagues in the Canadian study, is "backcasting." Robinson describes backcasting as "a method of analyzing alternative futures. Its major distinguishing characteristic is a concern with how desirable futures can be attained. It is thus explicitly normative, involving working backward from a desired future endpoint or set of goals to the present, in order to determine the physical feasibility of that future and the policy measures that would be necessary to reach that point."[18]

Backcasting does not have predicting the future as its goal; its goal is to explore the feasibility and costs of pursuing various policy goals. Values guide science toward the problems that must be explored using the scientific method. Further, these goals are stated for the future, thus stimulating an examination of what our values are, what we really want to create and sustain for the future. Robinson says of backcasting: "In order to permit time for futures significantly different than the present to come about, endpoints are usually chosen for a time twenty-five to fifty years in the future." Goals projected that far into the future will stimulate participants to ask what is really important to them today.[19] Values thus have a huge role in this process: they stand behind and give meaning to the goals we set as a society. Goals that emerge from a public process enhanced by backcasting reflect values held today, for the future. Science is then guided toward what is important to know, given the goals we have set for ourselves.

Participants *express* their values as they argue in favor of indicators and specific goals and management objectives. This pluralistic, cacophonous discourse, and the decisions it leads to, will define the bequest of the people in a place to their offspring and successors. And yet there is often room for compromise to find actual management objectives that can be supported. Hunters and birdwatchers, who place very different values on wildlife, can unite to pursue the goal of wetland and habitat protection. Backcasting can therefore be a very valuable tool for adaptive managers by shifting the debate about values from abstractions like intrinsic values versus commodity values, toward actual projections of possible scenarios describing how things might unfold in the future—what I have called development paths. Values, in other words, can remain highly diverse, provided specific shared goals can be articulated; scientists fit into this picture by focusing on questions that matter to action and by trying to translate their scientific work into more transparent, postdisciplinary demand models so that it can be examined and applied by interested parties.

We began this section with the question whether frankly mission-oriented science could be good science. Some, coming from the perspective that only value-neutral and "objective" science should guide management, will surely worry that the process I describe will promote biased science. These concerns about bias as a result of immersion in values are answered in section 9.3. The way to attack bias in mission-oriented science is not to cloister such science away from social values, but to encompass both science and the values of scientists and others within a single iterative discourse in which claims can be challenged and biases exposed. So I have here answered a slightly different question from the one we began with: I have answered the question whether mission-oriented science could be useful to adaptive managers. I think the examples given in this subsection and the preceding one support a positive answer to this question and suggest that there is a bright future for demand modeling as one important component of mission-oriented science.

For a pragmatist this is no dodge. I am not ignoring the fact that science undertaken in adaptive management processes will sometimes be biased; I am simply embedding the process of review of scientific findings within a broader management process. I am expecting the community of truth-seekers, who see the world from various perspectives, to recognize and point out bias. Provided experts and scientists are committed to transparency in the use of science, peer review will be supplemented with review by participants trained in multiple disciplines, as well as by participants with varied other interests and concerns. As long as these explorations are undertaken in a transparent way, there will be plenty of reviewers to point out biases. Science that is useful in adaptive management will be used; beliefs that are used in seeking community goals are open to adjustment in the face of experience. The question of bias is thus internalized into adaptive management because, when embedded in an open and inclusive process of participation, all claims, even scientific claims, can be disputed by the test of experience.

11.5 Environmental Evaluation: A Fresh Start in the World of What-If

Throughout this book, I have struggled against one of the deepest commitments of the Newtonian, modern age, the myth that science can be truly objective in its descriptions of the world. It cannot be so, at least not in the sense that the truths of science correspond to preexisting chunklets of the world. We have seen how this myth has disoriented and disorganized the process of blending social values and facts at EPA and elsewhere. The other side of this dubious claim, of course, was the suggestion that values, by contrast to objective facts, spring from a nonrational preference and are hardly worth studying. Economic science begins with the preferences already formed; its main tasks are in aggregation of given preferences. At best, the study of preference formation and alteration is considered something for cognitive psychologists to undertake; at worst, it is treated by economists as cognitively meaningless.[20] But if only the factual realm of science can be rational, and if values are banished from science, then we are doomed to making decisions based on subjectivity and nonrationality. The key point here is that, for economists, all values are preferences; and preference formation and reconsideration are treated as exogenous to environmental science, both natural and social. One goal of this book is to reassert the possibility of a robust and informative social scientific study of environmental values by sketching a new approach to environmental social science. I have called the new science the science of what-if. This approach is frankly value-laden and is committed to achieving cooperative behavior by fostering learning relationships. Social learning relationships, such as adaptive management processes, depend, of course, on communication, so contributions by social scientists and philosophers to improved communication about values could be crucial to the goals of adaptive management.

The social science of what-if approaches the whole question of environmental values from a new angle. Remember that, back in section 5.5, we noted that Economists and Intrinsic Value theorists both seemed obsessed with placing value on *objects*. When nonanthropocentric theorists extend intrinsic value language to ecosystems, they are sorely tempted to attribute a "good of their own" to them as entities or organisms.[21] I think we should sidestep this assumption and, following the suggestion of section 5.5, agree that the proper object of evaluation in a process of adaptive management is a development path or, more concisely, a scenario. By moving away from the idea that various objects or entities are the bearers of value, and by pluralizing the ways we talk about and evaluate possible future outcomes of proposed policies, we can avoid many of the dead-end discussions about the exact value of x, whether x is a commodity or a lump of nature that is purported to have intrinsic value. A development path, understood as a direction in which a community or "place" could evolve, is a scenario that the community might live out.

A development scenario is thus a possible path into the future that may come about in a place as a result of the dialectic between a human community and its environment. *Place,* here, refers not just to geographic location or to physical systems of nature, but rather to an evolving, interactive relationship between an ecological system and the people who occupy that ecosystem and use it as their habitat. So I am proposing that the "object" of evaluation be not an object at all, but processes of change and adaptation that emerge from a community's pursuit of various policies.

Scenarios, in keeping with pluralism, can be evaluated according to multiple criteria. Criteria, in turn, can be thought of as offering operationalizations and calibrations of various indicators. The criteria can be thought of as methodologically enhanced versions of the concerns and values that are felt in the community in question. So what we are doing, in an environmental evaluation, is judging a scenario, a possible unfolding of a local present into a future, according to multiple criteria {C1, C2, C3, . . .}. The criteria embody indicators {I1, I2, I3, . . .}, which are easily measurable variables that are hypothesized to be associated with the protection of important (unspecified) social values. These are the formal assumptions of the evaluative context; they should be thought of as apparatuses for expressing value.

Since we are involved in a *management* activity, we need a causal analysis so that policy responses will be understood as interventions in current causal dynamics. Admittedly, this is a point that might be challenged: can we, in fact, separate anthropogenic change from natural change? The answer is that sometimes we can and sometimes we cannot. But certainty is not expected when one embarks on an adaptive management process, so all we need is a hypothesis that a given impact is anthropogenic, in order to proceed to study it and to act to reduce uncertainty about it. Similarly, one need only assume that an impact is a consistently describable possibility to embed it in a possible scenario.

What is missing from this abstract model, of course, is choices. Choices as to what variables are important to a community and choices as to what levels of various indicators are "ideal" or "acceptable." These questions cannot be decided in abstraction from real communities, who must "fill in the blanks." Since we are talking about sustainability, we must also assume our community/the people (in our representative place) are concerned about the bequest they will leave for the future people who occupy their place (chapter 8); and since we assume that members of our community have short-term economic concerns, it follows that most communities for whom we might devise an evaluation plan will be pluralistic in that long-term impacts of proposed behaviors will also be of concern to them. What adaptive managers and participating scientists and stakeholders in adaptive management processes need is methods and heuristics that will help them to fill in the blanks in their own sustainability definition for their own place.

Dewey provided adaptive managers with the key idea of social learning; he offered a naturalistic approach by distinguishing in his ethics between saying that an outcome is "desired" and that it is "desirable." This distinction anchors Dewey's pragmatic method because it allows us to construe ethical statements as hypotheses that can be tested against experience. The distinction, in other words, is the key to Dewey's naturalism and his insistence that in the postmetaphysical, Darwinian world, there is a unitary logic by which both descriptive and prescriptive claims are held to the account of experience by active communities seeking cooperation.

The determination that an outcome is desired by an individual or a group is not the end of inquiry, but the beginning; it is a hypothesis that taking action to achieve the desired end will lead to satisfaction and recognition of success. It can be called into question if someone—an individual or an official body—acts on the goal of bringing about the outcome and instead gets an outcome that is widely rejected. By combining Dewey's distinction with some more recent innovations by pragmatist philosophers, especially Habermas and other discourse ethicists, I think we can envision a new direction for environmental social sciences, a direction we can refer to as the "world of what-if." Central to discourse ethics is the idea of a "discourse community," a community that uses discussion and public deliberation to seek cooperative action and which makes and evaluates claims and counterclaims according to practical reason. We can extended this thinking by hypothesizing an "ideal speech community," in which proposed actions and arguments for them are open to challenge by others, persons who perhaps have unique experience. Key to such a community, of course, would be a shared language in which communication is usually successful, but there are many additional conditions that could, arguably, be required to call a communal decision process ideal. Indeed, it may well be that no finite list of requirements would fully describe such an ideal community; nevertheless, we can imagine such an ideal, and I think we know roughly what would improve less-than-ideal discourse communities (the task of community improvement is explored in chapters 7 and 9).

Given this complex of assumptions applied to various concrete decision situations around the country, a new world of study now opens up for social scientists in three complementary directions. First, with regard to less-than-ideal public processes, including those that are failing to achieve progress toward consensus or social learning, one can ask, What conditions of public discourse and decision making would move this process closer to an ideal discourse? Answers might include both proposals of new terminology and new problem formulations, and also proposals for changing administrative structures and public participation venues. Second, by studying successful managerial processes that do achieve social learning and move toward cooperative action, one can ask what conditions support such progress; generalizations about what works and what does not work can help struggling communities and, eventually, provide the basis for empirical generalizations about the parameters of an idealized speech community designed to improve public deliberation in search of cooperative environmental action.

Social scientific progress in either of these directions would, in turn, open the door to a third world of information, the world of what-if. Assume that we have a community that sets out to develop a more cooperative approach; assume also, for the sake of concreteness, that the community develops an adaptive watershed management process to be guided by a citizens' advisory committee (CAC). Building on Dewey's distinction between desired and desirable, we can say that the community is fragmented into groups with many desires and demands, some compatible and others conflicting. The issue at hand is to determine which of these desires will indeed lead to outcomes that will be approved by (be desirable to) most participants. We have already admitted that we do not have a prior standard by which to judge which desires will turn out to have desirable outcomes. Worse, the inquiry begins amid ideology, disagreements about problem formulation, failures of communication, and rampant mistrust. Now we can ask, What goals would emerge from a process that is closer to an ideal speech situation? If there was a language adequate for communication and a forum for open deliberation, then conditions might emerge in which multidirectional communication between scientists, managers, stakeholders, and members of the general public would encourage social learning.

Social scientists, in such contexts, can create and manage situations that foster the emergence of improving, if not ideal, speech communities. For example, they might work with the CAC to create a science court or a jury; or they might institute monthly "informational" meetings or forums where groups could present their proposed goals (desires) for reaction from other interest groups. Or social scientists could participate in the development of shared "demand models" that would help to eliminate differences in mental models of participants and promote better communication. The point behind this research (besides improving our understanding of how public processes work or do not work) is not merely to try to learn what people in fact desire, but also to form a hypothesis about what they would find desirable *if* they looked at the problem from multiple perspectives (by listening to representatives of other interest groups), *if* they had better access to existing science, and *if* they felt empowered to undertake experiments in management to see what might be the outcome of pursuing proposed goals. This is the world of what-if: Community X is currently divided among interest groups who propose disparate and sometimes conflicting goals based on varied desires. What if members of this community were—contrary to fact—able to function as an ideal speech community committed to cooperative action?

This approach is referred to as the world of what-if because most social scientists and most communities simply do not have the resources or the widespread public interest necessary to involve all citizens cooperatively in an ideal community discourse that is well informed scientifically and engaged in public deliberation about goals and policies. For most communities, research on improved decision making by the whole community will be impossible; therefore, research into a counterfactual world of what-if is an acceptable substitute. Such research is not meaningless or hopeless, especially when combined with the first two types of research just mentioned. Based on this research on conditions conducive to cooperative agreement and on what characteristics of processes makes people more likely to achieve cooperative action, it is possible to begin studying groups already involved in the public debate, such as members of the CACs or science juries, treating them as microcosms of what the broader community would do if it was participating in an improved dialogue.

Such research may sound suspect to traditional, value-neutral social scientists who hope only to describe the desires of individuals and aggregate these for groups. For example, the use of participants in the process as subjects undermines most possibilities of generalizing from the participants tested to the general population because, far from being randomly chosen citizens, the subjects involved in the adaptive management processes are self-selected by their willingness to volunteer to work on management problems. But the point of such research is not to use sophisticated random sampling techniques to generalize about what is desired by the population prior to discussion, social learning, and public deliberation. The point, rather, is to determine what goals a subset of the population, representing some cross-section of the interest groups, would end up finding desirable if, in the face of the best available science and a few well-designed management experiments, they proceeded trustingly into an open-ended, iterative public dialogue about what to do.

In this chapter I have made one last tour through the disciplines, and I have tried to encourage a new way of relating them to one another, using the analogy of a rich stew. Whereas disciplines often vie to provide the basic vocabulary for analyzing environmental policy, this "turf war" approach has only reinforced towering. The central finding of linguistic philosophy, ranging from Peirce's theory of semiotics through the later Wittgenstein, ordinary language philosophers, modern pragmatists, and neopragmatists, is that the specialized languages of the disciplines cannot capture the richness of the world in which people communicate, deliberate, and cooperate. Repeated attempts to replace the sometimes-confusing vagueness of ordinary language with precise languages and algorithmic decision models have failed. In order to gain control over a small realm of nature or society, disciplines intentionally create precise but incomplete models of reality. Languages are more like spotlights than like floodlights: the choice of a precise language closes off important aspects of a situation, just as it highlights others, and no precise, specialized language can replace ordinary language with all its rich interconnections to communication and action, as the language of integration and choice.

And so I think of integration as the stew that is created when one adds the meat, potatoes, and vegetables of specialized disciplines into a rich broth of public discourse. It is the broth that unifies the dish and provides the stock for a rich, multidisciplinary discussion. The special sciences—the lumps—are invaluable when ordinary discourse bogs down and deliberators in the action phase cannot decide what to do next. We can, at such times, enlist the precise methods of a specialized science to resolve a disagreement or reduce crucial uncertainty as it occurs in ordinary discourse. When we face disagreement that is blocking cooperative action, we should try to formulate the disagreement as a testable hypothesis in one of the specialized sciences. If we can import just enough of the conceptual apparatus of that specialty discourse to create an experimental design, we can undertake "management experiments" to reduce uncertainty. For this technique of augmenting the practice of ordinary language with experimental tests to contribute to adaptive management processes, it will be necessary for the practitioners of the special discipline to connect their specialized concepts with ordinary language concepts by the use of bridge terms. In so doing, they will provide the broader discourse of the adaptive management community access to their scientific insights as introduced elements in ordinary discourse. At the same time, these bridges make possible the broader form of peer review—by both disciplines and stakeholders—necessary to integrate mission-oriented science into public discourse and into management plans.

12

INTEGRATED ENVIRONMENTAL ANALYSIS AND ACTION

12.1 Conservation: Moral Crusade or Environmental Public Philosophy?

In the preface I cited one of my heroes, John Muir, who initiated a moralistic and ideological rhetoric pitting good versus evil, God versus Mammon, and right versus wrong in the battles to save forests and Hetch Hetchy Valley. Muir's rhetoric was matched by that of the materialists Gifford Pinchot and his colleagues in the resource-use agencies of the government, who gladly counted board feet of timber and other resources extracted to fuel economic growth as measures of progress. Whatever the usefulness of Muir's theological rhetoric in kindling a movement to save nature, I have argued that these all-or-nothing rhetorics represent, today, the greatest barrier to cooperation in protecting the environment. The all-or-nothing rhetoric of the age of ideology, as detailed in this book, has blocked communication among participants in environmental discourse and has made compromises appear as deals with the Devil. Administratively, it has led to towering in the agencies as cadres of true believers in one ideology or another have created fiefdoms within bureaucracies. Politically, it has led to abrupt reversals of policies, even when there has been no change in scientists' basic understanding of a problem. All this and more has been visited upon the environmental movement as it has attempted to mature through the age of ideology.

A not-so-hidden agenda of this book has been to speculate whether it might have been different. Might the battle lines over conservation policy have been drawn in less black and white terms? Could conservation policy have been contested along a continuum of mores and lesses, rather than as all-or-nothing dilemmas? In answer to this question, I have shown, based on a classic 1973 paper by Rittel and Webber, that most environmental problems exhibit the characteristics of "wicked" problems and are better characterized and understood as involving competition among multiple goods and legitimate interests than as oppositions between right and wrong, good and bad, or optimal and suboptimal. As Rittel and Webber recognized, when wicked problems affecting conflicting values are treated as if they must have true or false answers, confusion reigns. If we can, instead, formulate an environmental problem as one of rationally allocating resources among competing values, the problem can be addressed in terms of more or less and in terms of balances among legitimate values. Such a move beyond ideology would go a long way toward reducing turf wars, eliminating fragmenting jargons, and encouraging the development of models designed to improve communication among stakeholders and other parties in an ongoing dialogue and decision process.

Rittel and Webber's argument accounts for the difficulty in agreeing on problem formulation in environmental policy, which in turn implies that we need public, iterative, deliberative processes if we are to respond reasonably to problems of great complexity. Following their argument has also encouraged us to embrace a pluralistic approach to values and to encourage the articulation of multiple values (pluralism) and multiple criteria for analyzing environmental policies, all within an open and deliberative public process. This combination of pluralism and process enables a new discourse about environmental values and policy objectives; it reframes controversies as questions of how to achieve multiple goals through cooperative action and how to encourage social learning through experimental action. If this reframing occurs, environmental ethics will give way to environmental public philosophy. Attention will shift from identifying centers of intrinsic value to improving linguistic and conceptual tools for relating the environmental sciences to one another and to improving verbal resources for communication about environmental values and policy goals. In this new discourse, a new kind of normative analysis will emerge. Better questions will be asked. It will be less important to know whether things of a particular kind have intrinsic value and more important to know how to choose improved policies, given multiple goals and values exemplified in the community. Debate will then shift from ideology to solving real problems.

Evaluation will concentrate on choosing development paths that promise the right mix, the right balance, of protections for various goods and interests. This shift to choice in action situations deflects attention away from ontological disputes in economics and ethics, and toward better methods of measuring and integrating values. In this new approach to evaluation, a theory will be judged not according to a priori arguments in ontology, but by the usefulness of ideas in promoting cooperation. Cooperation can lead to partnerships that support the monitoring of key trends, the setting of standards and goals with regard to those trends, experimenting, reducing uncertainty where possible, and seeking cooperative action. According to the pragmatist theory underlying adaptive management, this process can, through communication, lead to social learning. In this discourse, values are ever present; they affect the advocacy of various indicators and standards, they affect priorities accorded to certain criteria, and they may in some cases provoke great passion among advocates. As long as values support policies and standards, however, we can be pluralistic and inclusive, hoping to build coalitions behind integrative indicators that will track a range of values.

The choice between developing such an alternative discourse and the status quo is stark. Growth and development in more and more crowded space leads to complexifying problems. If policy responses are institutionalized in a fragmented fashion, they will be inefficient and contradictory. What has been offered here is an alternative, the alternative of formulating environmental problems as a competition among multiple legitimate goods, with the task at hand to be finding a reasonable balance in the pursuit of multiple goals formulated to apply to dynamics unfolding on different scales. Following this alternative, we try to focus deliberative attention on choosing indicators that are reflective of social goals and social values of the society. Once such an indicator is chosen, continued discussion must set specific goals, in the form of expectations and standards with respect to the indicators chosen. If a community can agree that a particular indicator can track important values that they have, the next stage in the deliberations will be a question of setting concrete goals with respect to the indicator. There will be legitimate grounds here for disagreement, but the disagreements will be about more or less, about how aggressively to pursue improvements in a given indicator, not about questions of right or wrong actions as judged by ideologies. Above all, this approach supports the possibility, by identifying policies that support multiple values associated with multiple scales and dynamics, to create win-win situations, which is the best strategy in the face of wicked problem.

The main point of this final chapter is to look to the future, but first I wish to summarize the costs of ideological environmentalism, which I understand as an environmental policy process in which theories and "first principles," usually borrowed from a single discipline, govern participants' views of environmental policy from the top down. It may be useful to divide these costs, prosaically, into three categories: academic-theoretical, practical-political, and socio-evaluative. These categories can be described briefly, since we have encountered many examples of each in our survey of environmental policy development and implementation.

First, let us consider the academic costs. As emphasized throughout this book, the two fields of academic study best suited to examining environmental values, environmental ethics and environmental economics, are mired in an all-or-nothing battle pitting nonanthropocentrists against market analyses, a battle tracing back to fundamental ideological differences and to basic differences in the terms used to evaluate changes in the environment. But this is only the beginning of the fragmentation. Economists, having adopted consumer sovereignty and being resolved to treat preferences as static snapshots of what people happen to want, have largely opted out of the broader question of how a community forms preferences and how these preferences change in the context of deliberation and cultural change across time. This position on the part of economists blocks the formation of bridges between economics, the cognitive sciences, and the other social sciences more generally.

I have also been critical of my own discipline, environmental ethics, because its practitioners have been seduced by Muir's engaging rhetoric and have cast problems of how to evaluate environmental change as problems of identifying objects that can be centers of intrinsic value, objects that demand to be considered morally. Identifying such centers of intrinsic value may bring comfort in waging moral war against the materialistic Philistines, but it also makes cooperative behavior in search of environmental solutions more difficult by perpetuating the myth that environmental problems can be addressed only through a moral crusade. My problem is not with the implication, shared by environmentalists from Thoreau to Leopold, that there is a deeply moral aspect of our treatment of nature and the broader, nonhuman communities we live within. The culprits, rather, are the a priori commitments that analysts and actors use to force all information about values into a monistic straitjacket in the service of specialized assumptions about what environmental value *must* be. These theoretical and preexperiential commitments of participants have blocked the development of a pluralistic and integrative approach. In environmental ethics, the requirement that values be expressed as comparable units of intrinsic value has an effect analogous to the effect of consumer sovereignty in environmental economics. Both channel attention away from changing, evolving values embedded in unfolding cultures, values being gradually adjusted in many particular situations, toward "chunks" of the world that can be assigned value, whether as commodities or as centers of intrinsic value.

I could go on, citing further cases of how ideological environmentalism reduces the likelihood of social learning and blocks improved problem formulation in academia. If we could get beyond ideology and address problems cooperatively, philosophy, economics, and other fields could lead the way in building a common, interdisciplinary language in which stakeholders and the public could deliberate about goals, objectives, and concretely proposed policies. We must take these contextual features of environmental policy—the uncertainty, the multiple competing values, the distrust among various interest groups—as givens and go on to design a process that can help communities to ask the right questions and to progress through incremental improvements toward common goals despite the challenges.

The political costs of ideological environmentalism are even more devastating than the academic ones, because all-or-nothing rhetoric causes environmental policy whiplash in federal and many state agencies each time there is a change in parties in the White House. With every switch in parties in presidential politics, the federal bureaucracy is subjected to a reversal in policy direction equivalent to the reversal of a battleship at sea. New administrations spend at least four years undoing or trying to undo what was accomplished, if anything, in the previous administration. Surely such rank inefficiency should at least encourage us to seek an alternative, more cooperative approach to environmental policy.

Without question, there are alternatives to the current ideological mess. Other countries have learned to transcend either-or ideology and have avoided the environmental policy whiplash that has been frequent in the age of ideology in America. Section 12.2 examines one such alternative, the system of environmental process and controls that has emerged in the Netherlands over the past two decades. The Dutch system of cooperation leads to far smoother transitions than does following all-or-nothing edicts implied by an ideological commitment to one form of right. Generally, shifts in electoral politics in the Netherlands result in changes, but they are changes of degree—slowing down or speeding up pursuit of an agreed-upon goal—rather than a matter of reversal of policies altogether. So we will set the stage for conclusions by examining one case study of a country that, like the United States, has a high standard of living, growing impacts on its natural environment from intense economic activity, and (broadly speaking) a democratic political process.

Then in section 12.3 we will revisit EPA and survey some documents that show that some people at EPA are raising the right questions and making very sensible recommendations, recommendations that would represent steps of various sizes toward the new approach to policy outlined here. If these proposals and recommendations were followed, huge improvements in process could occur. Unfortunately, during the George W. Bush administration these recommendations are languishing, as the administration has acted aggressively to reverse policies developed in the Clinton era. In the end we will conclude that, despite good advice, little progress has been made on the underlying problem—the lack of a shared and effective language for discussing environmental goals.

12.2 An Alternative: The Dutch System

One way of asking whether environmental policy discourse and problem formulation could be different is to ask whether there are nations that have avoided ideology and empty rhetoric in policy discourse and towering in their environmental agencies. Here I examine the case of the Netherlands, where initial encounters with environmental problems led to confrontation but where confrontation and towering in agencies with environmental responsibilities gave way to a concerted attempt to create a more rational and integrated process for addressing environmental problems as they emerge. The Dutch case proves that there is an alternative to the American ideological approach to public environmental discourse and to the towering that occurs at EPA.

We have seen that EPA, quite openly—indeed, in some cases, proudly—exhibits a tendency toward balancing antagonistic interests and groups against one another, which in turn reinforces towering at the agency as the minions of the various towers reinforce their defenses with more research that will be dismissed by opponents in other towers. The Balkanized agency, some would say, provides an opportunity for varied interest groups to establish a voice within the agency, ensuring that their viewpoint is registered. Less optimistically, other observers would argue that the present U.S. system, though messy and inefficient, is *inevitable,* given the fractured and partisan nature of American politics and the colorful history of the Muirites and the Pinchotians. Such assertions are disingenuous and question-begging, however. One could argue, to the contrary, that a more cooperative structure and process at EPA, one of the most contentious and politicized of all federal agencies, would go a long way toward reducing partisanship and fractured politics in the country.

At least in this one sense, there is an alternative. We could *try* to become less partisan and institute procedures likely to improve cooperation. We could *try* to measure values pluralistically, work to establish cooperative partnerships, and develop a process that promotes communication and social learning through iterative attempts to find policies that serve multiple goals. It would then be an empirical question whether such efforts would take root in American political soil. If we decide to take such steps, fortunately we are not without some guidance; other nations have been more successful than the United States in addressing environmental problems cooperatively and in innovating in the implementation of environmental policies. In fact, there are nations that have undertaken more integrative and cooperative approaches to environmental policy development, evaluation, and revision. The Netherlands provides perhaps the best example of a modern, industrialized nation with a high individual standard of living that has consciously and effectively created a more rational administrative structure and a more cooperative process for addressing environmental problems. I leave to the reader—and to the future—to determine whether the American political process could, given where it is today, impaled on a dilemma between two monistic value systems, evolve toward a more cooperative, integrated, and future-oriented process for setting environmental policies.

We can get a pretty good idea of how the Dutch system has worked by referring briefly to a recent book written by Paul De Jongh, who has served in several capacities in the evolving Dutch system of environmental management since 1977, and who set out to describe the Dutch search for integrated environmental policies in a book entitled *Our Common Journey*. The Netherlands, in response to growing environmental problems driven by high population and livestock density in a small, low-lying country, began in the 1960s and 1970s to address a catalog of problems including acid rain, urban smog, waste disposal, and deforestation. Early responses, De Jongh notes, concentrated on the *effects* of various polluting practices and problems and involved mounting an end-of-the-pipe response. During this early stage, a "war mentality" emerged as environmentalists (Greens) engaged in pitched battles against polluters. This old approach created and sustained numerous specialized institutions to deal with and regulate these effects. Programs piled on programs and new effects kept "popping up," demanding new responses. New legislation led to new regulations and new offices to administer them, and this trend fueled never-ending policy battles.[1]

So far, the story should sound familiar, because it parallels the creation of the U.S. system and its current condition. About the Dutch system, De Jongh says:

> Governments develop policies, and bureaucracies, to address each specific problem or problem area—such as waste, or pollution of the air, the water, the soil. After a while this leads to an inefficient duplication of effort. It resembles a hiking trip in which everyone is carrying his or her own equipment; there may be five tents or frying pans, when the group only needs one or two. They also have different maps and different ideas of how to reach the destination. In the policy field, that means separate bureaucracies for air, soil, water, toxics, and so on. Each has its own scientists and budgets; each issue has its own corresponding laws and regulations.[2]

Interestingly, De Jongh notes that the United States, though it was a pioneer in the environmental policy era, may have developed new and formative regulatory processes only to become trapped in the "confrontational culture" that was required for those early breakthroughs: "There are many drawbacks to being a pioneer. The US developed its environmental policy system in an era of intense conflict, in which every advance was won through hard political struggle. To force a change, activists pushed for strict, punitive laws, and litigation became a primary policy tool. The early battles of the US environmental movement established a confrontational culture which shaped public opinion, legislation and government institutions."[3] So, while the Dutch recognize their debt to pioneering policies and legislation in the United States, they can perhaps also look more objectively and critically at the results of pursuing the partisan approach that has become entrenched at EPA and throughout the U.S. environmental policy process.

Beginning in about 1982, the Dutch began an aggressive effort to rationalize the process of environmental policy discussion, debate, formation, and implementation. This policy process "was the product of extensive networking and consensus building. The approach was characterized by negotiations with important agents of economic change and oriented towards market forces; at the same time, however, it set out to safeguard the future of the environment in the national interest." And De Jongh emphasizes the importance of communication, just as I have done: "If they are to succeed, environmental policies must open new lines of communication on environmental issues between public and private agencies and governments. The aim must be to eliminate or reduce emissions at the source rather than at the end of production processes. Policy must focus on developing 'closed-loop' management of industrial processes to minimize waste and emissions and reduce energy needs. And it must emphasize the value of public education programmes to raise awareness of corporate and individual responsibility for cleaning up the environment."[4]

These were the goals. How does one get there from the contentious and disorganized "ad hoc" stage of environmental policy? In 1984 a working group of the Dutch Environment Department prepared a proposal, called simply the "nota," entitled "More Than the Sum of Its Parts," which contained a new philosophy for environmental policy and advocated integration of the work of government authorities, offices, and programs. The nota proposed a long-term National Environmental Policy Plan (NEPP) to be prepared by the four ministries with legal responsibilities for environmental policy. As interim steps, a series of multiyear programs was proposed, in which the working group would set out concrete goals for the first year and general principles to apply in the subsequent three years. At first there were problems as the various ministries and directorates, with overlapping and conflicting mandates, were loath to give up their "compartmentalized" approaches.[5] Gradually, however, participants stopped looking at the task as one of completing a plan and began to view it as a contribution to a larger process, which was to lead gradually to a redefinition of environmental problems.

The goal adopted was to institute NEPP by 1989, in order to track progress toward comprehensive, long-term goals. NEPP, it turned out, was not so much a set of fixed and substantive goals as a general approach to policy formation. One of its advocates described it as "70 per cent process and 30 per cent substance."[6] But NEPP, as it evolved, acquired several characteristics that are directly responsive to the problems addressed in this book. First, NEPP was intended to be *comprehensive;* from the beginning the plan was to avoid adding patches to patches as new problems arose. The aim was to reconstruct the policy process so that environmental problems would be understood and addressed in an integrated fashion. Second, NEPP was instituted with the full participation of the Dutch scientific community, and major, comprehensive studies were undertaken by RIVM (the National Institute for Health and the Environment), a respected national research institute. Scientists at RIVM, in turn, consulted scientists in universities and in other research institutes so that its reports were both comprehensive and widely reviewed. Third, NEPP was unabashedly future-oriented. Even though the process developed slowly, with interim plans and gradual implementation, both the science and the policy discussions were undertaken with a twenty-five-year horizon—one generation. Eventually it was realized that this generational focus tied the whole process directly to the growing interest, throughout Europe and the rest of the world, in "sustainable development" as a unifying policy goal capable of integrating economic and environmental concerns. The sustainability approach turned out to be quite successful, especially when expressed in twenty-five-year generational time spans. This period was long enough to make major changes conceivable to stakeholders, participants, and citizens, but short enough to imply some urgency. Once twenty-five-year chunks were associated with generations, it was easy for people to accept the goal of this generation planning to hand over a more livable world to their children.

The Dutch government instructed RIVM to conduct a comprehensive study of environmental quality and to involve a wide array of scientists so that the findings would appear as a scientific consensus. The study was to be cause-oriented: it was to trace environmental problems back to their causes. Whenever possible, quantitative measures of necessary reductions were to be stated. The study, like the planning aspects of NEPP, was to take a twenty-five-year time frame.

RIVM began working on two scenarios. One was based on a "business-as-usual" assumption, projecting the expected results of continuing current policies without change. The other was based on the assumption that the current, end-of-the-pipe programs would be intensified, applying all available cleanup technologies. The first scenario showed that current policies would fall far short of protecting environmental quality, especially in key areas such as acidification. More alarmingly, the intensified cleanup program of the second scenario was also going to fall far short of protection. Indeed, these models indicated that pollution levels would have to be reduced by 70–90 percent if environmental quality was to be protected for twenty-five years. The policymakers working on NEPP were stunned; they were seeking ways to cut pollution levels by as much as 20 percent, but now they realized that even best-case outcomes of their current proposals would be regarded as failures. These results, further discussed below, were at first disputed, but because they had been vetted so broadly prior to publication, few scientists challenged the results. This modeling exercise did much to focus attention on what came to be called the "third scenario." Nobody knew what such a scenario would look like in detail; but NEPP personnel did realize that the third scenario would necessarily move away from end-of-the-pipe cleanup technologies and toward prevention of pollution.

The third scenario also gained plausibility when economists at RIVM showed that both of the first two scenarios would result in spiraling costs to the economy; according to preliminary results, pollution prevention efforts might prove much less expensive. These results, derived by respected economists, had an important indirect effect. Some participants began to understand that pollution control need not be a zero-sum game in which every winner requires a loser. If, in fact, a new preventive approach was less expensive as well as more effective, industry and environmentalists could perhaps become allies rather than enemies.

An essential part of the RIVM modeling effort was the development of a multiscalar understanding of problems. The scientists believed that environmental problems were confusing to many laypeople because some environmental problems manifest themselves locally, whereas others are regional or global. So the RIVM team set out to classify environmental problems according to their scale, eventually developing a five-level system including local, regional, watershed, continental, and global scales. These scales were used to reorganize the available scientific data and to reorient assessment efforts. One of the leaders of the group explained: "Everything was there anyway, but we rearranged the stuff. As science, that was trivial, but for communication purposes it was an ideal approach."[7] Leopold's idea of thinking like a mountain has thus blossomed on European soil, with environmental problems and solutions articulated within a scalar approach to understanding interactions between humans and nature.

Once the working group was convinced that small, incremental changes to the pollution-control system would be overwhelmed by future developments, they began work on integration by redefining environmental problems; De Jongh, anticipating my search for a new language for talking through environmental problems, asserts that it is necessary to introduce a new language of integration into policy deliberations. In the Dutch case this meant developing a new way to classify environmental problems. The working group instituted the convention of classifying problems according to their source, rather than their impacts on media such as water and air. Instead of trying to regulate impacts on water, air, and so forth (the *effects* of pollutants) by tracing causes to their source, the new classification system focused on *sources* of pollutants, understood as types of economic activity, such as power-generating plants and pig farms. Once the emphasis was shifted from effects to causes, regulators were encouraged to integrate their activities with those of the various different ministries and departments: "Integration begins with a common language to describe environmental problems—the language of cause and effect. The different actors then divide their responsibilities according to where each can best influence the cause and effect chain. The actors do not give up their existing powers or authorities, but they work at using them in a way which is most effective for an integrated approach."[8]

This conventional decision to use new classifications opened up new lines of scientific thought, as modelers at RIVM produced computer models that showed rapid deterioration of the Dutch environment in a business-as-usual scenario, mainly owing to the cumulative impacts of the huge livestock industry in a relatively small area. These models showed, in particular, severe acidification of the environment and eutrophication of the waters as a result of excess nutrients. The excess nutrients arose from importing grains to feed livestock and exporting animal products, an economic process that leaves the manure within the country. Now the scientific models could shift from effects to causes, and scientists and policymakers began worrying less about media—water, soil, and so forth—and more about the source of the burdens affecting the media. Since economists at RIVM had, as noted above, produced models showing that the costs of business-as-usual would quickly overtake the costs of developing new, at-the-source reduction technologies, a shift occurred in the mental models that people used to think through pollution and responses to it.

What is impressive about the Dutch transformation, however, is that policymakers did not stop seeking better linguistic tools once the scientists and policymakers were comfortable with the cause-effect model for understanding problems. They used this accomplishment as a means to another, more democratic goal. They set out to articulate the problems in a way that all of the public could understand. One of the leaders of the new movement, Peter Winsemius, set the following standard: "If you can't explain an environmental problem to your next-door neighbor in five minutes, your story isn't good enough."[9] The cause-effect model was fine for scientists, policymakers, and environmental specialists, but they needed to be able to tell the moral of any environmental story in no more than five essential points! So the list of environmental themes, or problem areas, was reduced to five, each one labeled with the Dutch prefix *ver,* which signals a process that has gone too far. Translated, the five problems were acidification, eutrophication, dispersion, waste disposal, and local disturbance.

But the simplification and public-friendliness did not end there; for example, the term *eutrophication* seems to refer to an effect, the development of algae and the resultant lessening of suspended oxygen in bodies of water bodies. So the term *eutrophication* was replaced with *vermesting,* which could be translated as "manure-ification." This term focused attention on the economic processes—importing fodder, exporting meats and dairy products, and spreading manure on small plots of ground—that led to the problem, and it proved a far better educational tool than the technical, scientific, and neutral term *eutrophication.* The colorful term chosen thus had a dual impact. As an educational tool, it helped to gain support for new policies regarding animal waste, and at the same time it helped the public to see the value of the shift to a technical analysis based on causes rather than effects. This shift enabled the public to understand what values are at stake, given the negative associations of excrement and the implication of excess thereof. Eventually, the parliament accepted the proposed descriptions of environmental problems, and from that time forward there was a common vocabulary for discussing environmental problems.

The next step was to reorganize regulatory responsibilities. The managers did this by grouping tens of thousands of businesses into a few categories, such as the refinery industry, transportation, energy production, agriculture, and manufacturing. Interestingly, these groupings were accepted by the powerful Ministry of Economic Affairs, and the industries, rather than responding with paranoia for being singled out as causes, began to ask what changes they could make at least cost. Regulators thus opened up a dialogue with these main groups. The deputy secretary general of the Ministry of Economic Affairs summed up what was learned by this process: "After some discussion, there was a clear conviction that . . . if you do not involve the persons and the institutions and the companies that will have to do the job of cleaning up the environment, then you can just forget about it."[10]

Regulators now appeared not as enforcers so much as resources available to industry, and the contentious, warlike mentality of the ad hoc stage of policy enforcement began to pass. During the middle and late 1980s, although the Dutch economy was not doing well, the process of developing the National Environmental Policy Plan was gaining momentum. There was a real question whether the government would authorize the expenditure of 1.5 billion Dutch guilders necessary to reorganize environmental agencies and begin to address problems in an integrated fashion. A confrontation seemed inevitable. The question was whether the Central Economic Council and the cabinet would acknowledge the urgency implied by the RIVM study and appropriate the funds, thereby endorsing the ambitious NEPP approach. The issue was very much in doubt.

But the long-awaited RIVM study was published under the clever title *Zorgen voor Morgen* (Concern [care] for the future); it stated in strong and decisive language that environmental conditions would deteriorate unless a new approach was adopted. The report, though organized and published by RIVM, had drawn on the expertise of all the major scientific institutes in the country, so it appeared as a national consensus of Dutch scientists. This status discouraged "junk science" that repudiated the findings and greatly increased the comfort level of all participants in the process. But the plan had strong opponents on the Central Economic Council because of its initial cost and the weakness of the Dutch economy at the time.

Then, in a Christmas-day speech, the queen departed from the usual practice of giving nonpolitical holiday speeches and came out strongly in favor of decisive action, arguing that the Dutch people should accept "the challenge of finding a new relationship with nature, characterized by respect for ecological balance, caution, and careful management." The queen's speech carried the day, and several major trade organizations and unions endorsed the new approach. NEPP went forward.[11]

A political agreement between the government and industry emerged: industry accepted the long-term targets proposed in NEPP and the midterm projections regarding compliance costs. But the government had to accept a contingency plan: if the assumptions of the plan proved wrong, the whole deal would be renegotiated. A new organizational structure was gradually hammered out, and today most environmental issues in the Netherlands are addressed cooperatively, with orderly inputs from various stakeholder groups, public discussion of options, and gradual negotiations to find a policy that is acceptable to all parties. Negotiations with business shifted from enforcement of bureaucratic rules and standards to a "covenant" approach.[12] Problems were addressed by first finding points of agreement between all parties—the government, industrial representatives, and nongovernmental organizations—and getting consensus on the goals of the program. Over the next few years, the government negotiated covenants (business contracts) with all the major industrial sectors, generating support for environmental goals and trust among the parties. For dealing with local and regional governments, the scalar models proposed by RIVM provided a general guide to laying out responsibilities, and the Environment Ministry set about negotiating programs that addressed local and regional problems against the backdrop of national goals.

Although the Dutch still have some very serious environmental problems, what they have gained by a two-decade reexamination of their approach to environmental policy is a public process by which environmental problems are addressed cooperatively, with interest groups and stakeholders consulted throughout the regulatory process. This process includes the development of a new conception of problems; some new language for posing and discussing problems; a shift from command-and-control, end-of-the-pipe requirements to negotiated agreements; and even some new administrative structures for responding to problems. This cooperative strategy turns stakeholders into partners and focuses attention on finding least-cost solutions to problems, even as ongoing discussions about goals and improving institutions continues. Meanwhile, the Dutch are experiencing increased interest in restoration of natural systems, even abandoning some dikes and reestablishing ancient breeds of semiwild horses and cattle, and they have engaged the public in programs to appreciate "Second Nature"—nature that has been restored to something approaching wildness. These programs, which are very popular, show that the Dutch have accepted a commitment, undertaken as an obligation, to recreate natural areas where possible as a legacy to their future.[13]

Participating stakeholders are essential to the cooperative strategy in two ways. First, their participation increases the likelihood of buy-in, and buy-in has a way of increasing as trust develops: when all groups around the table recognize that everyone has made concessions—there are no "free riders"—and each one is willing to shoulder some of the burdens of achieving a common goal. Second, active representatives of interest groups are put in a difficult but productive position. On the one hand, as members of a negotiating team, they seek a successful agreement, recognizing that everyone must make concessions to achieve cooperation; on the other hand, they must justify the concessions they make when they consult with their constituencies. Their continued respect and support from their constituencies depends upon their being good educators, capable of explaining and justifying the negotiated package, even if it contains elements the constituency would originally have rejected.

Students of American environmental policy, upon reading this account of the Dutch transformation in the manner and content of environmental policy, may be justifiably skeptical. First, as noted, the Dutch have not solved their environmental problems in any final sense of the word—I am not claiming that the current Dutch system for protecting the environment is perfect. The Dutch have chosen to concentrate not on failures but on process. The Dutch system, for all its failings, can claim to have a relatively smooth, coordinated, and integrated process for addressing environmental problems.

Second, we must revisit the question of whether such a system—or any near relation of it—could possibly occur or succeed in the United States. As noted, if we take for granted the confrontational, warlike, interest-group partisanship that often characterizes U.S. politics and especially environmental politics, we might quickly conclude that such a transformation is impossible in the United States. And it would be easy to be defeatist in this situation if one compares the culture of Dutch politics to that of U.S. politics. For example, De Jongh makes much of the precedents in Dutch history of the polders and what is called the "polder model." Much of the land we now consider the Netherlands was reclaimed from the sea by a complex mixture of drainage canals, dikes, and windmills to create areas of land referred to as polders, which were managed by "water boards." These boards have managed these systems since the Middle Ages. Originally concerned to keep the sea out, the boards have recently and gradually accepted responsibility for the management of water quality also. As democratic and cooperative organizations, the water boards are in a sense expressive of the difficult situation faced, historically, by the Dutch, who occupy such low land. Holding back the sea has always been too big a task for any individual; cooperation was essential to survival. Now, faced with growing environmental problems, the Dutch are falling back on this multicentury experience in cooperative management and applying it to environmental problems.

There is another dissimilarity between attitudes in the United States and the Netherlands. The NEPP program is put forward as a program of "ecological modernization" that is mainly science-driven. I have argued that adaptive management must combine community participation *and* good science as it emerges within a management situation. In my visits to the Netherlands to learn about the Dutch system, I learned that the Dutch people tend to be much more trusting of bureaucrats and managers than Americans are; the Dutch are willing to leave more questions to experts to resolve. In the United States there is great distrust of government agencies, and it will, I expect, be more difficult to develop the trust and deference necessary for a community-driven process to inject good science into the decision process. Again, I can only say that, though this situation will present challenges to groups that try to engage in adaptive management and social learning, these challenges must be embraced if environmental policy is to move past ideology and conflict toward sustainability.

One could, of course, point out that citizens of the United States—a frontier country with rich land and water resources—have never had to learn the cooperative skills that the polder model encouraged and that the Dutch model is irrelevant here. Alternatively, we could say that the future of the United States is that of a highly complex, technological society with potentially overwhelming environmental problems; we could conclude that if we fail to learn to use something like the polder model, we will not survive. The point is that we do face a choice. We can continue to accept the current polarized approach to environmental problems, or we can—as the Dutch began to do in the early 1980s—attempt a major rethinking of the current fragmented and partisan approach. We could then undertake the search for a new process designed to improve communication and cooperation in environmental policy development and implementation.

12.3 EPA and Environmental Policy Today: A Report Card

Has there been any progress? Has there been a learning curve at EPA and at other agencies responsible for environmental policy? I have recounted numerous problems in policy analysis and formulation—what I have called towering—at EPA and elsewhere; I have also argued that these problems are symptomatic of deeper problems with environmental discourse in all its manifestations. Before considering more drastic alternatives, it is worth asking whether EPA is on course toward improvement. The good news is that EPA is now ensconced in its new building, which has straighter corridors. And there have been several interesting advisory publications, all pointing in the general direction I have advocated: they call for more participation, more explicit treatment of values, community-based and watershed-scale management, and a number of other tenets of the adaptive approach to management.

As of this writing, however, the George W. Bush administration, far from moving toward cooperative efforts involving all stakeholders, has embraced an ideological, free-market, antiregulation approach and has perfectly exemplified the pattern I have called environmental policy whiplash. It has rolled back protection for roadless areas in the national forests, voided critical habitat designations for endangered species, and reduced protections against pollutants of air and water, all of which had been hailed by environmentalists as progressive and long overdue policy improvements.

As was made clear in appointments and decisions after Bush took office, the ideology of an unfettered market rules. Environmentalists have been branded as extremists; any chance of compromise, consensus, and cooperation seems very remote at this writing. Lacking a shared vocabulary for evaluating environmental changes, and with little agreement about the substance of sustainability goals, environmentalists and the public at large see only chaos as the Bush administration has worked to reverse advances made by environmentalists in the Clinton years. In the process, the ideology of free markets and unfettered business has been accompanied by huge political contributions from polluters and resource exploiters. Tensions between environmentalists and probusiness forces intent upon minimizing regulation have never been higher.

My main point, however, goes deeper than a critique of the growth ideology of Bush and economically conservative Republicans; their ideological approach to environmental problems is only half the problem. Despite its fairly good record in achieving environmental objectives, the Clinton administration never communicated to the public a coherent message on environmental policy and too often resorted, on the other side, to rhetoric and ideology. And these retreats to ideology are symptomatic of a more fundamental failure: there exists no unified and transparent language in which to discuss environmental goals and objectives. This point can be made independently of administrations—independently, that is, of the pendulum swings from right to left and back again. My point cuts deeper than politics: whoever the political players are at a given moment, ideology will continue to reign until we have a coherent, experience-based language for discussing environmental problems and goals, a language that allows rational discussion of shared goals and objectives.

Since I have no crystal ball by which to foretell the effects of yet another swing of the pendulum toward the right or, for that matter, a swing back to the Democrats, and since I cannot even predict in detail what is in store for environmental policy development in the present administration, I will base my remarks on recent documents discussing policy directions and strategies. These documents were prepared during the Clinton administration, but they are nevertheless indicative of current expert thinking about the future at EPA on the part of nonpartisan scientific panels.

I will survey, first, some of the advice that EPA is getting from very distinguished sources, including the National Research Council and EPA's own Science Advisory Board, regarding the integration of management efforts. We will presently see that over the past few years, EPA has heard some excellent advice (by this I mean, of course, that it is hearing advice that is consonant with my own); up to this point, EPA administrators have been less than enthusiastic in their response to these recommendations. The first case in point has already been discussed in section 10.4; in 1996 the National Research Council published a report, *Understanding Risk,* which strongly qualified, even rescinded, the most basic principles that had been evoked in the council's influential Red Book (1983), which had served for almost fifteen years as the Bible of risk analysis at EPA. The 1996 report recommended that risk characterization should become a more decision-driven activity based on a broad and comprehensive assessment of a range of impacts; the authors advocated a combination of analytic and deliberative process that is mutual and recursive, that concentrates on problem articulation and refinement of our understanding of the decision situation, and that builds community-based institutions for the purpose of improving public decision making about risks to social values.

More recently, EPA's Science Advisory Board (SAB) has undertaken a study of current efforts and future possibilities for the agency to take a more integrated approach to environmental decision making. The SAB worked on this project—which was understood as an extension of the Integrated Risk Project begun in the 1990 report, *Reducing Risk*—at the direction of Carol Browner, administrator under Clinton. The report, called "Toward Integrated Environmental Decision-Making," recommends that

- EPA should continue development of integrated outcomes-based environmental protection, while maintaining the safeguards afforded by the current system.
- When evaluating risk reduction options, EPA should strive to weight the full range of advantages and disadvantages, both those measured in dollars as costs and benefits and those for which there may not be a comprehensive dollar measure, such as sustainability and equity.
- EPA should seek and develop methods to characterize public values and incorporate those values into goal-setting and decision-making.
- EPA should expand and develop new collaborative working relationships with other federal and non-governmental agencies.[14]

The SAB report closes, in a section called "Lessons Learned," by recommending a "culture change" at the agency, arguing that "decision makers will need to interact more extensively with scientific and technical analysts and the public in the course of developing integrated approaches to environmental risks." Further, experts must "recognize the legitimate role of values in establishing environmental goals and selecting management approaches." The culture change, these analysts conclude, must result from a growing "familiarity and experience with integrated environmental decision-making." There may be reason for optimism, because the authors of the report also seem to recognize the underlying problems of communication in the absence of a coherent and comprehensive language for evaluating environmental change. On the last page, they concisely state the very problem I have amplified in this book, saying that a major challenge that must be encountered on the way to an integrated approach to environmental management will be "problems of understanding arising from differences of terminology and outlook imbedded in the different disciplines and backgrounds of the participants in the process."[15]

EPA is thus getting very good advice, based on a deep understanding of the underlying problems. The agency is encouraged to pay more attention to problem formulation, to focus on values early and often in the policy process, and to begin to think of participation by interested parties as a continual process of multidirectional communication. If EPA was to follow this advice, the process of environmental policy formulation, deliberation, decision making, and implementation would be immeasurably improved. Such changes might eventually show us the way past the age of ideology, and forward into an age of adaptive community learning.

What this book has added to the urgings of sage advisers is a deeper analysis of this just-mentioned challenge, the lack of a common language for understanding and communicating about environmental goals and objectives. Organized into towers, riven by ideological divides, EPA's employees are finding it difficult to accept the good advice offered, because ideology and towering—enemies of communication, compromise, learning, and cooperation—are indigenous to the language we use to discuss problems. Polarized languages, those that unnecessarily pit human welfare values against the values of nature, encourage rhetoric and ideology and undermine the desire to learn. Fragmented and polarized linguistic forms allow communication within towers but block communication across towers, which is essential to cooperate in adaptive learning.

Our deeper analysis of the language of environmental policy discourse has led to the conclusion that in addition to improving process, we need a conscious effort to reformulate environmental problems by developing a pluralistic language of environmental evaluation. This pluralistic language will encourage the formulation of multiple indicators and competing measures of human impacts. Environmental evaluation, on this new, adaptive model of management, will become an interactive, communicative process, as information flow is increased.

Environmentalists in the United States can then follow the Europeans, who have undertaken a variety of national and regional experiments in participatory envisioning exercises that combine goal-setting and public participation with attempts at cooperative, mission-oriented science designed to resolve differences that block cooperative action, in comprehensive national and regional assessments and visioning processes. My intention here is to encourage, at the EPA and throughout the national discourse, a more pluralistic language and a more experimental attitude toward learning about how we value nature, which will open the doors of communication that are so essential to successful public discourse and deliberation. Along with the good advice being given to EPA, what is now needed is a concerted effort to learn how to improve the ways we *talk about* environmental problems and goals.

This book has offered a general strategy—that of using sustainability as a central concept capable of organizing our multilayered and multigenerational responsibilities—to address the underlying causes, ideology, towering, and the current linguistic poverty of environmental discourse. If environmentalists will explicitly address the problem of linguistic poverty by actively working to create a more comprehensive, pluralistic language of environmental values, it may be possible to see our way through the fog of ideologically driven failures of communication and beyond the age of ideology and paralysis resulting from towering in environmental policy formation and implementation. The above-mentioned recommendations of two respected panels of experts to EPA, which coincide with my own recommendations, should not come as a surprise to EPA personnel if they have been reading their own, internal reports. For example, a small group of EPA employees, variously organized, have advocated for years, with varying degrees of administrative support, the development and strengthening of a "community-based environmental programs" work group. Very similar recommendations have been emerging from the reports of this group for almost a decade, and the importance of involving stakeholders and community-based decision making was acknowledged in the 1999 annual report.[16] Unfortunately, the G. W. Bush administration, far from providing leadership in seeking compromise, seems uninterested in compromise with environmentalists.

Learning to live in a place—learning, Leopold would say, to think like a mountain—requires a rich, place-based language, a language in which culture and nature can be seen as intertwined and complementary. One aspect of learning to love, and to live in, a place is to learn its limitations and the ways in which natural systems are vulnerable to careless use such as overgrazing, as Leopold learned among the deer carcasses in the Southwest. I do not think that learning about such vulnerabilities is the whole story, however. If Americans are going to develop a new, more inspiring vision for environmental protection and create a rich language for expressing it, we must develop positive aspirations for a life integrated within, and co-evolved with, the local habitats that have become the "places" of America. The aspirations of each community will articulate a multigenerational idea of a good life in a good place. In the process, to return to the philosophical language developed in this book, community members will articulate, as ideals expressive of the identity of a community in a place, constitutive values, values that are make that place a home to those who live there.

12.4 Constitutive Values and Constitutional Environmentalism

I suppose my readers are by now rather impatient with my citing, mantralike, this idea of constitutive values; they are more than justified at this point in asking what exactly is meant by referring to a value, or set of values, of community members as constitutive of a place. I will respond to this challenge by offering a political analogy, an analogy that suggests a broad strategy for putting the question of obligations to the future on the political and policy agenda. Deciding how to develop a fair bequest for the future—filling in the blanks in a community's definition of sustainability—is, I will argue, more like writing a constitution than it is like anything else. I cannot claim originality for this strategy, because several respected writers have already developed the analogy between setting environmental policy goals and writing a constitution. This important literature provides the perfectly fitting keystone for my overarching theory of environmental value and for my practical and political strategy for developing long-term goals for environmental policy. The best example we have in our political culture of an act that expresses deep and abiding community values by a performative act of commitment is the adoption or amendment of a constitution.

Constitutive environmental values derive from the distinctiveness of a place, understood as an unfolding dialectic between a physical system and the cultures that have lived there, do live there, and will live there. The making of a constitution, like the making of a bequest for future generations, must be an ongoing process whereby a community collectively writes and rewrites—through amendments and applications to new situations—a document that can be the guiding force in a dynamic and responsive legal system. Several recent authors have developed this analogy of constitutions to planning for the future in two complementary ways. First, the idea of creating a constitution or a trust for the future has been explored as a useful analogy for understanding present-day deliberations about sustainability. Second, other authors have taken the analogy between writing a constitution and identifying essential elements of a bequest for the future quite literally, suggesting that modern democracies have good reason to actually pass a constitutional amendment to protect key resources such as biological diversity, to bind themselves to maintain certain elements of the legacy they inherited.

12.4.1 The Analogy between Rule of Law and Sustainability Bequests

It has been suggested that we should explore the analogy between creating a thoughtful bequest for the future and the process of writing and adjusting a constitution, and we can learn a great deal by exploring this analogy.[17] For the purposes of concreteness, I will sometimes refer in this discussion to events in the creation and alteration of the U.S. Constitution, but I intend all of my points to be generalizable to any case of constitution making. Several features of this analogy are illuminating.

First, there is no question about the importance of the legal difference between constitutional principles and particular legislative mandates. Both constitutions and bequests necessarily involve a multigenerational aspect, a discourse about principles and goals; this discourse transcends particular, day-to-day judgments of individuals. This feature of the constitutional process parallels our finding that sustainability goals have to function on a different level than do our day-to-day and individual goals and objectives. One might say that both writing a constitution and planning an environmental bequest require multiscalar thinking, thinking that takes into account long-term effects in a dynamic, changing system, as well as day-to-day decisions in which individual preferences are important.

Second, the constitutional process is similar to choosing a bequest package in the sense that chosen principles become "self-binding." A bequest package is a sort of contract—not with individuals of the future, because they do not exist, but rather with "our" community, the society Burke refers to as those who have lived, are living, and will live in the future. In adding to the existing heritage, generations will be successful if they are both faithful to the wisdom of the past and sensitive to new opportunities and constraints that emerge in a changing, evolving system. A commitment to sustain certain features of one's cultural and natural history is a commitment to an ongoing community and to the place it occupies. This obligation relates to questions of national and group identity, to an authentic sense of a place as a co-evolved creation of humans and nature.

Third, the process by which a people bind themselves with constitutional strictures is clearly forward-looking; a constitution expresses the aspirations of a people, their ideals, and their ultimate values as a culture, and it projects these values into the future. A constitution is a sort of birthright; it is at first bestowed by birth into a community; but then it must be "earned," as each generation matures and, over time, adds its own accretions to the culture and to the place. In this sense, constitutions and bequests respond to intergenerational continuities and dynamics that unite a culture across generations; successive cohorts of constitutional scholars and judges "interpret" the constitution in evolving situations. Similarly, a society intent upon creating an appropriate bequest for the future must draw on its deepest and longest-lasting values. It must choose its legacy based on the deeply unifying ideas of the culture, and yet each succeeding generation must interpret those deep-lying values in a new and evolving situation. Because most constitutions contain within them rules for their own amendment, a constitution is expected from the outset to be a living, changing process; grand principles and values are adjusted in response to day-to-day experience. A constitution and a bequest commitment to the future must both exhibit this tension between overarching goals and aspirations of a culture and the ongoing struggle to survive in changing circumstances. It exemplifies something very like the two-phase process of alternating between action and reflection.

Fourth, constitutional decision making requires a special framing of questions; it asks participants in the process to adopt an unusual and distinctive viewpoint and perspective. Persons chosen as delegates to a constitutional convention are asked to set aside their daily and personal commitments, to think as a citizen, and to think for the ages rather than for the present. Delegates who take personal business opportunities into account when writing constitutional principles, for example, have missed the point of a constitution and have disgraced the trust placed in them when they were chosen as delegates. In this sense one might say that both constitutional thinking and sustainability thinking require a shift to a multigenerational perspective in which the society is understood as a flow of multiple generations, rather than a synchronic aggregation of individually motivated "consumers."

Finally, the choice of a constitution embodies and articulates the crucial, underlying commitments—such as commitment to the rule of law—that will define the character of a society; it will even determine the nature of justice as it is practiced in that society. The constitution shapes the political environment in which people struggle to live together in a civil society. Analogously, when a generational cohort places its stamp on the landscape, when it makes choices entailing physical and ecological impacts, it is literally shaping the physical environment in which the members of future generations will face problems and challenges.

U.S. citizens, for example, might say that the value placed on the rule of law by the founding fathers was constitutive of our nation as a political entity. In the same sense, identification of features of the physical landscape for protection *constitutes* a community as a place, a place considered "home" by its inhabitants whose activities are embedded in a particular physical system, or habitat. The interaction between these two forces—the natural development of an ecological system and the cultural choices made by each generation—sculpts the landscape and determines the bequest that will be left to subsequent generations. A commitment to sustainable living, I would say, will institute a broader form of constitutional democracy, a democracy in which people of the future have a stake in the choices made today and in which today's decision makers must contemplate the long-term consequences of their choices.

The analogy of the rule of law provides a precise parallel for understanding the nature of constitutive community values as they pertain to a community's treatment of its ecological community, the habitat within which it is carving out a culture. Consider that the Founding Fathers, having emigrated from societies where birth and social ranking determined individual worth, reacted by endorsing the principle of equality before the law as the anchoring idea of their system of justice. Such a unifying principle can give both a constitution and a society integrity and strength. This embodiment of a political ideal within a constitution, a document that is one generation's political bequest to its successors, perfectly parallels Leopold's decision to think like a mountain; both involve acceptance of responsibility for the possibilities of the future; both involve multiscaled, multigenerational thinking.

12.4.2 A Constitutional Amendment as an Expression of Commitments to the Future

Some authors have gone beyond the use of the constitution as an analogy for understanding intergenerational commitments to suggest that we should adopt amendments to national constitutions that would protect natural assets, such as biological diversity.[18] This move would automatically give the future a sort of standing in courts of law, with environmental protection guaranteed to the extent of judicial discretion. I think this is a great idea, and I hope it will eventually be a topic of serious deliberation in public discourse about environment and politics. Although a constitutional amendment to protect the environment is perhaps a political impossibility at this point in history, working toward such an amendment, which would explicitly recognize an obligation for us to take the interests of the future into account, may be the best way of exploring possible means to learn about and fulfill our intergenerational obligations.

One problem to be overcome would be the writing of such an amendment. If it was too specific and binding, it probably would not pass muster as an amendment; but would a vague affirmation of intent be sufficient to do any good? Because I cannot yet imagine a set of words that would find a political balance between these extremes, I think it unlikely that a constitutional amendment is in the immediate offing. I definitely think we should be discussing the proposal, though, suggesting language, debating the merits of various formulations, trying to embody acceptance of responsibility for our impacts on the future. Doing so would improve public discourse about intertemporal relations, and such a dialogue may be just what is needed to spark some interest in the language we use to describe environmental goals, objectives, and values. A discourse like this would have as its object the further specification of many locally based sustainability definitions.

12.5 Problem-Solving Environmentalism

A second-generation pragmatist, the Italian Papini, once said that pragmatism is like a long hallway with many doors; from it one can access many "rooms."[19] This is an apt characterization, since pragmatism is more a method or a way of thinking than it is a settled doctrine or a set of "first principles"; I fear, however, that I have in this book explored pragmatism mainly within the several rooms, perhaps leaving the unfamiliar reader with the impression of a fragmented philosophy, a pragmatism that does not fully exhibit the long hallway connecting the rooms.

In the appendix pragmatism appears as a Peircean approach to language, an approach that eventually led to Carnap's conventionalist view that we can choose to institute linguistic improvements in order to improve communication, provided we are choosing within contexts where speakers have shared purposes. With shared purposes, we can ask the question, What linguistic forms and interpretations will improve communication in pursuit of our common goals? Answers to this question yield empirically testable hypotheses about the effectiveness of various linguistic forms and vocabularies. This view, in turn, gave rise to the science of the pragmatics of language, which is devoted to improving communication within cooperating groups with shared purposes. The linguistic "room" of pragmatism supports my ambitious project of developing a new linguistic framework, a constellation of terms and definitions that revolves around the term *sustainable* and its cohort *sustainable development.*

Pragmatists, I have argued, can show the way between the hopeless, Cartesian vision of a rationally organized system of beliefs, anchored in self-evident truths, and the specter of paralyzing skepticism. Pragmatists believe in the method of experience; every belief is open to challenge on the basis of contrary experience, but in any community of inquirers, there will be shared beliefs and norms that can serve as starting points for inquiry. Like the repairmen on a boat under constant sail, we stand on the solidest planks of our shared experience—areas where consensus, however temporary, has emerged—and proceed to repair weaker planks. The weaker ones, once repaired, can serve as solid footing from which to question, as necessary, what is accepted today. Although this method cannot offer certainty, it does offer the hope of controlling skepticism by limiting it to specified areas of disagreement—specifically, those disagreements that affect a proposed cooperative action in a given situation with active participants advocating various courses of action. This approach is ideal for adaptive managers who admit at the outset that their quest will be undertaken in a state of uncertainty both about particular facts and about the desirability of ultimate end points. This experience-based method, I have argued, is all that is required to start an adaptive management process, provided we have a community of inquirers with a shared language and some shared purposes.

When our attention turned, in part 2, to normative matters, the third-generation pragmatism of the German scholar Jürgen Habermas offered us an alternative, postmetaphysical way forward. This way bypasses the so-far fruitless search for a particular monistic and universal theory of ethics, embraces pluralism, and allows us to focus on deliberative processes. We found that by welcoming the plural voices of modern societies, encouraging expression of values in multiple vernaculars and based on varied worldviews, we could count on varied experience being brought to bear upon disagreements. Admittedly, such an approach will begin chaotically, as many participants will express their values in many tongues. But if we can assume that participants have already come together and are committed to seeking cooperative solutions to shared problems, then we have a way forward—deliberation, to be interspersed with limited action in the forms of experimental probing, reconsideration, more action, and on and on, iteratively. Our method of experience recognizes also the crucial role of communication. Communication in search of cooperative action stimulates the invocation of varied experiences by participants with different viewpoints, and the struggle to agree upon a shared problem formulation can inspire linguistic innovations and simplify chaotic discourse.

Speaking more broadly, the pragmatist's approach to community, especially as evidenced in our treatment of constitutive values as values associated with multiple generations, encourages a more organic, temporally unfolding community similar to a Burkean multigenerational community. Peirce's idea of truth as emergent in a community of inquirers, who will gradually tend toward agreement, can justify any given generation in seeing its choices in terms of commitments—performatives—choices regarding what values and what opportunities for experience are important enough to protect as the platform upon which the next generation (and subsequent generations) can build. Each accretion to this platform is a bridge from the past to the future. The choices of each generation will shape the free choices of the next generation, just as the choices of past inhabitants of a place affect our choice set.

Choosing a sustainable lifestyle and choosing a bequest for future generations, we noted in section 12.4, are similar to writing and implementing a constitution, and all three are similar in involving commitments made by the people of the present, with an eye to the past, including the natural history of the place, and projected into the future as a hypothesis about what will sustain the place, its people, and their evolving values. Such an enduring community, complete with a willingness to experiment and to submit the wisdom of the founders to new challenges from new experiences, can find new ways to express new aspirations as they emerge in new situations. It is this fluid, experimental aspect of pragmatism that provides the spirit guiding adaptive management. And so the development of communities and the development of integrated goals and integrated approaches to management illustrate in their own way the guiding ideas of pragmatism.

One might still ask, given that the pragmatist method of experience provides general guidance within the various "rooms" of policy formation and implementation, What is the unifying idea of pragmatism? What is the corridor, referred to by Papini, that leads from room to room? The answer, emphatically, is that pragmatism most basically stands for a problem-solving approach; pragmatism is a stubborn attempt to see knowledge and evaluation as contextual and conditioned by the agreements and disagreements existent in real situations with real people, real stakes, real dialogue, and real attempts at cooperative behavior. In this context, shared agreements provide the platform, and disagreements inspire the group to test their hypotheses and find new possible solutions. So it is ultimately problem orientation and the problem-solving perspective of pragmatism that drive the specific changes within the rooms and compartments of environmental thought and policy deliberation. Problem orientation, when coupled with the pragmatist's willingness to submit all beliefs and claims to counterclaims and challenges based on contrary experience, changes one's attitude toward disciplines and the theories propounded within them.

Much of this book has dealt with theories—theories of knowledge, theories of value, scientific theories—but I have also warned that theories developed within specific disciplinary contexts, if left to their disciplinary practitioners, are unlikely to provide a comprehensive picture. So we have tried to look at theory from the broader perspective of real-life problem orientation, as already shaped by strongly held beliefs and values; from this orientation, all factors are a part of the problematic situation. Any consideration of what to do in contexts such as this, once one adds a commitment to cooperative action, necessarily involves a holistic view of the problem situation. What is important in such contexts is not what theory tells us must be the case, but rather what works in the face of existing constraints and possibilities. In other words, adaptive management, if understood in Leopold's sense of experimental management within a value-laden, controversial social context, already starts with a complex, many-scaled problem. Since adaptive management is active and experimental and involves choosing to undertake certain pilot projects and probes of the system, we must see adaptive management as *political* activity. Adaptive management is thus a practice, a way of responding to environmental problems. From this perspective, the disciplines provide tools and methods for exploring possible solutions to problems as encountered in real contexts.

Similarly, adaptive management should change the way we think about scientific theories. Pragmatists think inductively; they see theory as flowing from experience, including both observation and controlled management experiments and pilot projects. Theory is emergent from practice and is at best a general guide to future practice. Every situation creates its own context; there are no one-size-fits-all solutions. The pragmatic attitude forces theory to confront experience in the presence of interested observers. In this way, our approach to policy can become "scientific" by continually submitting beliefs and claims to critical appraisal in the light of new and varied experience.

When one looks at theory and academic disciplines from the viewpoint of environmental managers seeking solutions to real environmental problems, communication among parties and communication across scientific disciplines become all-important. This means that, viewed from the perspective of action, our basic discourse about sustainability cannot be contained within the particular languages of the various disciplines; the assumptions of the practitioners, which get codified in the languages they use, make these languages ill suited to express a big-picture concept like sustainability.

In order to avoid restriction of discourse to specialists' very precise languages, which are too narrow to capture the full range of environmental concerns, I have insisted that the language of relevance to decision making is the ordinary language of people deliberating about what to do. And I have tried to guide discourse, especially the many local discourses centered in particular places, toward asking better questions by embedding heuristics in a self-consciously political discourse and deliberation about what to protect. Theories are best viewed as hypotheses, which must yield results in particular situations where action is demanded; they must contribute to problem-solving. If they do, we have every reason to weave these working consensuses into the broader consensus we call science. We consider these beliefs "true" until new experience, perhaps in a new context, calls them into question; in this case, inquiry can shine the spotlight of specialized sciences on the newly suspect hypotheses. And on and on within a community of inquirers, persistently engaged in problem-solving.

Problem-solving environmentalism does not eliminate disciplinary science; rather, it provides the multiple disciplines a forum for trying out, criticizing, and validating their theories. In the postdisciplinary era of problem-solving environmentalism, the necessary discipline derives from the need to respond adequately to real and persistent problems in a real place. The key to success in such responses is to recognize that problem formulation must be an integral part of problem solution, that dialogue and participation are essential for an activist program to embody the values and aspirations of the people who live in a place, and that problems, if addressed iteratively, are transformed: solutions interact with problems to create new formulations. The goal of adaptive management—and of the pragmatist methodology propounded here—is to help communities to better identify the real problems and to provide those communities better tools to address these problems. That is pragmatism in a nutshell.

12.6 Seeking Convergence

One problem with ideological environmentalism is that it begins by dividing people according to their preconceptions, rather than their experiences. If one starts with a preexperiential commitment to an ideology such as free marketeering or the moral intuition that nature has intrinsic value, then that preconception will manifest itself in the language one uses. In a community, each preconception-colored language will act as a barrier blocking communication with others who do not have the same preconception. In the interests of communication, we have concluded, it is far better to advocate indicators that *express* one's values and to seek cooperation and coalitions with other groups likely to support the indicators one thinks are important. In this way, a community concentrates on what should be measured and what can be done to improve performance of an indicator, without becoming bogged down in an all-or-nothing debate about the ultimate nature of environmental values. Choosing indicators, and standards with respect to those indicators, can thus turn the focus away from abstract theories and ontological commitments and can hold our language closer to real experience. We hope that, under favorable conditions in which inhabitants of a place have committed to cooperative action through a partnership or adaptive management process, they can focus attention on possible objectives that can unite people behind action in pursuit of shared goals.

Pragmatists believe that dichotomies and dualisms usually stand in for disguised continua and are best avoided—dissolved—when possible. What we need is integration through shared experience, not separations according to ideology hidden in choices between one version and another of polarized rhetorical language. This explains why I choose, thinking prospectively, an environmental philosophy based mainly in *intergenerational* morality rather than a nonanthropocentric morality. Nonanthropocentrism represents an incomplete rejection of a dualistic worldview. Extensionist proposals merely displace the dualism (those things with, and those without, intrinsic value) to a new point in the continuum of life, from including only humans as intrinsically valuable to including sentient animals, living organisms, and functioning ecosystems. The dichotomy, once moved, is preserved in a new form.

In some past publications—faced with the polarization around the question of anthropocentric reductionism of economists and the nonanthropocentric extensionism of environmental ethicists—I have introduced into the bifurcated discourse of environmental values what I call the convergence hypothesis.[20] The convergence hypothesis is most basically an empirical hypothesis about environmental policy, today and in the future. It states that if anthropocentric and nonanthropocentric theories are properly formulated, policies advocated by anthropocentrists (aware of the full range of human interests and the long-term obligations we have to the future) will be similar to those advocated by nonanthropocentrists. The convergence hypothesis says that, provided these antithetical theories are formulated in their most defensible form (defensible, that is, in their own lights as independent theories), applications of the two theories would approve many, perhaps all, of the same policies. Differing worldviews, when stated in this oppositional way, can still converge on shared policies because advocates of the opposed positions, when faced with shared problems, will often gravitate toward solutions that make sense on nonideological grounds.

The convergence hypothesis must be understood conditionally, in the sense in which it was proposed—in a context *assuming* dualistic categories. The hypothesis assumes, in this sense, the dualistic categories of human interests *versus* nonhuman interests and only makes sense in the context of that assumption. In this book I have adopted nondualistic, less ideological language to discuss environmental problems, and convergence of advocates of multiple values on shared indicators and policies is expected. Consequently, we see that the problem, as well as the solution, has its origin in dualistic assumptions.

I do not doubt, of course, that the interests of particular humans—including large categories such as stockholders in various large corporations—will find their selfish interests at odds with elements of nonhuman nature. We often read in the newspapers, for example, that a condominium development, a school, a telescope, or some other structure valued by groups of people can be built only if habitat for a threatened species is destroyed. But notice that these conflicts, no matter how sharp or how hotly contested, are conflicts between human individuals or groups—those who favor and those who oppose the proposed development. Development plans that threaten species and ecosystems proposed by one group of persons are often opposed by other human groups.

The central problem of sustainability is that in modern, technological societies, the most vulnerable among us, the young and the not-yet-born, must accept the consequences of risky chances taken by adults. I have argued that a society of persons—if it is to be understood as a true society in the Burkean sense of being a legitimate, multigenerational community—must include in some way the interests of future persons, persons who are not yet born but who might someday have an interest in a threatened site or the endangered habitat under pressure. It follows that a true community of humans, one that seeks to cooperatively pursue a rational environmental policy that protects legitimate human interests vested in it, will include articulate voices in defense of endangered species and threatened ecosystems as the birthright of future people. Multigenerational problems are community-scale problems, and acceptance of responsibility for them must be seen as a community responsibility; multigenerational problems therefore cannot be successfully addressed in purely individualistic terms. Individualism cannot, ultimately, provide voiceless future individuals a place at the bargaining table. Such long-term values are necessarily lodged in an evolving community. The task of finding a good policy to follow cooperatively is not a matter of identifying interests that are vested in nature and located outside the human community and enforcing those interests in opposition to the human community. The task is, rather, to develop a shared understanding and framework of language that will encourage the recognition of how much of nature we must preserve if we are to be fair to the not-yet-born.

Even when we see human individuals and groups proposing actions that are threatening to important elements of nature, there are other human individuals and groups who oppose those actions on the grounds that our generation owes subsequent generations intact ecosystems and access to the bounties of nature. Similarly, when humans propose changes to the landscape, perhaps altering habitat—whether the changes are explained as opportunities for human development or as restoration of degraded habitat—the human action may turn out to provide improved habitat for some species even as habitat for others is lost. So if one really wanted to test the idea that—on the whole—human interests are at odds with the interests of nature, one would have to first figure out some way to sum, aggregate, or characterize the "total" good of humans and set this sum in opposition to the "total" interests of nature. Given the hypothetical nature of the convergence hypothesis—it can only be formulated once one accepts that human and nonhuman interests can be opposed and contrasted—it becomes clear how difficult it would be to understand, much less support, its opposite, a *divergence* hypothesis. I therefore doubt that the statement "human interests diverge from the interests of nature" makes sense; at least I doubt that it could be the subject of a reasonable disagreement. Every time one might identify some change that would harm nature, someone on the other side could no doubt find some species that would benefit; similarly, every time one might identify some action as helpful to humans—though harmful to nature—there would be someone to argue that the action would actually harm humans in the long run.

Even if these internal problems of the divergence hypothesis were somehow avoided or solved, and even if one accepted for the sake of argument the dualistic assumptions hidden in the formulation of the issue, the divergence hypothesis seems unattractive when compared to its opposite, the convergence hypothesis. I prefer to seek convergence in actions, regardless of the value system or worldview of participants.

This view, emphasizing as it does the idea of placing humans within nature, leads me to embrace a useful formulation developed by the deep ecologists, who insist that we are not really separate from nature; our skin is but a permeable membrane. When I consider myself to belong to a place, a community, insults to my immediate environment are insults to the broader self I embrace as a member of a community. On this view, which I think is implicit in Leopold's simile of thinking like a mountain, there will be little use for the convergence hypothesis. In a nondualistic world, where humans are never severed from nature, the convergence hypothesis will wither away for lack of polarized interests to be brought together. In the world beyond dualism, we will care for the environment because it is *our* environment; even more, it is as much a part of us as we are a part of it, according to the useful reinterpretation by deep ecologists and, more formally, by hierarchy theorists. On the multiscalar, human perspective on evaluation—evaluation according to the multigenerational standard of sustainability—we must go beyond methodological individualism and protect communal values, values that are emergent on the scale of multiple human generations and on the scale and pace of whole landscapes. What remains, in the last section of this book, is to go the final step and explain how adaptive management can provide a general explanation of the goals of environmental management.

12.7 Ecology and Opportunity

America has often been referred to as the Land of Opportunity, as wave after wave of immigrants have arrived on its shores. Early on especially, but also now, the foundation of those opportunities is the unbelievably rich physical endowment of the continent. Included are living resources, like old-growth forests and rich soils, and also minerals and other natural sources of opportunity. These resources are the basis of America's great wealth. Further, this New World affords fantastic views and remarkable natural phenomena, places of great beauty and mystery. All of these characteristics have encouraged the development of local human cultures as contributions to a larger-than-human community, a "place." I will offer here a general theory of sustainable living in a place. My proposed theory draws heavily on ecology, especially on hierarchy theory, in order to pay homage to the irreplaceable role of natural systems and processes in the growth of human communities. The theory, however, cannot be just a physical theory with physical models—that would fail to accomplish a key purpose of this book, which is to unite the physical aspects of sustainability with intergenerational social values via a theory of environmental value that makes some sense of the pluralistic system of value that I have spoken of all along. If the theory captures what is truly at issue when we worry about sustainability, it may create a bridge between the putatively "objective," descriptive physical sciences and the humanities and social sciences.

Such a theory will support the constellation of concepts we are building around the term *sustainable* and encourage a more integrated discourse about uncertainty and about problems, goals, and values, Perhaps optimistically, I have assumed that an experimental approach to management is possible and that debate about environmental goals and indicators is already embedded in a rich public discourse. Given these assumptions, I have proposed a constellation of definitions, norms, and heuristics whereby communities can act in an integrated manner. In chapter 3 we noticed that the three axioms of adaptive management (which include the two axioms of hierarchy theory) provide an elegant schematic definition of sustainability. Each generation makes choices, as individuals, against the backdrop of a slower-changing context that can be thought of as the environment of those individuals; the environment is experienced by the individuals as a mixture of opportunities and constraints. Since the individuals, as smaller systems cycling more quickly within that environment, also *constitute* the environment itself, some combinations of actions by individuals in a particular generation—when viewed collectively—can irreversibly change the mix of options and constraints stored in the normally slow-changing environmental mix. When the actions of earlier generations narrow options and reduce opportunities for the future, their practices are unsustainable. When such actions maintain important options and opportunities and create no new and onerous constraints, the practices are deemed sustainable.

These options and opportunities, though important to individuals, are especially important to individuals viewed *as members of a community.* The community has both a human and a natural aspect, and both are important because the human community inhabits a natural place. I therefore concluded that obligations to protect these options and opportunities are better thought of as communal values that emerge as choices of whole communities among their options and opportunities, choices that define a way of life in a place. These choices are constitutive of meanings that resonate in that place and express the individual and collective identity of persons who live authentically there. These values can be comprehended only by understanding the many ways in which they interweave human actions and feelings with the natural dynamic that forms the "ecological" habitat of an authentic place. An authentic place retains its continuity and integrity through time and maintains the key options and opportunities that define it as a dynamic place to live, a place with a natural history.

We have had to admit, of course, that no society can achieve complete protection of all options—the position sometimes derisively referred to as "absurdly strong sustainability." The forces of change and development cannot be stopped; the dialectical relationship between human communities and natural communities must be a dynamic one. This recognition imposes an awesome responsibility upon societies such as our own that cast their lot with accelerating technological change, what Leopold referred to as "violence." We must constantly be asking ourselves, Which changes to our place are consistent with its integrity as a place, and which are not? To answer this question, a community must determine which new technologies and new conventions are consistent with, and supportive of, the community's deepest values. These are the values that define it as a cohesive, multigenerational, ongoing community.

The lesson of chapter 8 was that the requirement of weak sustainability cannot fully capture the obligations of prior generations to subsequent ones. Weak sustainability does not fully address the specific obligations of generations, such as our own, that are endowed with the power to transform the lives of those who come after them. If everything natural is replaceable by human-made substitutes, how can a culture maintain continuity? How can it perpetuate multigenerational practices and behaviors and develop "adaptations" that avoid the constraints and take advantage of the opportunities particular to their place? How can a culture, if everything is replaceable, maintain a shared sense of meaning and identity across time? How can such a culture expect to maintain the physical aspects of the relationship that has formed its "place" and endowed its relationships with meaning? If a society lets market forces decide which development path to pursue without considering communal values, that society runs the risk of replacing, with substitutes created by the generic outputs of global economic competition, everything that makes for a meaningful and authentic relationship with a home place.

The alternative for a democratic society is to deliberate regarding what is important to save.[21] I have characterized this difficult task as one of choosing, through a deliberative process, some indicators that are widely accepted as tracking important dynamics. If these dynamics are chosen through an iterative, deliberative process, they can be expected to reflect the social values that are cherished by participants. Participants, given a process that encourages social learning about ecological and other dynamics, will insist on monitoring and protecting those elements they see as important. By such a process they can democratically select a set of indicators and goals for each indicator, despite continued diversity in the values espoused by people in the society. A diverse community can arrive at a negotiated set of indicators and goals because if the right synoptic indicator is chosen, coalitions will support the indicator on the basis of varied values.

Pragmatism encourages us to begin with concrete choices and behaviors and work *toward* a general theory of sustainability values, rather than setting up universal principles and values prior to action and experience and deducing actions from principles. We have thus tried to think inductively, from experience toward theory. Theory can, at its best, guide action and the search for more experience. So far, however, I have not stated a positive theory of sustainability value, emphasizing instead local, bottom-up *processes*. The idea behind this approach is that if policy is chosen in an open, public deliberation in which the sciences and cultural voices are all expressed and heard—in a context, in short, that encourages social learning—then we can hope the community will be headed in the right general direction.

This process-oriented strategy, however, may still be thought to leave communities too little of substance as a guide to the difficult choices of what indicators to monitor, which options and opportunities to protect, and how, in general, to choose a suitable bequest package worthy of a commitment to live sustainably. The heuristics offered can only guide process. Am I going to end the book having said nothing, substantively, about what it is communities *should* try to save? Fortunately, I can say something about how we can associate our schematic definition of sustainability with a general theory about what is important to human communities and their members.

The theory advanced here is not a theory designed to shape experience. It is rather a conjecture about how we will come to understand and usefully describe experience in the future. Because our theory addresses the large-scale problem of how a society or culture adapts to its slower-changing environment, it must provide a linkage between social valuation and change on ecological scales. Multigenerational responsibilities, which we have seen to affect a community, its sense of place, and its sense of identity over multiple generations, must therefore be modeled in ecological time (the temporal scale on which species adapt and fill a niche). Adaptive managers advise that if the pursuit of economic growth is causing accelerating change in normally slower-paced systems, we must examine the impacts of rapid ecological change on the mix of opportunities and constraints that will be available in the future. Examples of situations that called for such examination include the dust bowl on the Great Plains and the chip mills in the Appalachians. In these examples the individual choices of actors, once they were writ large on the landscape, transformed the landscape and reduced opportunities. The basic idea of evaluating long-term effects of our actions points us toward a general theory of what should be sustained by a community committed to sustainable living. Sustainability is intimately tied to a particular aspect of human freedom. Commitments to sustainability are commitments to perpetuate that aspect of freedom.

Because of the overwhelming ambiguity attending the concepts of freedom in ordinary speech, it will help to separate three theoretical definitions of freedom. For simplicity of reference, I will refer to these as (1) radical freedom, (2) individual freedom, and (3) opportunity freedom.

1. Radical freedom, as articulated by existentialists, emphasizes the close relationship between action and responsibility. According to the advocates of radical freedom, each and every choice we face necessarily involves a form of self-definition; responsibility is the necessary concomitant of freedom. Advocates of this position also assert that in many situations, the choice not to act is itself a choice; a melancholy view that freedom and responsibility are inescapable could follow: we are "doomed" to be free. We have referred to this notion briefly above, noting that powerful technological societies, assuming some level of knowledge about the impacts of their current actions on life in the future, cannot escape responsibility for choices they make that seriously affect the future. Although this aspect of freedom informs our analysis, it has little to do with the theory I will develop to explain sustainability.

2. Individual freedom can be understood as the ability of an individual "rational" being to make a decision uncoerced by either physical compulsion, such as threats of physical violence, or mental compulsion. This idea has of course been controversial throughout most of the history of philosophy. I was reminded of this definition when, at a small conference, I first shared my theory that environmental sustainability values celebrate human freedom. One of the members of the audience took me to task for proposing such an obviously flawed theory; as he pointed out, the pursuit of individual freedoms, including the freedom to consume and the freedom of mobility, for example, are often the driving force behind serious environmental problems such as management of enormous amounts of waste and sprawling cities with air-quality problems. I could not deny his point—individual freedom, unqualified and untutored, is clearly not an environmental value. To avoid this reasonable criticism, I must distinguish between freedom to choose within a set of options—lack of coercion, that is—and freedom in the sense of having a particular option open to the chooser. In the first case, there is a set of choices open, and the question is whether the chooser is unimpeded in choosing a favored option.

This is the question of individual, uncoerced freedom, and the clear implication is that—no matter how irrational or misguided the actual choice might be—freedom requires the ability of the chooser to follow through with his or her most favored choice. This distinction, however, allows us to admit that individual freedom can be abused, if a chooser is irresponsible or oblivious to negative consequences of her or his acts. We must distinguish, then, even among free acts of individuals, separating those that are done with good reason and responsibly from those that are based on whim or greed. It is now possible to say unequivocally that the freedom so central to sustainable thinking does not, in any sense, endorse irresponsible behavior. Indeed, a central aspect of Leopold's thinking-like-a-mountain motif is the acceptance of responsibility for individually motivated actions that might harm the future. Such acceptance of responsibility necessarily involves accepting, also, some limitations on the exercise of individual freedom.

3. The theoretical connection I hope to establish between ecology and freedom is based on the opportunity aspect of freedom, "opportunity freedom." Experiencing opportunity freedom is facing a wide range of choices in a given choice situation. The difference between opportunity freedom and individual freedom can be illustrated with a simple example: if I say, "I never shop at store X because it has an inadequate selection," I am making a claim about the range of choices available *for* making a free choice. There is no hint, in the complaint against the store, that there is some limitation on the free choice *among* the options available there. Since it would not be unreasonable to say one prefers another store to X because X's poor selection limits one's freedom of choice in pursuit of a goal—whether that goal be to prepare an interesting meal or to find a good deal on an unusual piece of clothing—we see that the range of available choice is an important prerequisite of any degree of freedom and responsibility.

There are interesting parallels—and differences—between what I am saying here and the important work of the Nobel laureate economist and philosopher Amartya Sen.[22] Although Sen has concentrated more on intragenerational equity than intergenerational equity, he has developed a pluralistic, normative system of concepts for distinguishing "opulence" (the command one has of commodities) from capabilities (what one can achieve by choice) and well-being from choices of functions—the freedom one has to choose a way of life, suggesting that "the quality of life a person enjoys is not merely a matter of what he or she achieves, but also of what options the person has had the opportunity to choose from. In this view, the 'good life' is partly a life of genuine choice, and not one in which the person is forced into a particular life—however rich it might be in other respects."[23] Sen's notions, which he presents with appropriate suggestions about possible quantifications of key terms, closely parallel the idea of holding open options and opportunities for the future, and he introduces and defines the term *advantage* to capture this idea: "'Advantage' refers to the real opportunities that the person has, especially compared with others. . . . The freedom to achieve well-being is closer to the notion of advantage than well-being itself."[24] The concepts and measures that Sen has developed have considerable promise for replacing simple welfare comparisons with a more pluralistic approach to characterizing interpersonal comparisons of opportunity. These new concepts may lead the way to improved formulations of the trade-offs faced in a multicriteria system that emphasizes opportunity and freedom of choice, as well as well-being, in judging choice sets that present themselves to people. These conceptualizations might even be developed into a relatively comprehensive framework of evaluation for judging fairness, both intragenerational and intergenerational.

Sustainability, I theorize—building on the schematic definition developed on the basis of the axioms of adaptive management—is a function of the degree to which members of a future community experience no diminution of opportunity freedom in comparison to the opportunities open in earlier generations. This connection is illustrated, structurally, by a return to the core, Darwinian analogy that informs adaptive management. True "adaptation" must succeed on two scales. First, in genetic evolution, each "player" tries to survive long enough to reproduce; for long-term success, however, there must also be enough stability for the offspring of one's offspring to survive, a requirement that becomes more severe in highly social animals who necessarily have long periods of dependence because of the need to learn complex social behaviors associated with communication. Long-term adaptation of complex societies thus requires successful adaptations at the community as well as the individual level—what we call social learning. Social learning is analogous to the learning of a species over many generations and many "experiments" in varied environments.

The adaptation analogy, then, as applied to cultures and cultural practices, is ultimately the key to understanding sustainability. Adaptation—speaking culturally, adaptive management—must always be a function of making good choices both on the individual scale (analogous to individual fitness leading to reproduction) and on the larger, collective scale that affects the range of choices available to community members in the long run. Each generation, acting as individuals, makes choices based on the options and opportunities stored in the environment it encounters. On the one hand, short-term survival and thriving simply depend on making the right choices. Long-term survival, on the other hand, depends on making today's choices in such a way as to avoid destroying options that may be important in the future. Sustainability, then, is a relation between the choice-sets of individuals and communities that exist in different generations. It is clear, however, that we cannot directly operationalize the idea of an "important option" by simply counting options available at different times. Options are not comparable in value, and I have no simple plan to make them so.

The strategy of this book has been to place the burden of identifying, monitoring, and prioritizing options directly upon communities. If current choices cut future actors off from important options, those options that are essential to fulfilling a community's aspirations, the result will be a reduction of fitness of the community, vis-à-vis the emerging landscape. Here, of course, fitness refers not just to physical survival, but also to the flourishing of individuals as members of a cohesive community. To this strategy I have in this final section added some theoretical guidance to communities seeking to identify indicators that they wish to monitor and commitments to goals that they feel are essential to maintaining a healthy community in a place with continuing integrity. Success in sustainable living will occur when, over multiple generations, the pattern of individual choices, even when aggregated to a community scale, maintains the integrity and resilience of the place the community inhabits. A community will have fallen into unsustainability if the successors of its members, who carry on in their place, face a narrowed and poorer range of options than they do.

APPENDIX
JUSTIFYING THE METHOD

A.1 Philosophy's Abdication

The usual story is that philosophy in the twentieth century was an abject failure, that the obsession of Western philosophers with language, especially in the English-speaking world, left the field bereft of relevance to individuals or communities facing important decisions. Speaking only of language, philosophers apparently ignored the serious and substantive questions of their discipline, questions of the nature and meaning of reality and questions of defining right action. According to this standard story, the Queen of the Sciences had abdicated and was reduced to the status of a mere handmaiden to the all-powerful sciences. Philosophy today, though sometimes thought to be "interesting," is expected by very few to regain its potency and once more change the world. You can't get there (to relevant philosophy) from where we are (mired in the study of language), according to this usual story.

Or can you? This may be a case in which accepting received wisdom would be unwise. Given our diagnosis in chapter 1 of the failures of communication among environmentalists and their critics, the most effective antidote may be some clearheaded analysis of the language we use to understand and discuss environmental problems; and I know of no better way to challenge what everybody claims to know about philosophy's failures than to demonstrate that philosophy can offer a key ingredient in a creative rethinking of environmental policy. In chapter 1 I hypothesized that failures of communication and deadlocks over environmental policy choices are at least partly a result of the poverty of the language available to communities for discussion and deliberation about environmental goals. Here I hypothesize a remedy and sketch out a method. I believe that the techniques and ideas of analytic philosophy of language support the effort, undertaken in the main body of this book, to create more useful language for discussing environmental problems and to develop a more useful definition of *sustainability.* It's worth a try, at least. Before examining our hypotheses directly, however, it is necessary to revise somewhat the story that is usually told about linguistic philosophy in the last century and its relation to modern philosophy. In this appendix I place the progressive and pragmatist ideas sketched in section 2.3 in the broader context of modern philosophy and, in the process, explain the rational, philosophical arguments that undergird my use of the pragmatic method to analyze environmental policies.

We usually mark the beginning of the modern period of philosophy with René Descartes, inventor of solid geometry, whose most influential philosophical work was published in 1641. Descartes, much impressed with the empirical work of Galileo, wrested dominance from the Scholastic logicians and provided necessary support for the rising tide of observation-based science. As a geometer, Descartes offered a confident rationalism, according to which he claimed to have validated science and knowledge of the senses through a deduction from self-evident and indubitable premises traceable to God, who by definition could not be a deceiver. Over the decades and centuries to follow, however, Descartes's rationalism was gradually eroded, beginning with challenges by the great British empiricists, from John Locke (1689) forward, and continuing with more and more rationalist retrenchment and with the final abdication of philosophy in favor of mere linguistic analysis in the twentieth century.

The story of modern philosophy, then, first featuring giants like Descartes, Bacon, and Newton, is thus usually told as a story of loss: loss of faith, loss of a priori knowledge, loss of "foundations" for knowledge, loss of rationality, loss of human privilege, and so on. The modern period, since Descartes's confident embrace of science within a religiously based epistemology, seemed trapped in a downward spiral of confidence lost and increasing skepticism. This skepticism further undermines our confidence in what we thought we knew, leaving us unable to justify our actions on the basis of universal and unquestionable principles, or even on the basis of rationality more generally. When viewed as a progression forward of Descartes's confident rationalism, linguistic analysis appears to be the endpoint of modernism and encourages the judgment that linguistic philosophy is but the sad denouement of a noble but failed philosophical project. Since there was no less at stake in Descartes's crusade than the possibility of knowledge of the external world and the possibility of rational deliberation as a guide to action, recognition for the development of techniques to better analyze language does indeed seem like a booby prize. This story of loss can be found in any competent history of modern philosophy. And although I cannot question the facts as told in the usual story, I will here argue that this tale of loss is a one-sided and misleading version of the far more complex story of linguistic philosophy.

This same story, if reframed, can also be told as a story of liberation and reconstruction, and it is in this mode that I will briefly retell it here. There is no denying that the twentieth century was, on the whole, a bad century for philosophy. That century gave us, in addition to "linguistic idealism," existentialism, nihilism, positivism, and, in Europe, the scandal that the leading phenomenologist expressed an unseemly appreciation for Hitler's Third Reich. One result of this bad century has without question been a lowering of expectations regarding philosophy's contribution to both intellectual and real-world problems. Indeed, when I admit to being both a philosopher and an environmentalist, most people look at me as if I'd just said I was a bricklayer and a concert pianist. Usually it is easier not to challenge the assumption that philosophy is useless, and when wearing my activist shoes, I try to let my actions speak for themselves; but here, in a long book on environmental discourse, where I claim to offer a dose of philosophy of language as the antidote to the impoverished discourse of environmentalism, it seems reasonable to back up a bit into philosophical history and see if it is possible to tell a plausible story in which philosophy now reenters the mainstream of public discourse as a creative and positive force.

The first step in rehabilitating contemporary linguistic philosophy is to reject the view that places linguistic philosophy as the final denouement of modernism. It is more accurate to think of modernism as having collapsed under the epistemological weight of its own unsupportable assumptions, for reasons inherent in the modernist project. Descartes, after all, founded his system of empirical science and metaphysics on an a priori proof for God's existence; he attributed our knowledge of causation to an innate idea of reason; and he placed God, who he reasoned could not be a deceiver, at the center of modern science, as the very guarantor of human experience. Descartes's science, in other words, was a God-centered science; God's intentions give meaning and structure to the world we strive to understand so that we may do his will. Descartes bought a knowable world by paying the price of implicitly attributing purpose—teleology—to nature, making it more than the dead, inert nature that he claimed made up the physical universe. In so teleologizing nature, Descartes created an image of the modern world that, it turned out, could not be rebuilt from empirically observable atoms of experience. Put most simply, Descartes assumed that reason alone could validate belief in God, and belief in God gave meaning and structure to the universe because an omnipotent and benevolent God could not create a chaotic and unknowable universe. Human reason and knowledge could not be futile in a world governed by a benevolent Almighty. Attempts to reconstruct Descartes's rationally knowable universe from human experience proved impossible, however. Descartes's world, by its very constitution, was suffused with unobservable forces and meanings. Descartes had implicitly construed his "problem" so that a "solution" would necessarily include a supernatural element. When empiricists set out to offer another, nonsupernatural solution, they found that, as a result of the problem definition, their legitimately available empirical methods were simply unable to solve the problem as Descartes had formulated it. Meanwhile, in the broader society, the apparently unstoppable march of science and secularization gradually undermined the only foundation sufficient to support the Cartesian edifice of knowledge: innate or self-evident knowledge of God's benevolence.

If we see the stalemate between rationalists and empiricists as an inevitable outcome of these internal Cartesian contradictions, we then see Nietzsche, who proclaimed God dead as, for practical purposes, the end of the modernist story. Nietzsche announced the end of modernism; he saw that authority, even—or rather, especially—the authority of God himself, could no longer support a rational edifice of general knowledge and certainty. On this telling, the story of philosophy's loss of potency culminates in the nineteenth century, with Nietzsche in Europe and pragmatism and conventionalism advocated by Holmes and Peirce in North America (discussed in chapter 2). The story of philosophy in the twentieth century can thus be told as a story of new departures from a new starting point. Consistent empiricism could not indefinitely coexist with rationalism because rationalism's commitment to universality and certainty required divine epistemological support. In fact, however, philosophy and philosophical movements take time to grow, and they sometimes also take time to die. Indeed, modernism, as we shall see presently, survived Nietzsche's death sentence for decades and, like the mortally wounded prima donna, it came back to sing one last, brilliant aria, a virtuoso performance by great minds in logic and mathematical foundations. What the interplay between the rationalists and the empiricists eventually showed was that, for the rationalists' conception of the world to be meaningful and knowable, it must express the will of a divine and supernatural creator. The wounded prima donna may have pulled herself up from the catsup to sing one last, impassioned aria, but it was too late. The plot was sealed. Finis for modernism, however long or virtuosic the subsequent denouement.

The more positive story of linguistic philosophy begins, then, not with Descartes but with the irascible genius of American philosophy, Charles Sanders Peirce, born in 1839 and a major figure in our discussion of early pragmatism in chapter 2. Peirce, a mathematician and the most penetrating logical mind among the pragmatists, saw clearly that the modernist, dualistic system of existence was doomed. He was not alone in this recognition, but he provided a convincing reason to reject not just the answers of Descartes but also the central questions of Cartesianism and modernism. That question had been, (How) can we use human reason to validate knowledge of the external world? Peirce, by contrast, realized that the modern project of using reason to prop up the senses was a lost cause. Reason would ultimately lead back to an authoritarian basis for resolving disagreement, and since authority bears no internal, inevitable relation to the emergence of truth, Peirce gave up on the modernist question; all answers to it will be either question-begging or self-defeating. So Peirce asked a different question: How can experience be used to reconstruct a knowable and known world, in the absence of preexperiential knowledge of the forms and structure of nature? If linguistic philosophy is seen as developing a set of concepts and tools to respond to Peirce's question, then we can begin to see the more positive aspects of philosophy's turn toward linguistic analysis.

One reason this more positive story is often missed is that Peirce, besides being a mess personally, was also a terrible writer; he changed his terminology and philosophical "system" constantly throughout his career, often arguing in great detail against his own previous positions. Unlike most great philosophers, he never completed a single between-two-covers account of his system and its significance, leaving instead a legacy of a few published, semipopular essays and countless cartons of manuscripts in various stages of completion and revision. I think it is safe to say that at the time of his death nobody—perhaps including even Peirce himself—really understood Peirce's philosophical system or its historical significance.

Because the first great philosopher of communication was such a bad communicator, the positive story of the progress of linguistic philosophy travels in a circle, showing that philosophical progress, like scientific progress, seldom finds the shortest distance between points in history. Peirce, writing at the end of the nineteenth century, had found the answer to the key riddle of the twentieth century; but most other philosophers of his time were not able to recognize this because they were not yet asking the question for which Peirce had an answer. Like a mystery story that begins with its solution, twentieth-century philosophy thus begins and ends with Peirce. The story in between is one of learning to ask the right questions. The story's plot follows anything but a direct line, as widespread questioning of the Cartesian project had left philosophy in confusion with regard to its role and possibilities. And, not surprisingly, with the field in a state of confusion, a variety of ways forward were tried.[1] Let us look briefly at the several philosophical "movements" that emerged in the first three decades of the twentieth century. All of them were deeply influenced by important advances in symbolic logic and in the foundations of mathematics.

Linguistic analysis, which entered the philosophical stage during the sad denouement of Descartes's project, emerged in a confused philosophical situation in which both the rationalists and the empiricists seemed blocked in their search for an epistemological order. The rationalists apparently required a priori knowledge of important aspects of reality. The empiricists had in every case fallen far short of reconstructing key ideas—such as causation, substance, and so forth—from bits of experience. Descartes's rich endorsement of science as trustworthy, as in principle deducible from indubitable premises, could be based only on an invocation of God himself. This demand for the strongest form of justification, installed at the very heart of modernism by Descartes, inevitably clashed with science's demand for skepticism, which submits all assertions to the yardstick of experience and cannot abide "self-evident" truths. With modern science a consistent victor whenever scientific hypotheses have come into conflict with church pronouncements about how nature *must* be, it was only a matter of time before the modern synthesis of religion and science would collapse from its own internal contradictions.

My determination here to separate the Peircean way forward from the denouement of modernism is especially important. Not all philosophers, it turns out—especially not leading European philosophers—had yet given up on the Cartesian project. Consequently, because of an insufficient understanding of the internal contradictions of Cartesianism, and with European philosophers still hopeful of responding decisively to Descartes's challenge, the tools of logic were at first brought to bear upon the traditional problems of Western philosophy, questions of the nature and structure of reality. Enamored of important discoveries in logic and the foundations of mathematics, twentieth-century philosophy embarked upon one last wild goose chase, down one last blind alley, in its attempt to escape the collapse of the Cartesian project into skepticism and doubt. The growth in the study of logic and the philosophy of mathematics, especially the new techniques and concepts developed in these studies, would, in the process, create important tools of great importance to the postmodern, Peircean journey. Like the technological spin-offs from government-supported space-age technologies, the tools developed can acquire value even if the goal they were invented for is misguided.

Meanwhile, throughout the first half of the twentieth century, pragmatism maintained a few bastions of power and excellence in North American universities, especially the University of Chicago, and many pragmatist philosophers were public figures as well as scholars. In Europe, however, philosophy went in a much more abstract direction, examining the more subjective side of Descartes's subject-object dichotomy, a direction not obviously inconsistent with continuing the Cartesian quest. As a result, getting back to Peirce's insights as a starting point with wide-based support took most of the remainder of the twentieth century. Now, at the beginning of the new millennium, I would argue, the general direction forward for philosophy is becoming much clearer, and it may be time to give the new philosophical vehicle—linguistic analysis—a test spin through the environment.

Having sketched the plot, let us get on with telling the story, and telling it, hopefully, in just enough complexity to make it plausible and interesting. I admit to abject selectivity in telling the story. To retell it in detail, with complete arguments, would take another book at least; the most that can be accomplished here is to dust off a few of the signposts left by Peirce and his pragmatist followers and to sketch how linguistic philosophers gradually worked their way back to Peirce's insights and starting points. The point is that for most of a century, Peirce's arguments were ignored, obscured by logical razzle-dazzle based on new and exciting discoveries in the symbolic logic of deduction and subsequent studies in logic and mathematics.

As noted in chapter 2, pragmatists, as advocates of the "method of experience," came to see that the categories and kinds we use in our descriptions of nature are embodied in our language and gain their validity in human communities from the practices of communication and symbolic behavior. Our descriptions of nature and the categories we use in these descriptions must be validated, for pragmatists, within communicative communities that survive over time by learning, changing, and adapting in place. Language, now freed from the dead hand of religious orthodoxy and a priori doctrines about how nature *must* be, pragmatists thought, could now be analyzed functionally. On this view, language becomes a tool for communication, useful for achieving cooperative behavior in a larger social context. Utterances can be judged according to many purposes, which vary according to context, including human projects, goals, and objectives, which help to form these actual social contexts. This contextual approach to language encourages us to look at language as inseparable from a dynamic social context; in such a contextual analysis, "correspondence" to a preordained reality becomes less important than a deep understanding of human motives and actions.

The shift toward a more pragmatic and conventionalist view of language has, as emphasized throughout this book, great potential for reframing environmental problems less ideologically, by following Leopold's Darwinian-Hadleian approach to judging communities as more or less adapted to their natural environments. Leopold shifted the debate away from preexperiential commitments to prior philosophical principles, redirecting attention to the (at least in principle) empirical question of whether a given society, its practices, and its institutions are environmentally appropriate and "adaptive."

A.2 The Rise of Linguistic Philosophy: Its Inevitability and Meaning

We should start our story of linguistic philosophy by emphasizing developments in symbolic logic, the same developments that led eventually to digitalization and computing. It was in this area that philosophers developed great virtuosity in the manipulation and interpretation of symbols; but it was also an area that led to much nonsense and confusion. These studies had their origins in the late nineteenth century when the German philosopher Gottlob Frege and later Bertrand Russell, Alfred North Whitehead, and the German émigré Ludwig Wittgenstein, in England, began to develop new and powerful forms of symbolic and mathematical logic. These pioneers in logic, one might say, recognized with Peirce the ultimate importance of language and symbolic behavior, but they failed, at this early stage, to fully accept the insight of Peirce and Dewey that language must ultimately be understood as a social, community-based tool. They took the assertorial aspect of discourse—declarative sentences descriptive of the world—to represent the core of all linguistic meaning; they then removed these sentences from their natural habitat in ordinary discourse and concentrated their attention on language as a tool for inferences among sentences. This focus on descriptive language was reinforced by the strong trend at that time favoring science over all other forms of knowledge and wisdom. Historically, these scholars' view of psychology and language was much affected by that of Locke, who had treated words as signs for ideas in the human mind; so it was a small step to think of sentences as "picturing" reality. Moreover, not all philosophers, in the first decades of the twentieth century, saw the limits placed on philosophy by the recognition that neither philosophy nor any other search for truth can attain direct (i.e., presymbolic) access to a "real" world that validates our assertions.

So the twentieth century in Europe began with a bullish attitude toward philosophy, with Russell and later his brilliant but troubled student Wittgenstein proclaiming that the powerful methods of logic would ultimately resolve traditional philosophical problems. After flirting with idealism as a young man, multiplying concepts to answer to every linguistic form, Russell had by 1905 rejected idealism and its bloated world of shadow concepts and unrealized ideas. He adopted a broadly empiricist viewpoint and became convinced that analysis of the logical structure of sentences represented a new role for philosophers and logicians. The Europeans, in other words, were at first unaffected by the constructivist and conventionalist—relational—approach to communication of Peirce, or by Holmes's understanding of a legal system as a set of predictions rather than a set of formal principles. It was still assumed by many European philosophers that traditional philosophical problems could be solved by uncovering their "true" logical form, by rewriting them as other expressions that avoided embarrassing implications. In 1913 Russell, working with Whitehead, completed *Principia Mathematica,* a tour de force in logical reasoning in which they created the most sophisticated logical system ever constructed and used it to derive all of pure mathematics from a few definitions and axioms of formal logic and set theory.

This remarkable achievement showed, Russell thought, that mathematics was an expression of truths inherent in logical relationships, and he took such logical relationships to reveal the basic structure of the world. The best example of such solutions, one that had quite a strong influence on Russell and his followers, was his treatment of referring expressions and related philosophical problems; these referring expressions had troubled idealists, who were bound by their theory to designate an "idea" that embodied the meaning of each meaningful expression.[2] They struggled with nondenoting referring expressions. Russell considered the sentence "The present King of France is bald" and noted that since there was no king of France at the time, idealists would have to invent an entity to which the referring expression applied in order to declare the sentence meaningful. Using his devices from symbolic logic, Russell rewrote the sentence as a conjunction, creating what he called a "contextual definition" and allowing the substitution of the logical sentence for the troublesome one that seemed to create a whole realm of nonexistent entities. Russell rewrote the sentence as "There is one and only one King of France (in 1916), and he is bald." Since Russell considered this rewritten sentence false on the basis of the existence-claim it made, he felt he had eliminated the need to posit shadowy nonexistents in the conceptual world, and hence he considered his logic to have provided a breakthrough in philosophy's attempt to identify what actually exists. This analysis provided for Russell an example of how logical analysis can "solve" (or "dissolve") philosophical problems. In fact, of course, the analysis did not really solve the philosophical problem; what it did was to remove one apparently unavoidable argument for shadow concepts. But in his euphoria about this paradigm of successful philosophical analysis, Russell saw the example as providing an incipient philosophical method.

J. O. Urmson, in his astute chronicling of analytic philosophy between the two world wars, noted that the structure of the world would then resemble the structure of *Principia Mathematica.* Russell came to believe that a logically perfect language such as his PM logic reveals the structure of the world, with both language and the world being constructed of "atoms." Every meaningful sentence about the world, which will be made up of a reference to a bit of reality as having a particular quality or standing in a relation to other bits, must correspond to some collection of atomic facts. Simple sentences that are true correspond to simple "facts" in the world; and true complex sentences correspond to complexes of such atomic facts. False sentences are ones that lack a corresponding fact in the real world.[3]

The brilliant Wittgenstein, who was separated from Russell when he returned to his native Austria during World War I to fight on the side of the Austro-Hungarian Empire, built upon the work of Russell and other logicians, suggesting that language must be understood as atomic sentences that "show" something in the world, and that these elements are connected by truth-functional connectives. According to this "picture" theory of meaning and language, meaningful sentences represent concatenations of atomic facts structured and aggregated by "truth-functional" logical connectives. In complex sentences assembled by such connectives, one can substitute another sentence of the same truth value without affecting the truth value of the whole sentence. Such languages, called "extensional" languages, paved the way toward digitalization of information and also apparently provided Russell and his cohorts with an ideal device by which to assemble "atomic facts" into more and more complex constructions and perhaps to build a logic that would correspond to the structure of reality. From atomic sentences denoting atomic facts corresponding to them in the world, Russell thought he could reconstruct the world, supporting the atomic sentences as direct reports of sensations and using logical connectives to reconstruct the whole of knowledge from the atoms of nature arranged according to the patterns of a logically perfect language. At this point, early in the twentieth century, then, Russell believed that the goal of philosophy should be to construct a logically perfect language. In such a language, the logical relations would be built into the truth-functional structure of sentences, and atomic sentences would correspond—or fail to correspond—on a one-to-one basis with atomic facts. That such a language would be useless in human social life did not bother him. And, convinced that atomic facts are "out there" ready to be denoted by the propositions of his perfect language, he implicitly smuggled a form of essentialism into the foundations of his theory. Based on this crumbling support, he ridiculed and baited the pragmatists as enemies of philosophy and truth.

Although there were dissenters to almost every aspect of Russell's philosophy, younger philosophers were inspired at first by its elegant boldness and its promise to find the "true" structure of reality underlying confusing, everyday language and to project those findings onto the world, making the study of the true structure of language a means to fulfill the dream of precise knowledge of the true structure of the world behind our experience. Russell emphasized deductive logical relationships, and he and the other logicians had limited themselves to "formal" aspects of language. To traditional philosophers, these logical manipulations hardly qualified as uncovering the structure of the world. For those disappointed by the apparent failure of the Cartesian epistemological enterprise but still committed to its goals, however, Russell's formalisms were viewed as the only game in town. Formalism, as understood in this context, is a commitment to discuss only the "formal" properties of a linguistic expression, the study of syntax. Symbolic logic got its start in syntax because its originators were mainly interested in the foundations of mathematics and sought to characterize as much symbolic behavior as possible in terms that could be characterized merely in terms of reproducible shapes. This technical definition of formalism connects, more generally, with the formalism—obsession with rules and deduction from rules—that led Holmes, in reaction, to embrace the "predictive" theory of law.

Besides seeming appropriate for the study of mathematical foundations, this narrowing of the study of language and symbol systems was strongly motivated philosophically. It was thought by some philosophers that if one attempted to relate linguistic forms to nature itself, one might be accused of metaphysics, or considered to be making philosophical assumptions. So work in syntax seemed immune to objection from the extreme empiricists and science-oriented philosophers who were influential at that time. Given these two strong motivations, the early decades of symbolic logic concentrated heavily on merely formal aspects of language. Frege, Russell, Whitehead, Wittgenstein, and others succeeded in deriving all of the truths of logic and mathematics from a few axioms. From the viewpoint of symbolic logic, this was a very healthy direction. This strategy linked the logicians directly to the active field of mathematics, where discussions of foundations were spirited and highly creative during this period. Philosophically, however, this emphasis was all wrong, because the emphasis on deductive logic and its reducibility to syntax turned attention away from the weakness of Cartesianism, which was its inability to deal reasonably with skepticism, uncertainty, and inductive logic. This emphasis allowed philosophers, who were thinking of logical analysis as mainly a syntactical study, to avoid the critical problem of how logic, the study of the logical connective tissue of a language, relates to the world in which that language is spoken. And it allowed philosophers in Europe to ignore the cogent arguments of Peirce and Dewey for language as a conventionalistic tool that has its root in the functions of communication.

Wittgenstein, who studied briefly with Russell, had a truly enigmatic career as a philosopher and logician. Many books have been and will be written about this period of intense creativity in philosophy, which was stimulated by related creativity in symbolic logic and the foundations of mathematics. Wittgenstein's positions, and his influences, remain controversial today. What can be said without much controversy is that whatever Wittgenstein said, or meant to say, the impact of his difficult early writings was (1) to reinforce the connection between analysis of logical structures and philosophy—to promote, that is, Russell's program of addressing philosophical problems through symbolic logic; and (2) to reinforce a strong form of empiricism according to which primitive nonlogical expressions denote bits of experience from which the world could be rebuilt, thus collapsing the tasks of finding a logically perfect language with the separate, epistemological problem of grounding our beliefs firmly, in this case, firmly on sensory atoms.

Russell treated his development of a logically perfect language as a work in progress, using *Principia Mathematica* and its logical system as a starting point. His idea was that if one could first achieve a consistent and complete system of logical connectives, one might then systematically analyze troublesome words and phrases, avoid puzzles and paradoxes of language by translating troublesome phrases or sentences into the formal system, and inch closer to the ideal of a logically perfect language. If that language could be used to describe the world completely and consistently, Russell thought, it would not be unreasonable to infer that the process would be carrying us closer to understanding the structure of the world, as it is known by science and organized according to principles of symbolic logic.

When their story is told this way, Russell and his cohorts, including Wittgenstein, seem to have fallen into a harmless, intriguing, and possibly insight-yielding experiment in harnessing logic and philosophy together. The idea was that ordinary language is highly imprecise and far from the logically perfect language, and philosophical debates, carried out in imprecise ordinary language, made it impossible to resolve important philosophical disagreements. For Russell, these philosophical problems remained interesting in their own right, and he saw himself as discussing real and important philosophical issues. Indeed, he often used philosophical arguments to justify choices in logical structure. What had changed for Russell was that he thought he had a new and powerful tool, logical analysis, for solving philosophical problems. With confusions removed, who could doubt that philosophical problems would then yield to rational, philosophical debate, because of the clarity introduced by an analysis of the "true" logic of the sentences and phrases in question? The application of the tool of logical/linguistic analysis was, for Russell, part and parcel of the philosophers' task; and he expected his generation, armed with the new techniques of symbolic logic, to eventually resolve the traditional philosophical problems by rewriting troubling sentences in more perspicuous form and "solving" these problems, as in the problems of reference in sentences about the present king of France. In effect, the assumption that there is a logically perfect language that embodies the rules of logical deduction was just a new way to express, and for a time disguise, the Cartesian commitment to rationalism and certainty as the only acceptable outcome of analysis.

Wittgenstein, however, lacked Russell's intellectual patience. He was not content with Russell's small victories on the way to a noble goal. Wittgenstein was apparently influenced by the argument that because we can describe the workings of nature in the logically perfect language, the world must in fact have the broad logical structure of that language. He apparently began to see the logically perfect language not as an ideal or goal, but as the core structure any language must have if it is to successfully describe the world.

Wittgenstein was quicker than Russell, however, to recognize that the very idea of a logically perfect language poses a serious, perhaps fatal, dilemma for philosophy itself. To see this, suppose one has a logically perfect language and one uses this language to assert every true atomic sentence and deny every false one. Logic will have done its part by finding the perfect structure for aggregating and concatenating the sensory atoms; and science will have contributed by separating false from true atomic propositions. But what is left, if this level of success is achieved, for philosophy? Wittgenstein, the iconoclast, said in the last proposition of his first book, "Whereof one cannot speak, thereof one must be silent,"[4] retired from philosophy, and took up schoolteaching in remote villages in Austria.

Meanwhile, Wittgenstein's popularity grew immensely, even as he retired from the field, and his early work had a huge impact on the Vienna Circle, a group of scientifically oriented scientists, philosophers, and sociologists who met regularly in Vienna, and on a loosely connected but similarly disposed group that met in Berlin. Controversy still exists today regarding whether the Vienna Circle's ideas could be attributed to Wittgenstein, or whether his influence was mainly based on a misinterpretation. Either way, these young philosophers meeting in Vienna were to become the core of a new movement called logical positivism or, as they preferred to call it, logical empiricism, a movement that was destined to be at least as seriously misunderstood as Wittgenstein himself had been. The much-maligned movement of positivism, however, if properly understood, will lead the way back to Peirce and Peirce's new starting point. Rehabilitation of the positivists will require a section of its own (section A.3). But before going to Vienna, let's tie up a few loose ends of our revisionist story in Britain.

Russell's enthusiasm for formalism, and the impressive advances he and others made in the study of logic and the philosophy of mathematics, implicitly violated Peirce's earlier insight that the correction of our beliefs must be accomplished only through expansion of collective experience. Since the (real, underlying) structure of language was discovered by analysis, not experience, and since Russell inferred the structure of the world from this analytically derived structure of language, his approach to philosophy could work only if he had a source of extraexperiential knowledge about which analysis is the logically perfect one. Or, to put the same point slightly differently, Russell's reasoning was caught in a tight, if elegant, circle. If you asked him how he knew that a constituent bit of nature, such as an atomic fact, existed, he would say it must exist because the logically perfect language demands, in order for its elements to be meaningful, such a constituent to correspond to an element of its structure. If you then asked Russell how he knew a bit of logical code was an element of the logically perfect language, the only answer available was that writing the expression in this way solved a philosophical puzzle and must therefore correspond to the structure of reality. QED, but circularly. So it was soon questioned whether Russell's experiments in "logical metaphysics" could claim any status as a method by which to divine metaphysical truth. Nevertheless, Russell's image of philosophy as able to find the true structure of language remained a powerful ideal, illustrating what philosophers could hope to accomplish through linguistic analysis.

Notice, however, what is missing—what Russell needs if he is to break out of his circle. He needs an answer to Dewey's question, How do we anchor the categories that we use to describe reality? Russell had not seriously addressed this question.[5] Russell instead arrogantly dismissed the pragmatists as unworthy philosophers because of their willingness to accept a less rigorous standard of truth than correspondence to reality. But Dewey and Peirce were much closer to right than Russell was: to know something about the world—to actually use a language to describe the world—clearly requires some categories, some taxonomy, some sorting terms, and these are normally furnished by conventions of language formed in communities. Russell, because he adopted formalism as a methodological commitment, never had to talk seriously about the relationship between the formal marks of his syntactic system and the world itself. The system was treated as a series of marks representing variables and constants. The problem Russell's logical analysis hadn't addressed was, How (on what basis) do we fill in the variables? And this question left Russell and his followers struggling with a dilemma. Either they had to say that *they*, as philosophers, would answer this question—that philosophers have a method (symbolic logic, pure reason, or reading tea leaves) by which they can identify and articulate the categories of being—or to recognize that the categories we use to describe nature are themselves based on "experience." This latter admission was the poison pill that Russell could not bring himself to swallow. To do so would be to admit that philosophy has no distinctive subject matter and nothing to say beyond a scientific description of the world. So Russell, not convinced that it had already been shown to be untenable by Peirce, grasped the first horn of the dilemma, a course that led to the virtuoso, formalist arias of the dying modernist cast of characters.

Russell could not, ultimately, avoid the Cartesian dilemma: either the map of reality made by science and logic is the endpoint of all knowledge—in which case philosophy has no work left to do—or the map of science and logic is not complete, and philosophy has a role in uncovering truth, drawing the picture of nature. But what could this role be? Science had become the voice of experience; logic, purified and separated from everyday speech, was limited to empty tautologies as the only truths of logic. Russell would have to claim a priori and nontrivial knowledge that could bridge logic and the world in order to continue the Cartesian struggle. But that struggle, however valiant and noble—once stated in these terms—was simply hopeless. Indeed, what Russell needed to continue his project was exactly what Descartes needed: an authoritarian epistemological figure who anchors the nature and structure of the world itself. But Bertie was a well-known atheist, and he could not pull a self-evident epistemological rabbit out of his hat as Descartes had; over time, Russell's attempts at logical reconstructions became fewer, and his claims for his method less brazen; he ultimately drifted away from technical philosophy altogether, and into social criticism, experiments in education, and peace activism.

As for Wittgenstein, he ended his self-imposed exile from philosophy in 1929 and returned to Cambridge to launch a whole new approach toward linguistic analysis, an approach that, interestingly, had important affinities to the work of Peirce and James. Wittgenstein, apparently convinced that his prior emphasis on atomism, deduction, and formalism was all wrong, began to emphasize the use of language in social situations and communication. Thus, after carrying logical atomism to its most extreme form, Wittgenstein came back to philosophy on a new trajectory that led directly away from atomism and, at least generally speaking, in the direction of Peirce and Dewey. Like Peirce, however, Wittgenstein could not finish a synoptic book on his new ideas; indeed, he never published another book. Like Peirce, his legacy was to bewitch graduate students and assistant professors, who wrote dissertations and articles on what he must have meant to say in not-quite-finished manuscripts. In Wittgenstein's case, his intentions were to be the subject of much speculation, and there was a nasty competition to get access to the unpublished manuscripts. In the absence of evidence in the form of Wittgenstein's written word, and given the ambiguity of that which existed, Wittgenstein commanded at first a kind of mindless obeisance among some of his students and eventually a much-too-facile dismissal by philosophers, who tired of puzzling over his ambiguous utterances in the final quarter of the twentieth century.

At about the same time Wittgenstein returned to philosophy, the 1930s and the 1940s, the center of philosophical creativity in the United Kingdom shifted from Cambridge to Oxford, where a group of philosophers, mostly trained in the classics and impatient with the formalisms of logical analysis, turned their attention to the nature and structure of ordinary language. This group included Gilbert Ryle, Peter Strawson, J. O. Urmson, and especially J. L. Austin. Austin, though he died young, laid out the essentials of "speech act theory," which has since been developed by successors into a full-blown semantic theory, a theory of meaning that is articulated in terms of linguistic usage and communicative practices of functioning communities. Austin, speech act theories, and communicative practices appear at various points in this book as important tools for the analysis of language and communication. But since these "ordinary language analysts" were generally rather modest in the claims they made for philosophy, we will not dwell on their importance here, except to notice that when their work is coupled with the new direction of Wittgenstein in the 1930s and 1940s, we can begin to see a trend. Another of the side roads in our new story of twentieth century philosophy can thus be seen to be veering Peirce-ward. By the middle of the twentieth century, language was being examined not as a logically perfect, skeletal system of formal marks, but as a tool functioning in real communities where people must communicate to get along and to get on.

A.3 The Rise and Transformation of Logical Empiricism, aka Positivism

As was noted above, Russell—unable or unwilling to admit that philosophy as he knew it had no role—chose the first, essentially Cartesian, horn of the dilemma that presented itself in the death-throes of the Cartesian project. He left open the possibility that somehow one could find a role for philosophy in the application of symbolic logic to problems as they arise in ordinary language. I have described this strategy and the resulting displays of technical, logical prowess as a deathbed aria of the modernist era, because Russell was using the tools of logic to address traditional philosophical problems; Russell still addressed problems that were shaped by Descartes's philosophical project. He sought to replace our currently incoherent and confused system of beliefs, expressed in imprecise, ordinary language, with a reconstruction made up of directly observable atoms, of language corresponding to atomistic facts in the world, strung together with the latest in logical connective devices and providing a map of the world.

On the Continent, these same concerns and arguments were carried in another direction, one that was at first seen as striking at the heart of philosophy itself, as the previously mentioned groups of scientifically oriented philosophers and physicists, influenced by an earlier generation of European positivists including Ernst Mach, formed in Berlin and Vienna. These positivists were also influenced by the strict empiricism they thought they discerned in Wittgenstein's early work, and, excited by the advances in logic that they took to allow new and ever-more-powerful combinations of the atoms of knowledge, they enthusiastically grasped the anti-Cartesian horn of the dilemma. Confidently calling themselves logical empiricists, they admitted that most of traditional philosophy was "meaningless" and set about showing how a consistent application of empiricism and logical analysis would lay waste to the field of philosophy, reducing it to either psychology (the empirical study of people's attitudes and preferences) or "logic" (the manipulation of formal symbols in the proof of empty tautologies).

Although the positivists included several brilliant and subtle thinkers, what was communicated to the outside world, including other philosophers, unfortunately, was hardly subtle. According to the simple view communicated to nonpositivists, knowledge was simply the sum total of the findings of empirical science; logic and mathematics, important because they provided the connective tissue of science, allowing the aggregation and concatenation of facts, asserted only tautologies and could not enlighten us about the nature and structure of the world. Most of the positivists, then, advocated the "verifiability criterion of meaningfulness." According to this "criterion" of what was called "cognitive meaning," a sentence is cognitively meaningful (either true or false) if it is either (1) a truth or a contradiction of logic or mathematics (a tautology that provides no information about the world) or (2) a verifiable scientific assertion (one that could be "verified" by observation). No other sentence could be considered to be cognitively meaningful. Given this criterion, one would have thought that positivists should have, like Wittgenstein in 1920, deserted philosophy, since it was apparently left with nothing to say. The Viennese positivists, however, were far more subtle than this in practice—they carried on lively debates about how to incorporate values, poetry, and art—even parapsychology—into their system of thought, for example.[6] Further, in their careful work, they used empiricism as a guide to reconstructions rather than as a cleaver to banish interesting questions from discussion. In their heyday, positivists contributed greatly to advances in logic, including formal semantics; the philosophy of mathematics; and the philosophy of the social sciences.

This subtlety and openness was absent, however, from presentations of logical empiricism to the outside world, especially the English-speaking world. In fact, positivism was presented as an extremely "closed" and monolithic system, for reasons over which the members of the Vienna Circle had little control. The more careful work of the positivists was first written in German, with lags of time between publication and translation into English. Therefore, many English-speaking philosophers read only short tracts in English, obtaining most of their knowledge of positivism from secondary sources written by English speakers. In particular, most English speakers were introduced to logical positivism by a young Englishman, A. J. Ayer, who visited Vienna and studied the positivists and their philosophies. Ayer later became a respected professor of philosophy at Oxford and a very subtle thinker in his own right in subsequent decades. In this case, however, he wrote a short, concise, and very readable book, in the form of an antimetaphysical manifesto, called *Language, Truth, and Logic,* which he thought captured the essence of the new scientific positivism. The book, which Ayer later described as being "in every sense a young man's book,"[7] eventually became the text by which logical positivism was introduced to most young philosophers for the next half century. It articulated a very hard-line and simplistic version of logical positivism, one that emphasized uncompromising empiricism, accepting only logical tautologies and descriptive sentences based on empirical observations as cognitively meaningful utterances. In the remainder of the book, Ayer simply worked his way through the fields of philosophy, ethics, and theology, showing the meaninglessness of their "pseudo-sentences." Each philosophical subject *appeared* to encompass real problems that required rational deliberation, he said, but analysis of the sentences spoken by philosophers on these subjects showed that they could be resolved into either tautologies, simple descriptive statements, or logical concatenations of descriptive statements. All else—including all the traditional problems of philosophy—he concluded, was meaningless nonsense. These meaningless philosophical sentences were mistakenly thought to be meaningful only because of the imprecisions and confusions of everyday language.

It would seem that advocates of an emerging philosophical movement like logical positivism should welcome a brief, well-written book that lays out the movement's basic tenets in a simple, straightforward way—and Ayer's book was unquestionably a box-office success and disseminated the positivists' ideas widely. It turned out, however, to be a philosophical disaster for positivism, especially in Great Britain and America. The positivists' complex, subtle, and changing views—as well as their internal diversity and controversies—were all frozen in one monolithic description of a philosophy that must end all philosophy. The result was to polarize all the issues that might have been left open for debate; and with the literal battle lines forming for the World War II and European and English-speaking philosophers becoming increasingly isolated from each other, philosophical battle lines formed between positivists—thought by most philosophers to be traitors to philosophy itself—and more traditional philosophers who, like Russell, still hoped to protect a special role for philosophy. Most of these latter philosophers took the attitude that no philosophical progress could be made until this treason against philosophy was avenged and philosophy's reputation restored. The polarization of philosophy and the eruption of World War II—especially in the presence of Ayer's simplified manifesto—put an end to all opportunity for philosophical compromise and left precious little opportunity for incremental learning between positivists in Germany and critics in North America. Positivism was isolated from its critics both by the growing threat of war, which was in some cases repaired with emigration, as leading positivists fled the Nazi scourge and took jobs in American universities, and by Ayer's bombastic rhetoric, which was never repaired and which left most philosophers with the belief that positivism, a grotesque monster, had died without issue and amounted to a dead end in twentieth-century philosophy.

Once it was stripped of its controversies and variations and frozen in rhetorical language by Ayer, positivism was easily dismissed, hoist on its own philosophical petard. For decades, most teachers of Philosophy 101, assigning (some portion of) Ayer's overconfident manifesto, used the same tired, self-referential argument to dispatch logical positivism, so that they could regain the right and necessity to do philosophy. Positivism, the argument went, is inconsistent. It claims that all true sentences are either tautologies or reports of sensory data; otherwise they are meaningless nonsense. "But what status," the traditional professor says with a sly smile, "are we to attribute to the verifiability criterion itself?" Is this sentence intended as a tautology (and hence empty of descriptive content)? Is it a simple report of observations (in which case it would be obviously false, since many sentences actually uttered by humans have no clear connection to experience)? Or is the criterion itself meaningless nonsense? Now the sly smile turns triumphant. The positivists' criterion must, it seems to follow, have the status of a *philosophical* pronouncement; if so, it must be open to philosophical argument. Logical empiricism (at least as buttoned up by Ayer) became its own victim. It could banish philosophical discourse only by invoking philosophy and a philosophical argument. Philosophy was saved.

The question was whether it was worth saving, because philosophy was at this point directionless—caught between an inability to get over the Cartesian questions of the past and an inability to see the liberating aspects of addressing the real questions of the future. The Cartesian project of discovering the one true structure of reality had been doomed to failure for a century by Darwin's dynamism and the consequent undermining of essentialism. No other philosophical project even promised universal and necessary answers to philosophical questions. If positivists were by this lame argument forced back into the fold of philosophy, it was a pretty sad fold to be in. Neither they nor their critics could see how to escape from philosophy as Descartes conceived it; but there they were with no defensibly valid method, other than appeals to authority, by which to address Descartes's questions. The rationalists, on the one hand, offered an effective philosophical method, but it could be embraced only on the basis of theistic authority, which eventually flies in the face of experience-based science. The empiricists, on the other hand, now saw through the logical legerdemain of the Cartesians who managed to conjure certainty from question-begging premises. But nobody had been able to mount any respectable empiricist reconstruction of knowledge since the devastating empiricist analyses of causation, substance, and personal identity by the brilliant skeptic David Hume. Even the evil positivists could not avoid philosophy; but it was a hollow victory over them, since philosophers could choose only between the apparent, but false, potency of Cartesian rationalism and the well-established impotency of empiricist reconstructions.

By the early 1950s, common philosophical wisdom in North America was that the logical empiricists' attack on philosophy had failed; it was OK to "do philosophy" again. But if one asked what "doing philosophy" consisted in during the midcentury period, one might get more answers than the number of philosophy professors surveyed.[8] Some very good philosophers continued in Russell's tradition, doing logical analysis of troubling and important concepts, proposing alternative formulations; Wittgenstein's later philosophy became wildly popular in both Britain and North America, so there was increasing emphasis on language use and the communicative use of ordinary language. But Wittgenstein never gave up his antiphilosophical stance, and having gained respect for everyday discourse, he began to think of philosophical discourse as aberrant use of ordinary language, reversing Russell's ordering of philosophical reparsings of ordinary confusions. Often using the analogy of illness, he advocated "curing," rather than solving, philosophical problems.

Pragmatists retained a strong voice in America but often found themselves on the defensive with regard to both positivists and the positivists' critics, who often lumped the pragmatists and positivists together because of their shared emphasis on science and empiricism. In Europe the reaction against positivism expressed itself in phenomenology, existentialism, and later structuralism, poststructuralism, and deconstructionism. Philosophers seemed, at best, to be eclectic analysts, with a "tool kit" of logical and analytic techniques with which they addressed philosophical problems, usually piecemeal. The closest thing to a philosophical ideal seemed still to be Russell's idea that if we improve language, remove confusions here and there, we may be getting closer to the truth. But, as noted above, by midcentury this was a tattered and tarnished ideal, transparently circular in its most elegant form. Now that logical empiricism, viewed à la Ayer as a monolithic and inflexible empiricism, had failed and no clear alternatives had emerged, philosophy seemed to represent no more than the uneasy denouement—the death-throes—of modernism and Cartesianism.

Our positive story about linguistic analysis seems badly off track. How will we get to Peirce's new starting point through this impasse? I've already hinted at where we went off track. In the English-speaking world, most philosophers and most other philosophically sophisticated intellectuals accepted Ayer's simple, monolithic characterization of logical empiricism as a dogmatic, closed form of scientism and empiricism. As a result, the historically promising connection backward from early positivism and scientific philosophy to Peirce's theory of signs was severed; worse, philosophical acceptance of Peirce's way forward remained tentative and incomplete at best, even in America. Peirce's importance was that he identified knowledge-searching as a symbolic activity that occurs in human, communicative communities. To find our way from early positivism back to Peirce, it will be necessary to challenge Ayer's interpretation of positivism as a frozen, antimetaphysical doctrine. We must show that in fact the positivists adopted a more complex and dynamic approach to philosophy, an approach based on a deep faith in science but committed also to developing a number of important new, "postmetaphysical" projects. A careful look reveals that the group of philosophers usually thought of as core positivists—Moritz Schlick, Rudolf Carnap, Hans Reichenbach, Friedrich Waismann, and Herbert Feigl—were actually highly diverse, having a range of philosophical viewpoints. Wittgenstein is sometimes given credit for furnishing key, uniting ideas of the Vienna group; and he sometimes had conversations with Schlick, Carnap, and Waismann when he was in Vienna. But Wittgenstein was never a member of the group, and today many interpreters of Wittgenstein even doubt that he shared many of the views the positivists developed and attributed to him.

The diversity did not end with the philosophers; the group was also highly interdisciplinary and included leading scientists—especially physicists—mathematicians, and the prescient sociologist/economist Otto Neurath, who stands out as a central figure in our more promising story of philosophy's way forward, by furnishing a key epistemological analogy. Logical empiricism, as noted above, was at its most dynamic and creative in the early years of the movement, during the heyday of the Vienna Circle. The Circle had been convened in 1922 by the mathematician Hans Hahn and included Neurath, the physicist Philip Frank, and the newly arrived Moritz Schlick, formerly a student of Max Planck. Activity intensified in 1926 when Carnap came to Vienna to work with Schlick, and the group retained considerable cohesion until the assassination of Schlick by a revengeful student in 1935 and the scattering of the rest of the group by worries of war and social unrest in Germany during the middle thirties. During this early period, the Vienna Circle was an active center of discussion of all manner of social and political ideas as well as problems at the edges of science and philosophy, science and logic, and philosophy of science.

Although positivism has suffered withering criticisms from various quarters, many scientists, if questioned, will still endorse principles of positivism. So positivism remains tremendously significant today and deeply affects how scientists think about environmental policy. It is therefore important to see what was justifiably discarded from positivism, what influences still survive, and what should be emphasized in the future. I will concentrate on the careers of two of these positivists, the logician and philosopher Rudolf Carnap and Otto Neurath, the social scientist. These two friends were both competitive and complementary thinkers, and together they are illustrative of the range of opinions and ideas considered by the members of the Vienna Circle.

The positivists' aggressive antimetaphysical stance gained the most press and was undoubtedly a big-ticket philosophical issue throughout the 1920s, 1930s, and 1940s, but one cannot understand the positivists' contribution without emphasizing more complex ideas that survived attacks on verificationism. Thus, although it is correct that verificationism and the associated verificationist theory of meaning were decisively rejected in the 1970s and 1980s, this rejection disposes of logical positivism only if one inaccurately equates logical empiricism with Ayer's simplified doctrine.

To explain the development of positivism between 1920 and 1970, I will trace the career of Rudolf Carnap, born in 1891 and the most intellectually nimble of the famous positivists. Fortunately, I just happen to have written my dissertation and my first book on Carnap, so this is actually a field where I know what I'm talking about. Carnap was an open-minded, cooperative person, quick to change positions once convinced there was a better one and yet tenaciously holding to his views until better arguments emerged. He seemed to delight in the tension between Circle participants' tentative attempts to state common positions on philosophical matters, on the one hand, and the fierce intellectual independence of the individuals, on the other. Both Carnap and Neurath were socialists, but Carnap was mainly an intellectual, and he was not active politically. Their brand of socialism was not strongly Marxist and was mainly based on the need for state planning; Carnap was a pacifist and, despite a religiously devout childhood, a humanist.

As I learned when I worked on my dissertation, Carnap was actually motivated by two apparently independent "principles." First, he and other positivists espoused the empiricist theory of meaning—the view that the meaning of a sentence is its means of verification—and that theory's associated criterion of meaningfulness. But Carnap also proposed what he called the "principle of tolerance." Of the latter, he later said, "It might perhaps be called more exactly the 'principle of the conventionality of language forms.'"[9] At first Carnap saw no conflict between his empiricist principle, which he often used to argue that a variety of philosophical doctrines were "metaphysical," and his principle of tolerance. He held, and appealed to, both of these principles throughout the 1920s and 1930s, as if they were independent and complementary principles.

Carnap's early inability to see the tensions the two principles created was due to a technical detail of great importance. Originally, he and the other positivists, following Russell's commitment to formalism and syntax, thought that all that could be meaningfully said about language or logic was what Carnap called "syntactical." Carnap, roughly following Peirce's ideas of semiotics, divided the study of language into three categories, *syntax, semantics,* and *pragmatics.* As noted above, syntax—the study of language as a physical mark or sound—turned out to be surprisingly fertile; Frege and Russell both used pure syntax to articulate the deductive structure of logic and mathematics. Syntax raised no problems with empiricism because the shapes of signs qualified as observable, and it produced the stunning successes of reducing mathematics to logic and set theory and of axiomatizing a complete and consistent system of deductive logic, truth tables, and other important discoveries, which set in motion the digitalization of the contemporary world. Impressed with the developments in logical syntax, Carnap sought in his early work "to construct a theory about language, namely the geometry of the written pattern," and was convinced that all that could meaningfully be said about language could occur in this "formal" mode.[10]

Semantics, the study of the relationship of signs to their object of signification, had a much more troubled infancy and childhood. Many years later, Carnap said, "We read in Wittgenstein's book that certain things show themselves but cannot be said; for example the logical structure of sentences and the relation between the language and the world."[11] Wittgenstein, who used a "picture" theory of meaning—units of language get their meaning by picturing a fact—believed that some things could not be said. At first, Moritz Schlick and Carnap, as well as most of the other members, endorsed this view, which apparently implied that if one tried to discuss the relationship between a symbol and the object signified by it, one would have gone beyond the limits of language and fallen into metaphysics. To say, "The term 'brick' refers to this brick" seemed paradoxical to them because reference to bricks beyond language seemed to require reference to the world itself. Carnap, following Wittgenstein, in his early period adopted "methodological solipsism," the view that we must reconstruct the world from only our individual sensations plus the observable marks of deductive logic. This flight from semantics also limited the study of pragmatics; if one could not talk about meanings without falling into "metaphysics," pragmatics could not be, strictly speaking, a field in philosophy. If legitimate at all, it fell into the fields of the "psychology" or the "sociology" of language—the empirical study of language as a system of signs and their uses. Carnap later described his and Neurath's view, however, as evolving gradually away from Schlick's and Wittgenstein's: "first tentatively, then more and more clearly, our conception developed that it is possible to talk meaningfully about language and about the relation between a sentence and the fact described."[12]

In the early 1930s, Alfred Tarski, a Polish logician from Warsaw, visited the Vienna Circle several times and eventually convinced Carnap (about 1935) that a "semantic definition of truth" was possible, which Tarski had achieved by treating truth claims as assertions that a given sentence, such as "snow is white," is true if and only if snow is, in fact, white. This may, at first glance, sound trivial or like an endorsement of a correspondence theory of truth, but it is in fact a "semantic" theory. According to Tarski's understanding, and later Carnap's, because the subject of the assertion of truth is the sentence, truth is attributed to an English assertion of three words. Tarski and Carnap simply included the object language (first-order language) within the metalanguage, allowing them to assert the metalanguage truth that a given object language sentence is true. As a convention, the sentence enclosed in quotation marks was treated as a referring expression for the sentence itself, and the attribution "truth" was made at the "metalogical" or "metalanguage" level.

Having by then recognized the serious limitations of the syntactical theory as a complete theory of language, Carnap found Tarski's solution a liberating discovery that allowed discussion of both the truth and the meaning of expressions. Although some members of the Vienna Circle were skeptical, Carnap saw that this breakthrough could greatly expand the horizons of logical analysis, and he published two books on semantics in the 1940s. Later, he continued the trajectory toward more inclusive studies of language by developing a system of pragmatics. In effect, he ended up believing that there was no problem with talking about the relationship between language and the world, as long as one recognized that one must do so in a metalanguage—a language in which the subjects of sentences are linguistic expressions. It was permissible to discuss language, its relation to the world, and the social impacts of its communicative powers, provided one's arguments were articulated as arguments *about* language.

By the 1940s, though still calling himself a logical empiricist, Carnap began to play down the demands of empiricism and emphasize the principle of tolerance. Carnap had originally formulated the principle of tolerance in his earlier, "syntax only" phase, and the important technicality mentioned above was the key to understanding the emerging importance of this principle. The principle of tolerance formed the basis of Carnap's important midlife masterpiece *The Logical Syntax of Language,* in which Carnap "developed the idea of the logical syntax of a language as the purely analytic theory of the structure of its expressions." In the late 1920s and early 1930s, Carnap and his colleagues in the Vienna Circle were struggling to articulate a "theory of language structure," and Carnap was immersed in this problem, both as a personal challenge and as a cooperative venture. Carnap describes the conception of the book as follows. During a sleepless night in 1931, when he was ill and suffering from a fever, "the whole theory of language structure and its possible applications in philosophy came to me like a vision."[13] The next day, he wrote out forty-four pages of notes; they became the first version of *Logical Syntax*, which was eventually published in German in 1934. In this book Carnap introduced his "metalinguistic" techniques to be used in conjunction with formal object languages and put them to work at two quite distinct tasks.

First, Carnap saw his development of a "meta-logic" as the completion of Russell and Whitehead's task in *Principia Mathematica*. Russell and Whitehead had constructed a system that had certain provable properties, such as consistency, transitivity of deductive validity through rules of inference, and so forth. The proofs of these latter characteristics of systems were conducted in informal, ordinary language augmented with symbols and variables. Carnap's *Logical Syntax of Language* completed this project by formalizing the discourse about language and sentences, showing that these metaproofs can also be proved within syntax—purely formally, that is. The book was a monumental undertaking, requiring a tour de force in the application of the tools of deductive symbolic logic. In the larger portion of the book, Carnap set out to define, by brute force, a language by which, using formal considerations only, one could derive all of the concepts of the theory of deductive logic, such as provability, and also all of the substance of mathematics.

At the time Carnap's book was written—given his beliefs at the time—he thought he must accomplish all of this using syntactical analysis alone if he was to avoid metaphysics in the discussion of language. The *technical* achievement of the book was undiminished by Carnap's later realization that semantics and pragmatics were also legitimate ways to discuss language. It remains an important problem in formal logic to show what *can* be accomplished syntactically, even if, as Carnap and others came to believe, syntactic analysis is only one type of analysis among others. Nevertheless, the later expansion of the positivists' metatheory of language to include semantics and pragmatics reduced the apparent importance of the technical task of the book, and *Logical Syntax of Language* was largely ignored in the 1940s and 1950s. It was recognized as a brilliant achievement but considered largely an anachronism because of its narrow foundations for the study of logic and language.

Carnap, however, had undertaken a second, less technical task in *The Logical Syntax of Language*: to show that many traditional philosophical controversies could be understood as disputes about what linguistic forms were to be used for a particular purpose.[14] The less technical discussion in the last part of the book outlined this task as one of finding translations of philosophical problems into the metatheory of language. He says, "When in what follows it is shown that the logic of science is syntax, it is at the same time shown that the logic of science can be formulated, and formulated not in senseless, if practically indispensable, pseudo-sentences, but in perfectly correct sentences. The difference of opinion here indicated is not merely theoretical; it has an important influence on the practical form of philosophical investigations." So the task of creating a language and the task of creating a limited but nonmetaphysical role for philosophy were closely related. Technically speaking, Carnap had set up a metatheoretical "filter" for sorting philosophical problems into meaningless metaphysics, to be expunged, and real philosophical problems, which were, according to his filter, ones that could be translated into metatheoretical questions about linguistic forms. In Carnap's words: "Translatability into the formal mode of speech constitutes the touchstone for all philosophical sentences, or, more generally for all sentences which do not belong to the language of any one of the empirical sciences."[15] He says, "Metaphysical philosophy tries to go beyond the empirical scientific questions of a domain of science and to ask questions concerning the nature of the objects of the domain. These questions we hold to be pseudo-questions." But then he goes on to sketch a new possibility: "The non-metaphysical logic of science, also, takes a different point of view from that of empirical science, not, however, because it assumes any metaphysical transcendency, but because it makes the language-forms themselves the objects of a new investigation."[16]

Thus, besides doing his reductionist, virtuosic logic, Carnap was also developing a framework within which philosophy could be reborn as discussions about what language we should use to discuss the world. Progress on this task was far more important to Carnap than any technical breakthroughs in syntactical theory; and it was this idea to which he returned when he again addressed the big questions in the late 1940s. In the meantime, Carnap's expansion of his "theory of language structure" to include semantics and pragmatics also expanded the interest and power of his hypothesis about the nature of philosophical problems. The original purpose of *Logical Syntax* was to show the power of a purely *syntactical* system, but that emphasis became unimportant to Carnap in the next decade because his attitude toward the nonsyntactical study of language changed so dramatically. It is worth quoting Carnap (from his intellectual autobiography, written in the 1960s) at length on his changing views of the book.

> A few years after the publication of the book, I recognized that one of its main theses was formulated too narrowly. I had said that the problems of philosophy

> or of the philosophy of science are merely syntactical problems; I should have said in a more general way that these problems are metatheoretical problems. The narrower formulation is historically explained by the fact that the syntactical aspect of language had been the first to be investigated by exact means by Frege, Hilbert, the Polish logicians, and in my book. Later we saw that the metatheory must also include semantics and pragmatics; therefore the realm of philosophy must likewise be conceived as comprising these fields.[17]

We have now circled back to the topic of tolerance and conventionalism and can begin to see its immense importance in the development of a new approach to philosophical analysis. Whereas the principle of tolerance had originally been limited to the narrow, syntactical question of choosing linguistic "forms," it was now expanded to apply also to the relationship between words and their objects and, further, to the *use* of language in social situations. By this point at the end of his career, Carnap had grasped two key ideas that are characteristic of pragmatist thought and are also empiricist in nature. First, he had fully embraced conventionalism, the idea that languages are tools for understanding our experience, recognizing that our linguistic choices shape the world we describe. Language must henceforth be understood functionally, in terms of human purposes and goals. Second, he presented the principle of tolerance—which he also described as conventionalism—as a source of freedom and as a justification for experimenting with alternate linguistic forms. He therefore recognized a mandate to actively consider, and judge, languages according to their appropriateness for various purposes.

These expansions of the empiricists' considerations of language had gradually encompassed the whole of Peirce's semiotics, the general theory for the study of signs and their use. Carnap's originally narrow reduction of philosophy to logical syntax was thus replaced with a broader thesis: philosophical problems can fruitfully be understood and addressed as questions about how to construct and use languages in science and in solving social problems. It is this broader conception of philosophy that gradually reemerged as dominant in Carnap's mature philosophy, which is summarized in a 1950 article, "Empiricism, Semantics, and Ontology."[18] This article, in my view, shows the road back to Peirce and his almost-forgotten insights, and we will examine it presently.

Having spent the late 1930s and the 1940s working on semantics and pragmatics, Carnap returned to broader philosophical questions later in his life. By then the philosophical landscape had changed dramatically, and he found it necessary to rethink his approach to empiricism itself. In its earlier formulation, the principle of tolerance, applying only to syntax, seemed to reinforce empiricism, because the geometry of signs as physical marks allowed the meaningful discussion of logic, even under the most rigorous empiricist criterion. The Vienna Circle, impressed with Russell's reduction of mathematics to logic, applied the same reasoning to the truths of mathematics. For the empiricists of the early twentieth century, this new conception of logical truth and mathematics as tautologies solved a long-standing problem of empiricism: What is the nature of mathematical and logical truth, given that its sentences seem to be true, and known independent of experience? Carnap summarizes the solution accepted so eagerly by empiricists early in the twentieth century: "What was important in this conception from our point of view was the fact that it became possible for the first time to combine the basic tenet of empiricism with a satisfactory explanation of the nature of logic and mathematics."[19] The result that the truths of deductive logic could be understood as simply a geometric relationship between signs was important to empiricists, who had hitherto been faced with an embarrassing dilemma: Are the truths of mathematics and logic known by reason (in which case empiricists' claim to support all knowledge on the basis of experience is clearly inadequate)? Or are the truths of mathematics and logic simply "generalizations" from experience (a laughably inadequate view, since we don't consider changing our "logic" to include a fallacy just because most people commit it).

So the logical empiricists considered themselves worthy of the title. They were empiricists, and they had all the mathematics they needed, based on the reduction of mathematics to logic and logic to syntax. Science would be built up from atomistic reports of discrete sensations, and logic was represented in the logical structure of deductions allowed in aggregating and concatenating the molecular facts made up of linguistic atoms. This was the above-mentioned approach of "methodological solipsism," which reconstructed the world relying only on sensory phenomena and making no claims regarding the existence of an external world to correspond to reality. Carnap saw this project as important epistemologically, because it promised, in principle, to achieve certainty by treating atomic sensory data as indubitable and reconstructing our complex belief system by deduction alone. Notice, then, that in both the commitment to finding certainty and his use of deductive relationships, Carnap was still operating within the Cartesian conception of knowledge. This ideal served as a beacon for Carnap, a sort of interpretation of what it would be to have what Russell called a "logically perfect language."

Although this ideal was presented by Ayer to the world as a fait accompli, Carnap realized from the beginning that the ideal was problematic in several ways. Carnap had broken with Wittgenstein and Schlick by 1936 and had dropped the stringent limitation that all meaningful discussion of language must be in the "formal" mode; he had earlier chosen a sense-datum language for his reconstructions, but later he began using a physical object language. This kind of experimentalism was clearly incompatible with Wittgenstein and Russell's view of a uniquely descriptive, deductively constructed language. So Carnap could not see how philosophy could be entirely eliminated; if the world is not "given" to us with a single structure that can be inferred from experience of it, and if there are multiple "logics," for example, there is inevitably a question of which language "should" be chosen.

Carnap did not write about the philosophical implications of his broadening views until much later, however. Rather, freed from the onerous, self-imposed limitation that all linguistic arguments must fit into the narrow and awkward constructions of formal syntax, Carnap entered a highly creative period in which he developed basic theories of semantics and pragmatics, emphasizing most the "formal" aspects of these broadened disciplines, offering definitions and rules for constructed languages that could serve as the basis for empirical studies of language. In doing so, he openly discussed such questions as whether it was advisable to use sense-data or physical objects as the atoms of reconstruction—which he took to be the philosophically legitimate (metatheoretical) version of the traditional philosophical problem of realism versus phenomenalism. Likewise, he considered the question of whether to include references to classes and kinds, which he took to be the metatheoretical interpretation of the traditional problem of realism and nominalism regarding properties and qualities of objects. He came to see questions like, "Should we use a physical object language or a phenomenalistic, sense-data language to describe the world?" or "Should we understand the term *class* to refer to a collection of entities?" to be legitimate and important philosophical questions. So Carnap, freed of the commitment to syntacticism, began to "do philosophy" by discussing the most appropriate linguistic forms for various purposes. What is interesting is the methods and arguments he proposed to resolve these philosophical problems.

Carnap had left Vienna for the German University in Prague in 1931 but remained in fairly close contact with the Circle. In 1934 he met two American philosophers, Charles Morris, a very able pragmatist, of the University of Chicago, and the young Willard Van Orman Quine, of Harvard University. Quine arranged for Carnap to be a guest at the Harvard University's Tercentenary celebration in 1936; he was then offered a visiting position at the University of Chicago for the fall and winter of 1936, joining Morris on the faculty there. Recognizing the threat from the Nazis if he returned to Prague, Carnap accepted a permanent position at Chicago and became a U.S. citizen in 1941. He taught philosophy at the University of Chicago until 1952, when he accepted a position at UCLA, where he continued teaching until his retirement. He remained active, despite failing sight, until his death in 1970.

Carnap's migration and personal changes were hardly more sweeping than the changes in his philosophical views over this period. During this time he built, from the bottom up, formal definitions and rules for a scientific semantics and pragmatics, so he was mainly concerned with technical details. Carnap's philosophy, and especially his view of philosophy and its role, changed tremendously in this period. In particular, his earlier idea of the principle of tolerance reemerged as the dominant principle guiding his philosophical constructions. As long as questions about the structure of language were formulated as metalinguistic or metalogical questions about the language of some object discourse—a special science or mathematics, for example—they could be discussed and rationally answered. Even as he worked on the technicalities, however, Carnap was thinking about the big questions and published "Empiricism, Semantics, and Ontology" in 1950. This essay can for our purposes be treated as the capstone of Carnap's philosophy (or, as he would prefer to call it, his "meta-philosophy"). Long freed from his early syntacticism, and leaving aside much of his past jargon, Carnap sketched his new approach to philosophical problems as it had evolved in the decades since the heyday of verificationism and logical empiricism.

First, and most important, the reader realizes at the very outset of the essay that empiricism, though still a commitment in some sense, has been transformed from a brashly used weapon by which to banish traditional philosophical questions into a constraint within which one must work. The problem of the essay, as stated, is that a number of empiricists had criticized Carnap (and others) for referring freely to numbers and classes as entities, arguing that Carnap, in his semantics, had deserted empiricism and adopted a metaphysically based, Platonistic metaphysics.[20] In his response, Carnap never questions empiricism as a commitment, arguing instead that his critics misunderstood his metalinguistic methodology and that he was accepting these abstract entities only as "conventions"—they were created merely for the convenience of semantic description. The essay is an attempt to explain how conventions such as this are introduced and how they are justified and that in referring, by convention, to physical objects or classes, one is not making ultimate/absolute statements about the world, but rather one is "choosing" to use a linguistic form. Sentences about which linguistic rules, forms, and interpretations to use in descriptive science were considered important questions that should be addressed by scientists in conjunction with logicians and philosophers, and in some cases these were very important questions. Carnap recognized, however, that these sentences, though perhaps affected by empirical data, could not be decided by it; nor could they be simply "analytic" sentences, ones that must be true according to logic. His new approach did not fit smoothly into logicians' verifiability principle and its either-or thinking. Carnap sidestepped the problem by creating a special category of metalanguage sentences that must be justified "pragmatically"—as choices regarding simplicity, descriptive adequacy, and effectiveness of a proposed language in communication. So Carnap could claim consistency with the verifiability criterion of meaningfulness, which applied only to sentences that claimed to make statements, by saying that these metatheoretical statements were "choices," not assertions, and that they had no "theoretical" content.

Carnap, in the 1950 essay, responded to the concerns of empiricists, arguing that we must separate two types of metatheoretical questions, "internal" and "external."[21] Internal questions are asked once one has already chosen a "linguistic framework" (a set of definitions, syntactic rules, and interpretations for a constructed language) and should be thought of as simply reporting a convention that is in effect. To say, then, that "constants a, b, c, . . . refer to the real numbers" is simply to say that we are speaking in a system of arithmetic (a linguistic framework) that is powerful enough to express the relationships representative of the real numbers; in interpreting this language, we take these constants to refer to such numbers. Similarly, if a physicist begins a paper on light phenomena by stating that he or she will use a language and interpretation in which light is described in terms of particles, that physicist is merely reporting a choice in terminology, a convention. In neither case, Carnap argued, should the mathematician or the physicist be expected to provide a metaphysical or "deeper" explanation for the references to numbers or particles. Either one could be understood as making a metalinguistic statement about the rules and features of the language that is chosen; and this statement is trivially true or false in that it simply reflects the conventional rules of the language as chosen.

Carnap also, however, recognized another kind of question, which he called "external" to the linguistic framework; these questions, he says, are "raised neither by the man in the street nor by scientists, but only by philosophers." Philosophers, who ask about the reality of a system itself, fall into confusion. But then Carnap says these philosophers "have perhaps in mind not a theoretical question but rather a practical question, a matter of a practical decision concerning the structure of our language. We have to make the choice whether or not to accept and use the forms of expression in the framework in question."[22] So now it seems that there is a perfectly good and interesting function for philosophy: to examine and to help decide what basic descriptive categories and predicates should be chosen to describe the world, in conjunction with scientists who are engaged in such description. Thus, although we cannot *assert* that there are physical objects, we can *choose* to speak in a language that countenances physical objects. After discussing at length what it means to "accept a new kind of entities," he again insists that such an acceptance cannot be seen as an assertion that is true or false; rather it is a choice that should be judged to be "more or less expedient, fruitful, conducive to the aim for which the language was intended. Judgments of this kind supply the motivation for the decision of accepting or rejecting the kind of entities."[23]

The acceptance of any kind of linguistic forms, Carnap argued, "will finally be decided by their efficiency as instruments, the ratio of the results achieved to the amount and complexity of the efforts required." And how does one go about resolving these "practical" problems? "To decree dogmatic prohibition of certain linguistic forms instead of testing them by their successes or failure in practical use, is worse than futile; it is positively harmful because it may obstruct scientific progress."[24] So Carnap became, as his rich career was winding down, a pragmatist. He was a pragmatist in the sense that he saw language as providing tools of communication; and having jettisoned the view that there could be only one useful language, Carnap advocated experiments *in particular discourses* as the basis for choices of structure and of the type of entities discussed. Although he emphasized formal criteria like simplicity and clarity in choosing a linguistic framework, his unequivocal embrace of semantics entailed that he must interpret these choices very broadly, to include the choice of useful categories and structures attributed to nature.

From this point it is a tiny further step to say that to the extent that our scientific models are themselves conventions for understanding and interpreting experience, semantic choices literally give structure to our experience. Carnap's broadened conventionalism, then, encourages a view of philosophers as members of larger scientific teams, assembled to address a problem or sets of problems. Philosophers contribute to the team with their expertise in logic and linguistic analysis and bring these to bear upon conceptual models chosen to interpret reality. Because the development of such models is both creative (it requires creating new categories and relations through the development of semantic rules) and conventional (in the sense that these decisions are open to choice), Carnap's idea of choosing linguistic frameworks can apply to modeling and prototype building. Carnap, in his mature philosophy, espoused what can be called the "proposal view of philosophy."[25] Philosophy contributes to intellectual and scientific development by proposing and examining linguistic forms and associated structures for interpreting the world of experience. Philosophy is thus provided a role that is at once positive, creative, and critical. Furthermore, given the expansion of the understanding of modeling and prototyping as a main avenue to intellectual and technical innovation, this approach to philosophy provides a very creative role for philosophers; we will return to a deeper examination of this role in the last section of the appendix.

Here we have, at last, an answer to Dewey's question. In a world in which all beliefs are considered open for testing, so are proposals about what languages and what categories of entities will be used to encapsulate our experience. But since these questions will, when legitimately formulated in the metalanguage, be questions about how to make our language work better as a tool, Carnap provides an essentially pragmatist answer to Dewey's question: the categories we use to characterize and describe nature are questions that are worked out in the process of pursuing our goals and duties. We develop a set of categories that are useful for responding in ways that will help us to survive. And of course we survive as a community, so these will be questions that are answered within communities of people who seek the truth as an aid to survival. "Let us grant those who work in any special field of investigation," Carnap said, "the freedom to use any form of expression which seems useful to them; the work in the field will sooner or later lead to the elimination of those forms which have no useful function. *Let us be cautious in making assertions and critical in examining them, but tolerant in permitting linguistic forms.*"[26] He had, that is, given up the Cartesian ideal of a unique and knowable, prelinguistic reality and the set of philosophical problems associated with it, and he recognized that, ultimately, pragmatics is more basic than either syntax or semantics. For it is in pragmatics that we can analyze the problems of communication in real-world situations and appeal to shared goals of a community to justify choices about how to describe and, ultimately, how to *structure* our experience in particular problematic situations.

Carnap was, throughout his life, a supporter of "international languages," and he was an enthusiastic participant in an association promoting the use of Esperanto as a means to improve communication in Europe. Although he saw the parallels between efforts to improve communication in ordinary speech contexts and his technical, professional work of creating new and more precise languages for science and logic, he at first treated these problems as distinct. Eventually he saw that both spring from, and are supported by, his belief in tolerance. "Thus, in time," he said, "I came to recognize that our task is one of *planning* forms of languages. Planning means to envisage the general structure of a system and to make, at different points in the system, a choice among various possibilities, theoretically an infinity of possibilities, in such a way that the various features fit together and the resulting total language system fulfills certain given desiderata."[27] Once Carnap had admitted, as he gradually did, that languages must be judged as instruments of communication in particular social situations, he could not avoid the next step: since there are many and varied purposes, there will be many and varied useful descriptive languages.

This conception carried Carnap far beyond the Cartesian project, despite his earlier commitments to a rational, unique reconstruction of reality through deductive logic. Choosing a language as a convention is now *based on human communication in real communities.* The search for a single set of rules, terms, and categories that reveal the structure of reality was entirely discredited. Carnap had thus brought positivism to the place where it had an answer to Dewey's question about the origin of the categories we use to describe nature: they are generated as conventions, conventions that are useful for communicating and solving human problems, within communities, in a cooperative way—and this is simply a detailing of the pragmatist view of language developed by Peirce and Dewey.

Perhaps now it is clear what I meant, above, when I said one could tell the story of linguistic philosophy since Peirce as a story of liberation. Once one gets away from the old, essentialist view that the world is structured prior to language and communication, whole new fields of opportunity open up, including the planning, proposal, and development of languages that are well suited for communication about particular social problems. If, as in the case of the environment, such social problems require the mixing of multiple scientific discourses and integrating these into a management process that is designed to protect and foster public values, then one problem is to consciously develop a language that will minimize confusion and increase communication at this nexus where public values, science, and policy intersect.

Would it not then be a noble calling for philosophers and logicians to use their understanding of language to address real problems of communication, problems that occur in public discourse about what to do to protect the environment? And that, as I understand it, is a role that will fit very nicely into a broadly pragmatist and, especially, Deweyan approach to philosophy and social policy. We have finally worked our way around the circle, which started with the recognition by Peirce that if experience is to be our guide and if every belief is to be vulnerable to revision in the face of experience, then experience must be understood as a form of symbolic behavior occurring within communicative communities. If, with Peirce, we begin the study of science and its application to environmental problems within the context of a community facing real problems, the test of truth will not be some external, human-independent reality, but the shared and ever-expanding experience of a community of truth-seekers.

What Carnap added is a belief that, provided we distinguish decisions about how to talk and communicate effectively from assertions about reality, it is reasonable to expect that we can consciously attempt to *improve* our language as a tool of communication, to the extent that we become consciously aware of our linguistic conventions. Conventionalism—the resounding rejection of essentialism and a priori commitment to how the world *must* be—leads to linguistic liberation, the freedom to develop and construct new forms of language that are more useful in understanding, for example, environmental problems, forms that encourage communication within a process of open and democratic dialogue and deliberation. The role of philosophy, then, might be to encourage and participate in discussions of how to describe the environment and how to communicate more effectively about environmental goals. In the process, a whole new set of questions about how to structure the world (for a given purpose) through the choice of more effective languages for discussing environmental policy open up for discussion. These questions involve building shared models to understand environmental problems. One of these questions is the one we set for this book: Can we develop a framework of definitions and linguistic relationships that would rehabilitate the term *sustainable* and make it a central term in discussions about long-term environmental policies?

What our review of the development of Carnap's thought did for us, I hope, is to highlight the possibility of a creative approach to linguistic analysis. Carnap offers a breathtakingly simple answer to Dewey's post-Darwinian question about how to construct the categories by which we describe the world. The answer is that the categories we use are conventions adopted in order to enable communication. The categories we find in ordinary language have some presupposition of validity—they have stood the test of time and expanding experience so far. At the same time, every one of our beliefs, including beliefs about how the world is to be categorized, is open to challenge on the basis of experience. What Carnap did was to show how linguistic conventions, in which the dead husks of traditional goals and traditional ideals have been implicitly embodied, are open to challenge. He also showed that we can propose new ways to speak and to experiment using them, and gradually improve communication. This sets the stage, for example, for scientific and, more particularly, environmental modeling as exercises in communicating information and for articulating goals. And Dewey, the social activist, provides the idea of "social learning," the idea that communities can experiment and learn, reducing scientific uncertainties about the impacts of current actions and policies, and revise goals in the face of new evidence.

So Carnap, the positivist warrior, eventually embraced pragmatism, treating the problems of traditional philosophy as "pragmatic" problems of choosing the best languages for various social purposes. This convergence of rehabilitated, "tolerant" logical empiricism with pragmatism, I suggest, provides us with a new way forward to improve our linguistic resources for discussing environmental problems and provides a new role for philosophy. Philosophy's role, on this approach, will be to work with scientists and policymakers to develop new models for describing (and, simultaneously, new ways to understand, interpret, and evaluate) ecological processes as they interact with human choices to produce environments that are either valued or abhorred.

The languages we have traditionally used and the categories that have been developed with them are no longer "given." We should think of them as a cultural heritage, but one that is in constant need of rethinking in the light of new situations and new information. The various languages we use to talk about environmental values, goals, and policies are best thought of as starting points, as prototypes for better models and linguistic structures. What Carnap's odyssey illustrates is that, once freed of Cartesian enslavement to the belief that reality is imposed upon us by the world out there, we can think of language as a creative tool for characterizing and talking about problems and solutions. Once language is seen as a tool that communities can use to communicate and deliberate, rather than as a reflection of a nonhuman reality, we are well on our way to developing what Peirce envisaged, an experientially based approach to understanding the world. The key, as Peirce saw, was to understand the entire process of experience-gathering as an inherently open-ended process that takes place within a community. In this community, every belief is open to challenge on the basis of anybody's contrary experience. The assertions we make about nature reflect not some human-independent reality, but rather a socialized reality in which experience and communication are inevitably intertwined.

Carnap, coming late to pragmatism, in fact fell short in several ways of providing a fully adequate method for proceeding, mainly because he failed to develop a fully satisfactory account of how we might, in practice, discuss and resolve external, or metatheoretical, issues raised by the need for interdisciplinary and public dialogue about environmental policy goals. What he did do was to liberate our thinking about language and to open the possibility of a pragmatic examination of language. Further, in his technical work on the metatheory of language, Carnap created important tools for exploring these questions.

A.4 Pragmatism: The New Way Forward

Carnap, as we have seen, gradually recognized that if conventionalism is true, if Dewey was correct that the categories we use to study nature are not "given" in prelinguistic reality, then language itself must give structure to the world as we encounter it and reason about it. Having recognized this, and the corollary that the linguistic choices we make in describing the world provide the categories and the structure attributed to nature in our descriptions, Carnap pointed us toward a new consciousness about linguistic decisions and proposed some important new methods for addressing them. Carnap's method for examining linguistic conventions as useful tools of communication is relevant here because of the importance of language in public policy discourse and deliberation about the environment.

Carnap's late "conversion" to pragmatism was driven by his study of languages, both natural and constructed; he concluded that both must involve an important element of choice; such choices are obviously important, because they reflect the "structure" we impose upon reality when we use language to describe it. Our choices in ordinary discourse are usually made implicitly—indeed, most of them have been made in a forgotten past—and are seldom held up for explicit examination. In science, and in whatever other context we are especially attentive to the concepts and words we use, we can make explicit decisions about them. Today, this process of creating concepts, defining them, and making them precise, is usually embedded in a more general activity called modeling. When we create a model for understanding experienced phenomena, an important aspect of the activity is the choice of descriptors as exemplified in syntax, semantics, and pragmatics.

After a career of chasing the impossible dream of a single, picturelike language that depicted reality as it really was, Carnap finally—in his late fifties and early sixties—recognized the impossibility of such a unique structure. Many different languages can be constructed, and one's purposes in constructing them can be the most important aspects of such constructions. These purpose-driven choices provide the means by which we interpret and describe the experiential world we encounter. So Carnap's new approach raised interesting questions regarding how to innovate in developing new languages, whether methods could be developed for gradually improving natural languages, and how to choose among multiple languages when there are many possibilities. Such questions, Carnap saw (once he gave up the dream of a single, representational language), were highly contextual and depended at least partly on practical concerns and human motives and goals. These motives and goals, unavoidably, connect with human values. In this sense, one might say Carnap embraced pragmatism because it provided a convenient and appropriate "patch" to his earlier, more dogmatic philosophies.

Fortunately, Carnap's conversion to pragmatism did not require him to develop a new philosophy from scratch. He was already well acquainted with pragmatism through conversations with his colleague Morris at Chicago and his younger friend W. V. O. Quine. Carnap often commented on the similarity between pragmatism and logical empiricism throughout his career; what he finally acknowledged was that there were important philosophical questions that must be asked *about* language and that unique answers to these questions could not be derived from the structure of *the* language of science. Carnap therefore recognized a new type of "philosophical sentences": metatheoretical questions about what language to use. Such questions apparently had to do with the pragmatics of language, which could of course be studied in the sociology of language and other social science disciplines; but Carnap insisted that important linguistic choices were also involved and that, ultimately, *choice* and *commitment* to act had to be an inevitable part of such decisions; the decisions, then, clearly depended upon individual and shared social values. Although Carnap did not explore the values aspect deeply, he did identify a new starting point for the investigation of linguistic forms and usage; he saw clearly that such an investigation would be openly pragmatic and goal-oriented, because external questions are about how people with shared purposes can better communicate.

If you can accept my story of Carnap becoming a pragmatist, we have found our way back to Peirce's new starting point. This new starting point seeks convergence through communication, and it is based on deliberation, experiment, and gradual expansion of the community's collective experience. This is precisely what John Dewey called social learning; it represents the ability of communities, divided by differing interests and viewpoints but possessed of a shared interest in maintaining a livable and pleasant environment, to deliberate together and decide what goals to pursue as a community. By melding the linguistic expertise of the logicians and language analysts together with the more communitarian approach to truth advocated by the pragmatists, it should be possible to develop a method by which to improve communication within communities regarding environmental problems and goals.

One important goal, a central one for Carnap, was the advancement of science through cooperation; and since communication is essential to any successful cooperation, we can set certain standards for communication and evaluate the usefulness of languages for various purposes. Carnap could not bring himself to say such judgments were true or false, but he never denied that they are ultimately important. Since it is our decisions about what terms we use and how we talk about environmental problems that determine the structure of the world we will encounter, it is important that a thorough investigation be undertaken before choosing a language for a given purpose, and such an investigation is the proper role of the linguistic, or analytic, philosopher. So, for Carnap, philosophy has a legitimate domain as the "normative" study of language. Carnap thought, and I agree, that this was no denigration of philosophy; it is a noble calling to ask, and to encourage others to ask, what linguistic forms will maximize communication and sharing of experience. And this is the purpose of this book: to ask what language, what descriptive models, and what representations of values will encourage better communication and more cooperation in the discussion and formation of environmental policy.

Because of Carnap's late embrace of pragmatism, he did not work out all the details of these pragmatic investigations. Indeed, there are several areas where Carnap probably did not go far enough in rejecting positivist ideas or in liberalizing empiricism to allow metalinguistic discussion of the usefulness of language forms. To get all the way back to Peirce's starting point, we need to add some important qualifications and limitations that we now know must be appended to Carnap's final position.

A.4.1 The Analytic-Synthetic Distinction

Carnap never gave up his claim that in constructed languages, one can draw a sharp distinction between "analytic" sentences and "synthetic" sentences. He thought that analytic sentences expressed "necessary" truths, truths that follow from the rules of the language, whereas synthetic sentences were related logically to sentences describing sensory information. This doctrine, together with Carnap's hope for a "logical" construction of complex sentences out of atomistic sensory reports, continued to connect him to the older Cartesian idea of a system anchored by "necessary truths." Historically the concept of analyticity was connected epistemologically to the concept of necessary truth, with analysis being the method of a priori reasoning. The search for certainty always seemed in one way or another to connect to necessity or analyticity, and Carnap sometimes referred to the "necessity" of analytic truths. Quine—who became Carnap's ablest critic—pointed out in the 1950s and 1960s, however, that Carnap's solution, to identify analytic sentences as those that are true by virtue of the rules of language, was too artificial to carry any *epistemological* weight.[28] Conventions do not result in necessity, except in the most hypothetical sense. Quine argued that since the languages we use are "conventional," so are the choices of the rules. And if the rules are chosen conventionally, then so are designations as to which sentences will be true by virtue of the rules of that language. As Carnap had acknowledged, the choice of a particular linguistic form is always underdetermined by experience.[29] Quine therefore argued that Carnap's attempts to separate the logical from the empirical content in a language was itself relative to the language and the rules chosen. If one restated the rules of the language, one could—without changing the empirical content of the theory—change sentences from analytic to synthetic.

On one level, this point seemed obvious to Carnap, and when he was interested only in choosing an efficient language, this form of relativity was unimportant to him. What Quine pointed out, however, was that by maintaining a special role for "analytic" sentences and associating such sentences with the tradition of "necessary" truth, it was easy to be misled back into the modernist, Cartesian ideas that some truths *must be true* and that we can therefore use them as starting points in all analysis of language and truth. One could not base anything universal and necessary on the purpose-driven decisions by which we decide upon efficient means of communication in particular situations. Quine wanted to ensure that references to "analyticity," now understood as the result of linguistic conventions, could never be resurrected as the starting points for a rationalist epistemology. Furthermore, he thought Carnap overestimated the importance of the move to the "metalanguage," and although he agreed with Carnap that choices regarding linguistic rules and structures must be resolved "pragmatically," Quine believed the same was true for empirical statements.

Quine's arguments (Carnap had no answer to them that was relevant to the epistemological point Quine made) entailed that the distinction between analytic and synthetic sentences, itself, is conventional, perhaps useful in particular situations. References to analyticity and necessity, however, are always language-bound and could never be used as a foundation to say that all languages must reflect these foundational truths. It also appeared that, similarly, Carnap's internal-external distinction would likewise be relative to choices of linguistic rules. I think Carnap should have simply accepted Quine's point and stated that the distinction between analytic and synthetic sentences was useful as a tool for understanding sentences and their logical relations. And that the distinction between internal and external questions lacked prelinguistic importance. It too is a useful distinction, because when we are developing new models, it is important to pay special attention to linguistic forms, and it is therefore useful to articulate explicit rules of language by convention. He should therefore have acknowledged that neither distinction could play an important part in any *epistemological* justification and settled for an argument that his use of the distinctions is *pragmatically* justified. His distinctions, analytic versus synthetic and internal versus external, were important conventions because they help us as scientists and philosophers to ask the right questions and to avoid dangerous confusions. But these choices should not be seen as conferring necessity or incontrovertibility on sentences that are true by virtue of the rules of language.

What is at stake here may seem to be a technical point, but it is of the utmost importance, and it has to do with "foundationalism." Foundationalism, in the sense used here, is the view that in any adequate belief system there is some set of sentences that (1) can themselves be shown to be indubitable and (2) can serve, through the operations of deductive logic, as indubitable building blocks capable of supporting our beliefs about the world. Foundationalism is at the heart of the modern debates between the rationalists and the empiricists; but it is important to see that both major strains of modernism have generally accepted a form of foundationalism. They both, that is, accepted the Cartesian challenge of providing deductive proof from indubitable premises as the only truly legitimate answer to skepticism and doubt. Where they differed was with respect to the nature of the indubitable starting points. For Descartes, the foundations were composed of necessary truths, innately known and self-evident on the basis of pure reason. For empiricists, the indubitable starting points were atoms of experience. The search for a perspicuous way of expressing these atoms led to a proliferation of terminology, and some positivists suggested that the atoms of sense could be expressed as "red, here, now," as minimal linguistic acts that simply reflected the speakers' visual field. So foundational empiricists differ from rationalists in that they "found" all knowledge on indubitable sensory data. Both systems depended upon a faith that certainty is a possibility, and on isolating some set of sentences as indubitable starting points from which further truths could be deduced. This set of sentences, then, were the "foundation" for the rest of knowledge. As argued above, the rationalists could never justify their "self-evident" truths; and the empiricists never came close to rebuilding the world we inhabit and have beliefs about from little bits of "red-here-now" and "blue-there-now," augmented only with the tautologies of symbolic logic. Both programs failed.

What the two modernist positions have in common, however, is more important than their differences, because in sharing a commitment to deduction as the epistemologically relevant logical relationship between evidence and belief, they implicitly assumed that the unit of knowledge is the sentence (because deductibility is a relationship between sentences). What Quine successfully argued was that in the real rough-and-tumble of scientific theory formation and verification, it is impossible to assess the truth value, or the analyticity, of any sentence independently of other sentences in the language. But truly taking this insight to heart, as we shall see, wreaks havoc with foundationalism in the sense defined above.

The general point had been known by scientists since about 1900, when it was originally articulated by Pierre Duhem, the French physicist and mathematician who argued persuasively that there cannot be "crucial experiments" in science.[30] Whenever we set out to test a hypothesis, he reasoned, we actually must submit for test not only the hypothesis, but also the associated assumptions and definitions necessary to assert it and to submit it to test and to possible falsification. If we get an unexpected result in the experiment, we can either give up the hypothesis under test, or we can adjust our assumptions and the definitions we use, and so forth. This implies, however, that every attempt at falsification of a given hypothesis actually makes vulnerable not just the hypothesis but rather the whole complex of meanings and methods—and theories—necessary to bring about the test. Duhem's apparently innocent finding, if expanded, entails that all of our experience, including sensory evidence that we can adduce in an experiment, is theory-bound. We could not have *that* experience without having also the theoretical assumptions and beliefs that allow us to test it. Actually, then, this point is a corollary of Peirce's more general—but more difficult to establish—argument that all experience is symbolic in nature.

This point is of utmost importance for the way we understand empiricism and the pragmatic method, so it's worth looking at one case study in the historical logic of science. Consider, for example, the multicentury history of the term *whale* in ordinary-language discussions of zoological taxonomy. At one time in history, whales were defined as large fish, because like other fish, they lived under water. Given the linguistic rules in effect at the time, it would be contradictory to deny that a whale is a fish and "analytic" to affirm it. As modern, genealogical taxonomy was developed, in response to new observations, physiological characteristics became more important than behavioral ones in determining taxons. Over time, and with improvements in our taxonomic system, it became more and more difficult to fit whale species into fish taxons, because dissections and other evidence found more and more differences between fish and whales. If one looks at linguistic usage at two different periods of scientific development, then, the sentence, "All whales are fish" was once considered to be "analytic," just as "All brothers are male siblings" would have been considered. At a later time, "All whales are fish" is not even considered true—indeed, it appears to be contradictory, given current usage and the "rules of language." One can, of course, say that no real contradiction occurs because, in the intervening time, the meaning of the terms were changed. Quine would insist, however, that although the changes in meaning were driven by many unexpected experiences—observations, for example, that whales, unlike fish, have lungs, suckle their young, and so on—there was no single point at which it was contradictory to a particular experience to continue to call whales fish. His point, and it is a profound one, is that in everyday science, changes in the use of terms are usually, ultimately, made in response to new observations. Taxonomists who began to encounter more and more structural and physiological differences between whales and fish could, for a time, accommodate this new information, simply by treating the family of fish as more diverse than first thought. But at some point the preponderance of observation tipped the scales, making it more reasonable and efficient to change the definition of fish and reclassify whales as mammals. Quine's point is that different scientific observers, each with a different set of past observations, might choose quite differently how to respond to an unexpected experience. According to Quine, matters of simplicity, elegance, and even personal taste will affect the exact response of a scientist to new and unexpected experiences, such as an experiment that yields unforeseen results.

It is more accurate, Quine argued, to think of relations of truth, meaning, and inference as a vast web of interrelationships. New and unexpected experiences must be fitted into our webs of belief; but assimilating them will seldom be a matter of simply substituting one atomic sentence for another. When faced with an unexpected experience, each of us must choose a response, which might include doubting the datum itself—pleading that the light wasn't good—or changing our theories or, in some cases, redefining terms involved. Quine, given the difficulty, in principle, of matching experiences with synthetic sentences and equating "logical truths" with the historical idea of necessity, chose to scuttle the analytic-synthetic distinction. He concluded that the best we can do is to concentrate on gradual improvements in some portion of the web of our beliefs. He advocates, then, a form of pragmatic empiricism that treats decisions about how to revise our system of beliefs as strategic and purpose-related more than as a matter of picturing.

Quine's position thus implies that, strictly speaking, truth is not a property of sentences, as such; truth can really be attributed only at the level of an entire belief system. Or, to put this point as Quine did in his tremendously influential essay "Two Dogmas of Empiricism," any observer, armed with a collection of beliefs about the world but surprised by a new and unexpected experience, can react in a number of ways. There is no way to identify a particular sentence that must be revised; pragmatism's conception of our belief system is not as a list of independent, atomic sentences—atoms of information—and all the permutations and concatenations of these. Sentences function within a web of belief, and given the intersentential connections animating the web of belief, it is always dangerous to sever a sentence from its context and attribute truth to the sentence without reference to those connections. To express the point another way, two theories stated in different languages with a shared base of observation sentences could predict all the same observations but use them to support quite different theoretical constructs. This is Quine's thesis of the indeterminacy of translation as applied to theoretical sentences. Quine summarized the argument as follows:

> The Vienna Circle espoused a verification theory of meaning but did not take it seriously enough. If we recognize with Peirce that the meaning of a sentence turns purely on what would count as evidence for its truth, and if we recognize with Duhem that theoretical sentences have their evidence not as single sentences but only as larger blocks of theory, then the indeterminacy of translation of theoretical sentences is the natural conclusion. And most sentences, apart from observation sentences, are theoretical.[31]

On the point of analyticity and sharp separation of a logical from an empirical component of language, Carnap had disagreed in the 1920s and 1930s with his friend Neurath, the radical sociologist and economist. Neurath had followed Duhem and rejected sharp and categorical separations between logical and empirical truth decades earlier, and Carnap was well aware of the difficulties of his position. As noted at several points in my book, Neurath introduced a key simile that makes sense of the gradual improvement of knowledge that accrues from careful observation, experience, and revision. Neurath, embracing radical empiricism, said that the task of reconstructing human knowledge is like sailing on a boat, going from port to port with no opportunity to put the boat in dry dock for repairs. In this situation, the boat must be repaired while being sailed, and so the repairmen must stand on some parts of the boat in order to repair other parts. Eventually, every plank in the boat may be repaired in this way, and the resulting ship can continue to fulfill its functions as before. Philosophical reconstructionists, Neurath argued, face a similar problem. They must accept some beliefs as valid in order to examine others. No particular sentence can be declared "true no matter what experience we have," and yet we succeed in identifying weak links in our system of beliefs and, standing on stronger sections temporarily, work on improving the weakest links. In the end, it is possible that every one of our beliefs will be repaired (recall the empiricist commitment to submit, sooner or later, every belief to the arbitration of experience), and yet we never lack a "platform" on which to stand in particular situations. This argument, of course, derives directly from Peirce's approach, which he called "fallibilism," in which he rejected Descartes's universal doubt, arguing that in real situations doubt reflects only on particular beliefs that are important to action.

By this ongoing repair process it seems unlikely that we will achieve anything like the Cartesian, or the early positivist, ideal of certainty, and yet it can be argued that such a deliberation can, in principle, given favorable social conditions, generate a community-based process that incorporates an informative, deliberative public discourse about social values and environmental goals. By this way of thinking we can gain a certain kind of "objectivity" in a community if, whenever anyone calls some assumption into question, the questioned element of knowledge is examined (even though the examination cannot be completed without "standing" on some planks that are accepted by consensus). Neurath's analogy, then, showed how advances in understanding and knowledge can occur without designating any particular sentences as beyond empirical reproach; rejection of foundationalism need not result in skepticism about knowledge and advances in science.

A casualty of the new analogy, apparently, was the notion of truth determinable on a sentence-by-sentence basis: when faced with a rotted section of five planks, the repairmen might replace each plank, or they might use three broader planks; or they could redesign the whole area, substituting metal parts for the rotted planks. The assessment, in other words, is not based simply on a direct comparison, plank by plank, because the criterion of a good repair is not whether it "corresponds" to the earlier plan, but whether the new construction "holds water"; similarly, in epistemology, since the world does not appear to us in sentence-size bits, a good response to empirical evidence may not be specific substitution of a true for a false sentence but a reworking of a whole area of theory, including the proposal of new terminology.

Quine, more than any other English-speaking philosopher, wrote the epitaph for the old, Cartesian project, because the Cartesian project was based on deduction, and deduction is a relationship between sentences as bearers of truth and information. Through three centuries, modern philosophy had argued about how we know specific sentences and how we infer new sentences from ones we know. Quine, the logician, deeply influenced by both Russell and Carnap and their struggles to find a uniquely perfect language as a substitute for the order introduced into descriptive discourse by old-fashioned "innate ideas," finally saw the futility of it all. He did not reject only the Cartesian method for reconstructing our knowledge; he rejected as well the Cartesian formulation of the epistemological problem as demanding a world that has only one "rational" interpretation held together by deduction. Quine succeeded in formulating a simple, incontrovertible argument that the Cartesian project—including, most poignantly, the empiricists who had attempted to rebuild Descartes's world from experience—was wrongheaded. Quine saw clearly that neither rationalism nor empiricism could reconstruct our knowledge because they shared a false assumption about language—the apparently innocuous assumption that human knowledge can be assessed and tested on a sentence-by-sentence basis.

This point is important for our search for an improved language for discussing environmental problems because it emphasizes the open-ended and creative task involved in learning how best to describe and classify our experiences. Describing the world is more like painting a picture than like taking a photograph. This insight also connects Carnap and Quine's gradual development of a pragmatist conception of scientific development and change as a series of choices with another idea that has proved extremely influential today. In Carnap's idea of internal versus external questions, and in Quine's recognition that changes in belief are not *dictated* by particular observations, we have the essentials of Thomas Kuhn's idea of a paradigm. Kuhn introduced this concept—a set of definitions, assumptions, beliefs, and methods that constitute a scientific discipline—in order to distinguish between "normal" and "revolutionary" science. Kuhn argued that paradigms, which are larger systems of definitions and meanings that articulate the structure of the world as it is discussed in a particular scientific discipline, do not stand or fall on the basis of one or a few observations.[32] A major difference, of course, is that Kuhn applied these ideas in order to understand the historical development of sciences, rather than as a tool of logical analysis. But the point I wish to make here is that the concepts and tools developed by the positivists and postpositivists, if applied to the future of environmental policy science rather than to the history of normal, disciplinary sciences, provide a pragmatic method for addressing problems in understanding and communicating about environmental problems.[33]

A.4.2 The Many Functions of Language

Another point, which neither Carnap nor Quine fully grasped, was that the rejection of deductivism, atomism, and the belief in a single structure that was imposed upon language by either reason or experience would actually undermine the special role that they, as philosophers of science, had always given to the descriptive sciences. Because of their strong bias in favor of natural science, the positivists and other philosophers who retained modernist leanings, much like those of Russell and the logical atomists, assumed that declarative sentences were the most basic form of speech and that we should study the structure of true and false sentences to understand the true structure of language. In logic, then, other forms of speech, such as questions, exclamations, or imperatives, were thought of as mere modifications of descriptive sentences. If sentence S pictures reality, the corresponding question could be written as ?S.

Peirce's understanding that all of our experience is shaped by language and his clear recognition of the community-based nature of linguistic activity should encourage us to look beyond the reporting and describing function of language. In real situations, language also functions to persuade, to ask questions, to give commands, and to sound alarms. Although all of these functions may often involve some element of "reporting," the use of nondescriptive language may be more important than descriptions in actual social situations where action is necessary for survival. So a truly pragmatic approach, one that pays attention to the uses of language in real, communicative situations, would place much more emphasis on the use of language for purposes other than to describe the world; more attention would be given to language that evokes attitudes, persuades others, enters into commitments, and builds trust. All of these uses of language, in any community-based attempt to discuss and implement environmental policy, can be as important as the descriptive use of language.

A truly pragmatic approach to the investigation of language as a tool for developing environmental policy will therefore require that we pay attention to all of these functions of language, not just to description. Communication can fail, and often does fail, as a result of miscommunications that have little to do with failures in the descriptive functions of language, and that fact highlights the importance of the project of this book, to develop a more useful concept of sustainability. In developing this concept as a useful tool for communicating about the broad goals of environmentalism, we must realize that factors other than measurability and descriptive adequacy will be important in our choice of definitions and that speech other than declarative sentences can be important.

A.4.3 The Fact-Value Dichotomy and "Value-Neutral" Science

Neither Carnap, nor Quine, nor, for that matter, most scientists today, have recognized that the very same arguments that undermined the distinction between analytic and synthetic sentences also undermine the positivists' other important dichotomy, between facts and values. When Quine pointed out that a scientist, faced with an unexpected observation, can consistently follow several courses of action—disallowing the experience and redefining terms, and so forth, in addition to changing a belief—he opened the possibility that substantive human values are an important component of many scientific decisions. What pragmatism does, what distinguishes it from the early philosophy of logical empiricism, is to recognize the humanity—and the social context—of scientists as symbol-using and hence social beings. Although pragmatists support attempts to protect the "objectivity" of science (by avoiding bias, by reducing conflicts of interest, etc.), they do not think of objectivity as a correspondence of assertions with "facts" that exist prior to those assertions. They think of it, rather, as persistence through a process of continual deliberation, disagreement, and compromise.

This change in the image of science from a snapshot to a negotiation is just what is needed if science is to be integrated into public policy processes. A truly pragmatist approach to environmental policy cannot be dominated by the descriptive, snapshotlike aspects of science. The context, and the problems that constitute it, will ensure that values will be injected into the process. And the meaning given to scientific information is largely determined by its bearing upon disagreements that affect management decisions. Once we, following Quine, see our responses to new and unexpected data as an impact upon a belief system understood as a force field, it is necessary to admit that scientists' values, and their experience of values, will affect the choices that they make when faced with new data.

Quine has acknowledged that science can advance even when its practitioners do not follow an exact, deductive, sentential logic whereby evidentiary sentences logically entail that some particular belief must be changed. He advocates a "naturalistic" epistemology, according to which we collapse the question of how we justify our beliefs into the more tractable, empirical question of how we learn how to learn. But, I submit, anyone not embodying a positivist, proscientific bias would be forced, rationally, to include human purposes, goals, and values as important considerations of the "natural" logician. To answer "pragmatic" questions, such as the choice of which structure to impose upon a problem (what scientific model to use), one must be willing to provide normative as well as descriptive reasons and justifications. This is quite natural in sciences like medicine and dentistry. The criteria of good science in these normatively based areas of study require that science be relevant to curing sickness and discomfort, as well as being replicable. But relevance requires a normative judgment as to what is important, and in environmental policy discussions, these discussions of importance can involve important value commitments. Science, when undertaken in a real situation with a defined "problem," is more focused. Pragmatism places science in a larger context in which scientists are actors and communicators. However much scientists value their "objectivity," doing science in a context where real management decisions hang in the balance necessarily thrusts social values into scientific debates.

Much more has been said on these points throughout the book; for now it is important simply to recognize that rejection of the Cartesian worldview and epistemology, and along with it the modernist concept of objectivity, does not entail skepticism. In the process of rejecting Cartesianism and embracing pragmatism, one loses certainty, but one does not lose all ability to rank truths according to their likelihood. This postmodern conception of objectivity, then, admits that the world is contingent, but it nevertheless locates an external check—in the form of a community, including people with different perspectives and interests—on our subjective beliefs. This community-based approach to supporting our beliefs relies upon deliberation and disagreement to identify weak links in the web of our common beliefs as its participants seek ways to reach agreement on what to do within a diverse community of inquirers.

A.4.4 Logical Empiricism as a Subdiscipline of Pragmatism

Pragmatism, on the understanding we have developed, should be thought of as encompassing and incorporating a revised positivism, rather than vice versa. Carnap, until his death, considered himself a logical empiricist; in his later years he would have no doubt described himself as a pragmatic logical empiricist. Positivism/logical empiricism, once revised as necessary given the arguments summarized in this appendix, however, should be considered one aspect of a broader philosophical pragmatism. This is true because the arguments of Peirce and Quine are far more general—they apply to all experience and to all symbolic behavior—whereas the logical empiricists concerned themselves professionally only with the languages of science. Peirce recognized that scientific language is not somehow prior to other uses of language; it is one specialized use of language.

Keeping this relationship clear is important because language, in its most basic and general occurrences—in ordinary situations in which humans are cooperating and communicating in order to achieve a common purpose—is highly contextual. One way of thinking of the whole Cartesian project, and the logical empiricists' last-gasp effort to mimic the certainty Descartes claimed, is that it posited a "non-contextual" world, a world that exists independent of human conception, values, and objectivity. And for Cartesians and their opponents, this idea of a noncontextual reality, a reality that would shape every human action, enforced a univocal sense of "objectivity." To be objective, a belief must correspond to a bit of the unitary, prelinguistic, human-independent reality that could, in principle, coerce our beliefs into a unitary, objective belief system. Pragmatism, however, which treats science as one human activity among others, recognizes that all activities take place in contexts in which humans express and act on values and in which certain assumptions are made. Given this dynamic, action-based approach to deliberation and knowledge, it is expected that in specific situations, assumptions will be made that would be disallowed in other contexts. Thus, the idea of a single body of knowledge that somehow "represents" a single reality is rejected as a model for science. So although some aspects of positivism may still remain viable as approaches to science, any special epistemological requirements placed on science must be justified within a broader action-based, pragmatic philosophy. In this sense, positivism becomes a contributor to a broader pragmatic philosophy, rather than vice versa.

Pragmatism, though diverse even today, is unified by allegiance to Peirce's three starting points. First, Peirce believed that the only source of information by which to improve our belief system is experience; he had no use for a priori pronouncements that tell us how the world must be, independent of experience. The shape of the world, the dynamic Darwinian world we actually live within and experience, is not substantively given, independent of experience; it emerges at the intersection of a culture and its physical milieu. That culture is made up of social and political beings, beings who can use language to communicate and to record and pass on culture, and sometimes to make collective decisions. But in a constantly changing world without fixed essences, language and communication will reflect both the physical milieu and the ideas, values, and ideals of the culture located at that intersection. In this sense, Peirce was the first true empiricist, since he recognized that experience itself is constructed and that the building blocks of knowledge need not reflect a preexisting structure implicit in nonhuman nature.

Second, and this is perhaps Peirce's greatest intellectual achievement, Peirce advanced his obscure but prescient theory of signs. We have discussed Peirce's theory of the semiotic in more detail in the main text; here it is important only to realize that the theory of signs established that all linguistic and communicative activity is necessarily social in nature.

Third, Peirce went beyond the individualism that made truth a relationship between a lone asserter and a corresponding fact, and he made truth emergent within a community that uses language to communicate and compare experience. Rather than seeking truth and justification in correspondence of sentence with fact, Peirce saw that the conventions of speech and the processes of deliberation take on meaning in a community in which individuals and groups act, compete, and make deals. Peirce believed that, given our ability to gather more experience and to learn from experience by gradually weeding out error through the gradual development of better methods for approximating truth, skepticism could be avoided. What is conspicuously absent from this theory of meaning, communication, and "validation" of assertions is any reference to an independent reality, of things or preexperiential categories that exist beyond experience and validate assertions.

The epistemological power and advantages of Peirce's three tenets and starting point are discussed in section 3.4, where a working epistemology that rests squarely on Peirce's three-legged stool is developed. Here, suffice it to say that Peirce, though a difficult writer and a terrible marketer of his ideas, had by early in the twentieth century identified a path beyond the Cartesian conception of the epistemological problem, a conception that had turned modern philosophy into an ongoing flip-flop between certainty and skepticism. Unfortunately, most philosophers did not immediately follow Peirce's directions, and the idea that assertions cannot be judged independently of the social milieu and the "paradigm" in which they are made lay fallow until it was rediscovered in midcentury.

Peirce's way forward, augmented by a Carnapian commitment to "tolerance" for new conventions and new ways of speaking, prepares the way for a new philosophy of language. When it is put to work in the context of environmental problems, where failures of communication create situations of doubt and confusion, this philosophy of language may provide some tools to improve deliberation and decision making on environmental policy. This process, and the process of articulating, deliberating, and choosing common goals, can be the occasion of increased scientific understanding of environmental problems; but also, and especially, it can contribute to increased understanding of ways in which to articulate and measure values and to develop visions of new and possible, but not yet grasped, relationships between human communities and their physical contexts. What we have learned in this appendix is that in postmodern philosophy, the physical context cannot be treated independently of human conceptions of it. To understand the reality in which environmental problems unfold, one must understand physical relationships from within a particular, human perspective.

A.5 Pragmatism and Environmental Policy

Pragmatism today is in the midst of a huge revival, both in academic philosophy and in applications to social problems. Many leading philosophers now espouse one or another form of pragmatism. We already saw how Wittgenstein, the Oxford analysts, and logicians such as Quine embraced versions or aspects of pragmatism by the middle of the twentieth century, a trend that accelerated as the century progressed. Most current lists of "top-ten" philosophers in America would include a half dozen self-avowed pragmatists and others with pragmatist leanings. Leading names include, in the United States, Quine of course, recently deceased after an active career into his nineties, but also Donald Davidson, Hilary Putnam, Richard Rorty, Richard Bernstein, Nancy Fraser, Cornel West, Nicholas Rescher, and Ian Hacking. These are all philosophers who have made significant marks on the discipline, and all of them acknowledge a major debt to earlier pragmatists. In Europe the influence of pragmatism is spreading both in ethical theory and in philosophy of science and epistemology, as is discussed briefly in chapter 7. As the movement has blossomed and diversified, it has become more and more difficult to state any set of shared principles that unify pragmatists, but that's nothing new—Peirce, James, and Dewey disagreed robustly among themselves.

Perhaps this diversity is not surprising because, as noted, the twentieth century was mainly preoccupied with which questions to ask, the old, Cartesian ones, or a new set of questions that break free from the strictures of rationalist epistemological assumptions. Once those matters were sorted out and now that the pragmatists' arguments have begun to carry the day, it has become possible at the dawn of a new century to address constructively the questions that pragmatism identifies as legitimate and important philosophical questions. Two broad types have emerged as important. First, there are philosophical questions about language, especially linguistic choices that, once made, structure our thinking. So pragmatists retain a strong interest in linguistic construction and linguistic choice; and inspired by Peirce's approach to semiotics, pragmatists undertake the study of language without the straitjacket embodied in the modernist assumption that we must find a single language that describes the world as it *must* be. Accordingly, pragmatists have contributed by clarifying linguistic forms through analysis and by development of new and experimental languages that are proposed as useful for given purposes.

Second, pragmatists have every incentive, given the emphasis on practice in their philosophy, to pay attention to real-world problems. From the pragmatist theory that language finds its most basic expression when human beings use it collectively to solve problems affecting survival or well-being, it follows that a hands-on, community-based, teamwork approach to solving problems, an approach that emphasizes model-building as a tool of integration and communication, should be the philosopher's laboratory. Each such problem has its own social context, which often affects how facts are gathered and whose values will get priority in the face of conflicts of interest. If pragmatism is to be true to its theory, then, it must manifest itself in practice. And if the pragmatists are right about the role and possibilities of philosophy, it seems to follow that pragmatist philosophers must do much of their work in the trenches, often at a local level, where social problems are identified and addressed.

Pragmatism is not hostile to theory, though such a charge is often leveled against it; discussions of theoretical questions is welcomed. What has changed is that the arguments of pragmatism have reversed the flow of premises to conclusions. As Dewey noticed, in a pre-Darwinian conception of knowledge and reality, philosophers were encouraged to start with the big questions, using a priori reasoning, and then to "apply" this general reasoning in particular contexts. If we take the arguments of pragmatism seriously, the flow of argument from theory to application must be reversed; theory must emerge from many specific practices.

The differences among pragmatists regarding how to treat problems of objectivity remain a controversial subject; some would say, for instance, that neopragmatists such as Richard Rorty have gone too far toward relativism and subjectivism. Rorty's differences from other pragmatists are representative of a philosophical rift that had already begun showing between Peirce and William James. The disagreement centered on the possibility of using the pragmatists' methods to arrive at a single, univocal reality. This matter had an important impact on how pragmatists responded to big questions such as belief in God or free will. James, impressed with the extent to which available data always underdetermines such questions, expected each individual to decide these questions in a way that was consistent with available empirical evidence; but he also expected that, since key evidence is always missing, real human individuals would still accept such hypotheses, keeping an open mind, perhaps, but acting on hypotheses that went considerably beyond available data. James said: "We have a right to believe at our own risk any hypothesis that is live enough to tempt our will." He stated his hypothesis more precisely as follows: "Our passional nature not only lawfully may, but must, decide an option between propositions, whenever it is a genuine option that cannot by its nature be decided on intellectual grounds; for to say, under such circumstances, 'Do not decide, but leave the question open,' is itself a passional decision—just like deciding yes or no—and is attended with the same risk of losing the truth."[34] In other words, James was emphasizing that in day-to-day life it makes a difference in behavior, at least, whether we believe in God or in free will, for example, but it is also impossible to resolve such questions on strict rules of evidence applied to available information. James thus emphasized the ways in which action-oriented contexts will induce us to act on hypotheses even when experiential evidence is far from adequate, no matter how much of an empiricist we are. Belief is "forced" in the particular situation.

Peirce, in contrast, argued that, given an indefinitely long time for the community of truth-seekers to apply the method of experience to all questions, and eventually to submit them to experimental test, the community would arrive at a unitary truth. To put the point simply, Peirce, who emphasized the "logic" of ongoing experience, associated truth with the outcome of an indefinite process involving an ongoing community of inquirers, whereas James emphasized the particularities and uncertainties inevitably faced by individuals who happen (as we all in fact must do) to have to act on the inadequate evidence available to them today. Accordingly, Peirce was able to call himself a realist by identifying truth with the hypothetical endpoint of an indefinite process; James's emphasis seemed to invite a more pluralistic variety of pragmatism, recognizing the potential for different individuals and groups, living in very different situations, to develop belief sets peculiar to their own situation and unlikely to converge with those of others who believe and act in very different situations.

Rorty, like James, emphasizes the problem of reconciling individual experience—which nobody can dispute is highly idiosyncratic—with an ideal of truth. Rorty argues, based on arguments similar to those I have rehearsed above, that we must give up the Cartesian concept of a reality beyond perception, that we must, today, choose between "objectivity" and "solidarity." Objectivity rests upon a concept of truth as correspondence. Solidarity, by contrast, refers to a process of justifying oneself, criticizing others, and reaching common agreement—or at least tolerance of disagreements—as members of a shared community. Rorty thinks that to continue to insist on realism and objectivity in the postmodern period is pointless: "So it is hard to see what difference is made by the difference between saying 'there is only the dialogue' and saying 'there is also that to which the dialogue converges.'"[35]

Rorty says that today we must accept the Jamesian version of pragmatism and place less importance on the convergence of individuals and groups coming from diverse, ethnocentric beginning points. There remains at present, then, somewhat of a split among pragmatists, between Rorty and those who, like Hilary Putnam, would describe themselves as realists. Rorty emphasizes the importance of dialogue, communication, and the substitution of "civilized" (nonviolent, open, etc.) discourse for violence and intimidation, whereas the latter group sees the pragmatic method as capable of establishing a progression, of creating a more and more inclusive experiential basis for our expanding set of shared beliefs. This is the same split that, we just saw, separated Peirce and James. The argument comes down to whether one thinks of increasing experience as pushing human belief systems—taken as gradually evolving within increasingly inclusive communities of inquirers—to converge toward a single interpretation, or whether one believes that people and cultures, beginning from separate starting points, will continue to develop and grow in diverse directions.

It is evident in this book that I side with Putnam in believing there can be a reasonable sense of truth as a directional ideal, that set of beliefs toward which the inclusive community of inquirers would trend, given indefinite time and experience, but Rorty's work shows that some participants in the pragmatic tradition do not share this ideal. Fortunately, we do not have to resolve this issue in order to use the pragmatist epistemology in the project of this book. One advantage of the community-based, bottom-up approach to practical problems adopted here is that we do, as Rorty says, start where we are and try to improve communication and agreement within our community. If our interests turn out to intersect the interests of a neighboring community—as they so often do in environmental problems, which do not respect conventional political boundaries—then we should try to engage the other community in dialogue and attempt to create a larger communicative community to address common problems. This process is open-ended, and progress is measured by the plausibility of the stories we tell about our common interests. It is a process that can be ongoing and important, regardless of whether it is viewed as having a single, ideal endpoint. The controversy about the endpoint addresses issues left over from modernism's concern with skepticism; but the people with whom we are dealing, people already engaged in community processes and cooperative action, have already bought into their community and share at least some of that community's beliefs about reality.

There is another advantage of this approach to studying environmental problems by starting at the local level. If we set agreement that results from open discussion and deliberation within a somewhat cohesive community as our standard for choosing environmental policies, then the shared values and aspirations of that community become a resource for building coalitions and support for improved policies. As is shown in chapter 5, the very agreement of diverse stakeholders to meet and seek a mutually acceptable course of action—

their commitment to solve problems cooperatively—already generates the raw material for a working consensus. Whatever one believes about the convergence of various cultures upon a single descriptive account of nature, the pragmatic approach to building upward from local perspectives emphasizes commitment and shared values: solidarity; and this is all that is required for our philosophical analysis to proceed. As long as we can work within a community that commits to solving problems cooperatively through deliberation and democratic decision making, we have enough solidarity to begin the pragmatic analysis of existing goals, existing beliefs, arguments, and disagreements and to begin experimentation to reduce uncertainty. An important part of that deliberation is development and experimentation with new forms of language, a process that involves building locally based models that are problem-oriented. Building such models can be an important contribution to the improvement of communicative tools for discussing cooperative action.

The problem of convergence of belief systems still receives a lot of attention from the critics of pragmatism, who often appeal, explicitly or implicitly, to an "objective" reality "out there" beyond any human experience and accuse pragmatists of being "relativistic" by comparison. The pragmatic approach, based on expanding experience through an experimental method, does not answer to the Cartesian categories; it seeks convergence not in abstract belief systems but rather in deliberation and action. When the goal is cooperative management of shared problems, consensus on what to do is more important than consensus about why it is important to do it. If one must—and one usually must—act cooperatively in a situation with diversity of beliefs and values, the goal should be to seek robust policies, ones that apply across a fairly wide range of opinions, and ones that fulfill a variety of preferences and aspirations.

It is much more accurate to describe pragmatists as contextualists than as relativists. What this means is that the explorations and deliberations of pragmatists are undertaken within a given context, with special attention to particular situations and the problems and possibilities that present themselves in that context. This means that most assertions and beliefs are already tied to significant facts and to value commitments that are extant in the community. Assertions are examined and judged in a particular community context; no claims are made that the assertions are corollaries of universal or necessary truths. But contextualism is consistent with careful attempts at community expansion and a drive for larger communities based on broader solidarities. Contextualism states a starting point, but no endpoint, for inquiry. The question of endpoint is left open for the future to unravel.

The point is that regardless of whether a convergent endpoint of the process is posited, there are sufficient shared intellectual and experiential resources available in a particular community to allow dialogue, deliberation, and progress toward shared goals. Contextualism is a commitment to start with particular problems at a local level and work toward solutions within relevant local communities. If some communities find solutions and, in the process, propose hypotheses of broader application, then these will be tested in other and larger contexts. In this process one expects, given careful use of observation and experiment, that generalities will emerge. Whether one needs to be able to posit a single endpoint for the process, however, affects nothing, practically, in the way we proceed. It is a perfect example of what the pragmatists refer to as a question that can make no difference in actual situations.

In this appendix I have tried to tell a positive story about linguistic philosophy, arguing that although linguistic philosophers did not at first use their tools to address the right questions, they did sharpen their tools throughout the twentieth century. The resurgence of pragmatism has refocused philosophy on the questions Dewey asked in 1910, questions about how we can come to understand our world as structured, given that our linguistic categories and concepts are all conventional. I am suggesting that philosophy could contribute importantly to a number of questions that frustrate progress in dealing with social problems, especially environmental problems. If it is to do so, however, I believe it must come down from its ivory tower, tools of logical and linguistic analysis in hand, and address real communication problems that emerge in real situations.

Speaking more specifically, I see philosophers as having an active role in analyzing current environmental discourse and in proposing new descriptive and prescriptive models that bridge hitherto separate disciplines, such as ecology, economics, toxicology, and so forth, working as "conceptual engineers" at the boundaries of these disciplines. The goal, then, is for philosophy to function to encourage the development of a coherent and integrated framework for discussing and evaluating problems, and to do so by applying special skills in logic and linguistic analysis at the boundaries of scientific and policy disciplines. Philosophical pragmatists must, on this view, take an active attitude toward understanding the role of language in current discussions of environmental policy and, especially, toward choosing a new language and proposing new forms of speech that will improve communication in real environmental debates. The goal is to shed light on that nexus where information from the sciences, social values, and political constraints all get factored into a decision. This goal, however, is not pursued for its own sake. It is pursued because language shapes the world we encounter, understand, and manage, and thus it is a powerful force in public discourse about what to do. Indeed, our language deeply affects what we see as a problem and how we formulate problems for public discussion and deliberation. The goals of pragmatist analyses of language and its uses are to improve public communication about environmental problems, reduce confusion and misunderstanding, promote cooperation, and improve the process of environmental policy formation and implementation in real contexts involving disagreements about what to do. Pragmatism's emphasis on praxis ensures that intelligence will be understood as primarily instrumental, encouraging convergence on an *action* designed to protect the community and the interests of its members. Cooperative action is consistent with considerable disagreement about specific beliefs and specific values.

What we have seen is that twentieth-century philosophy was not so much a period of abdication as a period of confusion, confusion mainly about what questions philosophers can answer and what tools are appropriate to address those questions. I have argued that however indirect the route from Peirce's breakthrough understanding of inquiry as a symbolic activity and his recognition of the importance of signs and conventions, linguistic philosophy worked its way back to Peirce, providing a new way forward for philosophy and for applications of philosophy to real-world problems. Peirce was imperfectly understood, and it took nearly a century for philosophers to ask again the question to which Peirce had an answer, the question of how the world is structured in the absence of fixed essences and categories. If philosophers will finally reject the last vestiges of Cartesianism, the field can develop in a new direction, one that is *both* science-oriented *and* goal-oriented.

Most discussions of integration in environmental policy assume that a major task will be to survey the important disciplinary fields, such as ecology, economics, hydrology, and so forth, and then to somehow integrate this factual material into a satisfactory "model" of the environment and of the processes that create problems. If the argument of this appendix is correct, this approach is backward. The first step is to pose environmental problems as problems of cooperative behavior within human communities. The sciences, then, become instrumental disciplines—what I have called mission-oriented science—and their goal is to ascertain answers to questions that emerge as disagreements and uncertainties in the management process.[36] This management process is embedded in a larger social process. At all levels, however, in this age of postfoundationalism, we know that language is important, both because our language structures the world we experience and because language is the tool of both communication and public deliberation.

A.6 Philosophy's Role: An Epilogue

We should discuss one more consequence of the rejection of foundationalism in favor of the pursuit of a more concrete, local, and contextual kind of truth.

It concerns the question, alluded to several times above in our history of twentieth-century philosophy, of whether philosophy has some unique method or some special "transcendental" access to reality, a method not used by scientists and laypersons. Rorty points out that it is characteristic of societies such as our own Western society, with Greek intellectual roots, to believe the realist assumption that there is one and only one correct set of beliefs against which all of our current, temporary beliefs must ultimately be matched, and that there is access to that reality.[37] Rorty notes that it is *philosophers* who consider themselves, of all people in the culture, those called to discover this reality, creating what Rorty refers to as a self-appointed "priesthood." The claim of philosophers to privilege in this case is suspect, however, because it can be supported only if foundationalism itself is supportable. The foundationalist, as defined above, believes that we can identify certain sentences or bits of information that can be known for sure and that these beliefs, together with deductive connections, can provide support for the remainder of our belief system. We have already seen how this view is intimately connected with the idea of essentialism, which posits a basis in reality for the categories we use to describe our experience. It is belief in these real categories that makes possible the claim of the priesthood to have special access to them. If one gives up foundationalism and its atomistic approach to sentences and their truth, and adopts a Quinean web-of-belief interpretation of empiricism, then it is apparently impossible to maintain the view that philosophers have a special status in the search for better answers. Like other participants in the process, their interpretations of experience can and should be reported, their opinions should be considered, and their views should have the same weight as those of others who participate in public deliberation about what to do. This is not to say that philosophers have no special skills. They can act as specialists in interpretation, communication, and language, especially as these factors structure and give order to the world within communities and communicative situations.

But philosophers cannot, in the postfoundational era of philosophy we have now entered, claim any special source of knowledge, beyond experience and the interpretation of experience. Once the pretense of special, a priori knowledge is given up, philosophy shifts ground to the interstices of the disciplines, weighing the conceptual power of various disciplinary languages and models and examining apparent incommensurabilities across disciplines. If they are willing to immerse themselves in one or more specific disciplines—studying the "philosophy of" various scientific and humanistic disciplines—philosophers can in fact make significant contributions. The tools developed in the fields of analysis of syntax, semantics, and pragmatics allow us to characterize languages according to their assertive power and show equivalencies across disciplines and their models when they exist.

Philosophers can thus find a role in clarifying concepts within and across disciplines, by examining the special conceptual demands of an integrative, activist science and by understanding all of these as taking place within a larger political context in which stakeholders and the public discuss, deliberate, and lobby for their favored policy options. Pragmatist philosophers contribute not by overriding other participants in public deliberation by appeal to self-evident truths, but by exercising a method of careful use of language and of careful analysis of the relationship of data to hypothesis. Philosophers also should pay special attention to the methods that are used in various disciplines and to the development of better methods for the pursuit of agreement and consensus. These are the questions, seldom examined by practitioners of the special sciences, where social values, the social sciences, and natural sciences all come together in the formation of policy.

What is interesting about the field of professional philosophy is that despite the apparent emergence of pragmatism as the leading edge of philosophy at the turn of the twenty-first century, and despite the apparent implication that philosophers should refocus their attention from general truths pursued with a priori methods to particular social problems that emerge in real communities, contemporary philosophers have been very negative about attempts to engage real problems in real contexts. In particular, philosophers in the best philosophy departments denigrate what they call "applied" philosophy and devote few positions to it. The implication of this description-turned-epithet is to suggest that there is a special type of philosopher, a "pure" philosopher, who, using special methods, derives general truths. It is a lesser calling, those philosophers seem to think, to actually put such general ideas into application in particular contexts where communities face real problems—a task that can be left to nonphilosophers.

The first thing we can say about this attitude is that, like the old commitments to objectivity and to a priori categories of being, this attitude embodies a vestige of the commitment to a Cartesian worldview. True pragmatists who reject foundationalism also reject the reference to "applied philosophy" as misleading, since there are no a priori philosophical principles to "apply." Pragmatists thus prefer to refer to their work as practical philosophy. Practical philosophy refers, most basically, to a problem orientation and a commitment to address social problems within real contexts, where people have and express real values and where disagreements make a difference in real choices about how to act.

Much of the main text of this book explores heuristics that are designed to encourage those involved in community-based environmental management to ask the right questions and to expand the range of disagreements that can be submitted to empirical test. Practical philosophy does not claim to bring general, substantive principles about reality or right action to the negotiating table to apply to a local case; the practical philosopher instead tries to pay attention to the particular facts of particular contexts and to work within the broad assumptions agreed upon by activists on all sides. One important role, then, for philosophers is to tease out the logical and inferential relationships that animate public debate about what to do in particular, local contexts. In these contexts, it is of the utmost importance to know what is agreed upon by all parties and how various disagreements interact with each other to block, and in some cases to encourage, coalitions. Hopefully, examination of the case studies spread throughout the book will help the reader see how philosophical analysis and attention to language and semantic relationships may lead to progress in understanding and solving environmental problems in particular situations.

NOTES

Chapter 1

1. United States Environmental Protection Agency, *Reducing Risk: Setting Priorities and Strategies for Environmental Protection* (Washington, DC: Environmental Protection Agency, 1990).

2. For details, see United States Environmental Protection Agency, *Ecological Risk Assessment Issue Papers,* Office of Research and Development (EPA/630/R-94/009), November 1994 (Washington, DC: Risk Assessment Forum, U.S. Environmental Protection Agency, 1994); and United States Environmental Protection Agency, *Peer Review Workshop Report on Ecological Risk Assessment Issue Papers,* Office of Research and Development (EPA/630/R-94/008), November 1994 (Washington, DC: Risk Assessment Forum, U.S. Environmental Protection Agency, 1994).

3. Glenn Suter II, *Ecological Risk Assessment* (Boca Raton, FL: Lewis, 1993), v.

4. Ibid., 21.

5. Ibid., 22.

6. Ibid.

7. Marc Landy, Marc Roberts, and Stephen Thomas, *The Environmental Protection Agency: Asking the Wrong Questions from Nixon to Clinton,* exp. ed. (New York: Oxford University Press, 1994), 283.

8. A. M. Freeman, *The Measurement of Environmental and Resource Values: Theory and Methods* (Washington, DC: Resources for the Future, 1993), 485.

9. Landy, Roberts, and Thomas, *The Environmental Protection Agency.*

10. Samuel P. Hays, *Beauty, Health, and Permanence: Environmental Politics in the United States, 1955–1985* (Cambridge: Cambridge University Press, 1989).

11. Landy, Roberts, and Thomas, *The Environmental Protection Agency;* Hays, *Beauty, Health, and Permanence.*

12. Landy, Roberts, and Thomas, *The Environmental Protection Agency,* 23. The following account summarizes the more detailed description by Landy and colleagues.

13. Ibid., 10.

14. Ibid.

15. Ibid.

16. Ibid., 11.

17. The proposals of the Ecosystem Valuation Forum are summarized in G. Bingham, R. Bishop, M. Brody, D. Bromley, E. T. Clark, W. Cooper, R. Costanza, T. Hale, G. Hayden, S. Kellert, R. Norgaard, B. Norton, J. Payne, C. Russell, and G. Suter, "Issues in Ecosystem Valuation: Improving Information for Decision Making," *Ecological Economics* 14 (1995): 73–90.

18. National Research Council, Committee on Mitigating Wetland Losses, *Compensating for Wetland Losses under the Clean Water Act* (Washington, DC: National Academy Press, 2001).

19. Ibid., 6.

20. Dennis M. King, "Using Ecosystem Assessment Methods in Natural Resource Damage Assessment," report to the U.S. Department of Commerce, National Oceanic and Atmospheric Administration, Damage Assessment and Restoration Program, Silver Spring, MD, January 31, 1997, 4–5.

21. W. J. Mitsch and J. G. Gosselink, *Wetlands,* 2nd ed. (New York: Van Nostrand Reinhold, 1993), 542.

22. Ibid., 565.

23. Dennis M. King, "Wetland Values, Wetland Mitigation, and Sustainable Watershed Management," unpublished manuscript, 1997.

24. Mitsch and Gosselink, *Wetlands,* 535; Mitch and Gosselink are paraphrasing E. P. Odum, in "The Value of Wetlands: A Hierarchical Approach," in *Wetland Functions and Values: The State of Our Understanding,* ed. P. E. Greeson, J. R. Clark, and J. E. Clark (Bethesda, MD: American Water Resources Association, 1979), 1–25.

25. Dennis King, "Comparing Ecosystem Services and Values," report to the U.S. Department of Commerce, National Oceanic and Atmospheric Administration, Damage Assessment and Restoration Program, Silver Spring, MD, January 12, 1997, 6–7.

26. Mitsch and Gosselink, *Wetlands,* 508.

27. King, "Comparing Ecosystem Services and Values," 6.

28. King, "Wetland Values and Sustainable Watershed Management."

29. King, "Comparing Ecosystem Services and Values," 7.

30. King, "Using Ecosystem Assessment Methods," 4–5.

31. J. B. Zedler, "Ecological Issues in Wetland Mitigation: An Introduction to the Forum," *Ecological Applications* 6 (1996): 36.

32. C. A. Simenstad and R. M. Thom, "Function Equivalency Trajectories of the Restored Gog-Le-Hi-Te Estuarine Wetland," *Ecological Applications* 6 (1996): 38.

33. Aldo Leopold, *A Sand County Almanac and Sketches Here and There* (Oxford: Oxford University Press, 1949), 129–33.

34. National Research Council, *Compensating for Wetland Losses,* 159.

35. Ibid.

Chapter 2

1. K. Arrow, B. Bolin, R. Costanza, P. Dasgupta, C. Folke, C. S. Holling, B. Jansson, S. Levin, K. Maler, C. Perrings, and D. Pimentel, "Economic Growth, Carrying Capacity, and the Environment," *Science* 268 (1995): 520–21.

2. For a more detailed explanation of this point, see Bryan Norton, "Resilience and Options," *Ecological Economics* 15 (1995): 133–36.

3. Sagoff's talk has been published as "The Hedgehog, the Fox, and the Environment," in *The Moral Authority of Environmental Decision Making,* ed. J. M. Gillroy and J. Bowersox (Durham, NC: Duke University Press, 2002), 262–75, quotation on 262.

4. Ibid., 262.

5. Ibid., 263.

6. I argue for this conclusion at length in *Toward Unity among Environmentalists* (New York: Oxford University Press, 1991).

7. Ibid.

8. Max Oelschlager, *The Idea of Wilderness* (New Haven, CT: Yale University Press, 1991), 214.

9. I am here summarizing the fascinating account by Alan Menand in *The Metaphysical Club* (New York: Farrar, Straus and Giroux, 2002).

10. Quoted in ibid., 230.

11. Charles S. Peirce, "How to Make Our Ideas Clear," in *Pragmatism: A Reader,* ed. L. Menand (New York: Vintage Books, 1997), 26–48.

12. Oliver Wendell Holmes, "The Path of the Law," in *The Mind and Faith of Justice Holmes,* ed. Max Lerner (New York: Modern Library, 1943 [originally published in 1897]), 71–89, quotation on 71–72.

13. Ibid., 80.

14. Ibid., 83.

15. Ibid., 85.

16. Ibid., 87.

17. Morris Hadley, *Arthur Twining Hadley* (New Haven, CT: Yale University Press, 1948), 197.

18. Aldo Leopold, "Some Fundamentals of Conservation in the Southwest," *Environmental Ethics* 1 (1979): 140.

19. Ibid., 140–41.

20. Aldo Leopold, "The Civic Life of Albuquerque," unpublished lecture, 1.

21. Arthur Twining Hadley, *Some Influences in Modern Philosophic Thought* (New Haven, CT: Yale University Press, 1913), 122.

22. Ibid., 125.

23. Ibid., 126.

24. Ibid., 129.

25. Ibid., 130.

26. Ibid., 131.

27. Leopold, *Sand County Almanac,* 202.

28. Ibid., 203.

29. Henry Plotkin, *Darwin Machines and the Nature of Knowledge* (Cambridge, MA: Harvard University Press, 1993).

30. See Anthony Weston, "Beyond Intrinsic Value: Pragmatism in Environmental Ethics," *Environmental Ethics* 7 (1985): 321–39.

31. Bryan Norton, "Conservation and Preservation: A Conceptual Rehabilitation," *Environmental Ethics* 8 (1986): 195–220.

32. John Dewey, *The Influence of Darwin on Philosophy and Other Essays* (New York: Henry Holt, 1910).

33. John Dewey, *Logic: The Theory of Inquiry* (New York: Henry Holt, 1938), 19.

34. Ibid., 43.

35. Ibid., 47.

36. Ibid., 20.

37. Ibid., 66.

38. Ibid.

39. Maarten A. Hajer, *The Politics of Environmental Discourse: Ecological Modernization and the Policy Process* (Oxford: Clarendon Press, 1995), 26.

40. Ibid., 282n.

41. Ibid. Hajer's reference is to Ulrich Beck, *Politique in der Risikogesellschaft* (Frankfort am Main: Suhrkamp, 1991).

42. Hajer, *Politics of Environmental Discourse,* 58–59.

43. Cf. ibid., 280f., 293.

44. Ibid., 69f.

45. Aldo Leopold, "A Biotic View of Land," *Journal of Forestry* 37 (1939): 727.

Chapter 3

1. Aldo Leopold, quoted in *The Essential Aldo Leopold: Quotations and Commentaries,* ed. C. Meine and R. Knight (Madison: University of Wisconsin Press, 1999), 211. My essay in the same volume, "Leopold as Practical Moralist and Pragmatic Policy Analyst," develops the argument that Leopold should be thought of as the first adaptive manager.

2. Larry Hickman, "Nature Meets Culture: John Dewey's Pragmatic Naturalism," in *Environmental Pragmatism,* ed. A. Light and E. Katz (London: Routledge, 1996), 50–62.

3. C. S. Holling and Lance Gunderson, "Resilience and Adaptive Cycles," in *Panarchy: Understanding Transformations in Human and Natural Systems,* ed. C. S. Holling and L. Gunderson (Washington, DC: Island Press, 2002), 31–32.

4. Bruce Hannon and I have developed the argument summarized here in more detail in "Environmental Values: A Place-Based Theory," *Environmental Ethics* 19 (1997): 227–45; and in "Democracy and Sense of Place Values," *Philosophy and Geography III: Philosophies of Place* 3 (1998): 119–46.

5. This is the approach of Kai Lee in *Compass and Gyroscope* (Washington, DC: Island Press, 1993), which describes adaptive management as a negotiation within a political process. It should be noted that some who legitimately call themselves adaptive managers have a somewhat narrower conception, treating "learning" in a narrower, technical sense of updating subjective probability judgments using Bayes's rule. See, for example, Donald Ludwig, "Missed Opportunities in Natural Resource Management," *Natural Resource Modeling* 8 (1994): 111–17.

6. Aldo Leopold, "A Biotic View of Land," *Journal of Forestry* 37 (1939): 728.

7. M. Granger Morgan and Max Henrion, *Uncertainty: A Guide to Dealing with Uncertainty in Quantitative Risk and Policy Analysis* (New York: Cambridge University Press, 1990).

8. Sylvio O. Funtowicz and Jerome R. Ravetz, *Uncertainty and Quality in Science for Policy* (Dordrecht, Netherlands: Kluwer, 1990).

9. Malte Faber, Reiner Manstetten, and John Proops, "Toward an Open Future: Ignorance, Novelty, and Evolution," in *Ecosystem Health: New Goals for Environmental Management,* ed. Robert Costanza, Bryan G. Norton, and Benjamin D. Haskell (Washington, DC: Island Press, 1992), 72–96.

10. Charles S. Peirce, *Collected Papers* (Cambridge, MA: Harvard University Press, 1960), 5:565, 5:553, 5:407.

11. Henry David Thoreau, *Walden* (New York: New American Library, 1960), 217.

12. Ibid., 70–71.

13. Henry David Thoreau, *The Journal of Henry David Thoreau,* vol. 6, ed. B. Torrey and F. H. Allen (Salt Lake City: Peregrine Smith Books, 1984), 206.

14. Readers will note a considerable similarity between my argument, following, and Richard Norgaard's critique of modernism in *Development Betrayed: The End of Progress and a Coevolutionary Revisioning of the Future* (London: Routledge, 1994).

15. Cf. Aldo Leopold, *A Sand County Almanac and Sketches Here and There* (Oxford: Oxford University Press, 1949), 200.

16. Sylvio O. Funtowicz and Jerome R. Ravetz, "Science for the Post-Normal Age," in *Perspectives on Ecological Integrity,* ed. L. Westra and J. Lemons (Dordrecht, Netherlands: Kluwer Academic Publishers, 1995).

17. Ibid., 206.

18. A. P. Ushenko, *Power and Events* (Princeton, NJ: Princeton University Press, 1946), 21.

19. John E. Smith, *Purpose and Thought: The Meaning of Pragmatism* (Chicago: University of Chicago Press, 1978), 117–18. The following owes much to Smith's analysis; see also 52–53 in his book.

20. Christopher Hookway, *Peirce* (London: Routledge and Kegan Paul, 1985), 43.

21. Dewey drew a somewhat different distinction—between what he called "common sense" and "scientific inquiry." John Dewey, *Logic: The Theory of Inquiry* (New York: Henry Holt, 1938), 114–19. One interesting area for further research would be to explore these differences between Peirce and Dewey and their implications for the logic of policy inquiry.

22. Smith, *Purpose and Thought,* 118.

23. Lee, *Compass and Gyroscope.*

24. Lee addresses these questions in ibid., but I do not find Lee's brief discussions of value formation, expression, and revision convincing. For example, he introduces the ends-means distinction on the way to an explanation of social learning in Dewey (*Compass and Gyroscope,* 105–8), creating a muddle, since rejection of the ends-means distinction is a keystone of Dewey's philosophy.

25. Bryan Norton and Anne Steinemann, "Environmental Values and Adaptive Management," *Environmental Values* 10 (2001): 473–506.

26. Lee, *Compass and Gyroscope;* and P. M. Haas, *Saving the Mediterranean: The Politics of International Environmental Cooperation* (New York: Columbia University Press, 1990).

27. J. Baird Callicott, *In Defense of the Land Ethic: Essays in Environmental Philosophy* (Albany: State University of New York Press, 1989), esp. 165.

28. C. S. Holling, *Adaptive Environmental Assessment and Management* (New York: John Wiley, 1978).

29. Leopold, *Sand County Almanac,* 203.

30. Stuart Kauffman, *At Home in the Universe* (New York: Oxford University Press, 1995).

Chapter 4

1. Horst W. J. Rittel and Melvin M. Webber, "Dilemmas in a General Theory of Planning," *Policy Sciences* 4 (1973): 155–69. I am indebted to my friend and sometimes collaborator E. Nicholas Novakowski for convincing me of the importance of the arguments of Rittel and Webber.

2. Ibid., 161. The page numbers for items 2 through 10 are given parenthetically in the text.

3. Ibid., 161–62.

4. Ibid., 163.

5. Robin S. Gregory, S. Lichtenstein, and P. Slovic, "Valuing Environmental Resources: A Constructive Approach," *Journal of Risk and Uncertainty* 7 (1993): 177–97; R. L. Keeney, "Value-Focused Thinking: Identifying Decision Opportunities and Creating Alternatives," *European Journal of Operational Research* 92 (1996): 537–49; Detlef Von Winterfeldt and Ward Edwards, *Decision Analysis and Behavioral Research* (Cambridge: Cambridge University Press, 1986).

6. Sandra Batie and H. H. Shugart, "The Biological Consequences of Climate Changes: An Ecological and Economic Assessment," in *Greenhouse Warming: Abatement and Adaptation,* ed. N. J. Rosenberg, W. E. Easterling III, P. R. Crosson, and J. Darmstadter (Washington, DC: Resources for the Future, 1989).

7. National Research Council, *Risk Assessment in the Federal Government* (Washington, DC: National Academy Press, 1983).

8. See Dale S. Rothman and John B. Robinson, "Growing Pains: A Conceptual Framework for Considering Integrated Assessments," *Environmental Monitoring and Assessment* 46 (1997): 23–43, for a similar argument using different terminology.

9. C. S. Holling, *Adaptive Environmental Assessment and Management* (New York: John Wiley, 1978); C. Walters, *Adaptive Management of Renewable Resources* (New York: Macmillan, 1986); Kai Lee, *Compass and Gyroscope* (Washington, DC: Island Press, 1993); L. Gunderson, C. S. Holling, and A. Light, *Barriers and Bridges to the Renewal of Ecosystems and Institutions* (New York: Columbia University Press, 1995).

10. Robert N. Proctor, *Value-Free Science?* (Cambridge, MA: Harvard University Press, 1991).

Chapter 5

1. Alan Menand, *The Metaphysical Club* (New York: Farrar, Straus and Giroux, 2002).

2. See Bryan Norton, *Toward Unity among Environmentalists* (New York: Oxford University Press, 1991), chapter 1.

3. Lynn White Jr., "The Historical Roots of Our Ecologic Crisis," *Science* 155 (1967): 1203–7; See J. Baird Callicott, *In Defense of the Land Ethic: Essays in Environmental Philosophy*

(Albany: State University of New York Press, 1989), 229, for an endorsement of White's importance. Note also that Callicott favors the interpretation that I consider unfortunate.

4. John Passmore, *Man's Responsibility for Nature* (New York: Scribner's, 1974).

5. For example, Tom Regan, "The Nature and Possibility of an Environmental Ethic," *Environmental Ethics* 3 (1981): 19–34; and Callicott, *In Defense,* 157, have argued that the field of environmental ethics is limited to the study of intrinsic value in nature.

6. Robert C. Mitchell and Richard T. Carson, *Using Surveys to Value Public Goods: The Contingent Valuation Method* (Washington, DC: Resources for the Future, 1989), 9–10.

7. Ibid., 2.

8. Ibid., 4–5.

9. G. H. McClelland, W. D. Schulze, J. K. Lazo, D. M. Waldman, J. Doyle, S. R. Elliott, and J. R. Irwin, "Methods for Measuring Non-Use Values: A Contingent Valuation Study of Groundwater Cleanup," draft report to EPA, 1992.

10. As of this writing, legal appeals and other complications have protected Exxon from paying these damages.

11. G. H. McClelland, W. D. Schulze, and J. K. Lazo, "Memorandum to the Science Advisory Board Environmental Economics Advisory Committee: Additional Explication of Methods for Measuring Non-Use Values: A Contingent Valuation Study of Groundwater Clean-up," Center for Economic Analysis, Boulder, CO, June 29, 1993, 33.

12. Richard Bishop, "Endangered Species and Uncertainty: The Economics of Safe Minimum Standards," *American Journal of Agricultural Economics* 60 (1978): 1–18.

13. Paul Portney, foreword to A. Myrick Freeman, *The Measurement of Environmental and Resource Values: Theories and Methods* (Washington, DC: Resources for the Future, 1993).

14. Freeman, *Measurement of Environmental and Resource Values,* 493.

15. Arild Vatn and Daniel Bromley, "Choices without Prices without Apologies," *Journal of Environmental Economics and Management* 26 (1994): 129–47.

16. There are a number of different and overlapping reasons that economists do not confront the mythological nature of their claims to comprehensiveness. They include, most prominently, a tendency toward disciplinary imperialism based on a discredited theory that all values are "emotive," noncognitive, and nonrational in nature; hence, economics is the only science that can study human values "rationally." Second, economists have adopted a many-headed theoretical/methodological monster, also mythological, called "consumer sovereignty," which recommends, methodologically, that preferences of economic actors be taken at face value, that we take people's choices as indicative of their true interests. This methodological convenience, unfortunately, has become entangled with a number of other theories of questionable credibility. A full clarification of these issues would carry us far from the present argument—about the comprehensiveness of the "measurables" of environmental economics, and so I here refer the reader to Bryan Norton, "Thoreau's Insect Analogies; Or, Why Environmentalists Hate Mainstream Economists," *Environmental Ethics* 13 (1991): 235–51; B. Norton, "Economists' Preferences and the Preferences of Economists," *Environmental Values* 3 (1994): 311–32; and B. Norton, R. Costanza, and R. Bishop, "The Evolution of Preferences: Why 'Sovereign' Preferences May Not Lead to Sustainable Policies and What to Do about It," *Ecological Economics* 24 (1998): 193–212.

17. Gifford Pinchot, *Breaking New Ground* (1947; repr., Covelo, CA: Island Press, 1987), 323.

18. Heraclitus, it appears, was the first "New Ecologist," emphasizing the importance of change in nature. See Bryan Norton, "Change, Constancy, and Creativity: The New Ecology and Some Old Problems," *Duke Environmental Law and Policy Forum* 7 (1996): 49–70. See also G. S. Kirk and J. E. Raven, *The Presocratic Philosophers* (Cambridge: Cambridge University Press, 1957), 196–97.

19. See Stuart Pimm, *The Balance of Nature? Ecological Issues in the Conservation of Species and Communities* (Chicago: University of Chicago Press, 1991); and Bryan G. Norton, "A New Paradigm for Environmental Management," in *Ecosystem Health: New Goals for Environmental Management,* ed. R. Costanza, B. Norton, and B. Haskell (Washington, DC: Island Press, 1992), 23–41.

20. B. Norton and B. Hannon, "Democracy and Sense of Place Values," *Philosophy and Geography III: Philosophies of Place* 3 (1998): 119–46.

21. J. Baird Callicott, *Beyond the Land Ethic: More Essays in Environmental Philosophy* (Albany: State University of New York Press, 1999), 248.

22. Ilya Prigogene and Isabelle Stengers, *Order out of Chaos: Man's New Dialogue with Nature* (New York: Bantam, 1984).

23. Leading scientists have thus joined Heraclitus, Henri Bergson, and A. N. Whitehead in embracing a process-oriented understanding of natural systems. See Henri Bergson, *Creative Evolution,* trans. A. Mitchell (New York: Holt, Rinehart, and Winston, 1911); and A. N. Whitehead, *Process and Reality* (New York: Macmillan, 1929).

24. See Bryan Norton and Bruce Hannon, "Environmental Values: A Place-Based Theory," *Environmental Ethics* 19 (1997): 227–45; and Norton and Hannon, "Democracy and Sense of Place Values."

25. B. Norton and B. Minteer, "From Environmental Ethics to Environmental Public Policy: 1973–Future," in *The International Yearbook of Environmental & Resource Economics 2002/2003,* ed. Henk Folmer and Tom Teintenberg (Cheltenham, UK: Edward Elgar, 2002).

26. See Kai Lee, *Compass and Gyroscope* (Washington, DC: Island Press, 1993), esp. chapter 3.

Chapter 6

1. Samuel P. Hays, *Beauty, Health, and Permanence: Environmental Politics in the United States, 1955–1985* (Cambridge: Cambridge University Press, 1989).

2. Stanley V. Ciriacy-Wantrup, *Resource Conservation* (Berkeley: University of California Press, 1952); John Krutilla, "Conservation Reconsidered," *American Economic Review* 57 (1967): 777–86.

3. Ben A. Minteer and Robert E. Manning, "Pragmatism in Environmental Ethics: Democracy, Pluralism, and the Management of Nature," *Environmental Ethics* 21 (1999): 191–208.

4. See J. Baird Callicott, *In Defense of the Land Ethic: Essays in Environmental Philosophy* (Albany: State University of New York Press, 1989), 119–20.

5. David Hume, *Treatise on Human Nature,* ed. D. F. Norton and M. J. Norton (Oxford: Oxford University Press, 2000), 302.

6. I thank my former colleague, Andy Ward, for help in understanding and articulating Hume's law and its role in the argument regarding moral naturalism.

7. B. A. O. Williams, *Ethics and the Limits of Philosophy* (Cambridge, MA: Harvard University Press, 1985).

8. Kai Lee, *Compass and Gyroscope* (Washington, DC: Island Press, 1993).

9. William D. Leach, N. W. Pelkey, and P. A. Sabatier, "Making Watershed Partnerships Work: A Review of the Empirical Literature," *Journal of Water Resources Planning and Management* 6 (2001): 378–85; Mark Borsuck, Robert Clemen, Lynn Maguire, and Kenneth Reckhow, "Stakeholder Values and Scientific Modeling in the Neuse River Watershed," *Group Decision and Negotiation* 10 (2001): 355–73.

10. J. Burgess, C. M. Harrison, and P. Filius, "Environmental Communication and the Cultural Politics of Environmental Citizenship," *Environment and Planning A* 30 (1998): 1445–60; Adolph Gundersen, *The Environmental Promise of Democratic Deliberation* (Madison: University of Wisconsin Press, 1995); Mark Sagoff, "Aggregation and Deliberation in Valuing Environmental Public Goods: A Look beyond Contingent Pricing," *Ecological Economics* 24 (1998): 213–30.

11. Herbert Simon, *Administrative Behavior: A Study of Decision-Making Processes in Administrative Organization,* 3rd ed. (New York: Free Press, 1976).

12. Stephen Fox, *John Muir and His Legacy: The American Conservation Movement* (Boston: Little, Brown, 1981), 43.

13. Henry David Thoreau, *Walden* (New York: New American Library, 1960), 146. See Bryan Norton, "Thoreau's Insect Analogies; Or, Why Environmentalists Hate Mainstream Economists," *Environmental Ethics* 13 (1991): 235–51, for a more detailed discussion of this trend in Thoreau's thought.

14. Aldo Leopold, *A Sand County Almanac and Sketches Here and There* (Oxford: Oxford University Press, 1949), 130.

15. Ibid., 132.

16. I have explored Leopold's profound thoughts on the scalar nature of temporal relationships in "Thoreau and Leopold on Science and Values" in my *Searching for Sustainability* (New York: Cambridge University Press, 2003).

17. J. Baird Callicott, "Animal Liberation and Environmental Ethics: A Triangular Affair," in Callicott, *In Defense.*

18. Ibid., 25, 37, 29, 27.

19. J. Baird Callicott, "On the Intrinsic Value of Nonhuman Species," in Callicott, *In Defense,* 148–49.

20. Tom Regan, *The Case for Animal Rights* (Berkeley: University of California Press, 1983), 262.

21. Callicott, *In Defense,* 93–94; Callicott continues to struggle with this issue. See, for example, "Holistic Environmental Ethics and the Problem of Ecofascism," in *Beyond the Land Ethic: More Essays in Environmental Philosophy* (Albany: State University of New York Press, 1999), 59–76.

22. C. S. Holling, "Cross-Scale Morphology, Geometry, and Dynamics of Ecosystems," *Ecological Monographs* 62 (1992): 447–502.

23. Ibid., 484.

24. S. L. Pimm, *The Balance of Nature? Ecological Issues in the Conservation of Species and Communities* (Chicago: University of Chicago Press, 1991).

25. For an early exploration, see Arthur Koestler, *The Ghost in the Machine* (New York: Macmillan, 1967); T. F. H. Allen and T. B. Starr, *Hierarchy: Perspectives for Ecological Complexity* (Chicago: University of Chicago Press, 1982); For an excellent review, see Jianguo Wu and Orie L. Loucks, "From Balance of Nature to Hierarchical Patch Dynamics: A Paradigm Shift in Ecology," *Quarterly Review of Biology* 70 (1995): 439–66.

26. R. V. O'Neill, D. L. DeAngelis, J. B. Waide, and T. F. H. Allen, *A Hierarchical Concept of Ecosystems* (Princeton, NJ: Princeton University Press, 1986).

27. P. S. White, "Pattern, Process, and Natural Disturbance in Vegetation," *Botanical Review* 45 (1979): 229–99; S. T. A. Pickett and P. S. White, eds., *The Ecology of Natural Disturbance and Patch Dynamics* (San Diego: Academic Press, 1985); Wu and Loucks, "From Balance of Nature to Hierarchical Patch Dynamics."

28. Bryan G. Norton, "Context and Hierarchy in Aldo Leopold's Theory of Environmental Management," *Ecological Economics* 2 (1990): 119–27.

29. Curt Meine, *Aldo Leopold: His Life and Work* (Madison: University of Wisconsin Press, 1988), 459.

30. Holling, "Cross-Scale Morphology."

31. Allen and Starr, *Hierarchy;* T. F. H. Allen and T. W. Hoekstra, *Toward a Unified Ecology* (New York: Columbia University Press, 1991).

Chapter 7

1. Garrett Hardin, "The Tragedy of the Commons," *Science* 162 (1968): 1243–48.

2. See, for example, Elinor Ostrom, *Governing the Commons: The Evolution of Institutions for Collective Actions* (Cambridge: Cambridge University Press, 1990), and section 7.3 below.

3. Colin Clark, "The Economics of Over-Exploitation," *Science* 182 (1974): 630–34.

4. Ibid., 634.

5. Derek Parfit, *Reasons and Persons* (Oxford: Clarendon Press, 1984).

6. The view that all value is individual value is of course challenged, at least apparently, by some nonanthropocentrists, such as Baird Callicott and Holmes Rolston, who speak of ecosystems as having intrinsic value. Although this view may be held by quite a few environmental ethicists, it is not prevalent among ethicists of human interactions or among social scientists.

7. William Ophuls, *The Politics of Scarcity: A Prologue to a Political Theory of the Steady State* (San Francisco: Freeman, 1977); William Ophuls, *The Politics of Scarcity Revisited: The Unraveling of the American Dream* (New York: Freeman, 1992); Robert Heilbroner, *An Inquiry into the Human Prospect* (New York: Norton, 1974); Bruce Hannon, "World Shogun," *Journal of Social and Biological Structures* 8 (1985): 329–41.

8. Laura Westra, *Living in Integrity: A Global Ethic to Restore a Fragmented Earth* (Lanham, MD: Rowman and Littlefield, 1998), 57.

9. Ibid., 68.

10. Ibid., 72.

11. Ibid.

12. Herbert Simon, *The New Science of Management Decisions* (New York: Harper and Row, 1960), 49.

13. Kenneth Arrow, *Social Choice and Individual Values*, 2nd ed. (New York: John Wiley, 1963).

14. Tore Sager, "Planning and the Liberal Paradox: A Democratic Dilemma in Social Choice," *Journal of Planning Literature* 12 (1997): 16–29.

15. Alan Menand, *The Metaphysical Club* (New York: Farrar, Straus and Giroux, 2002).

16. Quoted in ibid., 172.

17. Quoted in ibid., 228.

18. Ibid., 200.

19. Robin Gregory and R. L. Keeney, "Creating Policy Alternatives Using Stakeholder Values," *Management Science* 40 (1994): 1035–48; Robin Gregory, "Incorporating Value Trade-offs into Community-Based Environmental Risk Decisions," *Environmental Values* 11 (2002): 461–88.

20. Robert Axelrod, "The Emergence of Cooperation among Egoists," *American Political Science Review* 75 (1980): 306–18: R. Axelrod, *The Evolution of Cooperation* (New York: Basic Books, 1984).

21. Thomas Hanne, *Intelligent Strategies for Meta Multiple Criteria Decision Making* (Boston: Kluwer, 2001), 3.

22. Ibid., 44.

23. Mark Sagoff, *The Economy of the Earth: Philosophy, Law, and the Environment* (Cambridge: Cambridge University Press, 1988), 7–14.

24. Jürgen Habermas, *The Structural Transformation of the Public Sphere: An Inquiry into a Category of Bourgeois Society*, trans. Thomas Burger (1962; repr., Cambridge, MA: MIT Press, 1998).

25. Jürgen Habermas, *Between Facts and Norms: Contributions to a Discourse Theory of Law and Democracy*, trans. William Rehg (Cambridge, MA: MIT Press, 1998), 13–15.

26. Jürgen Habermas, *Justification and Application: Remarks on Discourse Ethics*, trans. C. P. Cronin (Cambridge, MA: MIT Press, 1993), xxix.

27. Ibid., 28; Habermas, *Between Facts and Norms*, 18.

28. Habermas, *Between Facts and Norms*, 352.

29. Discourse ethics has been usefully applied to problems in the planning process, especially by John Forester. See especially his *Critical Theory, Public Policy, and Planning Practice: Toward a Critical Pragmatism* (Albany: State University of New York Press, 1993); also see Sager, "Planning and the Liberal Paradox."

30. Karl-Otto Appel, "Ethics, Utopia, and the Critique of Utopia," in *The Communicative Ethics Controversy*, ed. S. Benhabib and F. Dallmayr (Cambridge, MA: MIT Press, 1991), esp. 45–59.

31. The following summary is based on Habermas, *Justification and Application*, 1–17; and Habermas, *Between Facts and Norms*, 107–9.

32. Habermas, *Justification and Application*, 9; and Habermas, *Between Facts and Norms*, 107.

33. Habermas, *Between Facts and Norms*, 108–9.

34. Ibid., 107.

35. *Justification and Application*, 3.

36. Ibid., 8.

37. Ibid., 11, 12.

38. Habermas, *Between Facts and Norms,* 8.

39. See Kai Lee, *Compass and Gyroscope* (Washington, DC: Island Press, 1993).

40. S. O. Funtowicz and Jerome R. Ravetz, "Science for the Post-Normal Age," in *Perspectives on Ecological Integrity,* ed. L. Westra and J. Lemons (Dordrecht, Netherlands: Kluwer, 1995); and S. Funtowicz and J. Ravetz, "The Worth of a Songbird: Ecological Economics as a Post-Normal Science," *Ecological Economics* 10 (1994): 197–207.

41. B. G. Norton, *Why Preserve Natural Variety?* (Princeton, NJ: Princeton University Press, 1987).

42. See section 7.1 for full discussion and references.

43. William D. Leach, N. W. Pelkey, and P. A. Sabatier, "Making Watershed Partnerships Work: A Review of the Empirical Literature," *Journal of Water Resources Planning and Management* 6 (2001): 378–85; J. Burgess, C. M. Harrison, and P. Filius, "Environmental Communication and the Cultural Politics of Environmental Citizenship," *Environment and Planning A* 30 (1998): 1445–60; Mark Borsuck, Robert Clemen, Lynn Maguire, and Kenneth Reckhow, "Stakeholder Values and Scientific Modeling in the Neuse River Watershed," *Group Decision and Negotiation* 10 (2001): 355–73; and Julia Wondolleck and Steven Yaffee, *Making Collaboration Work: Lessons from Innovations in Natural Resource Management* (Washington, DC: Island Press, 2000).

44. Marc Landy, Marc Roberts, and Stephen Thomas, *The Environmental Protection Agency: Asking the Wrong Questions from Nixon to Clinton,* exp. ed. (New York: Oxford University Press, 1994).

45. Benjamin Barber, *Strong Democracy: Participatory Politics for a New Age* (Berkeley: University of California Press, 1985).

Chapter 8

1. See section 8.3 for a substantive discussion of Barry's writings on the subject.

2. Gretchen Dailey, ed., *Nature's Services: Societal Dependence on Natural Ecosystems* (Washington, DC: Island Press, 1997); Robert Costanza, R. d'Arge, R. de Groot, S. Farber, M. Grasso, B. Hannon, S. Naeem, K. Limburg, J. Paruelo, R. V. O'Neill, R. Raskin, P. Sutton, and M. van den Belt, "The Value of the World's Ecosystem Services and Natural Capital," *Nature* 387 (1997): 253–60.

3. See Robert Solow, "An Almost Practical Step toward Sustainability," invited lecture on the occasion of the fortieth anniversary of Resources for the Future, Washington, DC, October 8, 1992.

4. See, for example, John Pezzey, "Sustainability Constraints," *Land Economics* 73 (1997): 448–66.

5. See Richard Howarth, "Sustainability as Opportunity," *Land Economics* 73 (1997): 569–79, esp. 572.

6. Herman Daly, "On Wilfred Beckerman's Critique of Sustainable Development," *Environmental Values* 4 (1995): 49–55.

7. I have developed my concerns about economically based attempts to specify a stronger version of sustainability in much more detail in "Intergenerational Equity and Sustainability," in my *Searching for Sustainability* (New York: Cambridge University Press, 2003).

8. David Pearce and Edward Barbier, *Blueprint for a Sustainable Economy* (London: Earthscan, 2000), 245.

9. Ibid., 24.

10. Ibid., 83.

11. See, for example, Richard Howarth, "Sustainability under Uncertainty: A Deontological Approach," *Land Economics* 71 (1995): 417–27.

12. Robert M. Solow, "The Economics of Resources or the Resources of Economics," *American Economic Review Proceedings* 64 (1974): 1–14; Robert M. Solow, "On the Intergenerational Allocation of Natural Resources," *Scandinavian Journal of Economics* 88 (1986): 141–49; Solow, "An Almost Practical Step toward Sustainability"; Robert M. Solow, "Sustainability: An Economist's Perspective," in *Economics of the Environment: Selected Readings,* ed. Robert Dorfman and Nancy Dorfman (New York: Norton, 1993), quotations on 181.

13. Solow, "Sustainability," 180.

14. Ibid., 181.

15. A Rawlsian with important environmental concerns could perhaps build more stringent environmental duties into this broader aspect of a Rawlsian bequest. For example, one might argue that the task of building a just society would include institutions adequate to protect essential natural resources. See Talbot Page, *Conservation and Economic Efficiency* (Baltimore: Johns Hopkins University Press, for *Resources for the Future,* 1977); and Bryan Norton, "Intergenerational Equity and Environmental Decisions: A Model Using Rawls' Veil of Ignorance," *Ecological Economics* 1 (1989): 137–59.

16. John Passmore, *Man's Responsibility for Nature* (New York: Scribner's, 1974), 91.

17. Brian Barry, "Circumstances of Justice and Future Generations," in *Obligations to Future Generations,* ed. R. I. Sikora and Brian Barry (Philadelphia: Temple University Press, 1978), 243–44; see also Brian Barry, "The Ethics of Resource Depletion," in *Democracy, Power, and Justice,* by Brian Barry (Oxford: Clarendon Press, 1989), 515.

18. Barry, *Democracy, Power, and Justice,* 519–20.

19. Ibid., 593.

20. It should be said, however, that Barry's position is thus explicitly stated only for nonrenewable resources, which may still leave room for more specialized obligations such as protecting natural capital. As far as I am aware, Barry has nowhere in print taken an explicit position regarding the fungibility of renewable resources.

21. See Richard Norgaard, "Economic Indicators of Resource Scarcity: A Critical Essay," *Journal of Environmental Economics and Management* 19 (1990): 19–25.

22. D. W. Bromley, "Searching for Sustainability: The Poverty of Spontaneous Order," *Ecological Economics* 24 (1998): 231–40.

23. Derek Parfit, *Reasons and Persons* (Oxford: Clarendon Press, 1984), esp. 449–54.

24. See Talbot Page, "Intergenerational Justice as Opportunity," in *Energy and the Future,* ed. Douglas MacLean and Peter G. Brown (Totowa, NJ: Rowman and Littlefield, 1983), for an excellent discussion of this problem and a demonstration that criteria of economic efficiency cannot solve it. Also see Page, *Conservation and Economic Efficiency,* for a more detailed treatment in terms of material policy.

25. Presentism, stated as a moral principle, might be thought to justify discounting. But many authors have argued that applying discounting across generations begs important moral questions. See, for example, Talbot Page, "Intergenerational Equity and the Social Rate

of Discount," in *Environmental Resources and Applied Welfare Economics,* ed. V. K. Smith and J. Krutilla (Washington, DC: Resources for the Future, 1988).

26. See Gregory Kavka, "The Futurity Problem," in *Responsibilities to Future Generations,* ed. Ernest Partridge (Buffalo, NY: Prometheus Books, 1981), 111–13; Page, "Intergenerational Justice as Opportunity"; and Brian Barry, "Justice between Generations," in Barry, *Democracy, Power, and Justice,* 500, for convincing reasons that this strong ignorance premise is implausible.

27. A notable exception is John O'Neill, "Future Generations, Present Harms," *Philosophy* 68 (1993): 35–51.

28. This point is made well by Daniel Callahan, "What Obligations Do We Have to Future Generations," in *Responsibilities to Future Generations,* ed. Ernest Partridge (Buffalo, NY: Prometheus Books, 1981).

29. See, for example, Thomas C. Schelling, "Some Economics of Global Warming," *American Economic Review* 82 (1992): 1–15; and Robert Mendelsohn, William Nordhaus, and Daigee Shaw, "The Impact of Global Warming on Agriculture: A Ricardian Analysis," *American Economic Review* 84 (1994): 753–71.

30. Solow, "Sustainability," 182.

31. It is of course a simplification to refer to environmentalists as if they represented a monolithic group, especially given the pluralism and contentions of environmentalists over values already described in this book. I believe this simplification is justified because I refer to the uniting idea of people who espouse varied values but who all care about the values of the future. For convenience, then, I will refer to the activity of saving areas for future enjoyment, use, or contemplation as "nature protection" or "protection of natural places," and the people who advocate these activities as "environmentalists" or "nature protectionists."

32. This approach does not require recognition of "inherent" values in the object or place itself, provided one believes in the value of cultural and community values. See Bryan Norton, *Why Preserve Natural Variety?* (Princeton, NJ: Princeton University Press, 1987), chapter 10.

33. This point is eloquently made in Joseph Sax's book, *Mountains without Handrails* (Ann Arbor: University of Michigan Press, 1981); and also by Mark Sagoff, "On Preserving the Natural Environment," *Yale Law Journal* 81 (1974): 205–67.

34. Passmore, *Man's Responsibility for Nature,* 91.

35. Martin P. Golding, "Obligations to Future Generations," *Monist* 56 (1972): 86.

36. Passmore, *Man's Responsibility for Nature,* 90. The embedded quotation is from Golding, "Obligations to Future Generations," 71.

37. J. L. Austin, *How to Do Things with Words* (New York: Oxford University Press, 1962), 5–7.

38. Ibid., 19–20.

39. Edmund Burke, *Reflections on the Revolution in France* (London: Dent, 1910), 93–94.

40. See Henry David Thoreau, *Walden* (New York: New American Library, 1960), 144; and Aldo Leopold, *A Sand County Almanac and Sketches Here and There* (Oxford: Oxford University Press, 1949), 199–201.

41. O'Neill, "Future Generations, Present Harms."

42. Adam Smith, *The Theory of Moral Sentiments* (New York: Garland, 1971), originally published in 1776.

43. Mark Sagoff, *The Economy of the Earth: Philosophy, Law, and the Environment* (Cambridge: Cambridge University Press, 1988), 92–94.

44. See Edith Brown Weiss, *In Fairness to Future Generations* (Tokyo: United Nations University; Dobbs Ferry, NY: Transnational Publishers, 1989); Peter G. Brown, *Restoring the Public Trust: A Fresh Vision for Progressive Government in America* (Boston: Beacon Press, 1994); Peter G. Brown, *Ethics, Economics, and International Relations* (Edinburgh, UK: Edinburgh University Press, 2000); and Paul Wood, *Biodiversity and Democracy: Rethinking Society and Nature* (Vancouver: University of British Columbia Press, 2000), for alternative conceptions of intergenerational trusts. Richard Howarth, "Sustainability under Uncertainty." Also see Page, "Intergenerational Justice as Opportunity," for a particularly clear explanation of how one must go beyond economic analysis to formulate the important questions of environmental and energy policy, because the assumptions of economic analysis are insensitive to crucial questions affecting intergenerational justice.

45. Brown, *Ethics, Economics, and International Relations,* 20, 21 (quotation on 20).

46. Ibid., 24.

47. Ibid., 21.

48. See, for example, ibid., 21.

49. Weiss, *In Fairness,* 2, 96, 80.

50. Ibid., 104.

51. Howarth, "Sustainability as Opportunity."

52. I discuss this topic briefly in section 10.2.

53. This use of hierarchy theory to correlate values, via their associated indicators, with physical scales has potential to provide quantifiable measures that can identify key processes in the production of a social good. See Bryan Norton and Robert Ulanowicz, "Scale and Biodiversity Policy," *Ambio* 21 (1992): 244–49.

Chapter 9

1. Arild Vatn and Daniel Bromley, "Choices without Prices without Apologies," *Journal of Environmental Economics and Management* 26 (1994): 129–48.

2. This approach is inconsistent, of course, with the attitude of Economists and with some legislation and bureaucratic pronouncements. The role suggested here would not, for example, justify executive orders, legislation, or administrative rules that require policies or regulations to "pass" a cost-benefit analysis. Decrees such as this imply that only those environmental values that can be priced will be counted, flying in the face of the obvious fact that many important environmental values cannot be meaningfully priced at this time. Decrees such as this are inconsistent with our decision to let experience—as it emerges from local processes of deliberation—be the only ultimate determinant of our beliefs.

3. See, for example, Holmes Rolston III, *Conserving Natural Value* (Buffalo, NY: Prometheus Books, 1994), 195, where he says that nature is a "value-generating system," independent of humans.

4. J. Baird Callicott, "The Pragmatic Power and Promise of Theoretical Environmental Ethics: Forging a New Discourse," *Environmental Values* 11 (2002): 3–25.

5. Ibid., 21.

6. Ibid., 10.

7. Ibid., 13–14.

8. Aldo Leopold, "A Biotic View of Nature," *Journal of Forestry* 37 (1939): 727; Aldo Leopold, *A Sand County Almanac and Sketches Here and There* (Oxford: Oxford University Press, 1949).

9. Callicott, "Pragmatic Power and Promise," 14.

10. Ibid.

11. Nor does Callicott think so. See J. Baird Callicott, *In Defense of the Land Ethic: Essays in Environmental Philosophy* (Albany: State University of New York Press, 1989), 134–35, for a strong denial that rights, independent of practices, are effectual.

12. For a fascinating study of organisms and ecological scale, see C. S. Holling, "Cross-Scale Morphology, Geometry, and Dynamics of Ecosystems," *Ecological Monographs* 62 (1992): 447–502.

13. See Per Ariansen, "The Non-Utility Value of Nature: An Investigation into Biodiversity and the Value of Natural Wholes," *Communications of the Norwegian Forest Research Institute* (Meddelelser fra Skogforsk), vol. 47 (Aas: Agricultural University of Norway, 1997). Ariansen suggests that some choices we make with respect to protecting our environment represent "constitutive" values. Loving and protecting special places and special features of a place, on this view, may be constitutive of a person's sense of self and of community membership. Careless destruction of these special features, correlatively, might be considered a kind of "cultural suicide." Ariansen's insight provides one interesting direction for the explication of what has been called, above, "noncompensable harms." Also see Alan Holland and John O'Neill, "The Integrity of Nature over Time," Thingmount Working Paper TWP 96–08, 1986, Department of Philosophy, Lancaster University, Lancaster, UK; and Alan Holland and Kate Rawls, "Values in Conservation," *Ecos* 14 (1993): 14–19, for a useful discussion of the inseparability of cultural and ecological ideals.

14. See section 4.1 for a discussion of wicked problems.

Chapter 10

1. Detlef von Winterfeldt and Ward Edwards, *Decision Analysis and Behavioral Research* (Cambridge: Cambridge University Press, 1986), 562.

2. National Research Council, *Risk Assessment in the Federal Government* (Washington, DC: National Academy Press, 1983).

3. Von Winterfeldt and Edwards, *Decision Analysis and Behavioral Research.*

4. National Research Council, *Understanding Risk* (Washington, DC: National Academy Press, 1996).

5. National Research Council, *Risk Assessment in the Federal Government,* 3.

6. Ibid., 21.

7. Ibid., 28.

8. Von Winterfeldt and Edwards, *Decision Analysis and Behavioral Research.*

9. Ibid., xiii.

10. Harold Glasser, "Towards a Descriptive, Participatory Theory of Environmental Policy Analysis and Project Evaluation" (Ph.D. diss., University of California, Davis, 1995).

11. Von Winterfeldt and Edwards, *Decision Analysis and Behavioral Research*, 16.

12. Ibid., 2.

13. See the appendix for an account of these struggles.

14. From the original—later amended—charge to the National Research Council committee. National Research Council, *Understanding Risk,* x.

15. Ibid., x–xi.

16. Ibid., 11–12.

17. Ibid., 12.

18. Ibid., 13.

19. Ibid., 27.

20. Ibid., 34.

21. Robin Gregory and R. L. Keeney, "Creating Policy Alternatives Using Stakeholder Values," *Management Science* 40 (1994): 1035–48.

22. Ibid., 1036.

23. Tom Horton, "Remapping the Chesapeake," *New American Land,* September–October 1987, 7–8.

Chapter 11

1. See section 9.2 above.

2. National Research Council, Committee on Mitigating Wetland Losses, *Compensating for Wetland Losses under the Clean Water Act* (Washington, DC: National Academy Press, 2001).

3. Curt Meine, *Aldo Leopold: His Life and Work* (Madison: University of Wisconsin Press, 1988), 282–84.

4. T. F. H. Allen and T. B. Starr, *Hierarchy: Perspectives for Ecological Complexity* (Chicago: University of Chicago Press, 1982), 6.

5. Aldo Leopold, *A Sand County Almanac and Sketches Here and There* (Oxford: Oxford University Press, 1949), 99.

6. L. H. Gunderson, C. S. Holling, and A. Light, *Barriers and Bridges to the Renewal of Ecosystems and Institutions* (New York: Columbia University Press, 1995); L. H. Gunderson and C. S. Holling, *Panarchy: Understanding Transformations in Human and Natural Systems* (Washington, DC: Island Press, 2002); L. H. Gunderson and Lowell Pritchard Jr., eds., *Resilience and the Behavior of Large-Scale Systems* (Washington, DC: Island Press, 2002); Mark Borsuk, Robert Clemen, Lynn Maguire, and Kenneth Reckhow, "Stakeholder Values and Scientific Modeling in the Neuse River Watershed," *Group Decision and Negotiation* 10 (2001): 355–73.

7. Jan Rotmans, "Methods for Integrative Assessment: The Challenges and Opportunities Ahead," *Environmental Modelling and Assessment* 3 (1998): 155–79.

8. Dale Rothman and John Robinson, "Growing Pains: A Conceptual Framework for Considering Integrated Assessments," *Environmental Monitoring and Assessment* 46 (1997): 23–43, quotation on 33.

9. Jan Rotmans, Marjolein van Asselt, Chris Anastasi, Sandra Greeuw, Joanne Mellors, Simone Peters, Dale Rothman, and Nicole Rijkens, "Visions for a Sustainable Europe," *Futures* 32 (2000): 809–31, quotation on 813.

10. International Centre for Integrative Studies, "Integrated Visions for a Sustainable Europe: Visions Final Report," submitted to the Research and Development Directorate, European Commission (ENV4-CT97-0462), April 2001, 2.

11. Ibid., 3.

12. Ibid., 52.

13. Ibid., 52–53.

14. See, for example, Gunderson and Holling, *Panarchy;* and Borsuk et al., "Stakeholder Values and Scientific Modeling."

15. John Robinson, David Biggs, George Francis, Russel Legge, Sally Lerner, D. Scott Slocombe, and Caroline Van Bers, *Life in 2030* (Vancouver: University of British Columbia Press, 1996), 8.

16. Rothman and Robinson, "Growing Pains," 37.

17. Hanna Cortner, Mary Wallace, and Margaret Moote, "A Political Context for Bioregional Assessment," in *Bioregional Assessments: Science at the Crossroads of Management and Policy,* ed. K. N. Johnson, F. J. Swanson, and S. Greene (Washington, DC: Island Press, 1999), 71.

18. Robinson et al., *Life in 2030,* 8.

19. Ibid.

20. See, for example, Milton Friedman, *Capitalism and Freedom* (Chicago: University of Chicago Press, 1962); George Stigler and Gary Becker, "De Gustibus Non Est Disputandum," *American Economic Review* 67 (1977): 76–89, esp. 89.

21. See, for example, J. Baird Callicott, *In Defense of the Land Ethic: Essays in Environmental Philosophy* (Albany: State University of New York Press, 1989).

Chapter 12

1. Paul De Jongh, with Seán Captain, *Our Common Journey: A Pioneering Approach to Cooperative Environmental Management* (London: Zed Books, 1999), 4, 5.

2. Ibid., 7.

3. Ibid., 226.

4. Ibid., xxi.

5. Ibid., 42. The following discussion mainly follows De Jongh's account.

6. Ibid., 62.

7. Ibid., 69.

8. Ibid., 44.

9. Ibid., 45.

10. Ibid., 48.

11. Ibid., 106.

12. For a somewhat different, but consistent, view of these political changes, see Maarten A. Hajer, *The Politics of Environmental Discourse: Ecological Modernization and the Policy Process* (Oxford: Clarendon Press, 1995).

13. See, for example, Fred Boerselman and Frans Vera, *Nature Development: An Exploratory Study for the Construction of Ecological Networks* (The Hague, Netherlands: Ministry of Agriculture, Nature Management, and Fisheries, 1995).

14. Selected from a list of ten recommendations in U.S. Environmental Protection Agency, Science Advisory Board, "Toward Integrated Environmental Decision Making," EPA-SAB-EC-00.011, August 2000, 38.

15. Ibid., 43, 106.

16. U.S. Environmental Protection Agency, "Reinventing Environmental Protection: 1998 Annual Report," EPA 100-R-99-002, 1999.

17. Talbot Page, *Conservation and Economic Efficiency* (Baltimore: Johns Hopkins University Press, for *Resources for the Future,* 1977); Michael A. Toman, "Economics and 'Sustainability': Balancing Tradeoffs and Imperatives," *Land Economics* 70 (1994): 399–413. Also see my citations and discussion of trust doctrines in section 8.7.

18. See, for example, Rodger Schlickheisen, "Protecting Biodiversity for Future Generations: An Argument for a Constitutional Amendment," *Tulane Environmental Law Journal* 8 (1994): 181–221; and Paul Wood, *Biodiversity and Democracy: Rethinking Society and Nature* (Vancouver: University of British Columbia Press, 2000).

19. William James, "What Pragmatism Means," in *Essays in Pragmatism,* by William James (New York: Hafner, 1948), 146.

20. Bryan Norton, "Conservation and Preservation: A Conceptual Rehabilitation," *Environmental Ethics* 8 (1986): 195–220; Bryan Norton, *Toward Unity among Environmentalists* (New York: Oxford University Press, 1991); Bryan Norton, "Convergence and Contextualism: A Reply to Steverson," *Environmental Ethics* 19 (1997): 87–100.

21. D. W. Bromley, "Searching for Sustainability: The Poverty of Spontaneous Order," *Ecological Economics* 24 (1998): 231–40.

22. Amartya Sen, *Commodities and Capabilities* (New Delhi: Oxford University Press, 1999); Amartya Sen, *Development as Freedom* (New York: Random House, Anchor Books, 1999); and Amartya Sen, *Rationality and Freedom* (Cambridge, MA: Harvard University Press, 2002).

23. Sen, *Commodities,* 44–45.

24. Ibid., 3.

Appendix

1. For an excellent, brief account of this experimentation, see Richard Rorty, introduction to *The Linguistic Turn: Essays in Philosophical Method,* ed. R. Rorty (Chicago: University of Chicago Press, 1967).

2. See Bertrand Russell, "The Philosophy of Logical Atomism," in *Readings in Twentieth Century Philosophy,* ed. W. P. Alston and G. Nakhnikian (London: Free Press of Glencoe, 1963) (from a series of lectures delivered in 1918), for a readable discussion of Russell's theory of descriptions.

3. J. O. Urmson, *Philosophical Analysis: Its Development between the Two World Wars* (Oxford: Oxford University Press, 1956), 7. The following account owes much to Urmson's discussion.

4. Ludwig Wittgenstein, *Tractatus Logico-Philosophicus,* trans. C. K. Ogden (London: Routledge and Kegan Paul, 1922).

5. See section 2.4 for more on Dewey's answer to this question.

6. See Rudolf Carnap, "Intellectual Autobiography," in *The Philosophy of Rudolf Carnap,* ed. P. A. Schilpp (LaSalle, IL: Open Court, 1963), 22–24.

7. A. J. Ayer, *Language, Truth, and Logic,* 2nd ed. (New York: Dover, 1949), 5.

8. Rorty, introduction to *The Linguistic Turn.*

9. Carnap, "Intellectual Autobiography," 55.

10. Ibid., 29.

11. Ibid. Carnap is referring to Wittgenstein's *Tractatus-Logico-Philosophicus,* of 1921.

12. Ibid.

13. Ibid., 53, where Carnap is writing about *The Logical Syntax of Language,* trans. Amethe Smeaton (London: Routledge and Kegan Paul, 1937).

14. Carnap, "Intellectual Autobiography," 54.

15. Carnap, *Logical Syntax,* 283, 313.

16. Ibid., 331.

17. Carnap, "Intellectual Autobiography," 56.

18. Rudolf Carnap, "Empiricism, Semantics, and Ontology," in *Meaning and Necessity: A Study in Semantics and Model Logic,* by Rudolf Carnap, 2nd ed. (Chicago: University of Chicago Press, 1956), 205–21.

19. Carnap, "Intellectual Autobiography," 47.

20. "Empiricism, Semantics, and Ontology," 205–6.

21. Ibid., 206–16.

22. Ibid., 207.

23. Ibid., 214.

24. Ibid., 221.

25. Rorty, introduction to *The Linguistic Turn,* 34.

26. Carnap, "Empiricism, Semantics, and Ontology," 221.

27. Carnap, "Intellectual Autobiography," 68.

28. Willard Van Orman Quine, "Two Dogmas of Empiricism," in *From a Logical Point of View,* by Willard Van Orman Quine (New York: Harper and Row, 1953), 20–46.

29. Carnap, *Logical Syntax,* 320.

30. Pierre Duhem, "Physical Theory and Experiment," in *Scientific Knowledge,* 2nd ed., ed. J. Kourany (Belmont, CA: Wadsworth, 1997), 189–92.

31. Willard Van Orman Quine, "Epistemology Naturalized," in *Ontological Relativity and Other Essays,* by Willard Van Orman Quine (New York: Columbia University Press, 1969), 80–81.

32. Thomas Kuhn, *The Structure of Scientific Revolutions* (Chicago: University of Chicago Press, 1962, 1970). The similarities between the viewpoint of Carnap at the end of his career with Kuhn's view has been noticed by, among others, Kuhn himself. I have heard it said by acquaintances of Kuhn, for example, that he at least once said that if he had read Carnap's essay "Empiricism, Semantics, and Ontology," he might never have bothered to write *The Structure of Scientific Revolutions.*

33. Further, given Quine's arguments, it should also be cautioned that there can be no sharp distinction between questions asked within a paradigm and those asked from outside, unless one "freezes" the paradigm by assigning to some beliefs the role of principles and rules

that can be rejected only if one rejects the whole system. This qualification, though theoretically important, need not reduce the usefulness of Kuhn's idea, because the idea of a paradigm is usually used to refer to the common and informal discourse of scientists.

34. William James, *The Will to Believe,* reprinted in *Classic American Philosophers,* ed. M. Fish (New York: Appleton-Century-Crofts, 1951), 147, 141.

35. Richard Rorty, *Objectivism, Relativism, and Truth* (New York: Cambridge University Press, 1991), 270.

36. See, for example, Sylvio O. Funtowicz and Jerome R. Ravetz, "Science for the Post Normal Age," in *Perspectives on Ecological Integrity,* ed. L. Westra and J. Lemons (Dordrecht, Netherlands: Kluwer, 1995).

37. Rorty, *Objectivism, Relativism, and Truth,* 21.

INDEX